Springer Series in Computational Mathematics

41

For further volumes:
http://www.springer.com/series/797

Jie Shen · Tao Tang · Li-Lian Wang

Spectral Methods

Algorithms, Analysis and Applications

Springer

Jie Shen
Department of Mathematics
Purdue University
N. University St. 150
West Lafayette, IN 47907-2067
USA
shen@math.purdue.edu

Tao Tang
Department of Mathematics
Hong Kong Baptist University
Waterloo Road 224
Kowloon
Hong Kong SAR
ttang@hkbu.edu.hk

Li-Lian Wang
Division of Mathematical Sciences
School of Physical & Mathematical Sciences
Nanyang Technological University
21 Nanyang Link
637371
Singapore
lilian@ntu.edu.sg

ISSN 0179-3632
ISBN 978-3-540-71040-0 e-ISBN 978-3-540-71041-7
DOI 10.1007/978-3-540-71041-7
Springer Heidelberg Dordrecht London New York

Library of Congress Control Number: 2011934044

Mathematics Subject Classification (2010): 65M70, 65M12, 65N15, 65N35, 65N22, 65F05, 35J25, 35J40, 35K15, 42C05

Cover design: deblik, Berlin

Printed on acid-free paper

Springer is part of Springer Science+Business Media (www.springer.com)

Preface

This book is developed from lecture notes of graduate courses taught over the years by the authors at the Pennsylvania State University, Purdue University, Hong Kong Baptist University and Nanyang Technological University of Singapore.

The aim of the book is to provide

- A detailed presentation of basic spectral algorithms
- A systematical presentation of basic convergence theory and error analysis for spectral methods
- Some illustrative applications of spectral methods

For many basic algorithms presented in the book, we provide Matlab codes (which will be made available online) which contain additional programming details beyond the mathematical formulas, so that the readers can easily use or modify these codes to suite their need. We believe that these Matlab codes will help the readers to have a better understanding of these spectral algorithms and provide a useful starting point for developing their own application codes.

There are already quite a few monographs/books on spectral methods. The classical books by Gottlieb and Orszag (1977) and by Canuto et al. (1987)[1] were intended for researchers and advanced graduate students, and they are excellent references for the historical aspects of spectral methods as well as in depth presentations of various techniques and applications in computational fluid dynamics. The book by Boyd (2001) focused on the Fourier and Chebyshev methods with emphasis on implementations and applications. The book by Trefethen (2000) gave an excellent exposition on the spectral-collocation methods through a set of elegant Matlab routines. The books by Deville et al. (2002) and by Karniadakis and Sherwin (2005) concentrated on the spectral-element methods with details on parallel implementations and applications in fluid dynamics, while the more recent book by Hesthaven and Warburton (2008) focused on the discontinuous Galerkin methods with a nodal spectral-element approach. On the other hand, Hesthaven et al. (2007) focused on

[1] An updated and expanded version of Canuto et al. (1987) is recently published. This new version Canuto et al. (2006, 2007) incorporated many new developments made in the last 20 years and provided a more systematical treatment for spectral methods.

the spectral methods for time-dependent problems with a particular emphasis on hyperbolic equations and problems with non-smooth solutions. The book length article by Bernardi and Maday (1997) and their monograph in French Bernardi and Maday (1992a) provided an excellent exposition on the basic approximation theory of spectral methods with a particular emphasis on Stokes equations, while the monograph (Shen and Tang 2006) presented a basic introduction in a lecture note style to the implementation and analysis of spectral methods. The emphasis of the book by Guo (1998b), on the other hand, was on numerical analysis of spectral methods for nonlinear evolution problems. Finally, spectral methods have been playing a very significant role in dealing with stochastic differential equations and uncertainty quantifications, and we refer to the recent books by Le Maître and Knio (2010) and by Xiu (2010) on these emerging topics.

The current book attempts to provide a self-contained presentation for the construction, implementation and analysis of efficient spectral algorithms for some model equations, of elliptic, dispersive and parabolic type, which have wide applications in science and engineering. It strives to provide a systematical approach based on variational formulations for both algorithm development and numerical analysis. Some of the unique features of the current book are

- Our analysis is based on the non-uniformly weighted Sobolev spaces which lead to simplified analysis and more precise estimates, particularly for problems with corner singularities. We also advocate the use of the generalized Jacobi polynomials which are particularly useful for dealing with boundary value problems.
- We develop efficient spectral algorithms and present their error analysis for Volterra integral equations, higher-order differential equations, problems in unbounded domains and in high-dimensional domains. These topics have rarely been covered in detail in the existing books on spectral methods.
- We provide online a set of well structured Matlab codes which can be easily modified and expanded or rewritten in other programming languages.

The Matlab codes as well as corrections/updates to the book will be available at http://www.math.purdue.edu/~shen/STWbook. In case this site becomes unavailable due to unforeseen circumstances in the future, the readers are advised to check the Springer Web site for the updated Web link on the book.

We do not attempt to provide in this book an exhaustive account on the wide range of topics that spectral methods have had impact on. In particular, we do not include some important topics such as spectral methods for hyperbolic equations and spectral-element methods, partly because these topics do not fit well in our uniform framework, and mostly because there are already some excellent books mentioned above on these topics. As such, no attempt is made to provide a comprehensive list of references on the spectral methods. The cited references reflect the topics covered in the book, but inevitably, the authors' bias. While we strive for correctness, it is most likely that errors still exist. We welcome comments, suggestions and corrections.

The book can be used as a textbook for graduate students in both mathematics and other science/engineering. Mathematical analysis and applications are organized mostly at the end of each chapter and presented in such a way that they can be skipped without affecting the understanding of algorithms in the following chapters. The first four chapters and Sects. 8.1–8.4 provide the basic ingredients on Fourier and polynomial approximations and essential strategies for developing efficient spectral-Galerkin and spectral-collocation algorithms. Section 8.5 deals with sparse spectral methods for high-dimensional problems. The topics in Chaps. 5, 6 and 7 are independent of each other so the readers can choose according to their need. Applications covered in Chap. 9, except for a slight dependence on Sects. 9.4–9.5, are also independent of each other. For the readers' convenience, we provide in the Appendices some essential mathematical concepts, basic iterative algorithms and commonly used time discretization schemes.

The book is also intended as a reference for active practitioners and researchers of spectral methods. The prerequisite for the book includes standard entry-level graduate courses in Numerical Analysis, Functional Analysis and Partial Differential Equations (PDEs). Some knowledge on numerical approximations of PDEs will be helpful in understanding the convergence theory and error analysis but hardly necessary for understanding the numerical algorithms presented in this book.

The authors would like to thank all the people and organizations who have provided support for this endeavor. In particular, the authors acknowledge the general support over the years by NSF and AFOSR of USA, Purdue University; Hong Kong Research Grants Council, the National Natural Science Foundation of China, Hong Kong Baptist University; Singapore Ministry of Education and Nanyang Technological University. We are grateful to Mrs. Thanh-Ha Le Thi of Springer for her support and for tolerating our multiple delays, and to Ms. Xiaodan Zhao of Nanyang Technological University for carefully checking the manuscript. Last but not the least, we would like to thanks our wives and children for their love and support.

Indiana, USA — Jie Shen
Hong Kong, China — Tao Tang
Singapore — Li-Lian Wang

Contents

Symbol List

Common Notation

$\mathbb{C}$	Set of all complex numbers
$\mathbb{R}$	Set of all real numbers
$\mathbb{Z}$	Set of all integers
$\mathbb{N}$	Set of all nonnegative integers
P_N	Set of all real polynomials of degree $\leq N$
i	Complex unit, i.e., $\mathrm{i} = \sqrt{-1}$
δ_{mn}	Kronecker Delta symbol
Γ	Gamma function defined in (A.1)
$\cong$	$z_n \cong w_n$ means that for $w_n \neq 0,\ z_n/w_n \to 1$ as $n \to \infty$
$\sim$	$z_n \sim w_n$ means that for $w_n \neq 0,\ z_n/w_n \to C$ (independent of n) as $n \to \infty$
$\lesssim$	$z_n \lesssim w_n$ means that $z_n \leq C w_n$ with C independent of n

Orthogonal Polynomials/Functions

L_n	Legendre polynomial of degree n defined in (3.168)
T_n	Chebyshev polynomial of degree n defined in (3.207)
$J_n^{\alpha,\beta}$	Jacobi polynomial of degree n with parameter (α,β) defined in (3.110)
$J_n^{k,l}$	generalized Jacobi polynomial of degree n with $k,l \in \mathbb{Z}$ defined in (6.1)
$\mathscr{L}_n$	Laguerre polynomial of degree n defined in (7.4) with $\alpha = 0$
$\widehat{\mathscr{L}}_n$	Laguerre function of degree n defined in (7.16) with $\alpha = 0$
$\mathscr{L}_n^{(\alpha)}$	generalized Laguerre polynomial of degree n with parameter α defined in (7.4)
$\widehat{\mathscr{L}}_n^{(\alpha)}$	generalized Laguerre function of degree n with parameter α defined in (7.16)
H_n	Hermite polynomial of degree n defined in (7.58)
$\widehat{H}_n$	Hermite function of degree n defined in (7.71)

Weight Functions and Weighted Spaces of Functions

ω	A generic non-negative weight function
$\omega^{\alpha,\beta}$	Jacobi weight function: $\omega^{\alpha,\beta}(x) = (1-x)^{\alpha}(1+x)^{\beta}$
ω_α	Weight function associated with $\mathscr{L}_n^{(\alpha)}$, i.e., $\omega_\alpha(x) = x^\alpha e^{-x}$
$\hat{\omega}_\alpha$	Weight function associated with $\widehat{\mathscr{L}}_n^{(\alpha)}$, i.e., $\hat{\omega}_\alpha(x) = x^\alpha$
$L^p(\Omega)$	L^p-space on Ω with $1 \le p \le \infty$
$H^r(\Omega)$	Sobolev space on Ω
$H^r_\omega(\Omega)$	Weighted Sobolev space on Ω
$B^r_{\alpha,\beta}(I^d)$	Non-uniformly Jacobi-weighted Sobolev space defined in (3.251) ($d=1$) and in (8.125) with vector-valued $\boldsymbol{\alpha},\boldsymbol{\beta}$
$B^r_\alpha(\mathbb{R}_+)$	Non-uniformly weighted Sobolev space defined in (7.103)
$\hat{B}^r_\alpha(\mathbb{R}_+)$	Non-uniformly weighted Sobolev space defined in (7.110)
$\mathbb{K}^r_{\boldsymbol{\alpha},\boldsymbol{\beta}}(I^d)$	Jacobi-weighted Korobov-type space defined in (8.190)

Inner Products and Norms

$(\cdot,\cdot)_\omega$	Inner product of $L^2_\omega(\Omega)$
$(\cdot,\cdot)$	Inner product of $L^2(\Omega)$
$\|\cdot\|_\omega$	Norm of $L^2_\omega(\Omega)$
$\|\cdot\|_{r,\omega}$	Norm of $H^r_\omega(\Omega)$
$\vert\cdot\vert_{r,\omega}$	Semi-norm of $H^r_\omega(\Omega)$
$\|\cdot\|$	Norm of $L^2(\Omega)$
$\|\cdot\|_r$	Norm of $H^r(\Omega)$
$\vert\cdot\vert_r$	Semi-norm of $H^r(\Omega)$
$\|\cdot\|_\infty$	Norm of $L^\infty(\Omega)$
$\langle\cdot,\cdot\rangle_{N,\omega}$	Discrete inner product associated with a Gauss-type quadrature
$\langle\cdot,\cdot\rangle_N$	$\langle\cdot,\cdot\rangle_N = \langle\cdot,\cdot\rangle_{N,\omega}$ with $\omega \equiv 1$
$\|\cdot\|_{N,\omega}$	Discrete norm associated with $\langle\cdot,\cdot\rangle_{N,\omega}$

One-Dimensional Projection/Interpolation Operators

$\pi_N^{\alpha,\beta}$	$L^2_{\omega^{\alpha,\beta}}$-orthogonal projection operator defined in (3.249)
$\pi^1_{N,\alpha,\beta}$	$H^1_{\omega^{\alpha,\beta}}$-orthogonal projection operator defined in (3.269)
$\pi^{1,0}_{N,\alpha,\beta}$	$H^1_{0,\omega^{\alpha,\beta}}$-orthogonal projection operator defined in (3.290)
$I_N^{\alpha,\beta}$	Jacobi-Gauss-type interpolation operator
π_N, I_N	Operators $\pi_N^{\alpha,\beta}$, $I_N^{\alpha,\beta}$ with $\alpha=\beta=0$
π_N^c, I_N^c	Operators $\pi_N^{\alpha,\beta}$, $I_N^{\alpha,\beta}$ with $\alpha=\beta=-1/2$
$\Pi_{N,\alpha}$	Orthogonal projection operator in $L^2_{\omega_\alpha}(\mathbb{R}_+)$ defined in (7.102)
$\hat{\Pi}_{N,\alpha}$	Orthogonal projection operator in $L^2_{\hat{\omega}_\alpha}(\mathbb{R}_+)$ defined in (7.109)
Π_N	Orthogonal projection operator in $L^2_\omega(\mathbb{R})$ with $\omega = e^{-x^2}$ defined in (7.125)
$\hat{\Pi}_N$	Orthogonal projection operator defined in (7.128)
I_N^α, $\hat{I}_N^\alpha$	Laguerre-Gauss-type interpolation operators
I_N^h, $\hat{I}_N^h$	Hermite-Gauss interpolation operators

Chapter 1
Introduction

Numerical methods for partial differential equations can be classified into the *local* and *global* categories. The finite-difference and finite-element methods are based on local arguments, whereas the spectral method is global in character. In practice, finite-element methods are particularly well suited to problems in complex geometries, whereas spectral methods can provide superior accuracy, at the expense of domain flexibility. We emphasize that there are many numerical approaches, such as *hp* finite-elements and spectral-elements, which combine advantages of both the global and local methods. However in this book, we shall restrict our attentions to the *global* spectral methods.

Spectral methods, in the context of numerical schemes for differential equations, belong to the family of weighted residual methods (WRMs), which are traditionally regarded as the foundation of many numerical methods such as finite element, spectral, finite volume, boundary element (cf. Finlayson (1972)). WRMs represent a particular group of approximation techniques, in which the residuals (or errors) are minimized in a certain way and thereby leading to specific methods including Galerkin, Petrov-Galerkin, collocation and tau formulations.

The objective of this introductory chapter is to formulate spectral methods in a general way by using the notion of residual. Several important tools, such as *discrete transform* and *spectral differentiation*, will be introduced. These are basic ingredients for developing efficient spectral algorithms.

1.1 Weighted Residual Methods

Prior to introducing spectral methods, we first give a brief introduction to the WRM. Consider the general problem:

$$\partial_t u(x,t) - \mathscr{L}u(x,t) = \mathscr{N}(u)(x,t), \quad t > 0,\ x \in \Omega, \tag{1.1}$$

J. Shen et al., *Spectral Methods: Algorithms, Analysis and Applications*, Springer Series in Computational Mathematics 41, DOI 10.1007/978-3-540-71041-7_1,

where $\mathscr{L}$ is a leading spatial derivative operator, and $\mathscr{N}$ is a lower-order linear or nonlinear operator involving only spatial derivatives. Here, Ω denotes a bounded domain of $\mathbb{R}^d$, $d = 1,2$ or 3. Equation (1.1) is to be supplemented with an initial condition and suitable boundary conditions.

We shall only consider the WRM for the spatial discretization, and assume that the time derivative is discretized with a suitable time-stepping scheme. Among various time-stepping methods (cf. Appendix D), semi-implicit schemes or linearly-implicit schemes, in which the principal linear operators are treated *implicitly* to reduce the associated stability constraint, while the nonlinear terms are treated explicitly to avoid the expensive process of solving nonlinear equations at each time step, are most frequently used in the context of spectral methods.

Let τ be the time step size, and $u^k(\cdot)$ be an approximation of $u(\cdot,k\tau)$. As an example, we consider the Crank-Nicolson leap-frog scheme for (1.1):

$$\frac{u^{n+1}-u^{n-1}}{2\tau} - \mathscr{L}\Big(\frac{u^{n+1}+u^{n-1}}{2}\Big) = \mathscr{N}(u^n), \quad n \geq 1. \tag{1.2}$$

We can rewrite (1.2) as

$$\mathbf{L}u(x) := \alpha u(x) - \mathscr{L}u(x) = f(x), \quad x \in \Omega, \tag{1.3}$$

where, with a slight abuse of notation, $u = \frac{u^{n+1}+u^{n-1}}{2}$, $\alpha = \tau^{-1}$ and $f = \alpha u^{n-1} + \mathscr{N}(u^n)$. Hence, at each time step, we need to solve a steady-state problem of the form (1.3).

At this point, it is important to emphasize that the construction of efficient numerical solvers for some important equations in the form of (1.3), such as Poisson-type equations and advection-diffusion equations, is an essential step in solving general nonlinear PDEs. With this in mind, a particular emphasis of this book is to design and analyze efficient spectral algorithms for equations of the form (1.3) where $\mathscr{L}$ is a *linear elliptic* operator.

The starting point of the WRM is to approximate the solution u of (1.3) by a finite sum

$$u(x) \approx u_N(x) = \sum_{k=0}^{N} a_k \phi_k(x), \tag{1.4}$$

where $\{\phi_k\}$ are the *trial* (*or basis*) *functions*, and the expansion coefficients $\{a_k\}$ are to be determined. Substituting u_N for u in (1.3) leads to the *residual*:

$$\mathbf{R}_N(x) = \mathbf{L}u_N(x) - f(x) \neq 0, \quad x \in \Omega. \tag{1.5}$$

The notion of the WRM is to force the residual to zero by requiring

$$(\mathbf{R}_N, \psi_j)_\omega := \int_\Omega \mathbf{R}_N(x)\psi_j(x)\omega(x)dx = 0, \quad 0 \leq j \leq N, \tag{1.6}$$

where $\{\psi_j\}$ are the *test functions*, and ω is a positive weight function; or

$$\langle \mathbf{R}_N, \psi_j \rangle_{N,\omega} := \sum_{k=0}^{N} \mathbf{R}_N(x_k)\psi_j(x_k)\omega_k = 0, \quad 0 \le j \le N, \tag{1.7}$$

where $\{x_k\}_{k=0}^N$ are a set of preselected collocation points, and $\{\omega_k\}_{k=0}^N$ are the weights of a numerical quadrature formula.

The choice of trial/test functions is one of the main features that distinguishes spectral methods from finite-element and finite-difference methods. In the latter two methods, the trial/test functions are local in character with finite regularities. In contrast, spectral methods employ globally smooth functions as trial/test functions. The most commonly used trial/test functions are trigonometric functions or orthogonal polynomials (typically, the eigenfunctions of singular Sturm-Liouville problems), which include

- $\phi_k(x) = e^{ikx}$ (Fourier spectral method)
- $\phi_k(x) = T_k(x)$ (Chebyshev spectral method)
- $\phi_k(x) = L_k(x)$ (Legendre spectral method)
- $\phi_k(x) = \mathscr{L}_k(x)$ (Laguerre spectral method)
- $\phi_k(x) = H_k(x)$ (Hermite spectral method)

Here, $T_k, L_k, \mathscr{L}_k$ and H_k are the Chebyshev, Legendre, Laguerre and Hermite polynomials of degree k, respectively.

The choice of test functions distinguishes the following formulations:

- *Galerkin.* The test functions are the same as the trial ones (i.e., $\phi_k = \psi_k$ in (1.6) or (1.7)), assuming the boundary conditions are periodic or homogeneous.
- *Petrov-Galerkin.* The test functions are different from the trial ones.
- *Collocation.* The test functions $\{\psi_k\}$ in (1.7) are the Lagrange basis polynomials such that $\psi_k(x_j) = \delta_{jk}$, where $\{x_j\}$ are preassigned collocation points. Hence, the residual is forced to zero at $\{x_j\}$, i.e., $\mathbf{R}_N(x_j) = 0$.

Remark 1.1. *In the literature, the term of pseudo-spectral method is often used to describe any spectral method where some operations involve a collocation approach or a numerical quadrature which produces aliasing errors* (cf. Gottlieb and Orszag (1977)). *In this sense, almost all practical spectral methods are pseudo-spectral. In this book, we shall not classify a method as pseudo-spectral or spectral. Instead, it will be classified as Galerkin type or collocation type.*

Remark 1.2. *The so-called* tau *method is a particular class of Petrov-Galerkin method. While the* tau *method offers some advantages in certain situations, for most problems, it is usually better to use a well-designed Galerkin or Petrov-Galerkin method. So in this book, we shall not touch on this topic, and refer to* El-Daou and Ortiz (1998), Canuto et al. (2006) *and the references therein for a thorough discussion of this approach.*

In the forthcoming sections, we shall demonstrate how to construct spectral methods for solving differential equations by examining several spectral schemes based on Galerkin, Petrov-Galerkin and collocation formulations in a general manner. We shall revisit these illustrative examples in a more rigorous fashion in the main body of the book.

1.2 Spectral-Collocation Method

To fix the idea, we consider the following linear problem:

$$\begin{aligned} &\mathbf{L}u(x) = -u''(x) + p(x)u'(x) + q(x)u(x) = f(x), \quad x \in (-1,1), \\ &B_{\pm}u(\pm 1) = g_{\pm}, \end{aligned} \tag{1.8}$$

where $B_{\pm}$ are linear operators corresponding to Dirichlet, Neumann or Robin boundary conditions (see Sect. 4.1), and the data p, q, f and $g_{\pm}$ are given such that the above problem is well-posed.

As mentioned earlier, the collocation method forces the residual to vanish pointwisely at a set of preassigned points. More precisely, let $\{x_j\}_{j=0}^N$ (with $x_0 = -1$ and $x_N = 1$) be a set of Gauss-Lobatto points (see Chap. 3), and let P_N be the set of all real algebraic polynomials of degree $\le N$. The spectral-collocation method for (1.8) amounts to finding $u_N \in P_N$ such that (a) the residual $\mathbf{R}_N(x) = \mathbf{L}u_N(x) - f(x)$ equals to zero at the interior collocation points, namely,

$$\mathbf{R}_N(x_k) = \mathbf{L}u_N(x_k) - f(x_k) = 0, \quad 1 \le k \le N-1, \tag{1.9}$$

(b) u_N satisfies exactly the boundary conditions, i.e.,

$$B_- u_N(x_0) = g_-, \quad B_+ u_N(x_N) = g_+. \tag{1.10}$$

The spectral-collocation method is usually implemented in the physical space by seeking approximate solution in the form

$$u_N(x) = \sum_{j=0}^{N} u_N(x_j) h_j(x), \tag{1.11}$$

where $\{h_j\}$ are the Lagrange basis polynomials (also referred to as *nodal* basis functions), i.e., $h_j \in P_N$ and $h_j(x_k) = \delta_{kj}$. Hence, inserting (1.11) into (1.9)-(1.10) leads to the linear system

$$\begin{aligned} &\sum_{j=0}^{N} \big[\mathbf{L}h_j(x_k)\big] u_N(x_j) = f(x_k), \quad 1 \le k \le N-1, \\ &\sum_{j=0}^{N} \big[B_- h_j(x_0)\big] u_N(x_j) = g_-, \quad \sum_{j=0}^{N} \big[B_+ h_j(x_N)\big] u_N(x_j) = g_+. \end{aligned} \tag{1.12}$$

The above system contains $N+1$ equations and $N+1$ unknowns, so we can rewrite it in a matrix form. To fix the idea, we consider (1.8) with Dirichlet boundary conditions: $u(\pm 1) = g_{\pm}$. In this case, setting $u_N(x_0) = g_-$ and $u_N(x_N) = g_+$ in the first equation of (1.12), we find that the system (1.12) reduces to

$$\sum_{j=1}^{N-1} \left[\mathbf{L}h_j(x_k)\right] u_N(x_j) = f(x_k) - \left\{ \left[\mathbf{L}h_0(x_k)\right] g_- + \left[\mathbf{L}h_N(x_k)\right] g_+ \right\}, \tag{1.13}$$

for $1 \le k \le N-1$. Differentiating (1.11) m times leads to

$$u_N^{(m)}(x_k) = \sum_{j=0}^{N} d_{kj}^{(m)} u_N(x_j) \quad \text{where } d_{kj}^{(m)} = h_j^{(m)}(x_k). \tag{1.14}$$

The matrix $D^{(m)} = \left(d_{kj}^{(m)}\right)_{k,j=0,\dots,N}$ is called the differentiation matrix of order m relative to $\{x_j\}_{j=0}^N$. If we denote by $\mathbf{u}^{(m)}$ the vector whose components are the values of $u_N^{(m)}$ at the collocation points, it follows from (1.14) that

$$\mathbf{u}^{(m)} = D^{(m)}\mathbf{u}^{(0)}, \quad m \ge 1. \tag{1.15}$$

Hence, we have

$$\mathbf{L}h_j(x_k) = -d_{kj}^{(2)} + p(x_k)d_{kj}^{(1)} + q(x_k)\delta_{kj}. \tag{1.16}$$

Denote by $\mathbf{f}$ the vector with $N-1$ components given by the right-hand side of (1.13). Setting

$$\begin{aligned} &\widetilde{D}_m = \left(d_{kj}^{(m)}\right)_{k,j=1,\dots,N-1}, \quad m = 1,2, \\ &P = \operatorname{diag}\big(p(x_1),\dots,p(x_{N-1})\big), \quad Q = \operatorname{diag}\big(q(x_1),\dots,q(x_{N-1})\big), \end{aligned} \tag{1.17}$$

the system (1.13) reduces to

$$\left(-\widetilde{D}_2 + P\widetilde{D}_1 + Q\right)\mathbf{u}^{(0)} = \mathbf{f}. \tag{1.18}$$

Observe that the collocation method is easy to implement, once the differentiation matrices are precomputed. Moreover, it is very convenient for solving problems with variable coefficients and/or nonlinear problems, since we work in the physical space and derivatives can be evaluated by (1.14) directly. As a result, the collocation method has been extensively used in practice. However, three important issues should be considered in the implementation and analysis of a collocation method:

- The coefficient matrix of the collocation system is always full with a condition number behaving like $O(N^{2m})$ (m is the order of the differential equation).
- The choice of collocation points is crucial in terms of stability, accuracy and ease of dealing with boundary conditions. In general, they are chosen as nodes (typically, zeros of orthogonal polynomials) of Gauss-type quadrature formulas.
- The aforementioned collocation scheme is formulated in a *strong* form. In terms of error analysis, it is more convenient to reformulate it as a (but not always equivalent) *weak* form, see Sect. 1.3.3 and Chap. 4.

1.3 Spectral Methods of Galerkin Type

The collocation method described in the previous section is implemented in the physical space. In this section, we shall describe Galerkin-type spectral methods in the frequency space, and present the basic principles of the spectral-Galerkin method, spectral-Petrov-Galerkin method, and spectral-Galerkin method with numerical integration.

1.3.1 Galerkin Method

Without loss of generality, we consider (1.8) with $g_{\pm} = 0$. The non-homogeneous boundary conditions can be easily handled by considering $v = u - \tilde{u}$, where $\tilde{u}$ is a "simple" function satisfying the non-homogeneous boundary conditions (cf. Chap. 4).

Define the finite-dimensional approximation space:

$$X_N = \{\phi \in P_N \, : \, B_{\pm}\phi(\pm 1) = 0\} \;\Rightarrow\; \dim(X_N) = N-1.$$

Let $\{\phi_k\}_{k=0}^{N-2}$ be a set of basis functions of X_N. We expand the approximate solution as

$$u_N(x) = \sum_{k=0}^{N-2} \hat{u}_k \phi_k(x) \in X_N. \tag{1.19}$$

Then, the expansion coefficients $\{\hat{u}_k\}_{k=0}^{N-2}$ can be determined by the residual equation (1.6) with $\{\psi_j = \phi_j\}$:

$$\int_{-1}^{1} (\mathbf{L}u_N(x) - f(x))\,\phi_j(x)\omega(x)dx = 0, \quad 0 \le j \le N-2, \tag{1.20}$$

which is equivalent to

$$\begin{cases} \text{Find } u_N \in X_N \text{ such that} \\ \big(\mathbf{L}u_N, v_N\big)_\omega = \big(f, v_N\big)_\omega, \quad \forall v_N \in X_N. \end{cases} \tag{1.21}$$

Here, $(\cdot,\cdot)_\omega$ is the inner product of $L^2_\omega(-1,1)$ (cf. Appendix B).

The linear system of the above scheme is obtained by substituting (1.19) into (1.20). More precisely, setting

$$\mathbf{u} = \big(\hat{u}_0, \hat{u}_1, \ldots, \hat{u}_{N-2}\big)^T; \quad f_j = (f,\phi_j)_\omega, \quad \mathbf{f} = \big(f_0, f_1, \ldots, f_{N-2}\big)^T;$$
$$s_{jk} = \big(\mathbf{L}\phi_k, \phi_j\big)_\omega, \quad S = (s_{jk})_{j,k=0,\ldots,N-2},$$

the system (1.20) reduces to

$$S\mathbf{u} = \mathbf{f}. \tag{1.22}$$

Therefore, it is crucial to choose basis functions $\{\phi_j\}$ such that:

- The right-hand side $(f,\phi_j)_\omega$ can be computed efficiently.
- The linear system (1.22) can be solved efficiently.

The key idea is to use *compact combinations* of orthogonal polynomials or orthogonal functions to construct basis functions. To demonstrate the basic principle, we consider the Legendre spectral approximation (i.e., $\omega \equiv 1$ in (1.20)-(1.22)). Let $L_k(x)$ be the Legendre polynomial of degree k, and set

$$\phi_k(x) = L_k(x) + \alpha_k L_{k+1}(x) + \beta_k L_{k+2}(x), \quad k \geq 0, \tag{1.23}$$

where the constants α_k and β_k are uniquely determined by the boundary conditions: $B_\pm \phi_k(\pm 1) = 0$ (cf. Sect. 4.1). We shall refer to such basis functions as *modal* basis functions. Therefore, we have

$$X_N = \text{span}\{\phi_0, \phi_1, \ldots, \phi_{N-2}\}. \tag{1.24}$$

Using the properties of Legendre polynomials (cf. Sect. 3.3), one verifies easily that, *if $p(x)$ and $q(x)$ are constants*, the coefficient matrix S is *sparse* so the linear system (1.22) can be solved efficiently. However, for more general $p(x)$ and $q(x)$, the coefficient matrix S is full and one needs to resort to an iterative method (cf. Sect. 4.4).

In the above, we just considered the Legendre case. In fact, the construction of such a basis is also feasible for the Chebyshev, Laguerre and Hermite cases (see Chaps. 4–7). The notion of using compact combinations of orthogonal polynomials/functions to develop efficient spectral solvers will be repeatedly emphasized in this book.

We now consider the evaluation of $(f,\phi_j)_\omega$. In general, this term can not be computed exactly and is usually approximated by $(I_N f,\phi_j)_\omega$, where I_N is an interpolation operator upon P_N relative to the Gauss-Lobatto points. Thus, we can write

$$(I_N f)(x) = \sum_{k=0}^{N} \tilde{f}_k \varphi_k(x), \tag{1.25}$$

where $\{\varphi_k\}$ is an orthonormal polynomial basis of P_N (orthogonal with respect to ω, i.e., $(\varphi_k,\varphi_j)_\omega = \delta_{jk}$). Thanks to the orthogonality, the *discrete transforms* between the physical values $\{f(x_j)\}_{j=0}^N$ and the expansion coefficients $\{\tilde{f}_k\}_{k=0}^N$ can be computed efficiently. In particular, the computational complexity of the Fourier and Chebyshev discrete transforms can be reduced to $O(N\log_2 N)$ by using the fast Fourier transform (FFT). An approach for implementing discrete transforms relative to general orthogonal polynomials is given in Sect. 3.1.5.

It is important to point out that in solving time-dependent nonlinear problems, f usually contains nonlinear terms involving derivatives of the numerical solution u_N at previous time steps (cf. (1.3)). Hence, numerical differentiations in the frequency space and/or in the physical space are required. Differentiation techniques relative to general orthogonal polynomials are addressed in Sects. 3.1.6 and 3.1.7.

1.3.2 Petrov-Galerkin Method

As pointed out in Sect. 1.1, the use of different test and trial functions distinguishes the Petrov-Galerkin method from the Galerkin method. Thanks to this flexibility, the Petrov-Galerkin method can be very useful for some non-self-adjoint problems such as odd-order equations.

As an illustrative example, we consider the following third-order equation:

$$\begin{aligned} &\mathbf{L}u(x) := u'''(x) + u(x) = f(x), \quad x \in (-1,1), \\ &u(\pm 1) = u'(1) = 0. \end{aligned} \tag{1.26}$$

As with the Galerkin case, we enforce the boundary conditions on the approximate solution. So we set

$$X_N = \{\phi \in P_N : \phi(\pm 1) = \phi'(1) = 0\} \;\Rightarrow\; \dim(X_N) = N-2.$$

Assuming that $\{\phi_k\}_{k=0}^{N-3}$ is a basis of X_N, we expand the approximate solution as

$$u_N(x) = \sum_{k=0}^{N-3} \hat{u}_k \phi_k(x) \in X_N.$$

The expansion coefficients $\{\hat{u}_k\}_{k=0}^{N-3}$ are determined by the residual equation (1.6) (with $\omega = 1$):

$$\int_{-1}^{1} \big(\mathbf{L}u_N(x) - f(x)\big)\psi_j(x)dx = 0, \quad 0 \le j \le N-3. \tag{1.27}$$

Since the leading third-order operator is not self-adjoint, it is natural to use a Petrov-Galerkin method with the test function space:

$$X_N^* = \{\psi \in P_N : \psi(\pm 1) = \psi'(-1) = 0\} \;\Rightarrow\; \dim(X_N^*) = N-2.$$

Assume that $\{\psi_k\}_{k=0}^{N-3}$ is a basis of X_N^*. Then, (1.27) is equivalent to the variational formulation:

$$\begin{cases} \text{Find } u_N \in X_N \text{ such that} \\ \big(\mathbf{L}u_N, v_N\big) = \big(f, v_N\big), \quad \forall v_N \in X_N^*, \end{cases} \tag{1.28}$$

where $(\cdot,\cdot)$ is the inner product of the usual L^2-space.

The theoretical aspects of the above scheme will be examined in Chap. 6. We now consider its implementation. Setting

$$\begin{aligned} &\mathbf{u} = \big(\hat{u}_0, \hat{u}_1, \ldots, \hat{u}_{N-3}\big)^T; \quad f_j = (f, \psi_j), \quad \mathbf{f} = \big(f_0, f_1, \ldots, f_{N-3}\big)^T; \\ &s_{jk} = (\phi_k', \psi_j''), \quad S = \big(s_{jk}\big)_{j,k=0,\ldots,N-3}; \\ &m_{jk} = (\phi_k, \psi_j), \quad M = \big(m_{jk}\big)_{j,k=0,\ldots,N-3}, \end{aligned}$$

the linear system (1.28) becomes

$$(S+M)\mathbf{u}=\mathbf{f}. \tag{1.29}$$

As described in the previous section, we wish to construct basis functions for X_N and X_N^*, so that the linear system (1.29) can be inverted efficiently. Once again, this goal can be achieved by using compact combinations of orthogonal polynomials. It can be checked that for $0 \le k \le N-3$,

$$\begin{aligned} \phi_k &= L_k - \frac{2k+3}{2k+5}L_{k+1} - L_{k+2} + \frac{2k+3}{2k+5}L_{k+3} \in X_N; \\ \psi_k &= L_k + \frac{2k+3}{2k+5}L_{k+1} - L_{k+2} - \frac{2k+3}{2k+5}L_{k+3} \in X_N^*, \end{aligned} \tag{1.30}$$

where L_n is the Legendre polynomial of degree n (cf. Sect. 3.3). Hence, $\{\phi_k\}_{k=0}^{N-3}$ (resp. $\{\psi_j\}_{j=0}^{N-3}$) forms a basis of X_N (resp. X_N^*). Moreover, using the properties of the Legendre polynomials, one verifies easily that the matrix M is seven-diagonal, i.e., $m_{jk}=0$ for all $|j-k|>3$. More importantly, the matrix S is diagonal.

1.3.3 Galerkin Method with Numerical Integration

We considered previously Galerkin-type methods in the frequency space, which are well suited for linear problems with constant (or polynomial) coefficients. However, their implementations are not convenient for problems with general variable coefficients. On the other hand, the collocation method is easy to implement, but it can not always be reformulated as a suitable variational formulation (most convenient for error analysis). A combination of these two approaches leads to the so-called *Galerkin method with numerical integration*, or sometimes called the *collocation method in the weak form.*

The key idea of this approach is to *replace the continuous inner products in the Galerkin formulation by the discrete ones.* As an example, we consider again (1.8) with $g_\pm = 0$. The spectral-Galerkin method with numerical integration is

$$\begin{cases} \text{Find } u_N \in X_N := \{\phi \in P_N : B_\pm\phi(\pm 1) = 0\} \text{ such that} \\ a_N(u_N, v_N) := \langle Lu_N, v_N\rangle_N = \langle f, v_N\rangle_N, \quad \forall v_N \in X_N, \end{cases} \tag{1.31}$$

where the discrete inner product is defined by

$$\langle u,v\rangle_N = \sum_{j=0}^{N} u(x_j)v(x_j)\omega_j,$$

with $\{x_j,\omega_j\}_{j=0}^N$ being the set of Legendre-Gauss-Lobatto quadrature nodes and weights (cf. Theorem 3.29).

For problems with variable coefficients, the above method is easier to implement, thanks to the discrete inner product, than the spectral-Galerkin method (1.21). It is also more convenient for error analysis, thanks to the weak formulation, than the spectral-collocation method (1.12).

We note that in the particular case of homogeneous Dirichlet boundary conditions, i.e., $B_{\pm}u(\pm 1) = u(\pm 1) = 0$, by taking $v_N = h_j$, $1 \le j \le N-1$ in (1.31) and using the exactness of Legendre-Gauss-Lobatto quadrature, i.e.,

$$\langle u, v \rangle_N = (u, v), \quad \forall u \cdot v \in P_{2N-1}, \tag{1.32}$$

we find that the formulation (1.31) is equivalent to the collocation formulation (1.12). However, this is not true for general boundary conditions (see Chap. 4).

1.4 Fundamental Tools for Error Analysis

In the previous sections, we briefly described several families of spatial discretization schemes using the notion of weighted residual methods. In this section, we present some fundamental apparatuses for stability and convergence analysis of numerical schemes based on weak (or variational) formulations.

We consider the linear boundary value problem (1.3):

$$\mathbf{L}u = f, \quad \text{in } \Omega; \quad Bu = 0, \quad \text{on } \partial\Omega, \tag{1.33}$$

where $\mathbf{L}$ and B are linear operators, and f is a given function on Ω.

As shown before, the starting point is to reformulate (1.33) in a *weak formulation*:

$$\begin{cases} \text{Find } u \in X \text{ such that} \\ a(u, v) = F(v), \quad \forall v \in Y, \end{cases} \tag{1.34}$$

where X is the space of trial functions, Y is the space of test functions, and F is a linear functional on Y. The expression $a(u, v)$ defines a bilinear form on $X \times Y$. It is conventional to assume that X and Y are Hilbert spaces. We refer to Appendix B for basic functional analysis settings.

Now, we consider what conditions should be placed on (1.34) to guarantee its well-posedness in the sense that:

- *Existence-uniqueness:* There exists exactly one solution of the problem.
- *Stability:* The solution must be stable which means that it depends on the data continuously. In other words, a small change of the given data produces a small change of the solution correspondingly.

The first fundamental result concerning the existence-uniqueness and stability is known as the Lax-Milgram lemma (see Theorem B.1) related to the abstract problem (1.34) with $X = Y$, i.e.,

$$\begin{cases} \text{Find } u \in X \text{ such that} \\ a(u,v) = F(v), \quad \forall v \in X. \end{cases} \tag{1.35}$$

More precisely, if the bilinear form $a(\cdot,\cdot) : X \times X \to \mathbb{R}$ satisfies

- Continuity:

$$\exists C > 0 \quad \text{such that} \quad |a(u,v)| \le C\|u\|_X \|v\|_X, \tag{1.36}$$

- Coercivity:

$$\exists \alpha > 0 \quad \text{such that} \quad a(u,u) \ge \alpha \|u\|_X^2, \tag{1.37}$$

then for any $F \in X'$ (the dual space of X as defined in Appendix B), the problem (1.35) admits a unique solution $u \in X$, satisfying

$$\|u\|_X \le \frac{1}{\alpha} \|F\|_{X'}. \tag{1.38}$$

Remark 1.3. *The constant*

$$\alpha = \inf_{0 \ne u \in X} \frac{|a(u,u)|}{\|u\|_X^2} \tag{1.39}$$

is referred to as the ellipticity constant of (1.35).

The above result can only be applied to the problem (1.34) with $Y = X$. We now present a generalization of the Lax-Milgram lemma for the case $X \ne Y$ (see, e.g., Babuška and Aziz (1972)).

Theorem 1.1. *Let X and Y be two real Hilbert spaces, equipped with norms $\|\cdot\|_X$ and $\|\cdot\|_Y$, respectively. Assume that $a(\cdot,\cdot) : X \times Y \to \mathbb{R}$ is a bilinear form and $F(\cdot) : Y \to \mathbb{R}$ is a linear continuous functional, i.e., $F \in Y'$ (the dual space of Y) satisfying*

$$\|F\|_{Y'} = \sup_{0 \ne v \in Y} \frac{|F(v)|}{\|v\|_Y} < \infty. \tag{1.40}$$

Further, assume that $a(\cdot,\cdot)$ satisfies

- *Continuity:*

$$\exists C > 0 \quad \text{such that} \quad |a(u,v)| \le C\|u\|_X \|v\|_Y, \tag{1.41}$$

- *Inf-sup condition:*

$$\exists \beta > 0 \quad \text{such that} \quad \sup_{0 \ne v \in Y} \frac{|a(u,v)|}{\|u\|_X \|v\|_Y} \ge \beta, \quad \forall 0 \ne u \in X, \tag{1.42}$$

- *"Transposed" inf-sup condition:*

$$\sup_{0 \ne u \in X} |a(u,v)| > 0, \quad \forall 0 \ne v \in Y. \tag{1.43}$$

Then, for any $F \in Y'$, the problem (1.34) *admits a unique solution $u \in X$, which satisfies*

$$\|u\|_X \le \frac{1}{\beta}\|F\|_{Y'}. \tag{1.44}$$

Remark 1.4. *The condition* (1.42) *is also known as the Babuška-Brezzi inf-sup condition* (cf. Babuška (1973), Brezzi (1974)), *and the real number*

$$\beta = \inf_{0 \ne u \in X} \sup_{0 \ne v \in Y} \frac{|a(u,v)|}{\|u\|_X \|v\|_Y} \tag{1.45}$$

is called the inf-sup constant.

Remark 1.5. *Theorem* 1.1 *with $X = Y$ is not equivalent to the Lax-Milgram lemma. In fact, one can verify readily the relation between the ellipticity and inf-sup constants: $\alpha \le \beta$. Indeed, by* (1.37),

$$\alpha\|u\|_X \le \frac{|a(u,u)|}{\|u\|_X} \le \sup_{0 \ne v \in X} \frac{|a(u,v)|}{\|v\|_X}, \quad \forall 0 \ne u \in X,$$

which implies

$$\alpha \le \inf_{0 \ne u \in X} \sup_{0 \ne v \in X} \frac{|a(u,v)|}{\|u\|_X \|v\|_X} = \beta.$$

This means that one can have $\alpha = 0$ but $\beta > 0$. In other words, the bilinear form is not coercive, but satisfies the inf-sup condition.

We review below the fundamental theory on convergence analysis of numerical approximations to (1.34).

We first consider the case $X = Y$. Assume that $X_N \subseteq X$ and

$$\forall v \in X, \quad \inf_{v_N \in X_N} \|v - v_N\|_X \to 0 \quad \text{as} \quad N \to \infty. \tag{1.46}$$

The Galerkin approximation to (1.35) is

$$\begin{cases} \text{Find } u_N \in X_N \text{ such that} \\ a(u_N, v_N) = F(v_N), \quad \forall v_N \in X_N. \end{cases} \tag{1.47}$$

The stability and convergence of this scheme can be established by using the following lemma (cf. Céa (1964)):

Theorem 1.2. (Céa Lemma). *Under the assumptions of the Lax-Milgram lemma (see Theorem* B.1), *the problem* (1.47) *admits a unique solution $u_N \in X_N$ such that*

$$\|u_N\|_X \le \frac{1}{\alpha}\|F\|_{X'}. \tag{1.48}$$

Moreover, if u is the solution of (1.35), *we have*

$$\|u - u_N\|_X \le \frac{C}{\alpha} \inf_{v_N \in X_N} \|u - v_N\|_X. \tag{1.49}$$

Here, the constants C and α are given in (1.36) *and* (1.37), *respectively.*

Proof. Since X_N is a subspace of X, applying the Lax-Milgram lemma to (1.47) leads to the existence-uniqueness of u_N and the stability result (1.48). Now, taking $v = v_N$ in (1.35), and subtracting (1.47) from the resulting equation, we obtain the error equation

$$a(u-u_N, v_N) = 0, \quad \forall v_N \in X_N, \tag{1.50}$$

which, together with (1.36)-(1.37), implies

$$\begin{aligned}\alpha\|u-u_N\|_X^2 &\le a(u-u_N, u-u_N) = a(u-u_N, u-v_N)\\ &\le C\|u-u_N\|_X\|u-v_N\|_X, \quad \forall v_N \in X_N,\end{aligned}$$

from which (1.49) follows. □

Remark 1.6. *If, in addition, the bilinear form is symmetric, i.e., $a(u,v) = a(v,u)$, the Galerkin method is referred to as the Ritz method. In this case, the constant in the upper bound of* (1.49) *can be improved to $\sqrt{C\alpha^{-1}}$.*

Remark 1.7. *In performing error analysis of spectral methods, we usually take v_N in* (1.49) *to be a suitable orthogonal projection of u upon X_N, denoted by $\pi_N u$, which leads to*

$$\|u-u_N\|_X \le \frac{C}{\alpha}\|u-\pi_N u\|_X. \tag{1.51}$$

Hence, the error estimate follows from the approximation result on $\|u-\pi_N u\|_X$, which takes a typical form:

$$\|u-\pi_N u\|_X \le cN^{-\sigma(m)}\|u\|_{H^m}, \tag{1.52}$$

where c is a generic positive constant independent of N and any function, $\sigma(m) > 0$ is the so-called order of convergence in terms of the regularity index m, and H^m is a suitable Sobolev space with a norm involving derivatives of u up to m-th order. The establishment of such approximation results for each family of orthogonal polynomials/functions will be another emphasis of this book.

Typically, if u is sufficiently smooth, the estimate (1.52) *is valid for every m. However, for a finite-element method, the order of convergence is restricted by the order of local basis functions. The explicit dependence of the estimates of* (1.52) *type on the regularity index m will also be explored in this book.*

Observe that the bilinear form and the functional F in the discrete problem (1.47) are the same as those in the continuous problem (1.35). However, it is often convenient to use suitable approximate bilinear forms and/or functionals (see, for example, (1.31)). Hence, it is necessary to consider the following approximation to (1.35):

$$\begin{cases}\text{Find } u_N \in X_N \text{ such that}\\ a_N(u_N, v_N) = F_N(v_N), \quad \forall v_N \in X_N,\end{cases} \tag{1.53}$$

where X_N still satisfies (1.46), and $a_N(\cdot,\cdot)$ and $F_N(\cdot)$ are suitable approximations to $a(\cdot,\cdot)$ and $F(\cdot)$, respectively. In general, although X_N is a subspace of X, the properties of the discrete bilinear form can not carry over from those of the continuous one. Hence, they have to be derived separately.

The result below, known as the first *Strang lemma* (see, e.g., Strang and Fix (1973), Ciarlet (1978)), is a generalization of Theorem 1.2.

Theorem 1.3. (First Strang lemma). *Under the assumptions of the Lax-Milgram lemma, suppose further that the discrete forms $F_N(\cdot)$ and $a_N(\cdot,\cdot)$ satisfy the same properties in the subspace $X_N \subset X$, and $\exists\, \alpha_* > 0$, independent of N, such that*

$$a_N(v,v) \ge \alpha_* \|v\|_X^2, \quad \forall v \in X_N. \tag{1.54}$$

Then, the problem (1.53) *admits a unique solution $u_N \in X_N$, satisfying*

$$\|u_N\|_X \le \frac{1}{\alpha_*} \sup_{0 \ne v_N \in X_N} \frac{|F_N(v_N)|}{\|v_N\|_X}. \tag{1.55}$$

Moreover, if u is the solution of (1.35), *we have*

$$\begin{aligned}\|u-u_N\|_X \le \inf_{w_N \in X_N} \Big\{ \Big(1+\frac{C}{\alpha_*}\Big) \|u-w_N\|_X \\ + \frac{1}{\alpha_*} \sup_{0 \ne v_N \in X_N} \frac{|a(w_N,v_N)-a_N(w_N,v_N)|}{\|v_N\|_X} \Big\} \\ + \frac{1}{\alpha_*} \sup_{0 \ne v_N \in X_N} \frac{|F(v_N)-F_N(v_N)|}{\|v_N\|_X}.\end{aligned} \tag{1.56}$$

Here, the constant C is given in (1.36).

Proof. The existence-uniqueness and stability of (1.55) follow from the Lax-Milgram lemma. The proof of (1.56) is slightly different from that of (1.49). For any $w_N \in X_N$, let $e_N = u_N - w_N$. Using (1.54), (1.35) and (1.53) leads to

$$\begin{aligned}\alpha^* \|e_N\|_X^2 \le a_N(e_N,e_N) = a(u-w_N,e_N) + a(w_N,e_N) \\ - a_N(w_N,e_N) + F_N(e_N) - F(e_N).\end{aligned}$$

Since the result is trivial for $e_N = 0$, we derive from (1.36) that for $e_N \ne 0$,

$$\begin{aligned}\alpha^* \|e_N\|_X \le C\|u-w_N\|_X + \frac{|a(w_N,e_N)-a_N(w_N,e_N)|}{\|e_N\|_X} \\ + \frac{|F(e_N)-F_N(e_N)|}{\|e_N\|_X} \\ \le C\|u-w_N\|_X + \sup_{0 \ne v_N \in X_N} \frac{|a(w_N,v_N)-a_N(w_N,v_N)|}{\|v_N\|_X} \\ + \sup_{0 \ne v_N \in X_N} \frac{|F(v_N)-F_N(v_N)|}{\|v_N\|_X},\end{aligned}$$

which, together with the triangle inequality, yields

$$\|u - u_N\|_X \le \|u - w_N\|_X + \|e_N\|_X.$$

Finally, taking the infimum over $w_N \in X_N$ leads to the desired result. □

The previous discussions were restricted to approximations of the abstract problem (1.35) based on Galerkin-type formulations. Similar analysis can be done for the Petrov-Galerkin approximation of (1.34) by using Theorem 1.1. Indeed, let $X_N \subseteq X$ and $Y_N \subseteq Y$. Consider the approximation to (1.34):

$$\begin{cases} \text{Find } u_N \in X_N \text{ such that} \\ a(u_N, v_N) = F(v_N), \quad \forall v_N \in Y_N. \end{cases} \tag{1.57}$$

Unlike the coercivity property, the inf-sup property can not carry over from the whole space to the subspace. Indeed, the infimum in (1.39) will not decrease if it is taken on a subspace, whereas the supremum in the inf-sup constant (1.45), in general, becomes smaller on a subspace. Consequently, we have to prove

- Discrete inf-sup condition:

$$\exists \beta_* > 0 \quad \text{such that} \quad \sup_{0 \ne v_N \in Y_N} \frac{|a(u_N, v_N)|}{\|u_N\|_X \|v_N\|_Y} \ge \beta_*, \quad \forall 0 \ne u_N \in X_N, \tag{1.58}$$

- Discrete "transposed" inf-sup condition:

$$\sup_{0 \ne u_N \in X_N} |a(u_N, v_N)| > 0, \quad \forall 0 \ne v_N \in Y_N. \tag{1.59}$$

The following result, which is another generalization of Theorem 1.2, can be found in Babuška and Aziz (1972).

Theorem 1.4. *Under the assumptions of Theorem* 1.1, *assume further that* (1.58) *and* (1.59) *hold. Then the discrete problem* (1.57) *admits a unique solution* $u_N \in X_N$, *satisfying*

$$\|u_N\|_X \le \frac{1}{\beta_*} \|F\|_{Y'}. \tag{1.60}$$

Moreover, if u is the solution of (1.34), *we have*

$$\|u - u_N\|_X \le \left(1 + \frac{C}{\beta_*}\right) \inf_{v_N \in X_N} \|u - v_N\|_X, \tag{1.61}$$

where the constant C is given in (1.41).

Remark 1.8. *If we consider the following approximation to* (1.34):

$$\begin{cases} \text{Find } u_N \in X_N \text{ such that} \\ a_N(u_N, v_N) = F_N(v_N), \quad \forall v_N \in Y_N, \end{cases} \tag{1.62}$$

then a result similar to Theorem 1.3 *can be derived, provided that* (1.58) *and* (1.59) *hold in the subspaces* X_N *and* Y_N.

1.5 Comparative Numerical Examples

The aim of this section is to provide some illustrative numerical examples for a qualitative comparison of:

- Global versus local approximations
- Spectral-Galerkin versus spectral-collocation methods

in terms of accuracy, computational complexity and/or conditioning of the linear systems.

1.5.1 *Finite-Difference Versus Spectral-Collocation*

In order to illustrate the main differences between the finite-difference and spectral methods, we compare numerical differentiations of a periodic function u by using a fourth-order finite-difference method and a spectral-collocation method.

Given $h = \frac{2\pi}{N}$ and a uniform grid $\{x_0, x_1, \ldots, x_N\}$ with $x_j = jh$, and a set of physical values $\{u_0, u_1, \ldots, u_N\}$ with $u_j = u(x_j)$, a fourth-order centered finite-difference approximation to $u'(x_j)$ is

$$w_j := \frac{u_{j-2} - 8u_{j-1} + 8u_{j+1} - u_{j+2}}{12h}. \tag{1.63}$$

To account for periodicity of u, we set

$$u_{-2} = u_{N-1}, \quad u_{-1} = u_N, \quad u_0 = u_{N+1}, \quad u_1 = u_{N+2}.$$

Then, the differentiation process (1.63) can be expressed as

$$\begin{bmatrix} w_0 \\ w_1 \\ \vdots \\ \vdots \\ \vdots \\ \vdots \\ w_{N-1} \\ w_N \end{bmatrix} = \frac{1}{12h} \begin{bmatrix} & & \ddots & & & 1 & -8 \\ & & \ddots & -1 & & & 1 \\ & & \ddots & 8 & \ddots & & \\ & & \ddots & 0 & \ddots & & \\ & & \ddots & -8 & \ddots & & \\ -1 & & & 1 & \ddots & & \\ 8 & -1 & & & \ddots & & \end{bmatrix} \begin{bmatrix} u_0 \\ u_1 \\ \vdots \\ \vdots \\ \vdots \\ \vdots \\ u_{N-1} \\ u_N \end{bmatrix}. \tag{1.64}$$

Note that the coefficient matrix is sparse, reflecting the local nature of the finite-difference method.

On the other hand, the Fourier-collocation approximation of the function u is

$$\phi(x) = \sum_{k=0}^{N-1} h_k(x) u_k, \tag{1.65}$$

where $h_k(x_j) = \delta_{jk}$ and (cf. Lemma 2.2)

$$h_k(x) = \frac{1}{N} \frac{\sin\big(N(x-x_k)/2\big)}{\sin\big((x-x_k)/2\big)} \cos\big((x-x_k)/2\big). \tag{1.66}$$

Then, we approximate $u'(x_j)$ by

$$w_j = \phi'(x_j) = \sum_{k=0}^{N-1} h_k'(x_j) u_k, \quad j = 0, 1, \ldots, N-1, \tag{1.67}$$

where we have the explicit formula (cf. (2.34)):

$$h_k'(x_j) = \begin{cases} \frac{(-1)^{k+j}}{2} \cot\left[\frac{(j-k)\pi}{N}\right], & \text{if } j \neq k, \\ 0, & \text{if } j = k. \end{cases} \tag{1.68}$$

Thus, the matrix form of (1.67) becomes

$$\begin{bmatrix} w_0 \\ w_1 \\ \vdots \\ \vdots \\ \vdots \\ \vdots \\ w_{N-2} \\ w_{N-1} \end{bmatrix} = \frac{1}{2} \begin{bmatrix} \ddots & \vdots & \\ \ddots & -\cot\frac{2h}{2} & \\ \ddots & \cot\frac{h}{2} & \\ & 0 & \ddots \\ & -\cot\frac{h}{2} & \ddots \\ & \cot\frac{2h}{2} & \ddots \\ & \vdots & \end{bmatrix} \begin{bmatrix} u_0 \\ u_1 \\ \vdots \\ \vdots \\ \vdots \\ \vdots \\ u_{N-2} \\ u_{N-1} \end{bmatrix}. \tag{1.69}$$

Note that the coefficient matrix is full, reflecting the global nature of the spectral-collocation method. More detailed discussions of the Fourier method will be conducted in Chap. 2.

Next, we take $u(x) = \ln(2 + \sin x)$, which is 2π-periodic, and compare the exact derivative $u'(x) = \cos x/(2 + \sin x)$ with the numerical derivative $\{w_j\}$ obtained by the finite difference (1.64), and Fourier-collocation method (1.69) at the same grid. In Fig. 1.1, we plot the error $\max_{0 \le j \le N-1} |u'(x_j) - w_j|$ against various N. We observe a fourth-order convergence $O(h^4)$ (or $O(N^{-4})$) of the finite difference (1.64). We also observe that the Fourier-collocation method converges much faster than the

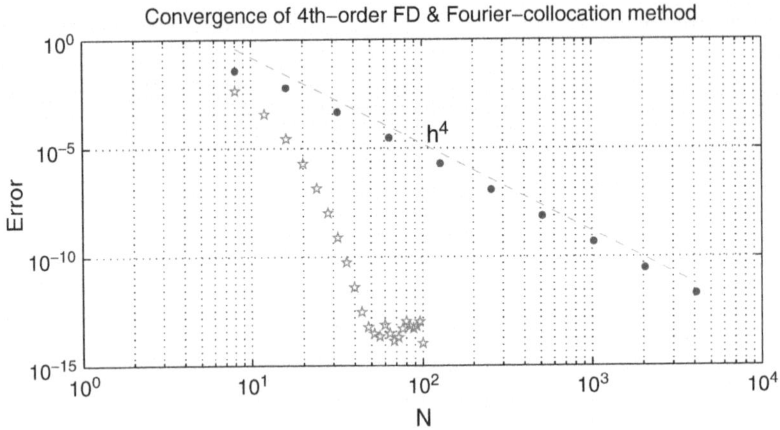

Fig. 1.1 Convergence of 4th-order finite difference (1.64) and Fourier-collocation (1.69) differentiation processes

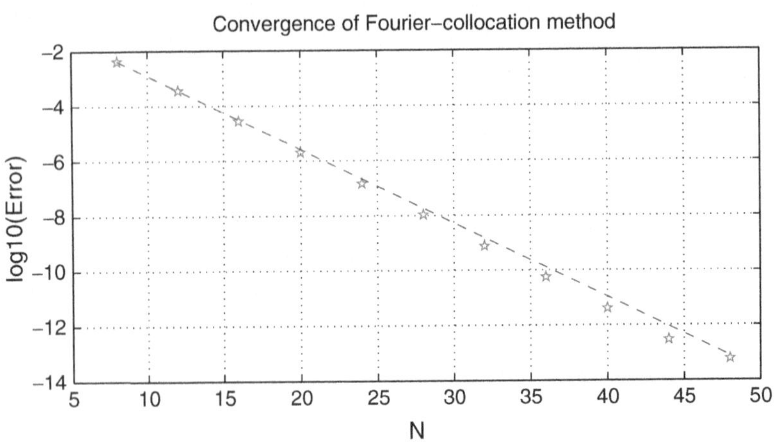

Fig. 1.2 Convergence of Fourier-collocation (1.69) differentiation process

finite difference method. To have a clearer picture of the convergence of the Fourier-collocation method (1.69), we plot in Fig. 1.2 the errors in the semi-log scale, which indicates an exponential convergence rate $O(e^{-cN})$ for some $c > 0$.

Remark 1.9. *The typical convergence behavior of a spectral method is $O(N^{-m})$ where m is a regularity index of the underlying function. In other words, its convergence rate is only limited by the regularity of the underlying function. A method exhibiting such a convergence behavior is often said to have spectral accuracy in the literature. On the other hand, the convergence rate of a finite element/finite difference method is limited by the order of the method, regardless of the regularity of the underlying function.*

A main advantage of a spectral method over a low-order finite element/finite difference method is that the former requires much fewer unknowns to resolve a given problem to a fixed accuracy, leading to potentially significant savings in storage and CPU time. For example, a rule of thumb (cf. Gottlieb and Orszag (1977)) *is that to achieve an engineering precision of* 1%, *a spectral method only needs* π *points per wave-length, as opposed to roughly ten points per wave-length required by a low-order method.*

Another important feature of spectral methods is that the derivatives of discrete functions are usually computed exactly (cf. (1.14)). *Therefore, spectral methods are usually free of phase errors, which can be very problematic for long-time integrations of partial differential equations.*

Remark 1.10. *If a function is analytic in a strip of width* 2β *(containing the underlying interval) in the complex plane, spectral approximations of such function can achieve an exponential convergence rate of* $O(e^{-\beta N})$. *We refer to* Davis (1975), Szegö (1975), *and* Gottlieb et al. (1992), Gottlieb and Shu (1997) *for such results on spectral projection errors, and to* Tadmor (1986) *for spectral differentiation errors (see* Reddy and Weideman (2005) *for a simpler analysis which also improved the estimates in* Tadmor (1986)). *Since the condition for an exponential convergence of order* $O(e^{-\beta N})$ *is quite generic, we shall not conduct analysis with exponential convergence in this book.*

1.5.2 Spectral-Galerkin Versus Spectral-Collocation

We compare in this section two versions of spectral methods: the Galerkin method in the frequency space and the collocation method in the physical space, in terms of the conditioning and round-off errors.

As an illustrative example, we consider the problem

$$u - u_{xx} = f, \quad u(\pm 1) = 0$$

with the exact solution $u(x) = \sin(10\pi x)$. In the comparison, the collocation solution is computed by (1.17)-(1.18) with $p = 0, q = \alpha$ and $g_{\pm} = 0$, while the Galerkin solution is obtained by solving (1.22) with the Legendre basis functions (cf. (1.23))

$$\phi_k(x) = \frac{1}{\sqrt{4k+6}}\big(L_k(x) - L_{k+2}(x)\big), \quad 0 \le k \le N-2.$$

Let us first examine the conditioning of the two linear systems. In Table 1.1, we list the condition numbers of the matrices resulted from the collocation method (COL) and the Galerkin method (GAL).

We see that for various α, the condition numbers of the GAL systems are all relatively small and independent of N, while those of the COL systems increase like $O(N^4)$.

Table 1.1 Condition numbers of COL and GAL

N	Method	$\alpha = 0$	$\alpha = 10$	$\alpha = 100$	$\alpha = 1,000$
32	COL	1.04E+04	2.05E+03	2.50E+02	2.64E+01
32	GAL	1.00	5.07	41.6	396
64	COL	1.71E+05	3.37E+04	4.09E+03	4.18E+02
64	GAL	1.00	5.07	41.6	407
128	COL	2.77E+06	5.47E+05	6.63E+04	6.78E+03
128	GAL	1.00	5.07	41.7	408
256	COL	4.46E+07	8.81E+06	1.07E+06	1.09E+05
256	GAL	1.00	5.07	41.7	408

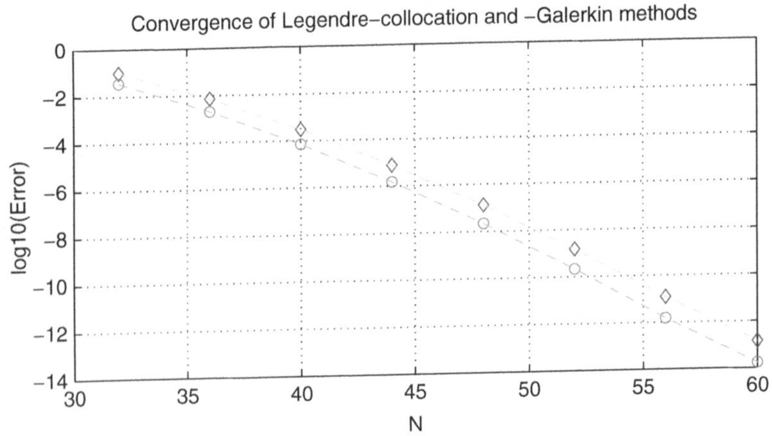

Fig. 1.3 Convergence: COL ("◇") vs. GAL ("○")

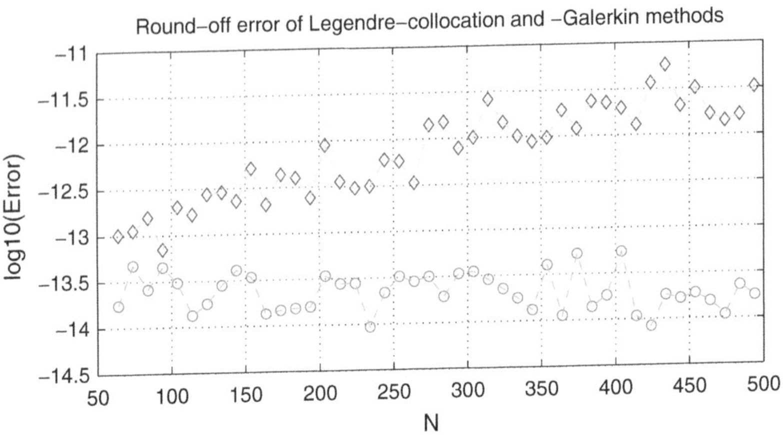

Fig. 1.4 Round-off errors: COL ("◇") vs. GAL ("○")

Next, we compare the effect of round-off errors. The maximum point-wise errors of two methods against various N are depicted in Figs. 1.3 and 1.4. We observe from Fig. 1.3 that, for relatively small N, both methods share essentially the same order of convergence rate. However, Fig. 1.4 indicates that the effect of roundoff errors may become severer in a collocation method as N becomes large.

The above comparison is performed on a simple one-dimensional model problem. It should be pointed out that similar behaviors can be expected for multidimensional and/or higher-order problems. Finally, we want to emphasize that in a collocation method, the choice of the collocation points (the quadrature nodes) should be in agreement with underlying differential equations and boundary conditions. For instance, the Gauss-Lobatto points are not suitable for third-order equations (cf. Huang and Sloan (1992), Merryfield and Shizgal (1993)). However, in a spectral-Galerkin method, the use of quadrature rules is merely to evaluate the integrals, so the usual Gauss-Lobatto quadrature works for the third-order equation as well.

Problems

1.1. Consider the heat equation

$$u_t(x,t) = u_{xx}(x,t), \ t > 0; \quad u(x,0) = u_0(x), \tag{1.70}$$

where $u_0(x)$ is 2π-periodic. We expand the periodic solution u in terms of Fourier series (cf. Sect. 2.1.1)

$$u(x,t) = \sum_{|k|=0}^{\infty} a_k(t)e^{\mathrm{i}kx} \text{ with } a_k(t) = \frac{1}{2\pi}\int_0^{2\pi} u(x,t)e^{-\mathrm{i}kx}dx, \tag{1.71}$$

where $\mathrm{i} = \sqrt{-1}$ is the complex unit.

(a) Show that

$$a_k(t) = e^{-k^2t}a_k(0), \quad \forall t \geq 0, \ k \in \mathbb{Z},$$

where

$$a_k(0) = \frac{1}{2\pi}\int_0^{2\pi} u_0(x)e^{-ikx}dx.$$

(b) Let

$$u_N(x,t) := \sum_{|k|=0}^{N-1} a_k(t)e^{ikx}.$$

Show that

$$\|(u-u_N)(\cdot,t)\|_\infty \leq ct^{-1/2}\|u_0\|_\infty \operatorname{erfc}(\sqrt{t}N),$$

where $\|v\|_\infty = \max_{x\in[0,2\pi]}|v(x)|$, and $\operatorname{erfc}(x)$ is the complementary error function defined by

$$\operatorname{erfc}(x) = \frac{2}{\sqrt{\pi}}\int_x^{\infty} e^{-y^2}dy.$$

(c) Use the property

$$\operatorname{erfc}(x) \cong \frac{e^{-x^2}}{\sqrt{\pi}x}, \quad x \gg 1,$$

to prove that

$$\|(u - u_N)(\cdot, t)\|_\infty \leq c t^{-3/2} N^{-1} e^{-tN^2}, \quad \forall t > 0.$$

Chapter 2
Fourier Spectral Methods for Periodic Problems

The spectral method was introduced in Orszag's pioneer work on using Fourier series for simulating incompressible flows about four decades ago (cf. Orszag (1971)). The word "spectral" was probably originated from the fact that the Fourier series are the eigenfunctions of the Laplace operator with periodic boundary conditions. This fact and the availability of the fast Fourier transform (FFT) are two major reasons for the extensive applications of Fourier methods to problems with periodic boundary conditions. In practice, a variety of physical problems exhibit periodicity. For instance, some problems are geometrically and physically periodic, such as crystal structures and homogeneous turbulence. On the other hand, many problems of scientific interest, such as the interaction of solitary waves and homogeneous turbulence, can be modeled by PDEs with periodic boundary conditions. Furthermore, even if an original problem is not periodic, the periodicity may be induced by using coordinate transforms, such as polar, spherical and cylindrical coordinates. Indeed, there are numerous circumstances where the problems are periodic in one or two directions, and non-periodic in other directions. In such cases, it is natural to use Fourier series in the periodic directions and other types of spectral expansions, such as Legendre or Chebyshev polynomials, in the non-periodic directions (cf. Chap. 7).

The objective of this chapter is to study some computational and theoretical aspects of Fourier spectral methods for periodic problems. In the first section, we introduce the continuous and discrete Fourier series, and examine the fundamental spectral techniques including discrete Fourier transforms, Fourier differentiation matrices and Fourier spectral differentiation based on FFT. The approximation properties of continuous and discrete Fourier series are surveyed in the second section. The applications of Fourier spectral methods to some linear and nonlinear problems are presented in the last section. For more detail and other aspects of Fourier approximations, we refer to Gottlieb and Orszag (1977), Gottlieb et al. (1984), Boyd (2001) and the references therein.

J. Shen et al., *Spectral Methods: Algorithms, Analysis and Applications*, Springer Series in Computational Mathematics 41, DOI 10.1007/978-3-540-71041-7_2,

2.1 Continuous and Discrete Fourier Transforms

This section is devoted to a brief review of the properties of Fourier series and Fourier transforms. Our focus is put on the discrete Fourier transforms and Fourier differentiation techniques, which play an important role in the Fourier spectral methods.

2.1.1 Continuous Fourier Series

We denote the complex exponentials by

$$E_k(x) := e^{\mathrm{i}kx} = \cos kx + \mathrm{i}\sin kx = (\cos x + \mathrm{i}\sin x)^k, \quad k \in \mathbb{Z},\ x \in \mathbb{R},$$

where $\mathrm{i} = \sqrt{-1}$. The set $\{e^{\mathrm{i}kx} : k \in \mathbb{Z}\}$ forms a complete orthogonal system in the complex Hilbert space $L^2(0,2\pi)$, equipped with the inner product and the norm

$$(u,v) = \frac{1}{2\pi}\int_0^{2\pi} u(x)\bar{v}(x)dx, \quad \|u\| = \sqrt{(u,u)}\,,$$

where $\bar{v}$ is the complex conjugate of v. The orthogonality of $\{E_k : k \in \mathbb{Z}\}$ reads

$$(E_k, E_m) = \frac{1}{2\pi}\int_0^{2\pi} e^{\mathrm{i}(k-m)x}dx = \delta_{km}, \tag{2.1}$$

where δ_{km} is the Kronecker Delta symbol.

For any complex-valued function $u \in L^2(0,2\pi)$, its *Fourier series* is defined by

$$u(x) \sim \mathscr{F}(u)(x) := \sum_{k=-\infty}^{\infty} \hat{u}_k e^{\mathrm{i}kx}, \tag{2.2}$$

where the *Fourier coefficients* are given by

$$\hat{u}_k = (u, e^{\mathrm{i}kx}) = \frac{1}{2\pi}\int_0^{2\pi} u(x)e^{-\mathrm{i}kx}dx. \tag{2.3}$$

It is clear that if u is a real-valued function, its Fourier coefficients satisfy

$$\hat{u}_{-k} = \bar{\hat{u}}_k, \quad k \in \mathbb{Z}, \tag{2.4}$$

and $\hat{u}_0$ is obviously real.

In fact, the Fourier series can be defined for general absolutely integrable functions in $(0,2\pi)$, and the convergence theory of Fourier expansions in different senses has been subjected to a rigorous and thorough investigation in Fourier analysis (see, e.g., Zygmund (2002), Stein and Shakarchi (2003)). It is well-known that, for any $u \in L^2(0,2\pi)$, its truncated Fourier series $\mathscr{F}_N(u) := \sum_{|k|\le N} \hat{u}_k e^{ikx}$ converges to u in the L^2-sense, and there holds the Parseval's identity:

$$\|u\|^2 = \sum_{k=-\infty}^{\infty} |\hat{u}_k|^2. \tag{2.5}$$

If u is continuous, periodic and of bounded variation on $[0,2\pi]$, then $\mathscr{F}_N(u)$ uniformly converges to u.

Notice that the truncated Fourier series can also be expressed in the convolution form, namely,

$$\mathscr{F}_N(u)(x) = (\mathscr{D}_N * u)(x) = \frac{1}{2\pi}\int_0^{2\pi} \mathscr{D}_N(x-t)u(t)dt, \tag{2.6}$$

where $\mathscr{D}_N$ is known as the *Dirichlet kernel* given by

$$\mathscr{D}_N(x) := \sum_{k=-N}^{N} e^{ikx} = 1 + 2\sum_{k=1}^{N} \cos kx = \frac{\sin((N+1/2)x)}{\sin(x/2)}. \tag{2.7}$$

It is sometimes convenient to express the Fourier series in terms of the trigonometric polynomials:

$$u(x) \sim \mathscr{S}(u)(x) = \frac{a_0}{2} + \sum_{k=1}^{\infty} \big(a_k \cos kx + b_k \sin kx\big), \tag{2.8}$$

where the expansion coefficients are

$$a_k = \frac{1}{\pi}\int_0^{2\pi} u(x)\cos kx dx, \quad b_k = \frac{1}{\pi}\int_0^{2\pi} u(x)\sin kx dx.$$

The coefficients of the two different representations (2.2) and (2.8) are related by

$$\hat{u}_0 = \frac{a_0}{2}, \quad \hat{u}_k = \begin{cases} \dfrac{a_k - ib_k}{2}, & \text{if } k \ge 1, \\[2mm] \dfrac{a_{-k} + ib_{-k}}{2}, & \text{if } k \le -1. \end{cases} \tag{2.9}$$

In particular, if u is a real-valued function, then

$$a_0 = 2\hat{u}_0, \quad a_k = 2\text{Re}(\hat{u}_k), \quad b_k = -2\text{Im}(\hat{u}_k), \quad k \ge 1. \tag{2.10}$$

2.1.2 Discrete Fourier Series

Given a positive integer N, let

$$x_j = jh = j\frac{2\pi}{N}, \quad 0 \le j \le N-1, \tag{2.11}$$

be the N-equispaced grids in $[0,2\pi)$, which are referred to as the Fourier collocation points. We define the discrete inner product by

$$\langle u,v\rangle_N = \frac{1}{N}\sum_{j=0}^{N-1} u(x_j)\bar{v}(x_j). \tag{2.12}$$

The following lemma is the discrete counterpart of (2.1).

Lemma 2.1. *Let $E_l(x) = e^{\mathrm{i}lx}$. For any integer $N \geq 1$, we have*

$$\langle E_k, E_m\rangle_N = \begin{cases} 1, & \text{if } k-m = lN,\ \forall l \in \mathbb{Z}, \\ 0, & \text{otherwise.} \end{cases} \tag{2.13}$$

Proof. Observe that if $k-m$ is not divisible by N, then

$$\begin{aligned}\langle E_k, E_m\rangle_N &= \frac{1}{N}\sum_{j=0}^{N-1} e^{\mathrm{i}(k-m)x_j} = \frac{1}{N}\sum_{j=0}^{N-1}\left(e^{2\pi\mathrm{i}(k-m)/N}\right)^j \\ &= \frac{1}{N}\frac{e^{2\pi\mathrm{i}(k-m)}-1}{e^{2\pi\mathrm{i}(k-m)/N}-1} = 0.\end{aligned}$$

If $k-m$ is divisible by N, we have $e^{2\pi\mathrm{i}(k-m)/N} = 1$, so the summation in the second line above equals to 1. □

In general, the Fourier coefficients $\{\hat{u}_k\}$ in (2.3) can not be evaluated exactly, so we have to resort to some quadrature formula. A simple and accurate quadrature formula for 2π-periodic functions is the rectangular rule

$$\frac{1}{2\pi}\int_0^{2\pi} v(x)dx \approx \frac{1}{N}\sum_{j=0}^{N-1} v(x_j), \quad \forall v \in C[0,2\pi), \tag{2.14}$$

which is exact for all

$$v \in \operatorname{span}\left\{e^{\mathrm{i}kx} : 0 \leq |k| \leq N-1\right\}.$$

Moreover, one verifies readily that (2.14) is also exact for $v = \sin(\pm Nx)$ but not for $v = \cos(\pm Nx)$.

Applying (2.14) to (2.3) leads to the approximation

$$\hat{u}_k \approx \tilde{u}_k := \frac{1}{N}\sum_{j=0}^{N-1} u(x_j)e^{-\mathrm{i}kx_j}, \quad k = 0, \pm 1, \ldots. \tag{2.15}$$

Note that $\{\tilde{u}_k\}$ are N-periodic, that is,

$$\tilde{u}_{k\pm N} = \frac{1}{N}\sum_{j=0}^{N-1} u(x_j)e^{-\mathrm{i}(k\pm N)x_j} = \frac{1}{N}\sum_{j=0}^{N-1} u(x_j)e^{-\mathrm{i}kx_j}e^{\mp 2\pi\mathrm{i}j} = \tilde{u}_k,$$

which implies that for even N, we have

$$\tilde{u}_{-N/2} = \tilde{u}_{N/2}. \tag{2.16}$$

Hence, for even N, the grids $\{x_j\}_{j=0}^{N-1}$ can not distinguish the modes: $k = \pm N/2$, since

$$e^{\mathrm{i}Nx_j/2} = e^{\mathrm{i}j\pi} = (-1)^j = e^{-\mathrm{i}Nx_j/2}, \quad 0 \le j \le N-1. \tag{2.17}$$

In other words, the two modes $k = \pm N/2$ are *aliased*.

In order to have an effective implementation of the discrete Fourier transform (DFT), it is preferable to use an even N, and accordingly, a symmetric finite set of modes: $-N/2 \le k \le N/2$ in the discrete Fourier series (cf. (2.20) below). In view of (2.16)–(2.17), we redefine the approximation (2.15) by modifying the two modes $k = \pm N/2$:

$$\hat{u}_k \approx \tilde{u}_k = \frac{1}{Nc_k} \sum_{j=0}^{N-1} u(x_j) e^{-\mathrm{i}kx_j}, \quad k = -N/2, \ldots, N/2, \tag{2.18}$$

where $c_k = 1$ for $|k| < N/2$, and $c_k = 2$ for $k = \pm N/2$. The expression (2.18) is referred to as the (forward) *discrete Fourier transform* of $u(x)$ associated with the grid points in (2.11).

Due to (2.16), there are only N independent coefficients. Hence, we set

$$\mathscr{T}_N = \Big\{ u = \sum_{k=-N/2}^{N/2} \tilde{u}_k e^{\mathrm{i}kx} : \tilde{u}_{-N/2} = \tilde{u}_{N/2} \Big\}, \tag{2.19}$$

and define the mapping $I_N : C[0, 2\pi) \to \mathscr{T}_N$ by

$$(I_N u)(x) = \sum_{k=-N/2}^{N/2} \tilde{u}_k e^{\mathrm{i}kx}, \tag{2.20}$$

with $\{\tilde{u}_k\}$ given by (2.18). The following lemma shows that I_N is the interpolation operator from $C[0, 2\pi)$ to $\mathscr{T}_N$ such that

$$(I_N u)(x_j) = u(x_j), \quad x_j = \frac{2\pi j}{N}, \quad 0 \le j \le N-1. \tag{2.21}$$

Lemma 2.2. *For any* $u \in C[0, 2\pi)$,

$$(I_N u)(x) = \sum_{j=0}^{N-1} u(x_j) h_j(x), \tag{2.22}$$

where

$$h_j(x) = \frac{1}{N} \sin\left[N \frac{x - x_j}{2}\right] \cot\left[\frac{x - x_j}{2}\right] \in \mathscr{T}_N \tag{2.23}$$

satisfying

$$h_j(x_k) = \delta_{jk}, \quad \forall j,k = 0,1,\ldots,N-1. \tag{2.24}$$

Proof. By (2.18) and (2.20),

$$\begin{aligned}
(I_N u)(x) &= \sum_{k=-N/2}^{N/2} \left(\frac{1}{Nc_k} \sum_{j=0}^{N-1} u(x_j) e^{-\mathrm{i}kx_j} \right) e^{\mathrm{i}kx} \\
&= \sum_{j=0}^{N-1} \left(\frac{1}{N} \sum_{k=-N/2}^{N/2} \frac{1}{c_k} e^{\mathrm{i}k(x-x_j)} \right) u(x_j) \\
&=: \sum_{j=0}^{N-1} h_j(x) u(x_j).
\end{aligned}$$

We derive from (2.7) and a direct calculation that

$$\begin{aligned}
h_j(x) &= \frac{1}{N} \sum_{k=-N/2}^{N/2} \frac{1}{c_k} e^{\mathrm{i}k(x-x_j)} \\
&= \frac{1}{N} \left(D_{N/2-1}(x-x_j) + \cos\left[N \frac{x-x_j}{2} \right] \right) \\
&= \frac{1}{N} \left(\frac{\sin\left[(N-1)\frac{x-x_j}{2} \right]}{\sin\frac{x-x_j}{2}} + \cos\left[N\frac{x-x_j}{2} \right] \right) \\
&= \frac{1}{N} \sin\left[N\frac{x-x_j}{2} \right] \cot\left[\frac{x-x_j}{2} \right].
\end{aligned} \tag{2.25}$$

Due to (2.17), we have $h_j(x) \in \mathscr{T}_N$, and it is clear that $h_j(x_i) = 0$ for $i \neq j$. Moreover, taking $x = x_j$ in the first identity of (2.25) yields $h_j(x_j) = 1$. □

Taking $x = x_j$ in (2.20) and using (2.21), leads to the *inverse (or backward) discrete transform*:

$$u(x_j) = \sum_{k=-N/2}^{N/2} \tilde{u}_k e^{\mathrm{i}kx_j}, \quad j = 0,1,\ldots,N-1. \tag{2.26}$$

It is obvious that the discrete Fourier transforms (2.18) and its inverse (2.26) can be carried out through matrix–vector multiplication with $O(N^2)$ operations. However, thanks to the fast Fourier transforms due to Cooley and Tukey (1965), such processes can be accomplished with $O(N\log_2 N)$ operations. Moreover, if u is a real valued function, then $\tilde{u}_{-k} = \overline{\tilde{u}}_k$, so only half of the coefficients in (2.26) need to be computed/stored.

The computational routines for FFT and IFFT are available in many software packages. Here, we restrict our attentions to their implementations in MATLAB. Given the data $\{\mathbf{v}(j) = u(x_{j-1})\}_{j=1}^{N}$ sampled at $\{x_k = 2\pi k/N\}_{k=0}^{N-1}$, the command "$\tilde{\mathbf{v}} = \mathtt{fft}(\mathbf{v})$" returns the vector $\{\tilde{\mathbf{v}}(k)\}_{k=1}^{N}$, defined by

$$\tilde{\mathbf{v}}(k) = \sum_{j=1}^{N} \mathbf{v}(j) e^{-2\pi \mathrm{i}(j-1)(k-1)/N}, \quad 1 \le k \le N, \tag{2.27}$$

while the inverse FFT can be computed with the command "$\mathbf{v} = \texttt{ifft}(\tilde{\mathbf{v}})$" which returns the physical values $\{\mathbf{v}(j)\}_{j=1}^{N}$ via

$$\mathbf{v}(j) = \frac{1}{N} \sum_{k=1}^{N} \tilde{\mathbf{v}}(k) e^{2\pi \mathrm{i}(j-1)(k-1)/N}, \quad 1 \le j \le N. \tag{2.28}$$

Notice that some care has to be taken for the ordering of the modes. To illustrate this, we examine the one-to-one correspondence of the transforms in (2.18) and (2.26). More precisely, let

$$u(x_j) = \mathbf{v}(j+1), \quad x_j = \frac{2\pi j}{N}, \quad 0 \le j \le N-1. \tag{2.29}$$

We find that

$$\begin{aligned} \tilde{u}_k &= \frac{1}{N}\tilde{\mathbf{v}}(k+1), \qquad 0 \le k \le \frac{N}{2} - 1, \\ \tilde{u}_k &= \frac{1}{N}\tilde{\mathbf{v}}(k+N+1), \quad -\frac{N}{2} + 1 \le k \le -1, \\ \tilde{u}_{-N/2} &= \tilde{u}_{N/2} = \frac{1}{2N}\tilde{\mathbf{v}}(N/2+1). \end{aligned} \tag{2.30}$$

A tabulated view of the above relations is given in the following table.

Table 2.1 Correspondence of DFT and FFT & IFFT in MATLAB

j	1	2	...	$N/2-1$	$N/2$	$N/2+1$	$N/2+2$	...	$N-1$	N
$\mathbf{u} = \mathbf{v}$	u_0	u_1	...	...	...	...	...	...	u_{N-2}	u_{N-1}
$\tilde{\mathbf{u}} = \tilde{\mathbf{v}}/N$	$\tilde{u}_0$	$\tilde{u}_1$	...	$\tilde{u}_{N/2-2}$	$2\tilde{u}_{N/2-1}$	$\tilde{u}_{N/2}$	$\tilde{u}_{-N/2+1}$	...	$\tilde{u}_{-2}$	$\tilde{u}_{-1}$
$\mathbf{k}$	0	1	...	$N/2-2$	$N/2-1$	0	$-N/2+1$	...	-2	-1

In the table, we denote $\{u_j = \mathbf{u}(j)\}_{j=0}^{N-1}$ and $\{\tilde{u}_k = \tilde{\mathbf{u}}(k)\}_{k=-N/2}^{N/2}$. The last row gives the frequency vector $\mathbf{k}$ for the Fourier-spectral differentiation based on FFT, see Sect. 2.1.4 below. Note that the frequency $-N/2$ is aliased with the frequency $N/2$ in the discrete Fourier transform.

2.1.3 *Differentiation in the Physical Space*

In a Fourier spectral method, differentiation can be performed in the physical space as well as in the frequency space.

We start with differentiation in the physical space. Let $\{x_j\}$ and $\{h_j\}$ be defined in (2.11) and (2.23), respectively. Setting

$$u(x) = \sum_{j=0}^{N-1} u(x_j)h_j(x), \tag{2.31}$$

and taking the m-th derivative, we get

$$u^{(m)}(x) = \sum_{j=0}^{N-1} u(x_j)h_j^{(m)}(x). \tag{2.32}$$

This process can be formulated as a matrix–vector multiplication

$$\mathbf{u}^{(m)} = D^{(m)}\mathbf{u}, \quad m \geq 0, \tag{2.33}$$

where

$$\begin{aligned} D^{(m)} &= \big(d_{kj}^{(m)} := h_j^{(m)}(x_k)\big)_{k,j=0,\ldots,N-1}, \\ \mathbf{u} &= \big(u(x_0), u(x_1), \ldots, u(x_{N-1})\big)^T, \\ \mathbf{u}^{(m)} &= \big(u^{(m)}(x_0), u^{(m)}(x_1), \ldots, u^{(m)}(x_{N-1})\big)^T. \end{aligned}$$

In particular, we denote $D = D^{(1)}$. The compact form of the first-order differentiation matrix is given below.

Lemma 2.3. *The entries of the first-order Fourier differentiation matrix D are determined by*

$$d_{kj}^{(1)} = h_j'(x_k) = \begin{cases} \frac{(-1)^{k+j}}{2}\cot\left[\frac{(k-j)\pi}{N}\right], & \text{if } k \neq j, \\ 0, & \text{if } k = j. \end{cases} \tag{2.34}$$

Proof. Differentiating the Lagrange basis in (2.23) directly gives

$$\begin{aligned} h_j'(x) =& \frac{1}{2}\cos\left[N\frac{x-x_j}{2}\right]\cot\left[\frac{x-x_j}{2}\right] \\ &- \frac{1}{2N}\sin\left[N\frac{x-x_j}{2}\right]\csc^2\left[\frac{x-x_j}{2}\right]. \end{aligned}$$

It is clear that if $x = x_k \neq x_j$, then the second term is 0 and the first term can be simplified into the desired expression in (2.34).

We now consider the case $k = j$. For convenience, let $\theta = (x - x_j)/2$, and rewrite the above formula as

$$h_j'(x) = \frac{1}{2}\frac{\cos(N\theta)\cos\theta\sin\theta - N^{-1}\sin(N\theta)}{\sin^2\theta}. \tag{2.35}$$

Using the Taylor expansion, we find

$$\cos(N\theta)\cos\theta\sin\theta = \theta + O(\theta^3), \quad N^{-1}\sin(N\theta) = \theta + O(\theta^3), \quad |\theta| \ll 1.$$

Hence, we derive from (2.35) that $h_j'(x_j) = \lim_{x\to x_j} h_j'(x) = 0$, since $\theta \to 0$ as $x \to x_j$.
□

Remark 2.1. *The first-order Fourier differentiation matrix has the following properties:*

- *D is a real and skew-symmetric matrix, since* $\cot(-x) = -\cot(x)$ *and* $d_{kk} = 0$.
- *D is a circulant Toeplitz matrix, since* $d_{kj} = d_{k+1,j+1}$.
- *The distinct eigenvalues of D are* $\{\mathrm{i}k : -N/2+1 \le k \le N/2-1\}$, *and the eigenvalue* 0 *has a multiplicity* 2.

The approximation of higher-order derivatives follows the same procedure. From the first relation in (2.25), we find

$$h_j^{(m)}(x_i) = \frac{1}{N}\sum_{k=-N/2}^{N/2} \frac{(\mathrm{i}k)^m}{c_k} e^{2\pi \mathrm{i}k(i-j)/N}. \tag{2.36}$$

In particular, the entries of the second-order differentiation matrix $D^{(2)}$ are given by

$$d_{kj}^{(2)} = h_j''(x_k) = \begin{cases} \dfrac{(-1)^{k+j}}{2}\sin^{-2}\left[\dfrac{(k-j)\pi}{N}\right], & \text{if } k \neq j, \\ -\dfrac{N^2}{12} - \dfrac{1}{6}, & \text{if } k = j. \end{cases} \tag{2.37}$$

It is worthwhile to point out that $D^{(2)} \neq D^2$. Indeed, we consider $u = \cos(Nx/2)$ and denote $\mathbf{u}$ the vector that samples u at $\{x_j\}_{j=0}^{N-1}$. Since $u(x_j) = (-1)^j$, one verifies readily that $D\mathbf{u} = \mathbf{0}$ and $D^2\mathbf{u} = \mathbf{0}$, while $D^{(2)}\mathbf{u} = -N^2\mathbf{u}/4$.

It is clear that the differentiation procedure through (2.33) requires $O(N^2)$ operations. We shall demonstrate below how to perform the differentiation in the frequency space with $O(N\log_2 N)$ operations using FFT.

2.1.4 Differentiation in the Frequency Space

For a function given by (2.31), we can rewrite it as a finite Fourier series

$$u(x) = \sum_{k=-N/2}^{N/2} \tilde{u}_k e^{\mathrm{i}kx}, \tag{2.38}$$

where $\tilde{u}_{N/2} = \tilde{u}_{-N/2}$ as before. Thus, we have

$$u'(x_j) = \sum_{k=-N/2}^{N/2} ik\tilde{u}_k e^{ikx_j}, \tag{2.39}$$

where $\{x_j = 2\pi j/N\}_{j=0}^{N-1}$ are the grids given by (2.11). Given the physical values $\{u(x_j)\}_{j=0}^{N-1}$, the approximation of the derivative values $\{w_j = u'(x_j)\}_{j=0}^{N-1}$ can be computed as follows:

- Call $\tilde{\mathbf{v}} = \mathtt{fft}(\mathbf{v})$, where the components of the input vector $\mathbf{v}$ are $v(j) = u(x_{j-1})$, $j = 1, \ldots, N$, and which returns the frequency vector:
$$\tilde{\mathbf{v}} = \big(\tilde{v}_1, \tilde{v}_2, \ldots, \tilde{v}_N\big).$$
- Compute the coefficients of the expansion of the derivative:
$$\begin{aligned}\tilde{\mathbf{v}}^{(1)} &= i\mathbf{k}.*\tilde{\mathbf{v}}\\ &= i\Big(0, \tilde{v}_2, \ldots, (N/2-1)\tilde{v}_{N/2}, 0, (-N/2+1)\tilde{v}_{N/2+2}, \ldots, -\tilde{v}_N\Big),\end{aligned}$$
where the multiplicative vector $\mathbf{k}$ is given in Table 2.1:
$$\mathbf{k} = \big(0, 1, \ldots, N/2-1, 0, -N/2+1, \ldots, -1\big). \tag{2.40}$$
- Call $\mathbf{w} = \mathtt{ifft}(\tilde{\mathbf{v}}^{(1)})$, which produces the desired derivative values $\{w_j\}_{j=0}^{N-1}$.

As a striking contrast to the differentiation process described in the previous section, the computational cost of the above procedure is $O\big(N\log_2 N\big)$. Moreover, higher-order derivatives can be computed using these three steps repeatedly. Multi-dimensional cases can be implemented similarly by using available routines such as `fft2.m` and `ifft2.m` in MATLAB.

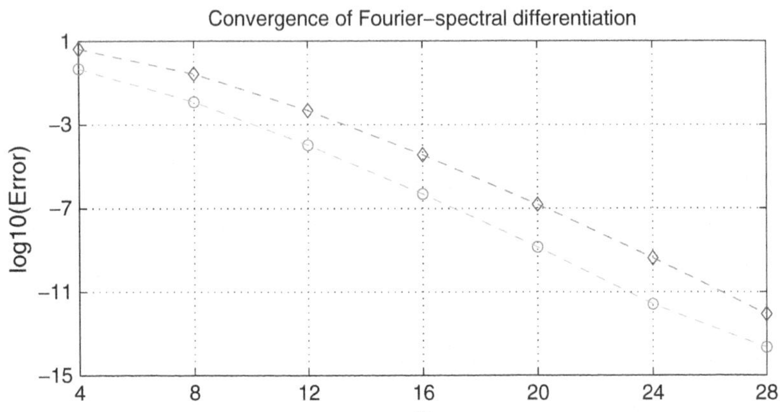

Fig. 2.1 Errors of Fourier-spectral differentiation of first-order ("∘") and second-order ("⋄")

As a numerical illustration, we consider the Fourier spectral differentiation of the 2π−periodic function $u(x) = e^{1+\sin x}$. In Fig. 2.1, we plot, in semi-log scale, the maximum point-wise errors of the first- and second-order derivatives against various N. The plots in Fig. 2.1 clearly indicate the exponential convergence of the Fourier spectral differentiation process.

2.2 Fourier Approximation

In this section, we summarize some fundamental results on the approximation of periodic functions by the continuous and discrete Fourier series.

2.2.1 Inverse Inequalities

Since all norms of a finite dimensional space are equivalent, we can bound a strong norm by a weaker one with bounding constants depending on the dimension of the space. This type of inequality is called inverse inequality. Our aim in this section is to find the optimal constants in such inequalities.

For notational convenience, we use $A \lesssim B$ to mean that there exists a generic positive constant c, which is independent of N and any function, such that $A \le cB$. We also use $\partial_x^m u$ or $u^{(m)}$ to denote the ordinary derivative $\frac{d^m u}{dx^m}$. Let $I := (0,2\pi)$, and define the complex $(2N+1)$-dimensional space

$$X_N := \text{span}\{e^{ikx} : -N \le k \le N\}. \tag{2.41}$$

The Banach space $L^p(I)$ with $1 \le p \le \infty$ and its norm $\|\cdot\|_{L^p}$ are defined as in Appendix B.4.

We first recall the following *Nikolski's* inequality.

Lemma 2.4. *For any $u \in X_N$ and $1 \le p \le q \le \infty$,*

$$\|u\|_{L^q} \le \left(\frac{Np_0+1}{2\pi}\right)^{\frac{1}{p}-\frac{1}{q}} \|u\|_{L^p}, \tag{2.42}$$

where p_0 is the least even integer $\ge p$.

Another type of inverse inequality, i.e., the so-called *Bernstein* inequality, relates the L^p-norms of a function and its derivatives.

Lemma 2.5. *For any $u \in X_N$ and $1 \le p \le \infty$,*

$$\|\partial_x^m u\|_{L^p} \lesssim N^m \|u\|_{L^p}, \quad m \ge 1. \tag{2.43}$$

In particular, for $p = 2$,

$$\|\partial_x^m u\| \lesssim N^m \|u\|. \tag{2.44}$$

The proofs of these inverse inequalities can be found in Butzer and Nessel (1971) (also see Guo (1998b)). In particular, the derivation of (2.44) is straightforward by using $\partial_x^m(e^{ikx}) = (ik)^m e^{ikx}$, and the orthogonality of the Fourier series.

2.2.2 Orthogonal Projection

Let $P_N : L^2(I) \to X_N$ be the L^2-orthogonal projection, defined by

$$\big(P_N u - u, v\big) = 0, \quad \forall v \in X_N. \tag{2.45}$$

It is obvious that $P_N u$ is the truncated Fourier series, namely,

$$(P_N u)(x) = \sum_{k=-N}^{N} \hat{u}_k e^{ikx},$$

where $\{\hat{u}_k\}$ are given by (2.3).

We next measure the errors between $P_N u$ and u in Sobolev spaces. For this purpose, we denote by $H_p^m(I)$ the subspace of $H^m(I)$ (cf. Appendix B.4), which consists of functions with derivatives of order up to $m-1$ being 2π-periodic. In view of the relation $(e^{ikx})' = ike^{ikx}$, the norm and semi-norm of $H_p^m(I)$ can be characterized in the frequency space by

$$\|u\|_m = \left(\sum_{k=-\infty}^{\infty} (1+k^2)^m |\hat{u}_k|^2\right)^{1/2}, \quad |u|_m = \left(\sum_{k=-\infty}^{\infty} |k|^{2m} |\hat{u}_k|^2\right)^{1/2}. \tag{2.46}$$

We see that the space $H_p^m(I)$ with fractional m is also well-defined.

Formally, for any $u \in H_p^m(I)$, we can differentiate the Fourier series term-wisely, and obtain

$$\partial_x^l u(x) = \sum_{k=-\infty}^{\infty} (ik)^l \hat{u}_k e^{ikx}, \quad 0 \le l \le m,$$

which implies the commutability of the derivative and projection operators:

$$\partial_x^l (P_N u) = P_N(\partial_x^l u), \quad 0 \le l \le m. \tag{2.47}$$

The main approximation result is stated below (cf. Kreiss and Oliger (1979), Canuto and Quarteroni (1982)).

Theorem 2.1. *For any $u \in H_p^m(I)$ and $0 \le \mu \le m$,*

$$\|P_N u - u\|_\mu \lesssim N^{\mu - m} |u|_m. \tag{2.48}$$

Proof. By (2.46),

$$\begin{aligned}\|P_N u - u\|_\mu^2 &= \sum_{|k|>N} (1+k^2)^\mu |\hat{u}_k|^2 \\ &\lesssim N^{2\mu-2m} \sum_{|k|>N} |k|^{2m-2\mu}(1+k^2)^\mu |\hat{u}_k|^2 \\ &\lesssim N^{2\mu-2m} \sum_{|k|>N} |k|^{2m}|\hat{u}_k|^2 \\ &\lesssim N^{2\mu-2m}|u|_m^2.\end{aligned}$$

This completes the proof. □

This theorem indicates that the projection $P_N u$ is the best approximation of u in all Sobolev spaces $H_p^m(I)$ $(m \ge 0)$.

The L^∞-estimate of the projection errors is stated as follows.

Theorem 2.2. *For any $u \in H_p^m(I)$ with $m > 1/2$,*

$$\max_{x\in[0,2\pi]} \big|(P_N u - u)(x)\big| \le \sqrt{\frac{1}{2m-1}} N^{1/2-m}|u|_m. \tag{2.49}$$

Proof. By the Cauchy–Schwarz inequality,

$$\begin{aligned}\big|(P_N u - u)(x)\big| &\le \sum_{|k|>N} |\hat{u}_k| \le \Big(\sum_{|k|>N} |k|^{-2m}\Big)^{1/2}\Big(\sum_{|k|>N} |k|^{2m}|\hat{u}_k|^2\Big)^{1/2} \\ &\le \sqrt{\frac{1}{2m-1}} N^{1/2-m}|u|_m.\end{aligned}$$

The last step is due to the fact that for $m > 1/2$,

$$\sum_{|k|>N} |k|^{-2m} \le \int_N^\infty x^{-2m} dx \le \frac{N^{1-2m}}{2m-1}.$$

This completes the proof. □

2.2.3 Interpolation

For the sake of consistency, we consider the Fourier interpolation on $2N$ collocation points $\{x_j = \pi j/N\}_{j=0}^{2N-1}$, but still denote the interpolation operator by I_N. That is,

$$(I_N u)(x) = \sum_{k=-N}^{N} \tilde{u}_k e^{ikx} \tag{2.50}$$

with $\tilde{u}_N = \tilde{u}_{-N}$ and

$$\tilde{u}_k = \frac{1}{2Nc_k} \sum_{j=0}^{2N-1} u(x_j) e^{-\mathrm{i}kx_j}, \quad -N \le k \le N. \tag{2.51}$$

The interpolation error: $I_N u - u$ is characterized by the following theorem (see also Kreiss and Oliger (1979), Canuto and Quarteroni (1982)):

Theorem 2.3. *For any $u \in H_p^m(I)$ with $m > 1/2$,*

$$\|\partial_x^l (I_N u - u)\| \lesssim N^{l-m} |u|_m, \quad 0 \le l \le m. \tag{2.52}$$

Proof. We first show that the expansion coefficients of the continuous (cf. (2.3)) and discrete Fourier series (cf. (2.51)) are connected by

$$c_k \tilde{u}_k = \hat{u}_k + \sum_{|p|>0}^{\infty} \hat{u}_{k+2pN}. \tag{2.53}$$

Indeed, plugging $u(x_j) = \sum_{|p|=0}^{\infty} \hat{u}_p e^{\mathrm{i}px_j}$ into (2.51) gives

$$c_k \tilde{u}_k = \frac{1}{2N} \sum_{j=0}^{2N-1} \Big(\sum_{|p|=0}^{\infty} \hat{u}_p e^{\mathrm{i}(p-k)x_j} \Big) = \frac{1}{2N} \sum_{|p|=0}^{\infty} \Big(\sum_{j=0}^{2N-1} e^{\mathrm{i}(p-k)x_j} \Big) \hat{u}_p,$$

where the constant $c_k = 1$ for $|k| < N$ and $c_k = 2$ for $k = \pm N$.

We deduce from Lemma 2.1 that $e^{\mathrm{i}(p-k)x_j} = 1$, if and only if $p - k = 2lN$ with $l \in \mathbb{Z}$, otherwise, it equals to zero. Hence, we have

$$c_k \tilde{u}_k = \sum_{|p|=0}^{\infty} \hat{u}_{k+2pN} = \hat{u}_k + \sum_{|p|>0}^{\infty} \hat{u}_{k+2pN},$$

which yields (2.53). Thus, a direct calculation leads to

$$\begin{aligned}
\|P_N u - I_N u\|^2 &= \sum_{|k|\le N} |\hat{u}_k - \tilde{u}_k|^2 \\
&= \sum_{|k|<N} |\hat{u}_k - \tilde{u}_k|^2 + \frac{1}{4} \sum_{k=\pm N} |2\hat{u}_k - 2\tilde{u}_k|^2 \\
&\le \sum_{|k|<N} |\hat{u}_k - \tilde{u}_k|^2 + \frac{1}{2} \sum_{k=\pm N} |\hat{u}_k - 2\tilde{u}_k|^2 + \frac{1}{2} \sum_{k=\pm N} |\hat{u}_k|^2 \\
&\le \sum_{|k|\le N} |\hat{u}_k - c_k \tilde{u}_k|^2 + \frac{1}{2} \sum_{k=\pm N} |\hat{u}_k|^2.
\end{aligned}$$

The last term is bounded by

$$|\hat{u}_N|^2 + |\hat{u}_{-N}|^2 \le N^{-2m} \sum_{|k|=N}^{\infty} |k|^{2m} |\hat{u}_k|^2 \le N^{-2m} |u|_m^2,$$

and the first term can be estimated by using the relation (2.53) and the Cauchy–Schwarz inequality:

$$\begin{aligned}\sum_{|k|\le N} |\hat{u}_k - c_k\tilde{u}_k|^2 &= \sum_{|k|\le N}\Big|\sum_{|p|>0}^{\infty} \hat{u}_{k+2pN}\Big|^2\\ &\le \sum_{|k|\le N}\Big\{\Big(\sum_{|p|>0}^{\infty}|k+2pN|^{-2m}\Big)\Big(\sum_{|p|>0}^{\infty}|k+2pN|^{2m}|\hat{u}_{k+2pN}|^2\Big)\Big\}\\ &\le \max_{|k|\le N}\Big\{\sum_{|p|>0}^{\infty}|k+2pN|^{-2m}\Big\}\Big(\sum_{|k|\le N}\sum_{|p|>0}^{\infty}|k+2pN|^{2m}|\hat{u}_{k+2pN}|^2\Big).\end{aligned}$$

It is clear that

$$\max_{|k|\le N}\Big\{\sum_{|p|>0}^{\infty}|k+2pN|^{-2m}\Big\} \le \frac{1}{N^{2m}}\sum_{|p|>0}^{\infty}\frac{1}{|2p-1|^{2m}} \lesssim N^{-2m},$$

and

$$\sum_{|k|\le N}\sum_{|p|>0}^{\infty}|k+2pN|^{2m}|\hat{u}_{k+2pN}|^2 \le 2|u|_m^2.$$

Hence, a combination of the above estimates leads to

$$\|P_N u - I_N u\| \lesssim N^{-m}|u|_m.$$

Moreover, by the inverse inequality (2.44),

$$\|\partial_x^l(P_N u - I_N u)\| \lesssim N^l\|P_N u - I_N u\| \lesssim N^{l-m}|u|_m.$$

Finally, using the triangle inequality and Theorem 2.1 yields

$$\|\partial_x^l(I_N u - u)\| \le \|\partial_x^l(P_N u - I_N u)\| + \|\partial_x^l(P_N u - u)\| \lesssim N^{l-m}|u|_m.$$

This ends the proof. □

We presented above some basic Fourier approximations in the Sobolev spaces. The interested readers are referred to the books on Fourier analysis (see, e.g., Körner (1988), Folland (1992)) for a thorough discussion on the Fourier approximations in different contexts.

2.3 Applications of Fourier Spectral Methods

In this section, we apply Fourier spectral methods to several nonlinear PDEs with periodic boundary conditions, including the Korteweg–de Vries (KdV) equation (cf. Korteweg and de Vries (1895)), the Kuramoto–Sivashinsky (KS) equation (cf. Kuramoto and Tsuzuki (1976)) and the Allen–Cahn equation (cf. Allen and Cahn (1979)). The emphasis will be put on the treatment for nonlinear terms and time discretizations.

2.3.1 Korteweg–de Vries (KdV) Equation

The KdV equation is a celebrated mathematical model of waves on shallow water surfaces. A fascinating property of the KdV equation is that it admits soliton-type solutions (cf. Zabusky and Galvin (1971)). Consider the KdV equation in the whole space:

$$\begin{aligned} &\partial_t u + u\partial_y u + \partial_y^3 u = 0, \quad y \in (-\infty,\infty), \ t > 0, \\ &u(y,0) = u_0(y), \quad y \in (-\infty,\infty), \end{aligned} \tag{2.54}$$

which has the exact soliton solution

$$u(y,t) = 12\kappa^2 \text{sech}^2\big(\kappa(y-y_0) - 4\kappa^3 t\big), \tag{2.55}$$

where y_0 is the center of the initial profile $u(y,0)$, and κ is a constant related to the traveling phase speed.

Since $u(y,t)$ decays exponentially to zero as $|y| \to \infty$, we can truncate the infinite interval to a finite one $(-\pi L, \pi L)$ with $L > 0$, and approximate the boundary conditions by the periodic boundary conditions on $(-\pi L, \pi L)$. It is expected that the initial-boundary valued problem (2.54) with periodic boundary conditions can provide a good approximation to the original initial-valued problem as long as the soliton does not reach the boundaries.

For convenience, we map the interval $[-\pi L, \pi L]$ to $[0, 2\pi]$ through the coordinate transform:

$$x = \frac{y}{L} + \pi, \quad y = L(x-\pi), \qquad x \in [0,2\pi], \ y \in [-\pi L, \pi L],$$

and denote

$$v(x,t) = u(y,t), \quad v_0(x) = u_0(y). \tag{2.56}$$

The transformed KdV equation reads

$$\begin{aligned} &\partial_t v + \frac{1}{L} v\partial_x v + \frac{1}{L^3}\partial_x^3 v = 0, \quad x \in (0,2\pi), \ t > 0, \\ &v(\cdot,t) \text{ periodic on } [0,2\pi), \quad t \geq 0; \quad v(x,0) = v_0(x), \quad x \in [0,2\pi]. \end{aligned} \tag{2.57}$$

Writing $v(x,t) = \sum_{|k|=0}^{\infty} \hat{v}_k(t) e^{ikx}$, taking the inner product of the first equation with e^{ikx}, and using (2.3) and the fact that $v\partial_x v = \frac{1}{2}\partial_x(v^2)$, we obtain that

$$\frac{d\hat{v}_k}{dt} - \frac{ik^3}{L^3}\hat{v}_k + \frac{ik}{2L}\widehat{(v^2)}_k = 0, \quad k = 0, \pm 1, \ldots, \tag{2.58}$$

with the initial condition

$$\hat{v}_k(0) = \frac{1}{2\pi}\int_0^{2\pi} v_0(x)e^{-ikx}dx. \tag{2.59}$$

The ODE systems (2.58)–(2.59) can be solved by various numerical methods such as the Runge–Kutta methods or the semi-implicit/linearly implicit schemes in which the nonlinear terms are treated explicitly while the leading linear term is treated implicitly.

Here, we use a combination of the integrating factor and Runge–Kutta methods as suggested in Trefethen (2000). More precisely, multiplying (2.58) by the integrating factor e^{-ik^3t/L^3}, we can rewrite the resulting equation as

$$\frac{d}{dt}\left[e^{-ik^3t/L^3}\hat{v}_k\right] = -\frac{ik}{2L}e^{-ik^3t/L^3}\widehat{(v^2)}_k, \quad k=0,\pm 1,\ldots. \tag{2.60}$$

Such a treatment makes the linear term disappear and can relax the stiffness of the system. The system (2.60) can then be solved by a standard ODE solver.

We now describe the Fourier approximation of (2.57) in MATLAB. Let **k** be the vector as in (2.40), and denote

$$\tilde{\mathbf{v}} = \left(\tilde{v}_0,\ldots,\tilde{v}_{N/2},\tilde{v}_{-N/2+1},\ldots,\tilde{v}_{-1}\right), \quad \mathbf{g} = e^{-i\mathbf{k}^3t/L^3}, \quad \tilde{\mathbf{u}} = \mathbf{g}.*\tilde{\mathbf{v}},$$

where the operations on the vectors are component-wise. Then, the Fourier approximation scheme based on (2.60) is as follows:

$$\frac{d\tilde{\mathbf{u}}}{dt} = -\frac{i\mathbf{k}}{2L}.*\mathbf{g}.*\mathtt{fft}\left(\left[\mathtt{ifft}\left(\mathbf{g}^{-1}.*\tilde{\mathbf{u}}\right)\right]^2\right), \quad t>0. \tag{2.61}$$

Therefore, a Runge–Kutta method, such as the fourth-order MATLAB routine rk4.m, can be directly applied to (2.61).

We present below some numerical results obtained by Program 27 (with some minor modifications) in Trefethen (2000). We first take $\kappa = 0.3$, $y_0 = -20$ and $L = 15$. On the left of Fig. 2.2, we plot the time evolution of the approximate solution (with $N = 256$, time step size $\tau = 0.01$ and $t \in [0,60]$), and on the right, we plot the maximum errors at $t = 1, 30, 60$ for various N with $\tau = 0.001$. Observe that the errors decay like $O(e^{-cN})$, which is typical for smooth solutions. The superior accuracy for this soliton solution indicates that the KdV equation on a finite interval can be used to effectively simulate the KdV equation on the whole space before the solitary wave reaches the boundaries.

In the next example, we consider the interaction of five solitary waves. More precisely, we consider the KdV equation (2.54) with the initial condition which consists of five solitary waves,

$$u_0(y) = \sum_{j=1}^{5} 12\kappa_j^2 \mathrm{sech}^2\left(\kappa_j(y-y_0)\right), \tag{2.62}$$

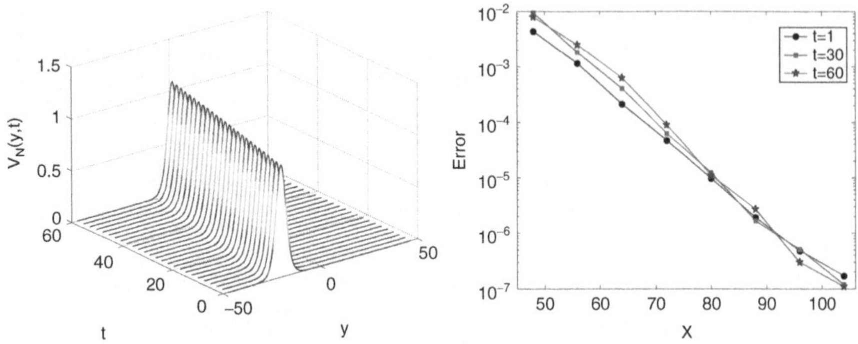

Fig. 2.2 *Left*: time evolution of the numerical solution; *right*: maximum errors vs. N

with

$$\begin{aligned} &\kappa_1 = 0.3, \ \ \kappa_2 = 0.25, \ \ \kappa_3 = 0.2, \ \ \kappa_4 = 0.15, \ \ \kappa_5 = 0.1, \\ &y_1 = -120, \ \ y_2 = -90, \ \ y_3 = -60, \ \ y_4 = -30, \ \ y_5 = 0. \end{aligned} \tag{2.63}$$

We fix $L = 50, N = 512$ and $\tau = 0.01$. In Fig. 2.3, we plot the time evolution of the approximate solution for $t \in [0, 600]$, and depict the initial profile and final profile at $t = 600$ in Fig. 2.4. We observe that the soliton with large amplitude travels with a faster speed, and the amplitudes of the five solitary waves are well preserved at the final time. This indicates the scheme has an excellent conservation property.

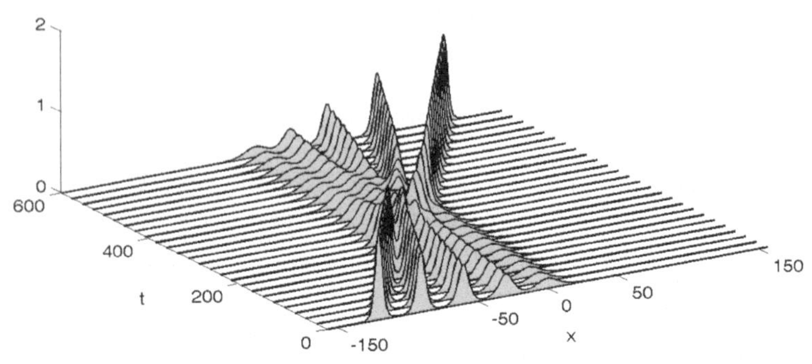

Fig. 2.3 Interaction of five solitary waves

2.3.2 Kuramoto–Sivashinsky (KS) Equation

The KS equation has been used in the study of a variety of reaction-diffusion systems (cf. Kuramoto and Tsuzuki (1976)), and is also an interesting dynamical PDE that can exhibit chaotic solutions (cf. Hyman and Nicolaenko (1986), Nicolaenko et al. (1985)).

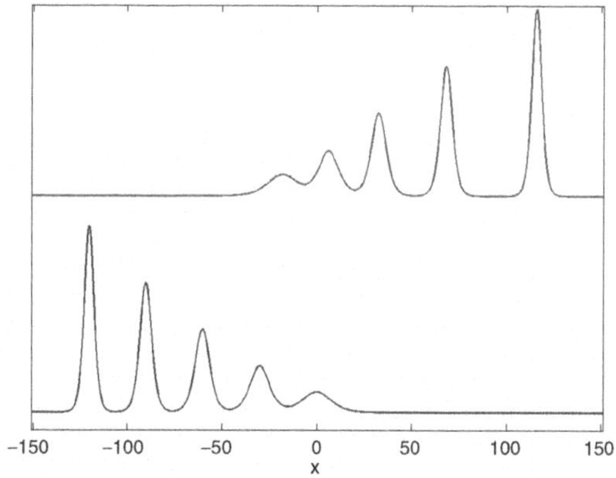

Fig. 2.4 Profiles at $t = 0, 600$

We consider the KS equation of the form

$$\begin{aligned} &\partial_t u + \partial_x^4 u + \partial_x^2 u + uu_x = 0, \quad x \in (-\infty, \infty), \ t > 0, \\ &u(x,t) = u(x + 2L\pi, t), \ \partial_x u(x,t) = \partial_x u(x + 2L\pi, t), \quad t \geq 0, \\ &u(x,0) = u_0(x), \quad x \in (-\infty, \infty), \end{aligned} \tag{2.64}$$

where the given function u_0 is $2L\pi$-periodic.

Thanks to the periodicity, it suffices to consider (2.64) in the reference interval $[0, 2L\pi]$. We discretize (2.64) in space by seeking the approximate solution

$$u_N(x,t) = \sum_{k=-N/2}^{N/2} \tilde{u}_k(t) e^{ikx/L}, \quad t > 0, \tag{2.65}$$

where $\tilde{u}_{N/2}(t) = \tilde{u}_{-N/2}(t)$. The N independent frequencies are determined by the scheme

$$\partial_t u_N + \partial_x^4 u_N + \partial_x^2 u_N = -\frac{1}{2} \partial_x I_N \big(u_N^2\big), \quad t > 0, \tag{2.66}$$

where I_N is the interpolation operator associated with the grids $\{x_j = 2L\pi j/N\}_{j=0}^{N-1}$. Thus, for each frequency k, we have

$$\tilde{u}_k'(t) + \Big(\frac{k^4}{L^4} - \frac{k^2}{L^2}\Big) \tilde{u}_k(t) = -\frac{1}{2L} ik\tilde{w}_k(t), \quad t > 0, \tag{2.67}$$

where $\tilde{w}_k(t)$ is the discrete Fourier coefficient of the nonlinear term, i.e.,

$$I_N \big(u_N^2\big) = \sum_{k=-N/2}^{N/2} \tilde{w}_k(t) e^{ikx/L}. \tag{2.68}$$

It is important to point out that, using the Fourier transform, linear operators with constant coefficients can always be diagonalized (i.e., the frequencies are separable) like (2.67). This leads to efficient time integrations for the resulting equation in the frequency space. We refer to Kassam and Trefethen (2005) for a review of various time-stepping schemes. Here, we use an exponential time-differencing (ETD) method as suggested in Kassam and Trefethen (2005).

Denote by $\tilde{\boldsymbol{u}}(t)$ the vector of the expansion coefficients as arranged in Table 2.1. Let $\boldsymbol{L}$ be the diagonal matrix with the diagonal $\boldsymbol{k}^2/L^2 - \boldsymbol{k}^4/L^4$, where $\boldsymbol{k}$ is the index vector given by Table 2.1, and $\boldsymbol{N}(t) := \boldsymbol{N}(\tilde{\boldsymbol{u}}, t)$ be the vector of the nonlinear term in (2.67)–(2.68). Then, we can rewrite (2.67)–(2.68) as a nonlinear ODE system

$$\tilde{\boldsymbol{u}}'(t) = \boldsymbol{L}\tilde{\boldsymbol{u}}(t) + \boldsymbol{N}(t), \quad t > 0. \tag{2.69}$$

Let τ be the time step size. It is clear that (2.69) is equivalent to

$$\tilde{\boldsymbol{u}}(t_n + \tau) = e^{\boldsymbol{L}\tau}\tilde{\boldsymbol{u}}(t_n) + e^{\boldsymbol{L}\tau}\int_0^{\tau} e^{-\boldsymbol{L}s}\boldsymbol{N}\big(\tilde{\boldsymbol{u}}(t_n + s), t_n + s\big)ds. \tag{2.70}$$

Based on how one approximates the integral, various ETD schemes may be constructed. For example, let $\tilde{\boldsymbol{u}}_n$ be the approximation of $\tilde{\boldsymbol{u}}(t_n)$. The following modified fourth-order ETD Runge–Kutta (ETDRK4) has been shown to be a very stable and accurate scheme for stiff equations (cf. Kassam and Trefethen (2005)):

$$\begin{aligned}
\boldsymbol{a}_n &= e^{\boldsymbol{L}\tau/2}\tilde{\boldsymbol{u}}_n + \boldsymbol{L}^{-1}\big(e^{\boldsymbol{L}\tau/2} - \boldsymbol{I}\big)\boldsymbol{N}(\tilde{\boldsymbol{u}}_n, t_n),\\
\boldsymbol{b}_n &= e^{\boldsymbol{L}\tau/2}\tilde{\boldsymbol{u}}_n + \boldsymbol{L}^{-1}\big(e^{\boldsymbol{L}\tau/2} - \boldsymbol{I}\big)\boldsymbol{N}(\boldsymbol{a}_n, t_n + \tau/2),\\
\boldsymbol{c}_n &= e^{\boldsymbol{L}\tau/2}\boldsymbol{a}_n + \boldsymbol{L}^{-1}\big(e^{\boldsymbol{L}\tau/2} - \boldsymbol{I}\big)\big[2\boldsymbol{N}(\boldsymbol{b}_n, t_n + \tau/2) - \boldsymbol{N}(\tilde{\boldsymbol{u}}_n, t_n)\big],\\
\tilde{\boldsymbol{u}}_{n+1} &= e^{\boldsymbol{L}\tau}\tilde{\boldsymbol{u}}_n + \Big\{\alpha\boldsymbol{N}(\tilde{\boldsymbol{u}}_n, t_n) + 2\beta\big[\boldsymbol{N}(\boldsymbol{a}_n, t_n + \tau/2)\\
&\qquad + \boldsymbol{N}(\boldsymbol{b}_n, t_n + \tau/2)\big] + \gamma\boldsymbol{N}(\boldsymbol{c}_n, t_n + \tau)\Big\},
\end{aligned} \tag{2.71}$$

where the coefficients

$$\begin{aligned}
\alpha &= \tau^{-2}\boldsymbol{L}^{-3}\Big[-4 - \boldsymbol{L}\tau + e^{\boldsymbol{L}\tau}\big(4 - 3\boldsymbol{L}\tau + (\boldsymbol{L}\tau)^2\big)\Big],\\
\beta &= \tau^{-2}\boldsymbol{L}^{-3}\Big[2 + \boldsymbol{L}\tau + e^{\boldsymbol{L}\tau}\big(-2 + \boldsymbol{L}\tau\big)\Big],\\
\gamma &= \tau^{-2}\boldsymbol{L}^{-3}\Big[-4 - 3\boldsymbol{L}\tau + e^{\boldsymbol{L}\tau}\big(4 - \boldsymbol{L}\tau\big) - (\boldsymbol{L}\tau)^2\Big].
\end{aligned} \tag{2.72}$$

In the following computations, we take $L = 16$ and impose the initial condition:

$$u_0(x) = \cos(x/L)(1 + \sin(x/L))$$

as in Kassam and Trefethen (2005). In Fig. 2.5, we depict the time evolution of the KS equation (2.64) obtained by the above algorithm with $\tau = 10^{-4}$ and $N = 128$. We plot in Fig. 2.6 the profiles of the numerical solution at various time in the waterfall format. We can observe the same pattern of the deterministic chaos as illustrated in Kassam and Trefethen (2005).

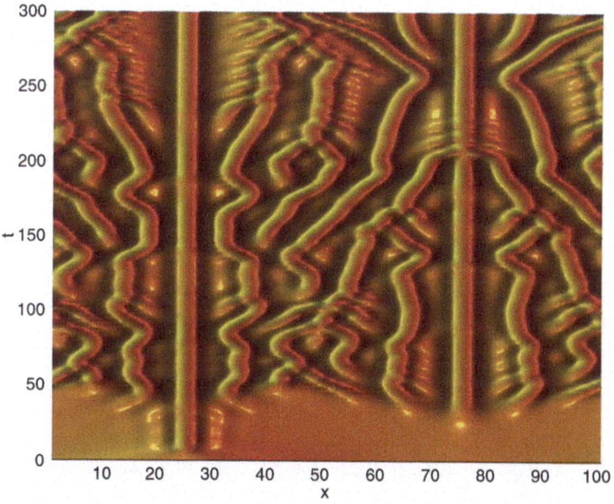

Fig. 2.5 Time evolution of the KS equation. Time runs from 0 at the bottom to 300 at the top

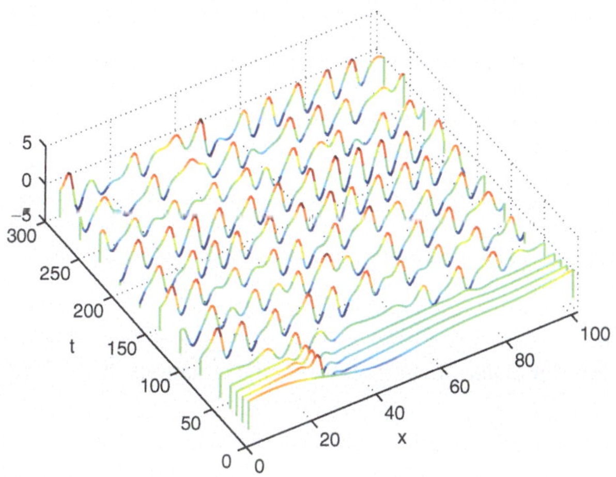

Fig. 2.6 Waterfall plot of the profiles of the numerical solution

2.3.3 Allen–Cahn Equation

The Allen–Cahn equation was originally introduced in Allen and Cahn (1979) to describe the motion of anti-phase boundaries in crystalline solids. It has been widely used in material science applications.

We consider the two-dimensional Allen–Cahn equation with periodic boundary conditions:

$$
\begin{aligned}
&\partial_t u - \varepsilon^2 \Delta u + u^3 - u = 0, \quad (x,y) \in \Omega = (-1,1)^2,\ t > 0,\\
&u(-1,y,t) = u(1,y,t), \quad u(x,-1,t) = u(x,1,t), \quad t \geq 0, \qquad (2.73)\\
&u(x,y,0) = u_0(x,y), \quad (x,y) \in \bar{\Omega},
\end{aligned}
$$

where ε is a small parameter which describes the inter-facial width. We refer to Sect. 9.3 for a more thorough discussion on the Allen–Cahn equation.

Let us write the Fourier approximation of the solution as

$$u_N(x,y,t) = \sum_{k,l=-N/2}^{N/2-1} \tilde{u}_{kl}(t) e^{\mathrm{i}(kx+ly)\pi}, \tag{2.74}$$

and denote by u_N^m the approximation of u_N at time $t_m = m\tau$ with τ being the time step size. Then a second-order stabilized semi-implicit scheme in time is (cf. Shen and Yang (2010)):

$$\begin{aligned} &\frac{3u_N^{m+1} - 4u_N^m + u_N^{m-1}}{2\tau} - \varepsilon^2 \Delta u_N^{m+1} + \left(2F_N(u_N^m) - F_N(u_N^{m-1})\right) \\ &\quad + s\left(u_N^{m+1} - 2u_N^m + u_N^{m-1}\right) = 0, \quad m = 1,2,\ldots, \end{aligned} \tag{2.75}$$

where $s > 0$ is an adjustable parameter, and $F_N(v) = I_N(v^3) - v$ with I_N being the two-dimensional tensorial interpolation operator on the computational grid. Notice that the extra dissipative term $s\left(u_N^{m+1} - 2u_N^m + u_N^{m-1}\right)$ (of order $s\tau^2$) is added to improve the stability while preserving the simplicity. At each time step, we only need to solve the linear problem

$$-2\tau\varepsilon^2 \Delta u_N^{m+1} + (3 + 2s\tau) u_N^{m+1} = \boldsymbol{N}\left(u_N^m, u_N^{m-1}\right), \tag{2.76}$$

where

$$\begin{aligned} \boldsymbol{N}\left(u_N^m, u_N^{m-1}\right) &= 4(1 + \tau s + \tau) u_N^m - (1 + 2\tau s + 2\tau) u_N^{m-1} \\ &\quad - 2\tau I_N\left[2(u_N^m)^3 - (u_N^{m-1})^3\right]. \end{aligned} \tag{2.77}$$

Applying the Fourier Galerkin method yields the equations in the frequency space:

$$\left(2\tau\varepsilon^2 (k^2 + l^2)\pi^2 + 3 + 2s\tau\right) \tilde{u}_{kl}^{m+1} = \tilde{w}_{kl}, \tag{2.78}$$

where $\{\tilde{w}_{kl}\}$ are the discrete Fourier coefficients of the nonlinear term $\boldsymbol{N}\left(u_N^m, u_N^{m-1}\right)$. The above semi-implicit scheme leads to an efficient implementation with the main cost coming from the treatment for the nonlinear term, which can be manipulated by FFT through a pseudo-spectral approach.

To test the numerical scheme, we consider the motion of a circular interface and impose the initial condition:

$$u_0(x,y) = \begin{cases} 1, & (x-0.5)^2 + (y-0.5)^2 < 1, \\ -1, & \text{otherwise.} \end{cases} \tag{2.79}$$

The motion of the interface is driven by the mean curvature of the circle, so the circle will shrink and eventually disappear, see Fig. 2.7. Note that the rate at which the diameter of the circle shrinks can be determined analytically (cf. Chen and Shen (1998)).

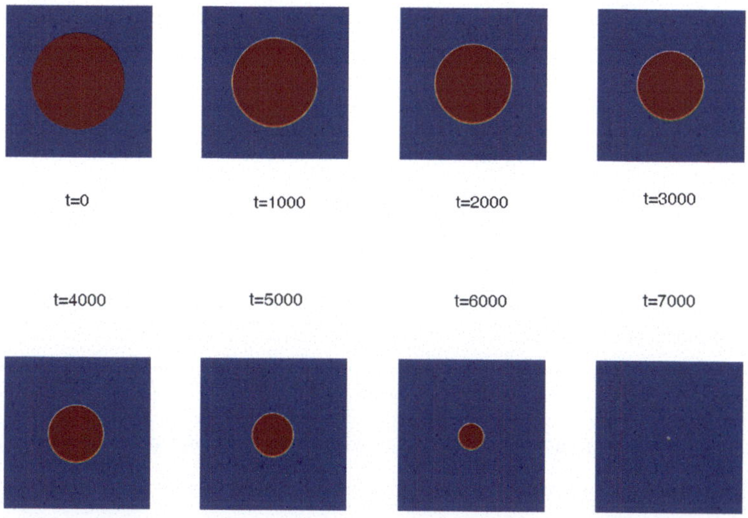

Fig. 2.7 Time evolution of a circular domain

Problems

2.1. Let $\mathscr{D}_N(x)$ be the Dirichlet kernel defined in (2.7).
(a) Show that $\mathscr{D}_N(x)$ is an even function, and it is symmetric about $x = 1/2$, namely,

$$\mathscr{D}_N(-x) = \mathscr{D}_N(x), \quad \mathscr{D}_N(1/2+x) = \mathscr{D}_N(1/2-x),$$

(b) Show that

$$\frac{1}{2\pi}\int_0^{2\pi} \mathscr{D}_N(x)dx = 1,$$

and

$$\frac{1}{2\pi}\int_0^{2\pi} \mathscr{D}_N^2(x)dx = 2N+1.$$

(c) Show that

$$\int_0^{2\pi} |\mathscr{D}_N(x)|dx \le c\ln N, \quad N \ge 2,$$

where c is a positive constant independent of N.
(d) Prove that for any $\phi \in X_N$ (defined in (2.41)),

$$\phi(x) = \frac{1}{2\pi}\int_0^{2\pi} \phi(t)\mathscr{D}_N(x-t)dt,$$

and

$$\|\phi\|_\infty \le \sqrt{2N+1}\|\phi\|.$$

2.2. The Fejér kernel is defined as the Nth arithmetic mean of the Dirichlet kernels:

$$F_N(x) = \frac{1}{N}\sum_{n=0}^{N-1} \mathscr{D}_n(x).$$

Show that

$$F_N(x) = \frac{\sin^2(Nx/2)}{N\sin^2(x/2)}.$$

2.3. Use the Sobolev inequality (B.33) and Theorem 2.1 to prove Theorem 2.2 with $m \geq 1$ in place of $m > 1/2$.

2.4. Determine $D^{(2)}$ and D^2 with $N = 4$ and confirm that $D^2 \neq D^{(2)}$.

2.5. Derive the formula (2.37) for the entries of the second-order differentiation matrix $D^{(2)}$.

2.6. Describe and implement a fourth-order Runge–Kutta and Fourier method for the Burger equation with periodic boundary conditions:

$$u_t = \varepsilon u_{xx} + uu_x, \quad x \in (-\pi, \pi); \quad u(x,0) = e^{-10\sin^2(x/2)},$$

with $\varepsilon = 0.03$ and the simulation running up to $t = 1$.

Chapter 3
Orthogonal Polynomials and Related Approximation Results

The Fourier spectral method is only appropriate for problems with periodic boundary conditions. If a Fourier method is applied to a non-periodic problem, it inevitably induces the so-called Gibbs phenomenon, and reduces the global convergence rate to $O(N^{-1})$ (cf. Gottlieb and Orszag (1977)). Consequently, one should not apply a Fourier method to problems with non-periodic boundary conditions. Instead, one should use orthogonal polynomials which are eigenfunctions of some singular Sturm-Liouville problems. The commonly used orthogonal polynomials include the Legendre, Chebyshev, Hermite and Laguerre polynomials.

The aim of this chapter is to present essential properties and fundamental approximation results related to orthogonal polynomials. These results serve as preparations for polynomial-based spectral methods in the forthcoming chapters. This chapter is organized as follows. In the first section, we present relevant properties of general orthogonal polynomials, and set up a general framework for the study of orthogonal polynomials. We then study the Jacobi polynomials in Sect. 3.2, Legendre polynomials in Sect. 3.3 and Chebyshev polynomials in Sect. 3.4. In Sect. 3.5, we present some general approximation results related to these families of orthogonal polynomials. We refer to Szegö (1975), Davis and Rabinowitz (1984) and Gautschi (2004) for other aspects of orthogonal polynomials.

3.1 Orthogonal Polynomials

Orthogonal polynomials play the most important role in spectral methods, so it is necessary to have a thorough study of their relevant properties. Our starting point is the generation of orthogonal polynomials by a three-term recurrence relation, which leads to some very useful formulas such as the Christoffel-Darboux formula. We then review some results on zeros of orthogonal polynomials, and present efficient algorithms for their computations. We also devote several sections to discussing some important topics such as Gauss-type quadrature formulas, polynomial interpolations, discrete transforms, and spectral differentiation techniques.

J. Shen et al., *Spectral Methods: Algorithms, Analysis and Applications*,
Springer Series in Computational Mathematics 41, DOI 10.1007/978-3-540-71041-7_3,

3.1.1 Existence and Uniqueness

Given an open interval $I := (a,b)$ $(-\infty \le a < b \le +\infty)$, and a generic weight function ω such that

$$\omega(x) > 0,\ \forall x \in I \text{ and } \omega \in L^1(I), \tag{3.1}$$

two functions f and g are said to be *orthogonal* to each other in $L^2_\omega(a,b)$ or orthogonal with respect to ω if

$$(f,g)_\omega := \int_a^b f(x)g(x)\omega(x)dx = 0.$$

An algebraic polynomial of degree n is denoted by

$$p_n(x) = k_n x^n + k_{n-1}x^{n-1} + \ldots + k_1 x + k_0, \quad k_n \neq 0, \tag{3.2}$$

where $\{k_i\}$ are real constants, and k_n is the leading coefficient of p_n.

A sequence of polynomials $\{p_n\}_{n=0}^\infty$ with $\deg(p_n) = n$ is said to be *orthogonal* in $L^2_\omega(a,b)$ if

$$(p_n, p_m)_\omega = \int_a^b p_n(x)p_m(x)\omega(x)dx = \gamma_n \delta_{mn}, \tag{3.3}$$

where the constant $\gamma_n = \|p_n\|^2_\omega$ is nonzero, and δ_{mn} is the Kronecker delta.

Throughout this section, $\{p_n\}$ denotes a sequence of polynomials orthogonal with respect to the weight function ω, and p_n is of degree n.

Denote by P_n the set of all algebraic polynomials of degree $\le n$, namely,

$$P_n := \text{span}\left\{1, x, x^2, \ldots, x^n\right\}. \tag{3.4}$$

By a dimension argument,

$$P_n = \text{span}\left\{p_0, p_1, \ldots, p_n\right\}. \tag{3.5}$$

A direct consequence is the following.

Lemma 3.1. *p_{n+1} is orthogonal to any polynomial $q \in P_n$.*

Proof. Thanks to (3.5), for any $q \in P_n$, we can write

$$q = b_n p_n + b_{n-1}p_{n-1} + \ldots + b_0 p_0.$$

Hence,

$$(p_{n+1}, q)_\omega = b_n(p_{n+1}, p_n)_\omega + b_{n-1}(p_{n+1}, p_{n-1})_\omega + \ldots + b_0(p_{n+1}, p_0)_\omega = 0,$$

where we have used the orthogonality (3.3). □

Hereafter, we denote the *monic* polynomial corresponding to p_n by

$$\bar{p}_n(x) := p_n(x)/k_n = x^n + a^{(n)}_{n-1}x^{n-1} + \ldots + a^{(n)}_0. \tag{3.6}$$

We show below that for any given weight function $\omega(x)$ defined in (a,b), there exists a unique family of monic orthogonal polynomials generated by a three-term recurrence formula.

Theorem 3.1. *For any given positive weight function $\omega \in L^1(I)$, there exists a unique sequence of monic orthogonal polynomials $\{\bar{p}_n\}$ with $\deg(\bar{p}_n) = n$, which can be constructed as follows*

$$\begin{aligned} &\bar{p}_0 = 1, \ \bar{p}_1 = x - \alpha_0, \\ &\bar{p}_{n+1} = (x - \alpha_n)\bar{p}_n - \beta_n \bar{p}_{n-1}, \quad n \geq 1, \end{aligned} \tag{3.7}$$

where

$$\alpha_n = \frac{(x\bar{p}_n, \bar{p}_n)_\omega}{\|\bar{p}_n\|_\omega^2}, \quad n \geq 0, \tag{3.8a}$$

$$\beta_n = \frac{\|\bar{p}_n\|_\omega^2}{\|\bar{p}_{n-1}\|_\omega^2}, \quad n \geq 1. \tag{3.8b}$$

Proof. It is clear that the first two polynomials are

$$\bar{p}_0(x) \equiv 1, \quad \bar{p}_1(x) = x - \alpha_0.$$

To determine α_0, we see that $(\bar{p}_0, \bar{p}_1)_\omega = 0$ if and only if

$$\alpha_0 = \int_a^b \omega(x) x dx \Big/ \int_a^b \omega(x) dx = \frac{(x\bar{p}_0, \bar{p}_0)_\omega}{\|\bar{p}_0\|_\omega^2},$$

where the denominator is positive due to (3.1).

We proceed with the proof by using an induction argument. Assuming that by a similar construction, we have derived a sequence of monic orthogonal polynomials $\{\bar{p}_k\}_{k=0}^n$. Next, we seek $\bar{p}_{n+1}$ of the form

$$\bar{p}_{n+1} = x\bar{p}_n - \alpha_n \bar{p}_n - \beta_n \bar{p}_{n-1} - \sum_{k=0}^{n-2} \gamma_k^{(n)} \bar{p}_k, \tag{3.9}$$

with α_n and β_n given by (3.8), and we require

$$\left(\bar{p}_{n+1}, \bar{p}_k\right)_\omega = 0, \quad 0 \leq k \leq n. \tag{3.10}$$

Taking the inner product with $\bar{p}_k$ on both sides of (3.9), and using the orthogonality of $\{\bar{p}_k\}_{k=0}^n$, we find that (3.10) is fulfilled if and only if

$$\begin{aligned} &(\bar{p}_{n+1}, \bar{p}_n)_\omega = (x\bar{p}_n, \bar{p}_n)_\omega - \alpha_n (\bar{p}_n, \bar{p}_n)_\omega = 0, \\ &(\bar{p}_{n+1}, \bar{p}_{n-1})_\omega = (x\bar{p}_n, \bar{p}_{n-1})_\omega - \beta_n (\bar{p}_{n-1}, \bar{p}_{n-1})_\omega = 0, \\ &(\bar{p}_{n+1}, \bar{p}_j)_\omega = (x\bar{p}_n, \bar{p}_j)_\omega - \sum_{k=0}^{n-2} \gamma_k^{(n)} (\bar{p}_k, \bar{p}_j)_\omega \\ &\quad \overset{(3.3)}{=} (x\bar{p}_n, \bar{p}_j)_\omega - \gamma_j^{(n)} \|\bar{p}_j\|_\omega^2 = 0, \quad 0 \leq j \leq n-2. \end{aligned} \tag{3.11}$$

Hence, the first equality implies (3.8a), and by the second one,

$$\beta_n = \frac{(x\bar{p}_n, \bar{p}_{n-1})_\omega}{\|\bar{p}_{n-1}\|_\omega^2} = \frac{(\bar{p}_n, x\bar{p}_{n-1})_\omega}{\|\bar{p}_{n-1}\|_\omega^2} = \frac{\|\bar{p}_n\|_\omega^2}{\|\bar{p}_{n-1}\|_\omega^2},$$

where we have used the fact

$$x\bar{p}_{n-1} = \bar{p}_n + \sum_{k=0}^{n-1} \delta_k^{(n)} \bar{p}_k,$$

and the orthogonality of $\{\bar{p}_k\}_{k=0}^n$ to deduce the last identity. It remains to show that the coefficients $\{\gamma_k^{(n)}\}_{k=0}^{n-2}$ in (3.9) are all zero. Indeed, we derive from Lemma 3.1 that

$$\left(x\bar{p}_n, \bar{p}_j\right)_\omega = (\bar{p}_n, x\bar{p}_j)_\omega = 0, \quad 0 \le j \le n-2,$$

which, together with the last equation of (3.11), implies $\gamma_k^{(n)} \equiv 0$ for $0 \le k \le n-2$, in (3.9). This completes the induction.

Next, we show that the polynomial sequence generated by (3.7)–(3.8) is unique. For this purpose, we assume that $\{\bar{q}_n\}_{n=0}^\infty$ is another sequence of monic orthogonal polynomials. Since $\bar{p}_n$, given by (3.7), is also monic, we have $\deg\left(\bar{p}_{n+1} - \bar{q}_{n+1}\right) \le n$. By Lemma 3.1,

$$(\bar{p}_{n+1}, \bar{p}_{n+1} - \bar{q}_{n+1})_\omega = 0, \quad (\bar{q}_{n+1}, \bar{p}_{n+1} - \bar{q}_{n+1})_\omega = 0,$$

which implies

$$\left(\bar{p}_{n+1} - \bar{q}_{n+1}, \bar{p}_{n+1} - \bar{q}_{n+1}\right)_\omega = 0 \quad \Rightarrow \quad \bar{p}_{n+1}(x) - \bar{q}_{n+1}(x) \equiv 0.$$

This proves the uniqueness. □

The above theorem provides a general three-term recurrence relation to construct orthogonal polynomials, and the constants α_n and β_n can be evaluated explicitly for the commonly used families. The three-term recurrence relation (3.7) is essential for deriving other properties of orthogonal polynomials. For convenience, we first extend it to the orthogonal polynomials $\{p_n\}$, which are not necessarily monic.

Corollary 3.1. *Let $\{p_n\}$ be a sequence of orthogonal polynomials with the leading coefficient $k_n \neq 0$. Then we have*

$$p_{n+1} = (a_n x - b_n) p_n - c_n p_{n-1}, \qquad n \ge 0, \tag{3.12}$$

with $p_{-1} := 0$, $p_0 = k_0$ and

$$a_n = \frac{k_{n+1}}{k_n}, \tag{3.13a}$$

$$b_n = \frac{k_{n+1}}{k_n} \frac{(xp_n, p_n)_\omega}{\|p_n\|_\omega^2}, \tag{3.13b}$$

$$c_n = \frac{k_{n-1}k_{n+1}}{k_n^2} \frac{\|p_n\|_\omega^2}{\|p_{n-1}\|_\omega^2}. \tag{3.13c}$$

Proof. This result follows directly from Theorem 3.1 by inserting $\bar{p}_l = p_l/k_l$ with $l = n-1, n, n+1$ into (3.7) and (3.8). □

The orthogonal polynomials $\{p_n\}$ with leading coefficients $\{k_n\}$ are uniquely determined by (3.12)–(3.13). Interestingly, the following result, which can be viewed as the converse of Corollary 3.1, also holds. We leave its proof as an exercise (see Problem 3.1).

Corollary 3.2. *Let* $\{k_n \neq 0\}$ *be a sequence of real numbers. The three-term recurrence relation* (3.12)–(3.13) *generates a sequence of polynomials satisfying the properties:*

- *the leading coefficient of* p_n *is* k_n *and* $deg(p_n) = n$;
- $\{p_n\}$ *are orthogonal with respect to the weight function* $\omega(x)$;
- *the* L^2_ω*-norm of* p_n *is given by*

$$\gamma_n = \|p_n\|^2_\omega = (a_0/a_n)c_1c_2\ldots c_n\gamma_0, \quad n \geq 0, \tag{3.14}$$

where $\gamma_0 = k_0^2 \int_a^b \omega(x)dx$.

An important consequence of the three-term recurrence formula (3.12)–(3.13) is the well-known *Christoff-Darboux formula.*

Corollary 3.3. *Let* $\{p_n\}$ *be a sequence of orthogonal polynomials with* $deg(p_n) = n$*. Then,*

$$\frac{p_{n+1}(x)p_n(y) - p_n(x)p_{n+1}(y)}{x-y} = \frac{k_{n+1}}{k_n}\sum_{j=0}^{n}\frac{\|p_n\|^2_\omega}{\|p_j\|^2_\omega}p_j(x)p_j(y), \tag{3.15}$$

and

$$p'_{n+1}(x)p_n(x) - p'_n(x)p_{n+1}(x) = \frac{k_{n+1}}{k_n}\sum_{j=0}^{n}\frac{\|p_n\|^2_\omega}{\|p_j\|^2_\omega}p_j^2(x). \tag{3.16}$$

Proof. We first prove (3.15). By Corollary 3.1,

$$\begin{aligned}
&p_{j+1}(x)p_j(y) - p_j(x)p_{j+1}(y)\\
&\quad= [(a_jx-b_j)p_j(x) - c_jp_{j-1}(x)]p_j(y)\\
&\qquad - p_j(x)[(a_jy-b_j)p_j(y) - c_jp_{j-1}(y)]\\
&\quad= a_j(x-y)p_j(x)p_j(y) + c_j[p_j(x)p_{j-1}(y) - p_{j-1}(x)p_j(y)].
\end{aligned}$$

Thus, by (3.13),

$$\begin{aligned}
&\frac{k_j}{k_{j+1}\|p_j\|^2_\omega}\frac{p_{j+1}(x)p_j(y) - p_j(x)p_{j+1}(y)}{x-y}\\
&\qquad - \frac{k_{j-1}}{k_j\|p_{j-1}\|^2_\omega}\frac{p_j(x)p_{j-1}(y) - p_{j-1}(x)p_j(y)}{x-y} = \frac{1}{\|p_j\|^2_\omega}p_j(x)p_j(y).
\end{aligned}$$

This relation also holds for $j = 0$ by defining $p_{-1} := 0$. Summing the above identities for $0 \leq j \leq n$ leads to (3.15).

To prove (3.16), we observe that

$$\frac{p_{n+1}(x)p_n(y)-p_n(x)p_{n+1}(y)}{x-y} = \frac{p_{n+1}(x)-p_{n+1}(y)}{x-y}p_n(y)-\frac{p_n(x)-p_n(y)}{x-y}p_{n+1}(y).$$

Consequently, letting $y \to x$, we obtain (3.16) from (3.15) and the definition of the derivative. □

Define the kernel

$$K_n(x,y)=\sum_{j=0}^{n}\frac{p_j(x)p_j(y)}{\|p_j\|_\omega^2}. \tag{3.17}$$

The Christoff-Darboux formula (3.15) can be rewritten as

$$K_n(x,y)=\frac{k_n}{k_{n+1}\|p_n\|_\omega^2}\frac{p_{n+1}(x)p_n(y)-p_n(x)p_{n+1}(y)}{x-y}. \tag{3.18}$$

A remarkable property of $\{K_n\}$ is stated in the following lemma.

Lemma 3.2. *There holds the integral equation*:

$$q(x)=\int_a^b q(t)K_n(x,t)\omega(t)dt, \quad \forall q\in P_n. \tag{3.19}$$

Moreover, the polynomial sequence $\{K_n(x,a)\}$ *(resp.* $\{K_n(x,b)\}$*) is orthogonal with respect to the weight function* $(x-a)\omega$ *(resp.* $(b-x)\omega$*).*

Proof. Thanks to (3.5), for any $q\in P_n$, we can write

$$q(x)=\sum_{j=0}^{n}\hat{q}_j p_j(x) \quad \text{with} \quad \hat{q}_j=\frac{1}{\|p_j\|_\omega^2}\int_a^b q(t)p_j(t)\omega(t)dt.$$

Thus, by definition (3.17),

$$q(x)=\sum_{j=0}^{n}\frac{1}{\|p_j\|_\omega^2}\int_a^b q(t)p_j(x)p_j(t)\omega(t)dt=\int_a^b q(t)K_n(x,t)\omega(t)dt,$$

which gives (3.19).

Next, taking $x=a$ and $q(t)=(t-a)r(t)$ for any $r\in P_{n-1}$ in (3.19) yields

$$0=q(a)=\int_a^b K_n(t,a)r(t)(t-a)\omega(t)dt, \quad \forall r\in P_{n-1},$$

which implies $\{K_n(x,a)\}$ is orthogonal with respect to $(x-a)\omega$.

Similarly, taking $x=b$ and $q(t)=(b-t)r(t)$ for any $r\in P_{n-1}$ in (3.19), we can show that $\{K_n(x,b)\}$ is orthogonal with respect to $(b-x)\omega$. □

3.1.2 Zeros of Orthogonal Polynomials

Zeros of orthogonal polynomials play a fundamental role in spectral methods. For example, they are chosen as the nodes of Gauss-type quadratures, and used to generate computational grids for spectral methods. Therefore, it is useful to derive their essential properties.

Again, let $\{p_n\}$ (with $\deg(p_n) = n$) be a sequence of polynomials orthogonal with respect to the weight function $\omega(x)$ in (a,b). The first important result about the zeros of orthogonal polynomials is the following:

Theorem 3.2. *The zeros of p_{n+1} are all real, simple, and lie in the interval (a,b).*

Proof. We first show that the zeros of p_{n+1} are all real. Assuming $\alpha \pm i\beta$ are a pair of complex roots of p_{n+1}. Then $p_{n+1}/((x-\alpha)^2+\beta^2) \in P_{n-1}$, and by Lemma 3.1,

$$0 = \int_a^b p_{n+1}\frac{p_{n+1}}{(x-\alpha)^2+\beta^2}\omega dx = \int_a^b ((x-\alpha)^2+\beta^2)\left|\frac{p_{n+1}}{(x-\alpha)^2+\beta^2}\right|^2 \omega dx,$$

which implies that $\beta = 0$. Hence, all zeros of p_{n+1} must be real.

Next, we prove that the $n+1$ zeros of p_{n+1} are simple, and lie in the interval (a,b). By the orthogonality,

$$\int_a^b p_{n+1}(x)\omega(x)dx = 0, \quad \forall n \geq 0,$$

there exists at least one zero of p_{n+1} in (a,b). In other words, $p_{n+1}(x)$ must change sign in (a,b). Let $x_0, x_1, \ldots, x_k$ be all such points in (a,b) at which $p_{n+1}(x)$ changes sign. If $k = n$, we are done, since $\{x_i\}_{i=0}^n$ are the $n+1$ simple real zeros of p_{n+1}. If $k < n$, we consider the polynomial

$$q(x) = (x-x_0)(x-x_1)\ldots(x-x_k).$$

Since $\deg(q) = k+1 < n+1$, by orthogonality, we derive

$$(p_{n+1}, q)_\omega = 0.$$

However, $p_{n+1}(x)q(x)$ cannot change sign on (a,b), since each sign change in $p_{n+1}(x)$ is canceled by a corresponding sign change in $q(x)$. It follows that

$$(p_{n+1}, q)_\omega \neq 0,$$

which leads to a contradiction. □

Another important property is known as the separation theorem.

Theorem 3.3. *Let $x_0 = a$, $x_{n+1} = b$ and $x_1 < x_2 < \ldots < x_n$ be the zeros of p_n. Then there exists exactly one zero of p_{n+1} in each subinterval (x_j, x_{j+1}), $j = 0, 1, \ldots, n$.*

Proof. It is obvious that the location of zeros is invariant with any nonzero constant multiple of p_n and p_{n+1}, so we assume that the leading coefficients $k_n, k_{n+1} > 0$.

We first show that each of the interior subintervals (x_j, x_{j+1}), $j = 1, 2, \ldots, n-1$, contains at least one zero of p_{n+1}, which is equivalent to proving

$$p_{n+1}(x_j)p_{n+1}(x_{j+1}) < 0, \quad 1 \le j \le n-1. \tag{3.20}$$

Since p_n can be written as

$$p_n(x) = k_n(x-x_1)(x-x_2)\ldots(x-x_n),$$

a direct calculation leads to

$$p_n'(x_j) = k_n \prod_{l=1}^{j-1}(x_j - x_l) \cdot \prod_{l=j+1}^{n}(x_j - x_l). \tag{3.21}$$

This implies

$$p_n'(x_j)p_n'(x_{j+1}) = D_{n,j} \times (-1)^{n-j} \times (-1)^{n-j-1} < 0, \tag{3.22}$$

where $D_{n,j}$ is a positive constant. On the other hand, using the facts that $p_n(x_j) = p_n(x_{j+1}) = 0$ and $k_n, k_{n+1} > 0$, we find from (3.16) that

$$-p_n'(x_j)p_{n+1}(x_j) > 0, \quad -p_n'(x_{j+1})p_{n+1}(x_{j+1}) > 0. \tag{3.23}$$

Consequently,

$$\left[p_n'(x_j)p_n'(x_{j+1})\right]\left[p_{n+1}(x_j)p_{n+1}(x_{j+1})\right] > 0 \overset{(3.22)}{\Longrightarrow} p_{n+1}(x_j)p_{n+1}(x_{j+1}) < 0.$$

Next, we prove that there exists at least one zero of p_{n+1} in each of the boundary subintervals (x_n, b) and (a, x_1). Since $p_n(x_n) = 0$ and $p_n'(x_n) > 0$ (cf. (3.21)), the use of (3.16) again gives $p_{n+1}(x_n) < 0$. On the other hand, due to $k_{n+1} > 0$, we have $p_{n+1}(b) > 0$. Therefore, $p_{n+1}(x_n)p_{n+1}(b) < 0$, which implies (x_n, b) contains at least one zero of p_{n+1}. The existence of at least one zero of p_{n+1} in (a, x_1) can be justified in a similar fashion.

In summary, we have shown that each of the $n+1$ subintervals (x_j, x_{j+1}), $0 \le j \le n$, contains at least one zero of p_{n+1}. By Theorem 3.2, p_{n+1} has exactly $n+1$ real zeros, so each subinterval contains exactly one zero of p_{n+1}. □

A direct consequence of (3.22) is the following.

Corollary 3.4. *Let $\{p_n\}$ be a sequence of orthogonal polynomials with* $\deg(p_n) = n$. *Then the zeros of p_n' are real and simple, and there exists exactly one zero of p_n' between two consecutive zeros of p_n.*

3.1.3 Computation of Zeros of Orthogonal Polynomials

We present below two efficient algorithms for computing zeros of orthogonal polynomials.

The first approach is the so-called *Eigenvalue Method.*

Theorem 3.4. *The zeros* $\{x_j\}_{j=0}^n$ *of the orthogonal polynomial* $p_{n+1}(x)$ *are eigenvalues of the following symmetric tridiagonal matrix:*

$$A_{n+1} = \begin{bmatrix} \alpha_0 & \beta_1 & & & \\ \beta_1 & \alpha_1 & \beta_2 & & \\ & \ddots & \ddots & \ddots & \\ & & \beta_{n-1} & \alpha_{n-1} & \beta_n \\ & & & \beta_n & \alpha_n \end{bmatrix}, \tag{3.24}$$

where

$$\alpha_j = \frac{b_j}{a_j}, \quad j \geq 0; \quad \beta_j = \frac{1}{a_{j-1}}\sqrt{\frac{a_{j-1}c_j}{a_j}}, \quad j \geq 1, \tag{3.25}$$

with $\{a_j, b_j, c_j\}$ *being the coefficients of the three-term recurrence relation* (3.12), *namely,*

$$p_{j+1}(x) = (a_j x - b_j)p_j(x) - c_j p_{j-1}(x), \quad j \geq 0, \tag{3.26}$$

with $p_{-1} := 0$.

Proof. We first normalize the orthogonal polynomials $\{p_j\}$ by defining

$$\tilde{p}_j(x) = \frac{1}{\sqrt{\gamma_j}} p_j(x) \quad \text{with} \quad \gamma_j = \|p_j\|_\omega^2 \;\Rightarrow\; \|\tilde{p}_j\|_\omega = 1.$$

Thus, we have

$$\begin{aligned} x\tilde{p}_j &\overset{(3.26)}{=} \frac{c_j}{a_j}\sqrt{\frac{\gamma_{j-1}}{\gamma_j}}\tilde{p}_{j-1} + \frac{b_j}{a_j}\tilde{p}_j + \frac{1}{a_j}\sqrt{\frac{\gamma_{j+1}}{\gamma_j}}\tilde{p}_{j+1} \\ &\overset{(3.13)}{=} \frac{1}{a_{j-1}}\sqrt{\frac{\gamma_j}{\gamma_{j-1}}}\tilde{p}_{j-1} + \frac{b_j}{a_j}\tilde{p}_j + \frac{1}{a_j}\sqrt{\frac{\gamma_{j+1}}{\gamma_j}}\tilde{p}_{j+1} \\ &= \beta_j \tilde{p}_{j-1}(x) + \alpha_j \tilde{p}_j(x) + \beta_{j+1}\tilde{p}_{j+1}(x), \quad j \geq 0, \end{aligned} \tag{3.27}$$

where we denote

$$\alpha_j := \frac{b_j}{a_j}, \quad \beta_j := \frac{1}{a_{j-1}}\sqrt{\frac{\gamma_j}{\gamma_{j-1}}}.$$

By (3.13),

$$\frac{\gamma_j}{\gamma_{j-1}} = \frac{a_{j-1}c_j}{a_j} > 0 \Rightarrow \beta_j = \frac{1}{a_{j-1}}\sqrt{\frac{a_{j-1}c_j}{a_j}}.$$

Then we rewrite the recurrence relation (3.27) as

$$x\tilde{p}_j(x) = \beta_j \tilde{p}_{j-1}(x) + \alpha_j \tilde{p}_j(x) + \beta_{j+1}\tilde{p}_{j+1}(x), \quad j \geq 0.$$

We now take $j = 0, 1, \ldots, n$ to form a system with the matrix form

$$x\widetilde{\mathbf{P}}(x) = A_{n+1}\widetilde{\mathbf{P}}(x) + \beta_{n+1}\tilde{p}_{n+1}(x)\mathbf{E}_{n+1}, \tag{3.28}$$

where A_{n+1} is given by (3.24), and

$$\widetilde{\mathbf{P}}(x) = \big(\tilde{p}_0(x), \tilde{p}_1(x), \ldots, \tilde{p}_n(x)\big)^T, \quad \mathbf{E}_{n+1} = (0, 0, \ldots, 0, 1)^T.$$

Since $\tilde{p}_{n+1}(x_j) = 0$, $0 \leq j \leq n$, the system (3.28) at $x = x_j$ becomes

$$x_j \widetilde{\mathbf{P}}(x_j) = A_{n+1}\widetilde{\mathbf{P}}(x_j), \quad 0 \leq j \leq n. \tag{3.29}$$

Hence, $\{x_j\}_{j=0}^n$ are eigenvalues of the symmetric tridiagonal matrix A_{n+1}. □

An alternative approach for finding zeros of orthogonal polynomials is to use an iterative procedure. More precisely, let x_j^0 be an initial approximation to the zero x_j of $p_{n+1}(x)$. Then, one can construct an iterative scheme in the general form:

$$\begin{cases} x_j^{k+1} = x_j^k + D(x_j^k), \ 0 \leq j \leq n, \ k \geq 0, \\ \text{given } \{x_j^0\}_{j=0}^n, \text{ and a termination rule.} \end{cases} \tag{3.30}$$

The deviation $D(\cdot)$ classifies different types of iterative schemes. For instance, the Newton method is of second-order with

$$D(x) = -\frac{p_{n+1}(x)}{p'_{n+1}(x)}, \tag{3.31}$$

while the Laguerre method is a third-order scheme with

$$D(x) = -\frac{p_{n+1}(x)}{p'_{n+1}(x)} - \frac{p_{n+1}(x)p''_{n+1}(x)}{2(p'_{n+1}(x))^2}. \tag{3.32}$$

Higher-order schemes can be constructed by using higher-order derivatives of p_{n+1}.

The success of an iterative method often depends on how good is the initial guess. If the initial approximation is not sufficiently close, the algorithm may converge to other unwanted values or even diverge. For a thorough discussion on how to find zeros of polynomials, we refer to Pan (1997) and the references therein.

3.1.4 Gauss-Type Quadratures

We now discuss the relations between orthogonal polynomials and Gauss-type integration formulas. The mechanism of a Gauss-type quadrature is to seek the best numerical approximation of an integral by selecting optimal nodes at which the integrand is evaluated. It belongs to the family of the numerical quadratures:

$$\int_a^b f(x)\omega(x)dx = \sum_{j=0}^N f(x_j)\omega_j + E_N[f], \tag{3.33}$$

where $\{x_j, \omega_j\}_{j=0}^N$ are the quadrature nodes and weights, and $E_N[f]$ is the quadrature error. If $E_N[f] \equiv 0$, we say the quadrature formula (3.33) is *exact* for f.

Hereafter, we assume that the nodes $\{x_j\}_{j=0}^N$ are distinct. If $f(x) \in C^{N+1}[a,b]$, we have (see, e.g., Davis and Rabinowitz (1984)):

$$E_N[f] = \frac{1}{(N+1)!}\int_a^b f^{(N+1)}(\xi(x))\prod_{i=0}^N (x-x_i)dx, \tag{3.34}$$

where $\xi(x) \in [a,b]$. The Lagrange basis polynomials associated with $\{x_j\}_{j=0}^N$ are given by

$$h_j(x) = \prod_{i=0;i\neq j}^N \frac{x-x_i}{x_j-x_i}, \quad 0 \le j \le N, \tag{3.35}$$

so taking $f(x) = h_j$ in (3.33) and using (3.34), we find the quadrature weights:

$$\omega_j = \int_a^b h_j(x)\omega(x)dx, \quad 0 \le j \le N. \tag{3.36}$$

We say that the integration formula (3.33)–(3.36) has a degree of precision (DOP) m, if there holds

$$E_N[p] = 0, \ \forall p \in P_m \ \text{ but } \exists q \in P_{m+1} \text{ such that } E_N[q] \neq 0. \tag{3.37}$$

In general, for any $N+1$ distinct nodes $\{x_j\}_{j=0}^N \subseteq (a,b)$, the DOP of (3.33)–(3.36) is between N and $2N+1$. Moreover, if the nodes $\{x_k\}_{k=0}^N$ are chosen as zeros of the polynomial p_{N+1} orthogonal with respect to ω, this rule enjoys the maximum degree of precision $2N+1$. Such a rule is known as the Gauss quadrature.

Theorem 3.5. (Gauss quadrature) *Let $\{x_j\}_{j=0}^N$ be the set of zeros of the orthogonal polynomial p_{N+1}. Then there exists a unique set of quadrature weights $\{\omega_j\}_{j=0}^N$, defined by* (3.36), *such that*

$$\int_a^b p(x)\omega(x)dx = \sum_{j=0}^N p(x_j)\omega_j, \quad \forall p \in P_{2N+1}, \tag{3.38}$$

where the quadrature weights are all positive and given by

$$\omega_j = \frac{k_{N+1}}{k_N} \frac{\|p_N\|_\omega^2}{p_N(x_j)p'_{N+1}(x_j)}, \quad 0 \le j \le N, \tag{3.39}$$

where k_j is the leading coefficient of the polynomial p_j.

Proof. Let $\{h_j\}_{j=0}^N$ be the Lagrange basis polynomials defined in (3.35). It is clear that

$$P_N = \text{span}\{h_j : 0 \le j \le N\} \;\Rightarrow\; p(x) = \sum_{j=0}^N p(x_j)h_j(x), \;\; \forall p \in P_N.$$

Hence,

$$\int_a^b p(x)\omega(x)dx = \sum_{j=0}^N p(x_j) \int_a^b h_j(x)\omega(x)dx \overset{(3.36)}{=} \sum_{j=0}^N p(x_j)\omega_j, \tag{3.40}$$

which implies (3.38) is exact for any $p \in P_N$. In other words, the DOP of this rule is not less than N.

Next, for any $p \in P_{2N+1}$, we can write $p = rp_{N+1} + s$ where $r,s \in P_N$. In view of $p_{N+1}(x_j) = 0$, we have $p(x_j) = s(x_j)$ for $0 \le j \le N$. Since p_{N+1} is orthogonal to r (cf. Lemma 3.1) and $s \in P_N$, we find

$$\begin{aligned} \int_a^b p(x)\omega(x)dx &= \int_a^b s(x)\omega(x)dx \\ &= \sum_{j=0}^N s(x_j)\omega_j \overset{(3.40)}{=} \sum_{j=0}^N p(x_j)\omega_j, \quad \forall p \in P_{2N+1}, \end{aligned} \tag{3.41}$$

which leads to (3.38).

Now, we show that $\omega_j > 0$ for $0 \le j \le N$. Taking $p(x) = h_j^2(x) \in P_{2N}$ in (3.41) leads to

$$0 < \int_a^b h_j^2(x)\omega(x)dx = \sum_{k=0}^N h_j^2(x_k)\omega_k = \omega_j, \quad 0 \le j \le N.$$

It remains to establish (3.39). Since $p_{N+1}(x_j) = 0$, taking $y = x_j$ and $n = N$ in the Christoff-Darboux formula (3.15) yields

$$p_N(x_j)\frac{p_{N+1}(x)}{x - x_j} = \frac{k_{N+1}}{k_N} \sum_{i=0}^N \frac{\|p_N\|_\omega^2}{\|p_i\|_\omega^2} p_i(x_j)p_i(x).$$

Multiplying the above formula by $\omega(x)$ and integrating the resulting identity over (a,b), we deduce from the orthogonality $(p_i, 1)_\omega = 0$, $i \ge 1$, that

$$\begin{aligned} & p_N(x_j) \int_a^b \frac{p_{N+1}(x)}{x - x_j}\omega(x)dx \\ &= \frac{k_{N+1}}{k_N}\|p_N\|_\omega^2 \frac{(p_0(x_j), p_0)_\omega}{\|p_0\|_\omega^2} = \frac{k_{N+1}}{k_N}\|p_N\|_\omega^2. \end{aligned} \tag{3.42}$$

Note that the Lagrange basis polynomial $h_j(x)$ in (3.35) can be expressed as

$$h_j(x) = \frac{p_{N+1}(x)}{p'_{N+1}(x_j)(x-x_j)}. \tag{3.43}$$

Plugging it into (3.42) gives

$$\begin{aligned} p_N(x_j)\int_a^b \frac{p_{N+1}(x)}{x-x_j}\omega(x)dx &= p_N(x_j)p'_{N+1}(x_j)\int_a^b h_j(x)\omega(x)dx \\ &= p_N(x_j)p'_{N+1}(x_j)\omega_j = \frac{k_{N+1}}{k_N}\|p_N\|_\omega^2, \end{aligned} \tag{3.44}$$

which implies (3.39). □

The above fundamental theorem reveals that the optimal abscissas of the Gauss quadrature formula are precisely the zeros of the orthogonal polynomial for the same interval and weight function. The Gauss quadrature is optimal because it fits all polynomials up to degree $2N+1$ exactly, and it is impossible to find any set of $\{x_j,\ \omega_j\}_{j=0}^N$ such that (3.38) holds for all $p \in P_{2N+2}$ (see Problem 3.3).

With the exception of a few special cases, like the Chebyshev polynomials, no explicit expressions of the quadrature nodes and weights are available. Theorem 3.4 provides an efficient approach to compute the nodes $\{x_j\}_{j=0}^N$, through finding the eigenvalues of the symmetric tridiagonal matrix A_{N+1} defined in (3.24). The following result indicates that the weights $\{\omega_j\}_{j=0}^N$ can be computed from the first component of the eigenvectors of A_{N+1}.

Theorem 3.6. *Let*

$$\mathbf{Q}(x_j) = \big(Q_0(x_j), Q_1(x_j), \ldots, Q_N(x_j)\big)^T$$

be the orthonormal eigenvector of A_{N+1} corresponding to the eigenvalue x_j, i.e.,

$$A_{N+1}\mathbf{Q}(x_j) = x_j\mathbf{Q}(x_j) \quad \textit{with} \quad \mathbf{Q}(x_j)^T\mathbf{Q}(x_j) = 1.$$

Then the weights $\{\omega_j\}_{j=0}^N$ can be computed from the first component of the eigenvector $\mathbf{Q}(x_j)$ by using the formula:

$$\omega_j = \big[Q_0(x_j)\big]^2\int_a^b \omega(x)dx, \quad 0 \le j \le N. \tag{3.45}$$

Proof. Using the Christoffel-Darboux formula (3.16) and the fact that $p_{N+1}(x_j) = 0$, we derive from (3.39) the following alternative expression of the weights:

$$\begin{aligned} \omega_j^{-1} &\overset{(3.39)}{=} \frac{k_N}{k_{N+1}}\frac{p_N(x_j)p'_{N+1}(x_j)}{\|p_N\|_\omega^2} \overset{(3.16)}{=} \sum_{n=0}^N \frac{p_n^2(x_j)}{\|p_n\|_\omega^2} \\ &= \widetilde{\mathbf{P}}(x_j)^T\widetilde{\mathbf{P}}(x_j), \quad 0 \le j \le N, \end{aligned} \tag{3.46}$$

where
$$\widetilde{\mathbf{P}}(x_j) = \big(\tilde{p}_0(x_j), \tilde{p}_1(x_j), \ldots, \tilde{p}_N(x_j)\big)^T \text{ with } \tilde{p}_n = \frac{p_n}{\|p_n\|_\omega}.$$

The identity (3.46) can be rewritten as

$$\omega_j \widetilde{\mathbf{P}}(x_j)^T \widetilde{\mathbf{P}}(x_j) = 1, \quad 0 \le j \le N.$$

On the other hand, we deduce from (3.29) that $\widetilde{\mathbf{P}}(x_j)$ is an eigenvector corresponding to the eigenvalue x_j. Therefore,

$$\mathbf{Q}(x_j) = \sqrt{\omega_j}\,\widetilde{\mathbf{P}}(x_j), \quad 0 \le j \le N, \tag{3.47}$$

is the unit eigenvector corresponding to the eigenvalue x_j. Equating the first components (3.47) yields

$$\omega_j = \left[\frac{Q_0(x_j)}{\tilde{p}_0(x_j)}\right]^2 = \frac{\|p_0\|_\omega^2}{\big[p_0(x_j)\big]^2}\big[Q_0(x_j)\big]^2 = \big[Q_0(x_j)\big]^2 \int_a^b \omega(x)dx, \quad 0 \le j \le N.$$

This completes the proof. □

Notice that all the nodes of the Gauss formula lie in the interior of the interval (a,b). This makes it difficult to impose boundary conditions. Below, we consider the Gauss-Radau or Gauss-Lobatto quadratures which include either one or both endpoints as a node(s).

We start with the Gauss-Radau quadrature. Assuming we would like to include the left endpoint $x = a$ in the quadrature, we define

$$q_N(x) = \frac{p_{N+1}(x) + \alpha_N p_N(x)}{x-a} \quad \text{with } \alpha_N = -\frac{p_{N+1}(a)}{p_N(a)}. \tag{3.48}$$

It is obvious that $q_N \in P_N$, and for any $r_{N-1} \in P_{N-1}$, we derive from Lemma 3.1 that

$$\begin{aligned} &\int_a^b q_N(x) r_{N-1}(x)\omega(x)(x-a)dx \\ &= \int_a^b \big(p_{N+1}(x) + \alpha_N p_N(x)\big) r_{N-1}(x)\omega(x)dx = 0. \end{aligned} \tag{3.49}$$

Hence, $\{q_N : N \ge 0\}$ defines a sequence of polynomials orthogonal with respect to the weight function $\tilde{\omega}(x) := \omega(x)(x-a)$, and the leading coefficient of q_N is k_{N+1}.

Theorem 3.7. (Gauss-Radau quadrature) *Let $x_0 = a$ and $\{x_j\}_{j=1}^N$ be the zeros of q_N defined in* (3.48)*. Then there exists a unique set of quadrature weights $\{\omega_j\}_{j=0}^N$, defined by* (3.36)*, such that*

$$\int_a^b p(x)\omega(x)dx = \sum_{j=0}^N p(x_j)\omega_j, \quad \forall p \in P_{2N}. \tag{3.50}$$

Moreover, the quadrature weights are all positive and can be expressed as

$$\omega_0 = \frac{1}{q_N(a)} \int_a^b q_N(x)\omega(x)dx, \tag{3.51a}$$

$$\omega_j = \frac{1}{x_j - a} \frac{k_{N+1}}{k_N} \frac{\|q_{N-1}\|_{\tilde{\omega}}^2}{q_{N-1}(x_j)q_N'(x_j)}, \quad 1 \le j \le N. \tag{3.51b}$$

Proof. The proof is similar to that of Theorem 3.5, so we shall only sketch it below.
Obviously, for any $p \in P_N$,

$$\int_a^b p(x)\omega(x)dx = \sum_{j=0}^{N} p(x_j) \int_a^b h_j(x)\omega(x)dx \overset{(3.36)}{=} \sum_{j=0}^{N} p(x_j)\omega_j. \tag{3.52}$$

Hence, the DOP is at least N.

Next, for any $p \in P_{2N}$, we write

$$p = (x-a)rq_N + s, \quad r \in P_{N-1},\ s \in P_N.$$

Since $(x-a)q_N(x)\big|_{x=x_j} = 0$, we have $p(x_j) = s(x_j)$ for $0 \le j < N$. Therefore, we deduce from (3.49) that

$$\begin{aligned} \int_a^b p(x)\omega(x)dx &= \int_a^b s(x)\omega(x)dx \\ &= \sum_{j=0}^{N} s(x_j)\omega_j = \sum_{j=0}^{N} p(x_j)\omega_j, \quad \forall p \in P_{2N}. \end{aligned}$$

Taking $p(x) = h_k^2(x) \in P_{2N}$ in the above identities, we conclude that $\omega_k > 0$ for $0 \le k \le N$.

Note that the Lagrange basis polynomials take the form

$$\begin{aligned} h_j(x) &= \frac{(x-a)q_N(x)}{((x-a)q_N(x))'\big|_{x=x_j}(x-x_j)} \\ &= \frac{(x-a)q_N(x)}{(q_N(x_j) + (x_j - a)q_N'(x_j))(x-x_j)}, \quad 0 \le j \le N. \end{aligned}$$

Hence, letting $j = 0$, we derive (3.51a) from the definition of ω_0, and for $1 \le j \le N$,

$$\omega_j = \int_a^b h_j(x)\omega(x)dx = \frac{1}{x_j - a} \int_a^b \frac{q_N(x)}{q_N'(x_j)(x-x_j)} \tilde{\omega}(x)dx.$$

Recall that $\{q_n\}$ are orthogonal with respect to $\tilde{\omega}$, so the integral part turns out to be the weight of the Gauss quadrature associated with N nodes being the zeros of $q_N(x)$. Hence, (3.51b) follows from the formula (3.39). □

Remark 3.1. *Similarly, a second Gauss-Radau quadrature can be constructed if we want to include the right endpoint $x = b$ instead of the left endpoint $x = a$.*

We now turn to the Gauss-Lobatto quadrature, whose nodes include two endpoints $x = a, b$. In this case, we choose α_N and β_N such that

$$p_{N+1}(x) + \alpha_N p_N(x) + \beta_N p_{N-1}(x) = 0 \text{ for } x = a, b, \tag{3.53}$$

and set

$$z_{N-1}(x) = \frac{p_{N+1}(x) + \alpha_N p_N(x) + \beta_N p_{N-1}(x)}{(x-a)(b-x)}. \tag{3.54}$$

It is clear that $z_{N-1} \in P_{N-1}$ and for any $r_{N-2} \in P_{N-2}$, we derive from Lemma 3.1 that

$$\begin{aligned} &\int_a^b z_{N-1} r_{N-2}(x-a)(b-x)\omega dx \\ &= \int_a^b (p_{N+1} + \alpha_N p_N + \beta_N p_{N-1}) r_{N-2} \omega dx = 0. \end{aligned} \tag{3.55}$$

Hence, $\{z_{N-1} : N \geq 1\}$ defines a sequence of polynomials orthogonal with respect to the weight function $\hat{\omega}(x) := (x-a)(b-x)\omega(x)$, and the leading coefficient of z_{N-1} is $-k_{N+1}$.

Theorem 3.8. (Gauss-Lobatto quadrature) *Let $x_0 = a$, $x_N = b$ and $\{x_j\}_{j=1}^{N-1}$ be the zeros of z_{N-1} in (3.53)–(3.54). Then there exists a unique set of quadrature weights $\{\omega_j\}_{j=0}^N$, defined by (3.36), such that*

$$\int_a^b p(x)\omega(x)dx = \sum_{j=0}^N p(x_j)\omega_j, \quad \forall p \in P_{2N-1}, \tag{3.56}$$

where the quadrature weights are expressed as

$$\omega_0 = \frac{1}{(b-a)z_{N-1}(a)} \int_a^b (b-x) z_{N-1}(x)\omega(x)dx, \tag{3.57a}$$

$$\omega_j = \frac{1}{(x_j-a)(b-x_j)} \frac{k_{N+1}}{k_N} \frac{\|z_{N-2}\|_{\hat{\omega}}^2}{z_{N-2}(x_j) z'_{N-1}(x_j)}, \quad 1 \leq j \leq N-1, \tag{3.57b}$$

$$\omega_N = \frac{1}{(b-a)z_{N-1}(b)} \int_a^b (x-a) z_{N-1}(x)\omega(x)dx. \tag{3.57c}$$

Moreover, we have $\omega_j > 0$ for $1 \leq j \leq N-1$.

Proof. The exactness (3.56) and the formulas of the weights can be derived in a similar fashion as in Theorem 3.7, so we skip the details. Here, we just verify $\omega_j > 0$ for $1 \leq j \leq N-1$ by using a different approach. Since $\{z_{N-1}\}$ are orthogonal with respect to the weight function $\hat{\omega}$, and $z_{N-1}(x_j) = 0$ for $1 \le j \le N-1$, we obtain from the Christoff-Darboux formula (3.16) that

$$\frac{k_N}{k_{N+1}} z_{N-2}(x_j) z'_{N-1}(x_j) = \sum_{j=0}^{N-2} \frac{\|z_{N-2}\|^2_{\hat{\omega}}}{\|z_j\|^2_{\hat{\omega}}} z_j^2(x_j) > 0, \ \ 1 \le j \le N-1.$$

Inserting it into the formula (3.57b) leads to $\omega_j > 0$ for $1 \le j \le N-1$. □

The Gauss-type quadrature formulas provide powerful tools for evaluating integrals and inner products in a spectral method. They also play an important role in spectral differentiations as to be shown later.

3.1.5 Interpolation and Discrete Transforms

Let $\{x_j, \omega_j\}_{j=0}^N$ be a set of Gauss, Gauss-Radau or Gauss-Lobatto quadrature nodes and weights. We define the corresponding discrete inner product and norm as

$$\langle u, v \rangle_{N,\omega} := \sum_{j=0}^{N} u(x_j) v(x_j) \omega_j, \quad \|u\|_{N,\omega} := \sqrt{\langle u, u \rangle_{N,\omega}}. \tag{3.58}$$

Note that $\langle \cdot, \ \cdot \rangle_{N,\omega}$ is an approximation to the continuous inner product $(\cdot,\cdot)_\omega$, and the exactness of Gauss-type quadrature formulas implies

$$\langle u, v \rangle_{N,\omega} = (u, v)_\omega, \quad \forall u \cdot v \in P_{2N+\delta}, \tag{3.59}$$

where $\delta = 1, 0$ and -1 for the Gauss, Gauss-Radau and Gauss-Lobatto quadrature, respectively.

Definition 3.1. *For any $u \in C(\Lambda)$, we define the interpolation operator $I_N : C(\Lambda) \to P_N$ such that*

$$(I_N u)(x_j) = u(x_j), \qquad 0 \le j \le N, \tag{3.60}$$

where $\Lambda = (a,b)$, $[a,b)$, $[a,b]$ for the Gauss, Gauss-Radau and Gauss-Lobatto quadrature, respectively.

The interpolation condition (3.60) implies that $I_N p = p$ for all $p \in P_N$. On the other hand, since $I_N u \in P_N$, we can write

$$(I_N u)(x) = \sum_{n=0}^{N} \tilde{u}_n p_n(x), \tag{3.61}$$

which is the counterpart of the discrete Fourier series (2.20) and may be referred to as the *discrete polynomial series*. By taking the discrete inner product of (3.61) with $\{p_k\}_{k=0}^N$, we can determine the coefficients $\{\tilde{u}_n\}$ by using (3.60) and (3.59). More precisely, we have

Theorem 3.9.

$$\tilde{u}_n = \frac{1}{\gamma_n} \sum_{j=0}^{N} u(x_j) p_n(x_j) \omega_j, \quad 0 \le n \le N, \tag{3.62}$$

where $\gamma_n = \|p_n\|_\omega^2$ *for* $0 \le n \le N-1$, *and*

$$\gamma_N = \begin{cases} \|p_N\|_\omega^2, & \textit{for Gauss and Gauss-Radau}, \\ \langle p_N, p_N \rangle_{N,\omega}, & \textit{for Gauss-Lobatto}. \end{cases} \tag{3.63}$$

The formula (3.62)-(3.63) defines the *forward discrete polynomial transform* as in the Fourier case, which transforms the physical values $\{u(x_j)\}_{j=0}^N$ to the expansion coefficients $\{\tilde{u}_n\}_{n=0}^N$. Conversely, the *backward (or inverse) discrete polynomial transform* is formulated by

$$u(x_j) = (I_N u)(x_j) = \sum_{n=0}^{N} \tilde{u}_n p_n(x_j), \quad 0 \le j \le N, \tag{3.64}$$

which takes the expansion coefficients $\{\tilde{u}_n\}_{n=0}^N$ to the physical values $\{u(x_j)\}_{j=0}^N$.

We see that if the matrices $\big(p_n(x_j)\big)_{0 \le n,j \le N}$ and/or $\big(\gamma_n^{-1} p_n(x_j)\omega_j\big)_{0 \le n,j \le N}$ are precomputed, then the discrete transforms (3.62) and (3.64) can be manipulated directly by a standard matrix–vector multiplication routine in about N^2 flops. Since discrete transforms are frequently used in spectral codes, it is desirable to reduce the computational complexity, especially for multidimensional cases. In particular, the Fast Fourier Transform (FFT) (cf. Cooley and Tukey (1965)) and discrete Chebyshev transform (treated as a Fourier-cosine transform) can be accomplished by $O(N \log_2 N)$ operations. However, with the advent of more powerful computers, this aspect should not be a big concern for moderate scale problems.

3.1.6 Differentiation in the Physical Space

Now, we are ready to address an important issue – polynomial-based spectral differentiation techniques. As with the Fourier cases, they can be performed in either the physical space or the frequency space.

Let us start with the implementation in the physical space. Assume that $u \in P_N$ is an approximation of the unknown solution U. Let $\{h_j\}_{j=0}^N$ be the Lagrange basis polynomials associated with a set of Gauss-type points $\{x_j\}_{j=0}^N$. Clearly,

$$u(x) = \sum_{j=0}^{N} u(x_j) h_j(x). \tag{3.65}$$

Hence, differentiating it m times leads to

$$u^{(m)}(x_k) = \sum_{j=0}^{N} h_j^{(m)}(x_k) u(x_j), \quad 0 \le k \le N. \tag{3.66}$$

Let us denote

$$\mathbf{u}^{(m)} := \left(u^{(m)}(x_0), u^{(m)}(x_1), \ldots, u^{(m)}(x_N)\right)^T, \quad \mathbf{u} := \mathbf{u}^{(0)};$$
$$D^{(m)} := \left(d_{kj}^{(m)}\right)_{0\le k,j\le N}, \quad D := D^{(1)}. \tag{3.67}$$

Different from the Fourier case, the higher-order differentiation matrix in this context can be computed by a product of the first-order one.

Theorem 3.10.

$$D^{(m)} = DD\ldots D = D^m, \quad m \ge 1, \tag{3.68}$$

and

$$\mathbf{u}^{(m)} = D^m \mathbf{u}, \quad m \ge 1. \tag{3.69}$$

Proof. Differentiating (3.65) gives

$$u'(x) = \sum_{l=0}^{N} u(x_l) h_l'(x). \tag{3.70}$$

Taking $u = h_j' \in P_{N-1}$ in the above equation leads to

$$h_j''(x) = \sum_{l=0}^{N} h_l'(x) h_j'(x_l).$$

Hence,

$$d_{kj}^{(2)} = h_j''(x_k) = \sum_{l=0}^{N} h_l'(x_k) h_j'(x_l) = \sum_{l=0}^{N} d_{kl}^{(1)} d_{lj}^{(1)},$$

which implies

$$D^{(2)} = DD = D^2. \tag{3.71}$$

Similarly, taking $u = h_j^{(i)}$ in (3.70) leads to

$$d_{kj}^{(i+1)} = h_j^{(i+1)}(x_k) = \sum_{l=0}^{N} h_l'(x_k) h_j^{(i)}(x_l) = \sum_{l=0}^{N} d_{kl}^{(1)} d_{lj}^{(i)}.$$

Therefore,

$$D^{(i+1)} = DD^{(i)}, \quad i \ge 1, \tag{3.72}$$

which yields (3.68).

Finally, (3.69) can be written in matrix form as in (3.66). □

Thanks to Theorem 3.10, it suffices to compute the first-order differentiation matrix D. We present below the explicit formulas for the entries of D.

Theorem 3.11. *The entries of D are determined by*

$$d_{kj} = h_j'(x_k) = \begin{cases} \dfrac{Q'(x_k)}{Q'(x_j)} \dfrac{1}{x_k - x_j}, & \text{if } k \neq j, \\ \dfrac{Q''(x_k)}{2Q'(x_k)}, & \text{if } k = j, \end{cases} \tag{3.73}$$

where

$$Q(x) = p_{N+1}(x), \ (x-a)q_N(x), \ (x-a)(b-x)z_{N-1}(x) \tag{3.74}$$

are the quadrature polynomials (cf. (3.48) *and* (3.54)) *of the Gauss, Gauss-Radau and Gauss-Lobatto quadrature, respectively.*

Proof. The Lagrange basis polynomials can be expressed as

$$h_j(x) = \frac{Q(x)}{Q'(x_j)(x-x_j)}, \quad 0 \le j \le N. \tag{3.75}$$

Differentiating (3.75) and using the fact $Q(x_j) = 0$ lead to

$$d_{kj} = h_j'(x_k) = \frac{Q'(x_k)}{Q'(x_j)} \frac{1}{x_k - x_j}, \quad \forall k \neq j.$$

Applying the L'Hopital's rule twice yields

$$d_{kk} = \lim_{x \to x_k} h_k'(x) = \frac{1}{Q'(x_k)} \lim_{x \to x_k} \frac{Q'(x)(x-x_k) - Q(x)}{(x-x_k)^2} = \frac{Q''(x_k)}{2Q'(x_k)}.$$

This completes the proof. □

Therefore, having precomputed the first-order differentiation matrix, the differentiation in the physical space can be carried out through matrix–matrix and matrix–vector multiplications.

3.1.7 Differentiation in the Frequency Space

Differentiation in the frequency space is to express the expansion coefficients of the derivatives of a function in terms of expansion coefficients of the function itself. More precisely, given $u \in P_N$, instead of using the Lagrange basis polynomials, we expand u in terms of the orthogonal polynomials:

$$u(x) = \sum_{n=0}^{N} \hat{u}_n p_n(x). \tag{3.76}$$

Using the orthogonality, we can determine the coefficients by

$$\hat{u}_n = \frac{1}{\|p_n\|_\omega^2} \int_a^b u(x) p_n(x) \omega(x) dx, \quad 0 \le n \le N. \tag{3.77}$$

Since $u' \in P_{N-1}$, we have

$$u'(x) = \sum_{n=0}^{N} \hat{u}_n^{(1)} p_n(x) \ \text{ with } \ \hat{u}_N^{(1)} = 0. \tag{3.78}$$

In order to express $\{\hat{u}_n^{(1)}\}_{n=0}^N$ in terms of $\{\hat{u}_n\}_{n=0}^N$, we assume that $\{p_n'\}$ are also orthogonal. Indeed, this property holds for the classical orthogonal polynomials such as the Legendre, Chebyshev, Jacobi, Laguerre and Hermite polynomials. In other words, $\{p_n'\}$ satisfy the three-term recurrence relation due to Corollary 3.1:

$$p_{n+1}'(x) = (a_n^{(1)} x - b_n^{(1)}) p_n'(x) - c_n^{(1)} p_{n-1}'(x). \tag{3.79}$$

Differentiating the three-term recurrence relation (3.12) and using (3.79), we derive

$$p_n(x) = \tilde{a}_n p_{n-1}'(x) + \tilde{b}_n p_n'(x) + \tilde{c}_n p_{n+1}'(x). \tag{3.80}$$

The coefficients $\{\hat{u}_n^{(1)}\}$ in (3.78) can be computed by the following backward recurrence formulas.

Theorem 3.12.

$$\begin{aligned} \hat{u}_{n-1}^{(1)} &= \frac{1}{\tilde{c}_{n-1}} \left[\hat{u}_n - \tilde{b}_n \hat{u}_n^{(1)} - \tilde{a}_{n+1} \hat{u}_{n+1}^{(1)} \right], \ n = N-1, \ldots, 1, \\ \hat{u}_N^{(1)} &= 0, \ \hat{u}_{N-1}^{(1)} = \frac{1}{\tilde{c}_{N-1}} \hat{u}_N. \end{aligned} \tag{3.81}$$

Proof. By (3.78) and (3.80),

$$\begin{aligned} u' &= \sum_{n=0}^{N-1} \hat{u}_n^{(1)} p_n = \sum_{n=0}^{N-1} \hat{u}_n^{(1)} [\tilde{a}_n p_{n-1}' + \tilde{b}_n p_n' + \tilde{c}_n p_{n+1}'] \\ &= \sum_{n=1}^{N-1} [\tilde{c}_{n-1} \hat{u}_{n-1}^{(1)} + \tilde{b}_n \hat{u}_n^{(1)} + \tilde{a}_{n+1} \hat{u}_{n+1}^{(1)}] p_n' + \tilde{c}_{N-1} \hat{u}_{N-1}^{(1)} p_N'. \end{aligned}$$

On the other hand, by (3.76),

$$u'(x) = \sum_{n=1}^{N} \hat{u}_n p_n'(x).$$

By the (assumed) orthogonality of $\{p_n'\}$, we are able to equate the coefficients of p_n' in the above two expressions, which leads to (3.81). □

Higher-order differentiations in the frequency space can be carried out by using the formula (3.81) repeatedly. It is important to point out that *spectral differentiations* together with *discrete transforms* form the basic ingredients for the so-called "*pseudo-spectral technique*" (particularly useful for nonlinear problems): *the differentiations* are manipulated in the frequency space, the inner products are computed in the physical space, and both spaces are communicated through discrete transforms.

3.1.8 Approximability of Orthogonal Polynomials

We now briefly review some general polynomial approximation results. One can find their proofs from standard books on approximation theory (see, for instance, Timan (1994), Cheney (1998)).

The first fundamental result is the remarkable *Weierstrass Theorem*, which states that any continuous function in a finite interval can be uniformly approximated by an algebraic polynomial.

Theorem 3.13. *Let* (a,b) *be a finite interval. Then for any* $u \in C[a,b]$, *and any* $\varepsilon > 0$, *there exist* $n \in \mathbb{N}$ *and* $p_n \in P_n$ *such that*

$$\|u - p_n\|_{L^\infty(a,b)} < \varepsilon. \tag{3.82}$$

This theorem forms the cornerstone of the classical polynomial approximation theory. The construction of p_n essentially relies on the solution of the best approximation problem:

$$\begin{cases} \text{Given a fixed } n \in \mathbb{N}, \text{ find } p_n^* \in P_n, \text{ such that} \\ \|u - p_n^*\|_{L^\infty(a,b)} = \inf\limits_{p_n \in P_n} \|u - p_n\|_{L^\infty(a,b)}. \end{cases} \tag{3.83}$$

This problem admits a unique solution, and as a consequence of Theorem 3.13, p_n^* uniformly converges to u as $n \to \infty$. However, the derivation of the best uniform approximation polynomial p_n^* is nontrivial, since a strong uniform norm is involved in (3.83), whereas the best approximation problem in the L^2-sense is easier to solve.

Theorem 3.14. *Let* $I = (a,b)$ *be a finite or an infinite interval. Then for any* $u \in L^2_\omega(I)$ *and* $n \in \mathbb{N}$, *there exists a unique* $q_n^* \in P_n$, *such that*

$$\|u - q_n^*\|_\omega = \inf_{q_n \in P_n} \|u - q_n\|_\omega, \tag{3.84}$$

where

$$q_n^*(x) = \sum_{k=0}^{n} \hat{u}_k p_k(x) \quad with \quad \hat{u}_k = \frac{(u, p_k)_\omega}{\|p_k\|_\omega^2}, \tag{3.85}$$

and $\{p_k\}_{k=0}^n$ *forms an* L^2_ω*-orthogonal basis of* P_n.

In particular, we denote the best approximation polynomial q_n^* by $\pi_n u$, which is the L^2_ω-orthogonal projection of u, and is characterized by the projection theorem

$$\|u - \pi_n u\|_\omega = \inf_{q_n \in P_n} \|u - q_n\|_\omega. \tag{3.86}$$

Equivalently, the L^2_ω-orthogonal projection can be defined by

$$(u - \pi_n u, \phi)_\omega = 0, \quad \forall \phi \in P_n, \tag{3.87}$$

so $\pi_n u$ is the first $n+1$-term truncation of the series $u = \sum_{k=0}^\infty \hat{u}_k p_k(x)$.

It is interesting to notice that a result similar to the Weierstrass theorem holds on infinite intervals, if suitable conditions are imposed on the growth of the given function u (cf. Funaro (1992)).

Theorem 3.15. *If* $u \in C[0,\infty)$ *and for certain* $\delta > 0$, u *satisfies*

$$u(x)e^{-\delta x} \to 0, \quad as \ x \to \infty,$$

then for any $\varepsilon > 0$, *there exist an* $n \in \mathbb{N}$ *and* $p_n \in P_n$ *such that*

$$|u(x) - p_n(x)|e^{-\delta x} \le \varepsilon, \quad \forall x \in [0,\infty).$$

Similar result holds on $(-\infty,\infty)$, *if we replace* $e^{-\delta x}$ *by* $e^{-\delta x^2}$.

3.1.8.1 A Short Summary of this Section

We presented some basic knowledge of orthogonal polynomials, which is mostly relevant to spectral approximations. We also set up a general framework for the study of each specific family of orthogonal polynomials to be presented in the forthcoming sections as tabulated in Table 3.1.

Table 3.1 List of orthogonal polynomials

	Symbol	Interval	Weight function	Section
Jacobi	$J_n^{\alpha,\beta}$	$(-1,1)$	$(1-x)^\alpha(1+x)^\beta, \alpha,\beta > -1$	3.2
Legendre	L_n	$(-1,1)$	1	3.3
Chebyshev	T_n	$(-1,1)$	$1/\sqrt{1-x^2}$	3.4
Laguerre	$\mathscr{L}_n^{(\alpha)}$	$(0,+\infty)$	$x^\alpha e^{-x}, \alpha > -1$	7.1
Hermite	H_n	$(-\infty,+\infty)$	e^{-x^2}	7.2

3.2 Jacobi Polynomials

3.2.1 Basic Properties

The Jacobi polynomials, denoted by $J_n^{\alpha,\beta}(x)$, are orthogonal with respect to the Jacobi weight function $\omega^{\alpha,\beta}(x) := (1-x)^\alpha(1+x)^\beta$ over $I := (-1,1)$, namely,

$$\int_{-1}^{1} J_n^{\alpha,\beta}(x)J_m^{\alpha,\beta}(x)\omega^{\alpha,\beta}(x)dx = \gamma_n^{\alpha,\beta}\delta_{mn}, \tag{3.88}$$

where $\gamma_n^{\alpha,\beta} = \|J_n^{\alpha,\beta}\|^2_{\omega^{\alpha,\beta}}$. The weight function $\omega^{\alpha,\beta}$ belongs to $L^1(I)$ if and only if $\alpha,\beta > -1$ (to be assumed throughout this section).

Let $k_n^{\alpha,\beta}$ be the leading coefficient of $J_n^{\alpha,\beta}(x)$. According to Theorem 3.1, there exists a unique sequence of monic orthogonal polynomials $\{J_n^{\alpha,\beta}(x)/k_n^{\alpha,\beta}\}$.

This class of Jacobi weight functions leads to Jacobi polynomials with many attractive properties that are not shared by general orthogonal polynomials.

3.2.1.1 Sturm-Liouville Equation

We first show that the Jacobi polynomials are the eigenfunctions of a singular Sturm-Liouville operator defined by

$$\begin{aligned}\mathscr{L}_{\alpha,\beta}u :&= -(1-x)^{-\alpha}(1+x)^{-\beta}\partial_x\big((1-x)^{\alpha+1}(1+x)^{\beta+1}\partial_x u(x)\big)\\ &= (x^2-1)\partial_x^2 u(x) + \{\alpha-\beta+(\alpha+\beta+2)x\}\partial_x u(x).\end{aligned} \tag{3.89}$$

More precisely, we have

Theorem 3.16. *The Jacobi polynomials are the eigenfunctions of the singular Sturm-Liouville problem:*

$$\mathscr{L}_{\alpha,\beta}J_n^{\alpha,\beta}(x) = \lambda_n^{\alpha,\beta}J_n^{\alpha,\beta}(x), \tag{3.90}$$

and the corresponding eigenvalues are

$$\lambda_n^{\alpha,\beta} = n(n+\alpha+\beta+1). \tag{3.91}$$

Proof. For any $u \in P_n$, we have $\mathscr{L}_{\alpha,\beta}u \in P_n$. Using integration by parts twice, we find that for any $\phi \in P_{n-1}$,

$$\big(\mathscr{L}_{\alpha,\beta}J_n^{\alpha,\beta},\phi\big)_{\omega^{\alpha,\beta}} = \big(\partial_x J_n^{\alpha,\beta},\partial_x\phi\big)_{\omega^{\alpha+1,\beta+1}} = \big(J_n^{\alpha,\beta},\mathscr{L}_{\alpha,\beta}\phi\big)_{\omega^{\alpha,\beta}} \overset{(3.88)}{=} 0.$$

Since $\mathscr{L}_{\alpha,\beta}J_n^{\alpha,\beta} \in P_n$, the uniqueness of orthogonal polynomials implies that there exists a constant $\lambda_n^{\alpha,\beta}$ such that

$$\mathscr{L}_{\alpha,\beta}J_n^{\alpha,\beta} = \lambda_n^{\alpha,\beta}J_n^{\alpha,\beta}.$$

To determine $\lambda_n^{\alpha,\beta}$, we compare the coefficient of the leading term x^n on both sides, and find

$$k_n^{\alpha,\beta} n(n+\alpha+\beta+1) = k_n^{\alpha,\beta} \lambda_n^{\alpha,\beta},$$

where $k_n^{\alpha,\beta}$ is the leading coefficient of $J_n^{\alpha,\beta}$. Hence, we have $\lambda_n^{\alpha,\beta} = n(n+\alpha+\beta+1)$. □

Remark 3.2. *Observe from integration by parts that the Sturm-Liouville operator $\mathscr{L}_{\alpha,\beta}$ is self-adjoint with respect to the inner product $(\cdot,\cdot)_{\omega^{\alpha,\beta}}$, i.e.,*

$$(\mathscr{L}_{\alpha,\beta}\phi, \psi)_{\omega^{\alpha,\beta}} = (\phi, \mathscr{L}_{\alpha,\beta}\psi)_{\omega^{\alpha,\beta}}, \tag{3.92}$$

for any $\phi, \psi \in \{u : \mathscr{L}_{\alpha,\beta}u \in L^2_{\omega^{\alpha,\beta}}(I)\}$.

As pointed out in Theorem 4.2.2 of Szegö (1975), the differential equation

$$\mathscr{L}_{\alpha,\beta}u = \lambda u,$$

has a polynomial solution not identically zero if and only if λ has the form $n(n+\alpha+\beta+1)$. This solution is $J_n^{\alpha,\beta}(x)$ (up to a constant), and no solution which is linearly independent of $J_n^{\alpha,\beta}(x)$ can be a polynomial. Moreover, we can show that

$$J_n^{\alpha,\beta}(x) = \sum_{k=0}^{n} a_k^n (x-1)^k,$$

where

$$\frac{a_{k+1}^n}{a_k^n} = \frac{\gamma_n^{\alpha,\beta} - k(k+\alpha+\beta+1)}{2(k+1)(k+\alpha+1)}. \tag{3.93}$$

Assume that the Jacobi polynomials are normalized such that

$$a_0^n = J_n^{\alpha,\beta}(1) = \binom{n+\alpha}{n} = \frac{\Gamma(n+\alpha+1)}{n!\Gamma(\alpha+1)}, \tag{3.94}$$

where $\Gamma(\cdot)$ is the Gamma function (cf. Appendix A). We can derive from (3.93) the leading coefficient

$$a_n^n = k_n^{\alpha,\beta} = \frac{\Gamma(2n+\alpha+\beta+1)}{2^n n!\Gamma(n+\alpha+\beta+1)}. \tag{3.95}$$

Moreover, working out $\{a_k^n\}$ by using (3.93), we find

$$J_n^{\alpha,\beta}(x) = \frac{\Gamma(n+\alpha+1)}{n!\Gamma(n+\alpha+\beta+1)} \sum_{k=0}^{n} \binom{n}{k} \frac{\Gamma(n+k+\alpha+\beta+1)}{\Gamma(k+\alpha+1)} \left(\frac{x-1}{2}\right)^k. \tag{3.96}$$

A direct consequence of Theorem 3.16 is the orthogonality of $\{\partial_x J_n^{\alpha,\beta}\}$.

Corollary 3.5.

$$\int_{-1}^{1} \partial_x J_n^{\alpha,\beta} \partial_x J_m^{\alpha,\beta} \omega^{\alpha+1,\beta+1} dx = \lambda_n^{\alpha,\beta} \gamma_n^{\alpha,\beta} \delta_{nm}. \tag{3.97}$$

Proof. Using integration by parts, Theorem 3.16 and the orthogonality of $\{J_n^{\alpha,\beta}\}$, we obtain

$$\left(\partial_x J_n^{\alpha,\beta}, \partial_x J_m^{\alpha,\beta}\right)_{\omega^{\alpha+1,\beta+1}} = \left(J_n^{\alpha,\beta}, \mathscr{L}_{\alpha,\beta} J_m^{\alpha,\beta}\right)_{\omega^{\alpha,\beta}} \overset{(3.90)}{=} \lambda_n^{\alpha,\beta} \|J_n^{\alpha,\beta}\|^2_{\omega^{\alpha,\beta}} \delta_{nm}.$$

This ends the proof. □

Since $\{\partial_x J_n^{\alpha,\beta}\}$ is orthogonal with respect to the weight $\omega^{\alpha+1,\beta+1}$, by Theorem 3.1, $\partial_x J_n^{\alpha,\beta}$ must be proportional to $J_{n-1}^{\alpha+1,\beta+1}$, namely,

$$\partial_x J_n^{\alpha,\beta}(x) = \mu_n^{\alpha,\beta} J_{n-1}^{\alpha+1,\beta+1}(x). \tag{3.98}$$

Comparing the leading coefficients on both sides leads to the proportionality constant:

$$\mu_n^{\alpha,\beta} = \frac{n k_n^{\alpha,\beta}}{k_{n-1}^{\alpha+1,\beta+1}} \overset{(3.95)}{=} \frac{1}{2}(n+\alpha+\beta+1). \tag{3.99}$$

This gives the following important derivative relation:

$$\partial_x J_n^{\alpha,\beta}(x) = \frac{1}{2}(n+\alpha+\beta+1) J_{n-1}^{\alpha+1,\beta+1}(x). \tag{3.100}$$

Applying this formula recursively yields

$$\partial_x^k J_n^{\alpha,\beta}(x) = d_{n,k}^{\alpha,\beta} J_{n-k}^{\alpha+k,\beta+k}(x), \quad n \ge k, \tag{3.101}$$

where

$$d_{n,k}^{\alpha,\beta} = \frac{\Gamma(n+k+\alpha+\beta+1)}{2^k \Gamma(n+\alpha+\beta+1)}. \tag{3.102}$$

3.2.1.2 Rodrigues' Formula

The Rodrigues' formula for the Jacobi polynomials is stated below.

Theorem 3.17.

$$(1-x)^{\alpha}(1+x)^{\beta} J_n^{\alpha,\beta}(x) = \frac{(-1)^n}{2^n n!} \frac{d^n}{dx^n}\left[(1-x)^{n+\alpha}(1+x)^{n+\beta}\right]. \tag{3.103}$$

Proof. For any $\phi \in P_{n-1}$, using integration by parts leads to

$$\begin{aligned}\int_{-1}^{1} \partial_x^n \left((1-x)^{n+\alpha}(1+x)^{n+\beta}\right) \phi dx &= \ldots \\ &= (-1)^n \int_{-1}^{1} \left((1-x)^{n+\alpha}(1+x)^{n+\beta}\right) \partial_x^n \phi dx = 0.\end{aligned}$$

Hence, by Theorem 3.1, there exists a constant c_n such that

$$\partial_x^n \left((1-x)^{n+\alpha}(1+x)^{n+\beta} \right) = c_n (1-x)^{\alpha}(1+x)^{\beta} J_n^{\alpha,\beta}(x). \tag{3.104}$$

Letting $x \to 1$ and using (3.94) leads to

$$\begin{aligned} c_n &= \frac{1}{J_n^{\alpha,\beta}(1)} \left\{ \frac{1}{(1-x)^{\alpha}(1+x)^{\beta}} \partial_x^n \left((1-x)^{n+\alpha}(1+x)^{n+\beta} \right) \right\} \Big|_{x=1} \\ &= (-1)^n n! 2^n. \end{aligned}$$

The proof is complete. □

We now present some consequences of the Rodrigues' formula. First, expanding the nth-order derivative in (3.103) yields the explicit formula

$$J_n^{\alpha,\beta}(x) = 2^{-n} \sum_{j=0}^{n} \binom{n+\alpha}{j} \binom{n+\beta}{n-j} (x-1)^{n-j}(x+1)^j. \tag{3.105}$$

Second, replacing x by $-x$ in (3.103) immediately leads to the symmetric relation

$$J_n^{\alpha,\beta}(-x) = (-1)^n J_n^{\beta,\alpha}(x). \tag{3.106}$$

Therefore, the special Jacobi polynomial $J_n^{\alpha,\alpha}(x)$ (up to a constant, is referred to as the Gegenbauer or ultra-spherical polynomial), is an odd function for odd n and an even function for even n. Moreover, using (3.94) and (3.106) leads to

$$J_n^{\alpha,\beta}(-1) = (-1)^n \frac{\Gamma(n+\beta+1)}{n!\Gamma(\beta+1)}, \tag{3.107}$$

and by the Stirling's formula (A.7),

$$J_n^{\alpha,\beta}(1) \sim n^{\alpha} \text{ and } |J_n^{\alpha,\beta}(-1)| \sim n^{\beta} \quad \text{for } n \gg 1. \tag{3.108}$$

As another consequence of (3.103), we derive the explicit formula of the normalization constant $\gamma_n^{\alpha,\beta}$ in (3.88).

Corollary 3.6.

$$\begin{aligned} \int_{-1}^{1} \left[J_n^{\alpha,\beta}(x) \right]^2 \omega^{\alpha,\beta}(x) dx &= \gamma_n^{\alpha,\beta} \\ &= \frac{2^{\alpha+\beta+1}\Gamma(n+\alpha+1)\Gamma(n+\beta+1)}{(2n+\alpha+\beta+1) n! \Gamma(n+\alpha+\beta+1)}. \end{aligned} \tag{3.109}$$

Proof. Multiplying (3.103) by $J_n^{\alpha,\beta}$ and integrating the resulting equality over $(-1,1)$, we derive from integration by parts that

$$\begin{aligned}
&\int_{-1}^{1}(1-x)^{\alpha}(1+x)^{\beta}[J_n^{\alpha,\beta}(x)]^2dx \\
&= \frac{(-1)^n}{2^n n!}\int_{-1}^{1}\partial_x^n\Big\{(1-x)^{n+\alpha}(1+x)^{n+\beta}\Big\}J_n^{\alpha,\beta}(x)dx \\
&= \frac{(-1)^{2n}}{2^n n!}\int_{-1}^{1}(1-x)^{n+\alpha}(1+x)^{n+\beta}\partial_x^n J_n^{\alpha,\beta}(x)dx \\
&= \frac{k_n^{\alpha,\beta}}{2^n}\int_{-1}^{1}(1-x)^{n+\alpha}(1+x)^{n+\beta}dx \\
&\overset{(3.95)}{\underset{(A.6)}{=}} \frac{2^{\alpha+\beta+1}\Gamma(n+\alpha+1)\Gamma(n+\beta+1)}{(2n+\alpha+\beta+1)n!\Gamma(n+\alpha+\beta+1)}.
\end{aligned}$$

This ends the proof. □

3.2.1.3 Recurrence Formulas

The Jacobi polynomials are generated by the three-term recurrence relation:

$$\begin{aligned}
&J_{n+1}^{\alpha,\beta}(x) = \big(a_n^{\alpha,\beta}x - b_n^{\alpha,\beta}\big)J_n^{\alpha,\beta}(x) - c_n^{\alpha,\beta}J_{n-1}^{\alpha,\beta}(x), \;\; n\ge 1, \\
&J_0^{\alpha,\beta}(x)=1, \;\; J_1^{\alpha,\beta}(x)=\frac{1}{2}(\alpha+\beta+2)x+\frac{1}{2}(\alpha-\beta),
\end{aligned} \tag{3.110}$$

where

$$a_n^{\alpha,\beta} = \frac{(2n+\alpha+\beta+1)(2n+\alpha+\beta+2)}{2(n+1)(n+\alpha+\beta+1)}, \tag{3.111a}$$

$$b_n^{\alpha,\beta} = \frac{(\beta^2-\alpha^2)(2n+\alpha+\beta+1)}{2(n+1)(n+\alpha+\beta+1)(2n+\alpha+\beta)}, \tag{3.111b}$$

$$c_n^{\alpha,\beta} = \frac{(n+\alpha)(n+\beta)(2n+\alpha+\beta+2)}{(n+1)(n+\alpha+\beta+1)(2n+\alpha+\beta)}. \tag{3.111c}$$

This relation allows us to evaluate the Jacobi polynomials at any given abscissa $x\in[-1,1]$, and it is the starting point to derive other properties.

Next, we state several useful recurrence formulas involving different pairs of (α,β).

Theorem 3.18. *The Jacobi polynomial* $J_n^{\alpha+1,\beta}(x)$ *is a linear combination of* $J_l^{\alpha,\beta}(x)$, $l=0,1,\ldots,n$, *i.e.,*

$$
\begin{aligned}
J_n^{\alpha+1,\beta}(x) = & \frac{\Gamma(n+\beta+1)}{\Gamma(n+\alpha+\beta+2)} \times \\
& \sum_{l=0}^{n} \frac{(2l+\alpha+\beta+1)\Gamma(l+\alpha+\beta+1)}{\Gamma(l+\beta+1)} J_l^{\alpha,\beta}(x).
\end{aligned} \tag{3.112}
$$

Proof. In the Jacobi case, the kernel polynomial (3.17) takes the form

$$
K_n(x,y) = \sum_{l=0}^{n} \frac{1}{\gamma_l^{\alpha,\beta}} J_l^{\alpha,\beta}(x) J_l^{\alpha,\beta}(y). \tag{3.113}
$$

By Lemma 3.2, $\{K_n(x,1)\}$ are orthogonal with respect to $\omega^{\alpha+1,\beta}$. By the uniqueness of orthogonal polynomials (cf. Theorem 3.1), $K_n(x,1)$ must be proportional to $J_n^{\alpha+1,\beta}$, i.e.,

$$
K_n(x,1) = \sum_{l=0}^{n} \frac{J_l^{\alpha,\beta}(1)}{\gamma_l^{\alpha,\beta}} J_l^{\alpha,\beta}(x) = d_n^{\alpha,\beta} J_n^{\alpha+1,\beta}(x). \tag{3.114}
$$

The proportionality constant $d_n^{\alpha,\beta}$ is determined by comparing the leading coefficients of both sides of (3.114) and working out the constants, namely,

$$
d_n^{\alpha,\beta} = \frac{k_n^{\alpha,\beta} J_n^{\alpha,\beta}(1)}{k_n^{\alpha+1,\beta} \gamma_n^{\alpha,\beta}} = 2^{-\alpha-\beta-1} \frac{\Gamma(n+\alpha+\beta+2)}{\Gamma(\alpha+1)\Gamma(n+\beta+1)}.
$$

Inserting this constant into (3.114), we obtain (3.112) directly from (3.94) and (3.109). □

Remark 3.3. *Thanks to* (3.106), *it follows from* (3.112) *that*

$$
\begin{aligned}
J_n^{\alpha,\beta+1}(x) = & \frac{\Gamma(n+\alpha+1)}{\Gamma(n+\alpha+\beta+2)} \times \\
& \sum_{l=0}^{n} (-1)^{n-l} \frac{(2l+\alpha+\beta+1)\Gamma(l+\alpha+\beta+1)}{\Gamma(l+\alpha+1)} J_l^{\alpha,\beta}(x).
\end{aligned} \tag{3.115}
$$

Theorem 3.19. *The Jacobi polynomials satisfy*

$$
J_n^{\alpha+1,\beta} = \frac{2}{2n+\alpha+\beta+2} \frac{(n+\alpha+1)J_n^{\alpha,\beta} - (n+1)J_{n+1}^{\alpha,\beta}}{1-x}, \tag{3.116a}
$$

$$
J_n^{\alpha,\beta+1} = \frac{2}{2n+\alpha+\beta+2} \frac{(n+\beta+1)J_n^{\alpha,\beta} + (n+1)J_{n+1}^{\alpha,\beta}}{1+x}. \tag{3.116b}
$$

Proof. In the Jacobi case, the Christoffel-Darboux formula (3.15) reads

$$
K_n(x,y) = \frac{k_n^{\alpha,\beta}}{k_{n+1}^{\alpha,\beta}\gamma_n^{\alpha,\beta}} \frac{J_{n+1}^{\alpha,\beta}(x)J_n^{\alpha,\beta}(y) - J_n^{\alpha,\beta}(x)J_{n+1}^{\alpha,\beta}(y)}{x-y}, \tag{3.117}
$$

which, together with (3.114), leads to

$$J_n^{\alpha+1,\beta}(x) = \frac{1}{d_n^{\alpha,\beta}} K_n(x,1)$$
$$= \frac{k_n^{\alpha,\beta}}{d_n^{\alpha,\beta} k_{n+1}^{\alpha,\beta} \gamma_n^{\alpha,\beta}} \frac{J_{n+1}^{\alpha,\beta}(x) J_n^{\alpha,\beta}(1) - J_n^{\alpha,\beta}(x) J_{n+1}^{\alpha,\beta}(1)}{x-1}.$$

Working out the constants yields (3.116a).

Replacing x in (3.116a) by $-x$ and using the symmetric property (3.106), we derive (3.116b) immediately. □

We state below two useful formulas and leave their derivation as an excise (see Problem 3.7).

Theorem 3.20.

$$J_{n-1}^{\alpha,\beta}(x) = J_n^{\alpha,\beta-1}(x) - J_n^{\alpha-1,\beta}(x), \tag{3.118a}$$

$$J_n^{\alpha,\beta}(x) = \frac{1}{n+\alpha+\beta}\big[(n+\beta) J_n^{\alpha,\beta-1}(x) + (n+\alpha) J_n^{\alpha-1,\beta}(x)\big]. \tag{3.118b}$$

More generally, we can express $J_n^{\alpha,\beta}$ in terms of $\{J_k^{a,b}\}_{k=0}^n$, where the expansion coefficients are known as the connection coefficients.

Theorem 3.21. *Suppose that*

$$J_n^{\alpha,\beta}(x) = \sum_{k=0}^{n} \hat{c}_k^n J_k^{a,b}(x), \quad a,b,\alpha,\beta > -1. \tag{3.119}$$

Then

$$\begin{aligned} \hat{c}_k^n = &\frac{\Gamma(n+\alpha+1)}{\Gamma(n+\alpha+\beta+1)} \frac{(2k+a+b+1)\Gamma(k+a+b+1)}{\Gamma(k+a+1)} \\ &\times \sum_{m=0}^{n-k} \frac{(-1)^m \Gamma(n+k+m+\alpha+\beta+1)\Gamma(m+k+a+1)}{m!(n-k-m)!\Gamma(k+m+\alpha+1)\Gamma(m+2k+a+b+2)}. \end{aligned} \tag{3.120}$$

Proof. By the Rodrigues' formula and integration by parts,

$$\begin{aligned} \hat{c}_k^n &= \frac{1}{\gamma_k^{a,b}} \int_{-1}^{1} J_n^{\alpha,\beta}(x) J_k^{a,b}(x) \omega^{a,b}(x) dx \\ &= \frac{(-1)^k}{2^k k! \gamma_k^{a,b}} \int_{-1}^{1} J_n^{\alpha,\beta}(x) \partial_x^k \big[\omega^{a+k,b+k}(x)\big] dx \\ &= \frac{1}{2^k k! \gamma_k^{a,b}} \int_{-1}^{1} \partial_x^k J_n^{\alpha,\beta}(x) \omega^{a+k,b+k}(x) dx. \end{aligned}$$

Using (3.101) and (3.96) yields

$$\hat{c}_k^n = \frac{d_{n,k}^{\alpha,\beta}}{2^k k! \gamma_k^{a,b}} \int_{-1}^{1} J_{n-k}^{\alpha+k,\beta+k}(x)\omega^{a+k,b+k}(x)dx$$
$$= \frac{d_{n,k}^{\alpha,\beta}\Gamma(n+\alpha+1)}{2^k k! \gamma_k^{a,b}\Gamma(n+k+\alpha+\beta+1)}$$
$$\times \sum_{m=0}^{n-k} \frac{(-1)^m \Gamma(n+k+m+\alpha+\beta+1)}{2^m m!(n-k-m)!\Gamma(k+m+\alpha+1)} \int_{-1}^{1} \omega^{a+m+k,b+k} dx.$$

Working out $\gamma_k^{a,b}, d_{n,k}^{\alpha,\beta}$ and the integral respectively by (3.109), (3.102) and (A.6) leads to (3.120). □

Next, we derive some recurrence formulas between $\{J_n^{\alpha,\beta}\}$ and $\{\partial_x J_n^{\alpha,\beta}\}$.

Theorem 3.22. *The Jacobi polynomials satisfy*

$$(1-x^2)\partial_x J_n^{\alpha,\beta} = A_n^{\alpha,\beta} J_{n-1}^{\alpha,\beta} + B_n^{\alpha,\beta} J_n^{\alpha,\beta} + C_n^{\alpha,\beta} J_{n+1}^{\alpha,\beta}, \tag{3.121}$$

where

$$A_n^{\alpha,\beta} = \frac{2(n+\alpha)(n+\beta)(n+\alpha+\beta+1)}{(2n+\alpha+\beta)(2n+\alpha+\beta+1)}, \tag{3.122a}$$

$$B_n^{\alpha,\beta} = (\alpha-\beta)\frac{2n(n+\alpha+\beta+1)}{(2n+\alpha+\beta)(2n+\alpha+\beta+2)}, \tag{3.122b}$$

$$C_n^{\alpha,\beta} = -\frac{2n(n+1)(n+\alpha+\beta+1)}{(2n+\alpha+\beta+1)(2n+\alpha+\beta+2)}. \tag{3.122c}$$

Proof. This formula follows from (3.100) and (3.116) directly. □

In the Jacobi case, the relation (3.80) takes the following form.

Theorem 3.23.

$$J_n^{\alpha,\beta} = \widehat{A}_n^{\alpha,\beta} \partial_x J_{n-1}^{\alpha,\beta} + \widehat{B}_n^{\alpha,\beta} \partial_x J_n^{\alpha,\beta} + \widehat{C}_n^{\alpha,\beta} \partial_x J_{n+1}^{\alpha,\beta}, \tag{3.123}$$

where

$$\widehat{A}_n^{\alpha,\beta} = \frac{-2(n+\alpha)(n+\beta)}{(n+\alpha+\beta)(2n+\alpha+\beta)(2n+\alpha+\beta+1)}, \tag{3.124a}$$

$$\widehat{B}_n^{\alpha,\beta} = \frac{2(\alpha-\beta)}{(2n+\alpha+\beta)(2n+\alpha+\beta+2)}, \tag{3.124b}$$

$$\widehat{C}_n^{\alpha,\beta} = \frac{2(n+\alpha+\beta+1)}{(2n+\alpha+\beta+1)(2n+\alpha+\beta+2)}. \tag{3.124c}$$

Proof. We observe from Corollary 3.16 that $\{\partial_x J_l^{\alpha,\beta}\}_{l=1}^{n+1}$ forms an orthogonal basis of P_n. Hence, we can express $J_n^{\alpha,\beta}(x)$ as

$$J_n^{\alpha,\beta}(x) = \sum_{l=1}^{n+1} e_l^{\alpha,\beta} \partial_x J_l^{\alpha,\beta}(x),$$

where

$$e_l^{\alpha,\beta} = \frac{1}{\gamma_l^{\alpha,\beta}\lambda_l^{\alpha,\beta}} \int_{-1}^{1} J_n^{\alpha,\beta}(x)(1-x^2)\partial_x J_l^{\alpha,\beta}(x)\omega^{\alpha,\beta}(x)dx.$$

Inserting (3.121) into the above integral and using the orthogonality of $\{J_n^{\alpha,\beta}\}$, we find that

$$\widehat{C}_n^{\alpha,\beta} = e_{n+1}^{\alpha,\beta} = \frac{A_{n+1}^{\alpha,\beta}\gamma_n^{\alpha,\beta}}{\gamma_{n+1}^{\alpha,\beta}\lambda_{n+1}^{\alpha,\beta}}, \qquad \widehat{B}_n^{\alpha,\beta} = e_n^{\alpha,\beta} = \frac{B_n^{\alpha,\beta}}{\lambda_n^{\alpha,\beta}},$$

$$\widehat{A}_n^{\alpha,\beta} = e_{n-1}^{\alpha,\beta} = \frac{C_{n-1}^{\alpha,\beta}\gamma_n^{\alpha,\beta}}{\gamma_{n-1}^{\alpha,\beta}\lambda_{n-1}^{\alpha,\beta}}, \qquad e_l^{\alpha,\beta} = 0, \ \ 0 \le l \le n-2.$$

Working out the constants yields the coefficients in (3.124). □

3.2.1.4 Maximum Value

Theorem 3.24. *For $\alpha, \beta > -1$, set*

$$x_0 = \frac{\beta-\alpha}{\alpha+\beta+1}, \qquad q = \max(\alpha,\beta).$$

Then we have

$$\max_{|x|\le 1} |J_n^{\alpha,\beta}(x)| = \begin{cases} \max\left\{ \left|J_n^{\alpha,\beta}(\pm 1)\right| \right\} \sim n^q, & \text{if } q \ge -\frac{1}{2}, \\ |J_n^{\alpha,\beta}(x')| \sim n^{-\frac{1}{2}}, & \text{if } q < -\frac{1}{2}, \end{cases} \tag{3.125}$$

where x' is one of the two maximum points nearest x_0.

Proof. Define

$$f_n(x) := \left[J_n^{\alpha,\beta}(x)\right]^2 + \frac{1}{\lambda_n^{\alpha,\beta}}(1-x^2)\left[\partial_x J_n^{\alpha,\beta}(x)\right]^2, \quad n \ge 1. \tag{3.126}$$

A direct calculation by using (3.90) leads to

$$\begin{aligned} f_n'(x) &= \frac{2}{\lambda_n^{\alpha,\beta}}\{(\alpha-\beta)+(\alpha+\beta+1)x\}\left[\partial_x J_n^{\alpha,\beta}(x)\right]^2 \\ &= \frac{2}{\lambda_n^{\alpha,\beta}}(\alpha+\beta+1)(x-x_0)\left[\partial_x J_n^{\alpha,\beta}(x)\right]^2. \end{aligned}$$

Notice that we have the equivalence

$$-1 < x_0 < 1 \iff \left(\alpha+\frac{1}{2}\right)\left(\beta+\frac{1}{2}\right) > 0.$$

We proceed by dividing the parameter range of (α,β) into four different cases.

- *Case I* : $\alpha,\beta > -\frac{1}{2}$. In this case, $f_n'(x) \le 0$ (resp. $f_n'(x) \ge 0$) for all $x \in [-1,x_0]$ (resp. $x \in [x_0,1]$). Hence, $f_n(x)$ attains its maximum at $x = \pm 1$, so we have

$$\max_{|x|\le 1}|J_n^{\alpha,\beta}(x)| = \max\left\{\left|J_n^{\alpha,\beta}(\pm 1)\right|\right\} \overset{(3.108)}{\underset{(3.106)}{\sim}} n^{\max\{\alpha,\beta\}}, \quad \alpha,\beta > -\frac{1}{2}. \tag{3.127}$$

- *Case II* : $\alpha \ge -\frac{1}{2}$ *and* $-1 < \beta < -\frac{1}{2}$. In this case, the linear function

$$(\alpha-\beta)+(\alpha+\beta+1)x \ge 0, \quad \forall x \in [-1,1],$$

which implies $f_n'(x) \ge 0$ for all $x \in [-1,1]$. Hence, we have

$$\max_{|x|\le 1}|J_n^{\alpha,\beta}(x)| = |J_n^{\alpha,\beta}(1)| \sim n^{\alpha}, \ \alpha \ge -\frac{1}{2}, \ -1 < \beta \le -\frac{1}{2}. \tag{3.128}$$

- *Case III* : $-1 < \alpha \le -\frac{1}{2}$ *and* $\beta \ge -\frac{1}{2}$. This situation is opposite to Case II, i.e., $(\alpha-\beta)+(\alpha+\beta+1)x \le 0$ and $f_n'(x) \le 0$ for all $x \in [-1,1]$. Thus, we have

$$\max_{|x|\le 1}|J_n^{\alpha,\beta}(x)| = |J_n^{\alpha,\beta}(-1)| \sim n^{\beta}, \ -1 < \alpha \le -\frac{1}{2}, \ \beta \ge -\frac{1}{2}. \tag{3.129}$$

- *Case IV* : $-1 < \alpha < -\frac{1}{2}$ *and* $-1 < \beta < -\frac{1}{2}$. In this case, we have $-1 < x_0 < 1$, and $f_n'(x) \ge 0$ (resp. $f_n'(x) \le 0$) for all $x \in [-1,x_0]$ (resp. $x \in [x_0,1]$). Therefore, the maximum of $f_n(x)$ is attained at x_0. Notice that the extreme point of $J_n^{\alpha,\beta}(x)$ in $(-1,1)$ is the zero of $\partial_x J_n^{\alpha,\beta}(x)$. Thus, we find from (3.126) that the maximum of $|J_n^{\alpha,\beta}(x)|$ can be attained at one of the zero of $\partial_x J_n^{\alpha,\beta}(x)$ nearest x_0 on the left or on the right of x_0.

The proof is complete. □

In Fig. 3.1, we plot the first six Jacobi polynomials $J_n^{1,1}(x)$ and $J_n^{1,0}(x)$. It is seen that the maximum values are attained at the endpoints. We also observe that $J_n^{1,1}(x)$ is an odd (resp. even) function for odd (resp. even) n, while the non-symmetric Jacobi polynomial $J_n^{1,0}(x)$ does not have this property.

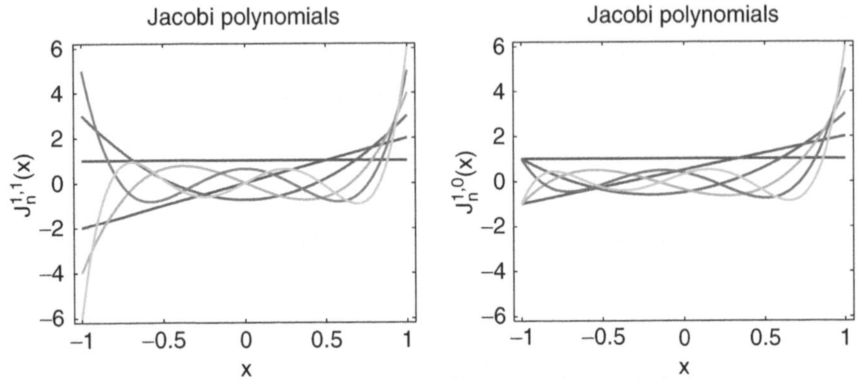

Fig. 3.1 Jacobi polynomials $J_n^{1,1}(x)$ (left) and $J_n^{1,0}(x)$ (right) with $n = 0, 1, \ldots, 5$

3.2.2 Jacobi-Gauss-Type Quadratures

It is straightforward to derive the Jacobi-Gauss-type (i.e., Jacobi-Gauss (JG), Jacobi-Gauss-Radau (JGR) and Jacobi-Gauss-Lobatto (JGL)) integration formulas from the general rules in Sect. 3.1.4. In the Jacobi case, the general quadrature formula (3.33) reads

$$\int_{-1}^{1} p(x)\,\omega^{\alpha,\beta}(x)dx = \sum_{j=0}^{N} p(x_j)\omega_j + E_N[p]. \tag{3.130}$$

Recall that if the quadrature error $E_N[p] = 0$, we say (3.130) is exact for p.

Theorem 3.25. (Jacobi-Gauss quadrature) *The JG quadrature formula* (3.130) *is exact for any* $p \in P_{2N+1}$ *with the JG nodes* $\{x_j\}_{j=0}^{N}$ *being the zeros of* $J_{N+1}^{\alpha,\beta}(x)$ *and the corresponding weights given by*

$$\omega_j = \frac{G_N^{\alpha,\beta}}{J_N^{\alpha,\beta}(x_j)\partial_x J_{N+1}^{\alpha,\beta}(x_j)} \tag{3.131a}$$

$$= \frac{\widetilde{G}_N^{\alpha,\beta}}{(1-x_j^2)\left[\partial_x J_{N+1}^{\alpha,\beta}(x_j)\right]^2}, \tag{3.131b}$$

where

$$G_N^{\alpha,\beta} = \frac{2^{\alpha+\beta}(2N+\alpha+\beta+2)\Gamma(N+\alpha+1)\Gamma(N+\beta+1)}{(N+1)!\Gamma(N+\alpha+\beta+2)}, \tag{3.132a}$$

$$\widetilde{G}_N^{\alpha,\beta} = \frac{2^{\alpha+\beta+1}\Gamma(N+\alpha+2)\Gamma(N+\beta+2)}{(N+1)!\Gamma(N+\alpha+\beta+2)}. \tag{3.132b}$$

Proof. The formula (3.131a) with (3.132a) follows directly from (3.39), and the constant

$$G_N^{\alpha,\beta} = \frac{k_{N+1}^{\alpha,\beta}}{k_N^{\alpha,\beta}} \gamma_N^{\alpha,\beta}$$

can be worked out by using (3.95) and (3.109).

In order to derive the alternative formula (3.131b) with (3.132b), we first use (3.110) and (3.121) to obtain the recurrence relation

$$\begin{aligned}
&(2N+\alpha+\beta+2)(1-x^2)\partial_x J_{N+1}^{\alpha,\beta}(x) \\
&= -(N+1)\big[(2N+\alpha+\beta+2)x+\beta-\alpha\big]J_{N+1}^{\alpha,\beta}(x) \\
&\quad + 2(N+\alpha+1)(N+\beta+1)J_N^{\alpha,\beta}(x).
\end{aligned} \tag{3.133}$$

Using the fact $J_{N+1}^{\alpha,\beta}(x_j) = 0$, yields

$$J_N^{\alpha,\beta}(x_j) = \frac{2N+\alpha+\beta+2}{2(N+\alpha+1)(N+\beta+1)}(1-x_j^2)\partial_x J_{N+1}^{\alpha,\beta}(x_j).$$

Plugging it into (3.131a) leads to (3.131b). □

We now consider the Jacobi-Gauss-Radau (JGR) quadrature with the fixed end-point $x_0 = -1$.

Theorem 3.26. (Jacobi-Gauss-Radau quadrature) *Let $x_0 = -1$ and $\{x_j\}_{j=1}^N$ be the zeros of $J_N^{\alpha,\beta+1}(x)$, and*

$$\omega_0 = \frac{2^{\alpha+\beta+1}(\beta+1)\Gamma^2(\beta+1)N!\Gamma(N+\alpha+1)}{\Gamma(N+\beta+2)\Gamma(N+\alpha+\beta+2)}, \tag{3.134a}$$

$$\begin{aligned}
\omega_j &= \frac{1}{1+x_j}\frac{G_{N-1}^{\alpha,\beta+1}}{J_{N-1}^{\alpha,\beta+1}(x_j)\partial_x J_N^{\alpha,\beta+1}(x_j)}, \\
&= \frac{1}{(1-x_j)(1+x_j)^2}\frac{\tilde{G}_{N-1}^{\alpha,\beta+1}}{[\partial_x J_N^{\alpha,\beta+1}(x_j)]^2}, \quad 1 \le j \le N.
\end{aligned} \tag{3.134b}$$

where the constants $G_{N-1}^{\alpha,\beta+1}$ and $\tilde{G}_{N-1}^{\alpha,\beta+1}$ are defined in (3.132). *Then, the quadrature formula* (3.130) *is exact for any $p \in P_{2N}$.*

Proof. In the Jacobi case, the quadrature polynomial q_N defined in (3.48) is orthogonal with respect to the weight function $\omega^{\alpha,\beta+1}$, so it must be proportional to $J_N^{\alpha,\beta+1}$. Therefore, the interior nodes $\{x_j\}_{j=1}^N$ are the zeros of $J_N^{\alpha,\beta+1}$.

We now prove (3.134a). The general formula (3.51a) in the Jacobi case reads

$$\omega_0 = \frac{1}{J_N^{\alpha,\beta+1}(-1)}\int_{-1}^{1} J_N^{\alpha,\beta+1}(x)\omega^{\alpha,\beta}(x)dy. \tag{3.135}$$

The formula (3.115) implies

$$J_N^{\alpha,\beta+1}(x) = a_{N,0}^{\alpha,\beta} J_0^{\alpha,\beta}(x) + \{\text{linear combination of } \{J_l^{\alpha,\beta}\}_{l=1}^N\}, \qquad (3.136)$$

where

$$a_{N,0}^{\alpha,\beta} = (-1)^N \frac{\Gamma(\alpha+\beta+2)\Gamma(N+\alpha+1)}{\Gamma(\alpha+1)\Gamma(N+\alpha+\beta+2)}.$$

In view of $J_0^{\alpha,\beta}(x) \equiv 1$, we find from the orthogonality (3.88) that

$$\omega_0 = \frac{a_{N,0}^{\alpha,\beta}\gamma_0^{\alpha,\beta}}{J_N^{\alpha,\beta+1}(-1)} = \frac{2^{\alpha+\beta+1}(\beta+1)\Gamma^2(\beta+1)N!\Gamma(N+\alpha+1)}{\Gamma(N+\beta+2)\Gamma(N+\alpha+\beta+2)},$$

where we have worked out the constants by using (3.107) and (3.109).

We next prove (3.134b). The Lagrange basis polynomial related to x_j is

$$\begin{aligned} h_j(x) &= \frac{(1+x)J_N^{\alpha,\beta+1}(x)}{\partial_x\big[(1+x)J_N^{\alpha,\beta+1}(x)\big]\big|_{x=x_j}(x-x_j)} \\ &= \frac{(1+x)J_N^{\alpha,\beta+1}(x)}{(1+x_j)\partial_x J_N^{\alpha,\beta+1}(x_j)(x-x_j)} \\ &= \frac{1+x}{1+x_j}\tilde{h}_j(x), \quad 1 \le j \le N, \end{aligned} \qquad (3.137)$$

where $\{\tilde{h}_j\}_{j=1}^N$ are the Lagrange basis polynomials associated with the Jacobi-Gauss points $\{x_j\}_{j=1}^N$ (zeros of $J_N^{\alpha,\beta+1}$) with the parameters $(\alpha,\beta+1)$. Replacing N and β in (3.131a) and (3.132a) by $N-1$ and $\beta+1$, yields

$$\begin{aligned} \omega_j &= \int_{-1}^1 h_j(x)\omega^{\alpha,\beta}(x)dx = \frac{1}{1+x_j}\int_{-1}^1 \tilde{h}_j(x)\omega^{\alpha,\beta+1}(x)dx \\ &= \frac{1}{1+x_j}\frac{G_{N-1}^{\alpha,\beta+1}}{J_{N-1}^{\alpha,\beta+1}(x_j)\partial_x J_N^{\alpha,\beta+1}(x_j)} \\ &= \frac{1}{(1-x_j)(1+x_j)^2}\frac{\widetilde{G}_{N-1}^{\alpha,\beta+1}}{[\partial_x J_N^{\alpha,\beta+1}(x_j)]^2}, \quad 1 \le j \le N. \end{aligned} \qquad (3.138)$$

This ends the proof. □

Remark 3.4. *A second Jacobi-Gauss-Radau quadrature with a fixed right endpoint* $x_N = 1$ *can be established in a similar manner.*

Finally, we consider the Jacobi-Gauss-Lobatto quadrature, which includes two endpoints $x = \pm 1$ as the nodes.

Theorem 3.27. (Jacobi-Gauss-Lobatto quadrature) *Let* $x_0 = -1,\ \ x_N = 1$ *and* $\{x_j\}_{j=1}^{N-1}$ *be the zeros of* $\partial_x J_N^{\alpha,\beta}(x)$, *and let*

$$\omega_0 = \frac{2^{\alpha+\beta+1}(\beta+1)\Gamma^2(\beta+1)\Gamma(N)\Gamma(N+\alpha+1)}{\Gamma(N+\beta+1)\Gamma(N+\alpha+\beta+2)}, \tag{3.139a}$$

$$\omega_N = \frac{2^{\alpha+\beta+1}(\alpha+1)\Gamma^2(\alpha+1)\Gamma(N)\Gamma(N+\beta+1)}{\Gamma(N+\alpha+1)\Gamma(N+\alpha+\beta+2)}, \tag{3.139b}$$

$$\begin{aligned}\omega_j &= \frac{1}{1-x_j^2}\frac{G_{N-2}^{\alpha+1,\beta+1}}{J_{N-2}^{\alpha+1,\beta+1}(x_j)\partial_x J_{N-1}^{\alpha+1,\beta+1}(x_j)},\\ &= \frac{1}{(1-x_j^2)^2}\frac{\widetilde{G}_{N-2}^{\alpha+1,\beta+1}}{\left[\partial_x J_{N-1}^{\alpha+1,\beta+1}(x_j)\right]^2}, \quad 1 \le j \le N-1,\end{aligned} \tag{3.139c}$$

where the constants $G_{N-1}^{\alpha,\beta+1}$ *and* $\widetilde{G}_{N-1}^{\alpha,\beta+1}$ *are defined in* (3.132). *Then, the quadrature formula* (3.130) *is exact for any* $p \in P_{2N-1}$.

The proof is similar to that of Theorem 3.26 and is left as an exercise (see Problem 3.8).

Remark 3.5. *The quadrature nodes and weights of these three types of Gaussian formulas have close relations. Indeed, denote by* $\{\xi_{Z,N,j}^{\alpha,\beta}, \omega_{Z,N,j}^{\alpha,\beta}\}_{j=0}^{N}$ *with* $Z = G,R,L$ *the Jacobi-Gauss, Jacobi-Gauss-Radau and Jacobi-Gauss-Lobatto quadrature nodes and weights, respectively. Then there hold*

$$\xi_{R,N,j}^{\alpha,\beta} = \xi_{G,N-1,j-1}^{\alpha,\beta+1}, \quad \omega_{R,N,j}^{\alpha,\beta} = \frac{\omega_{G,N-1,j-1}^{\alpha,\beta+1}}{1+\xi_{G,N-1,j-1}^{\alpha,\beta+1}}, \quad 1 \le j \le N, \tag{3.140}$$

and

$$\xi_{L,N,j}^{\alpha,\beta} = \xi_{G,N-2,j-1}^{\alpha+1,\beta+1}, \quad \omega_{L,N,j}^{\alpha,\beta} = \frac{\omega_{G,N-2,j-1}^{\alpha+1,\beta+1}}{1-\left(\xi_{G,N-2,j-1}^{\alpha+1,\beta+1}\right)^2}, \quad 1 \le j \le N-1. \tag{3.141}$$

This connection allows us to compute the interior nodes and weights of the JGR and JGL quadratures from the JG rule. Moreover, it makes the analysis of JGR and JGL (e.g., the interpolation error) easier by extending the results for JG case.

3.2.3 Computation of Nodes and Weights

Except for the Chebyshev case (see Sect. 3.4), the explicit expressions of the nodes and weights of the general Jacobi-Gauss quadrature are not available, so they have to be computed by numerical means. An efficient algorithm is to use the eigenvalue method described in Theorems 3.4 and 3.6.

Thanks to the relations (3.140) and (3.141), it suffices to compute the Jacobi-Gauss nodes and weights. Indeed, as a direct consequence of Theorem 3.4, the zeros of the Jacobi polynomial $J_{N+1}^{\alpha,\beta}$ are the eigenvalues of the following symmetric tridiagonal matrix

$$A_{N+1} = \begin{bmatrix} a_0 & \sqrt{b_1} & & & \\ \sqrt{b_1} & a_1 & \sqrt{b_2} & & \\ & \ddots & \ddots & \ddots & \\ & & \sqrt{b_{N-1}} & a_{N-1} & \sqrt{b_N} \\ & & & \sqrt{b_N} & a_N \end{bmatrix}, \tag{3.142}$$

where the entries are derived from (3.25) and the three-term recurrence relation (3.110):

$$a_j = \frac{\beta^2 - \alpha^2}{(2j+\alpha+\beta)(2j+\alpha+\beta+2)}, \tag{3.143a}$$

$$b_j = \frac{4j(j+\alpha)(j+\beta)(j+\alpha+\beta)}{(2j+\alpha+\beta-1)(2j+\alpha+\beta)^2(2j+\alpha+\beta+1)}. \tag{3.143b}$$

Moreover, by Theorem 3.6, the Jacobi-Gauss weights $\{\omega_j\}_{j=0}^N$ can be obtained by computing the eigenvectors of A_{N+1}, namely,

$$\omega_j = \gamma_0^{\alpha,\beta}\left[Q_0(x_j)\right]^2 = \frac{2^{\alpha+\beta+1}\Gamma(\alpha+1)\Gamma(\beta+1)}{\Gamma(\alpha+\beta+2)}\left[Q_0(x_j)\right]^2, \tag{3.144}$$

where $Q_0(x_j)$ is the first component of the orthonormal eigenvector corresponding to the eigenvalue x_j. Notice that weights $\{\omega_j\}_{j=0}^N$ may also be computed by using the formula (3.131).

Alternatively, the zeros of the Jacobi polynomials can be computed by the Newton's iteration method described in (3.30) and (3.31). The initial approximation can be chosen as some estimates presented below, see, e.g., (3.145).

We depict in Fig. 3.2 the distributions of zeros of some sample Jacobi polynomials:

- In (a), the zeros of $J_N^{1,1}(x)$ with various N
- In (b), the zeros $\left\{\theta_j = \cos^{-1} x_j\right\}_{j=0}^{N-1}$ of $J_N^{1,1}(\cos\theta)$ with various N
- In (c), the zeros of $J_{15}^{\alpha,\alpha}(x)$ with various α
- In (d), the zeros of $J_{15}^{\alpha,0}(x)$ with various α

We observe from (a) and (b) in Fig. 3.2 that the zeros $\{x_j\}$ (arranged in descending order) of the Jacobi polynomials are nonuniformly distributed in $(-1,1)$, while $\{\theta_j = \cos^{-1} x_j\}$ are nearly equidistantly located in $(0,\pi)$. More precisely, the nodes (in x) cluster near the endpoints with spacing density like $O(N^{-2})$, and are considerably sparser in the inner part with spacing $O(N^{-1})$. This feature is quantitatively characterized by Theorem 8.9.1 of Szegö (1975), which states that for $\alpha, \beta > -1$,

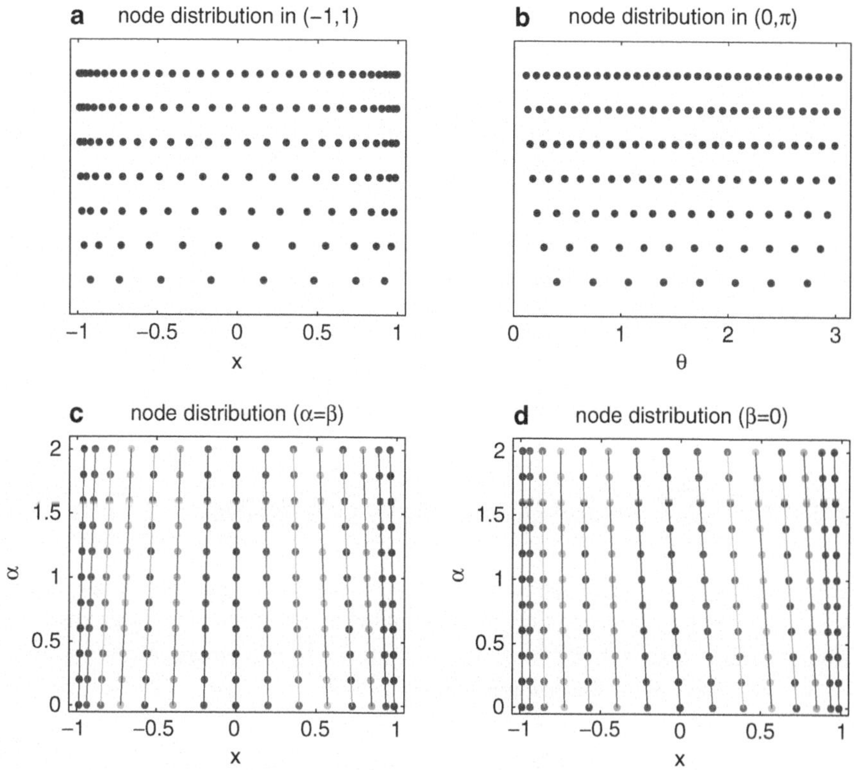

Fig. 3.2 Distributions of Jacobi-Gauss quadrature nodes

$$\cos^{-1} x_j = \theta_j = \frac{1}{N+1}\big((j+1)\pi + O(1)\big), \quad j = 0, 1, \ldots, N, \tag{3.145}$$

where $O(1)$ is uniformly bounded for all values $j = 0, 1, \ldots, N$, and $N = 1, 2, 3, \ldots$. We see that near the endpoints $x = \pm 1$ (i.e., $\theta = 0, \pi$),

$$1 - x_j^2 = \sin^2 \theta_j = O(N^{-2}), \quad j = 0, N.$$

Hence, the node spacing in the neighborhood of $x = \pm 1$ behaves like $O(N^{-2})$. In particular, for the case

$$-\frac{1}{2} \le \alpha \le \frac{1}{2}, \quad -\frac{1}{2} \le \beta \le \frac{1}{2}, \tag{3.146}$$

Theorem 6.21.2 of Szegö (1975) provides the bounds

$$\frac{2j+1}{2N+3} \le \theta_j \le \frac{2j+2}{2N+3}, \quad 0 \le j \le N, \tag{3.147}$$

where the equality holds only when $\alpha = -\beta = -\frac{1}{2}$ or $\alpha = -\beta = \frac{1}{2}$.

For a fixed j, we can view $x_j = x_j(N;\alpha,\beta)$ as a function of N, α and β, and observe from (c) and (d) in Fig. 3.2 that for a given N, the nodes exhibit a tendency to move towards the center of the interval as α and/or β increases. This is predicted by Theorem 6.21.1 of Szegö (1975):

$$\frac{\partial x_j}{\partial \alpha} < 0, \quad \frac{\partial x_j}{\partial \beta} > 0, \quad 0 \le j \le N. \tag{3.148}$$

In particular, if $\alpha = \beta$,

$$\frac{\partial x_j}{\partial \alpha} < 0, \quad j = 0, 1, \ldots, [N/2]. \tag{3.149}$$

3.2.4 Interpolation and Discrete Jacobi Transforms

Let $\{x_j, \omega_j\}_{j=0}^N$ be a set of Jacobi-Gauss-type nodes and weights. As in Sect. 3.1.5, we can define the corresponding interpolation operator, discrete inner product and discrete norm, denoted by $I_N^{\alpha,\beta}$, $\langle \cdot,\cdot \rangle_{N,\omega^{\alpha,\beta}}$ and $\|\cdot\|_{N,\omega^{\alpha,\beta}}$, respectively.

The exactness of the quadratures implies

$$\langle u, v \rangle_{N,\omega^{\alpha,\beta}} = (u,v)_{\omega^{\alpha,\beta}}, \quad \forall u \cdot v \in P_{2N+\delta}, \tag{3.150}$$

where $\delta = 1, 0, -1$ for JG, JGR and JGL, respectively. Accordingly, we have

$$\|u\|_{N,\omega^{\alpha,\beta}} = \|u\|_{\omega^{\alpha,\beta}}, \quad \forall u \in P_N, \text{ for JG and JGR.} \tag{3.151}$$

Although the above identity does not hold for the JGL case, we have the following equivalence.

Lemma 3.3.

$$\|u\|_{\omega^{\alpha,\beta}} \le \|u\|_{N,\omega^{\alpha,\beta}} \le \sqrt{2 + \frac{\alpha+\beta+1}{N}}\, \|u\|_{\omega^{\alpha,\beta}}, \quad \forall u \in P_N. \tag{3.152}$$

Proof. For any $u \in P_N$, we write

$$u(x) = \sum_{l=0}^{N} \hat{u}_l J_l^{\alpha,\beta}(x), \quad \text{with } \hat{u}_l = \frac{1}{\gamma_l^{\alpha,\beta}} \left(u, J_l^{\alpha,\beta}\right)_{\omega^{\alpha,\beta}}.$$

By the orthogonality of the Jacobi polynomials and the exactness (3.150),

$$\begin{aligned} \|u\|^2_{\omega^{\alpha,\beta}} &= \sum_{l=0}^{N} \hat{u}_l^2 \gamma_l^{\alpha,\beta}, \\ \|u\|^2_{N,\omega^{\alpha,\beta}} &= \sum_{l=0}^{N-1} \hat{u}_l^2 \gamma_l^{\alpha,\beta} + \hat{u}_N^2 \langle J_N^{\alpha,\beta}, J_N^{\alpha,\beta} \rangle_{N,\omega^{\alpha,\beta}}. \end{aligned} \tag{3.153}$$

To estimate the last term, we define

$$\psi(x) = [J_N^{\alpha,\beta}(x)]^2 + \frac{1}{N^2}(1-x^2)\left[\partial_x J_N^{\alpha,\beta}(x)\right]^2.$$

One verifies readily that $\psi \in P_{2N-1}$, since the leading term x^{2N} cancels out. Therefore, using the fact $(1-x_j^2)\partial_x J_N^{\alpha,\beta}(x_j) = 0$, and the exactness (3.150), we derive

$$\begin{aligned} \langle J_N^{\alpha,\beta}, J_N^{\alpha,\beta} \rangle_{N,\omega^{\alpha,\beta}} &= \langle 1, \psi \rangle_{N,\omega^{\alpha,\beta}} = (1, \psi)_{\omega^{\alpha,\beta}} = \left(J_N^{\alpha,\beta}, J_N^{\alpha,\beta}\right)_{\omega^{\alpha,\beta}} \\ &+ \frac{1}{N^2}\left(\partial_x J_N^{\alpha,\beta}, \partial_x J_N^{\alpha,\beta}\right)_{\omega^{\alpha+1,\beta+1}} \overset{(3.97)}{=} \left[1 + \frac{\lambda_N^{\alpha,\beta}}{N^2}\right]\gamma_N^{\alpha,\beta}. \end{aligned}$$

Hence, by (3.91),

$$\langle J_N^{\alpha,\beta}, J_N^{\alpha,\beta} \rangle_{N,\omega^{\alpha,\beta}} = \left(2 + \frac{\alpha+\beta+1}{N}\right)\gamma_N^{\alpha,\beta}. \tag{3.154}$$

Inserting it into (3.153) leads to the desired result. □

We now turn to the discrete Jacobi transforms. Since the interpolation polynomial $I_N^{\alpha,\beta} u \in P_N$, we write

$$\left(I_N^{\alpha,\beta} u\right)(x) = \sum_{n=0}^{N} \tilde{u}_n^{\alpha,\beta} J_n^{\alpha,\beta}(x), \tag{3.155}$$

where the coefficients $\{\tilde{u}_n^{\alpha,\beta}\}_{n=0}^N$ are determined by the *forward discrete Jacobi transform.*

Theorem 3.28.

$$\tilde{u}_n^{\alpha,\beta} = \frac{1}{\delta_n^{\alpha,\beta}} \sum_{j=0}^{N} u(x_j) J_n^{\alpha,\beta}(x_j)\omega_j, \tag{3.156}$$

where $\delta_n^{\alpha,\beta} = \gamma_n^{\alpha,\beta}$ *for* $0 \le n \le N-1$, *and*

$$\delta_N^{\alpha,\beta} = \begin{cases} \gamma_N^{\alpha,\beta}, & \textit{for JG and JGR}, \\ \left(2 + \frac{\alpha+\beta+1}{N}\right)\gamma_N^{\alpha,\beta}, & \textit{for JGL}. \end{cases}$$

Proof. This formula follows directly from Theorem 3.9 and (3.154). □

By taking $x = x_j$ in (3.155), the *backward discrete Jacobi transform* is carried out by

$$u(x_j) = (I_N^{\alpha,\beta} u)(x_j) = \sum_{n=0}^{N} \tilde{u}_n^{\alpha,\beta} J_n^{\alpha,\beta}(x_j), \quad 0 \le j \le N. \tag{3.157}$$

In general, the discrete transforms (3.156)-(3.157) can be performed by a matrix–vector multiplication routine in about N^2 flops. Some techniques to reduce the computational complexity to $N(\log N)^\alpha$ (with some positive α) are suggested in Potts et al. (1998), Tygert (2010).

3.2.5 Differentiation in the Physical Space

Let $\{x_j\}_{j=0}^N$ be a set of Jacobi-Gauss-type points, and let $\{h_j\}_{j=0}^N$ be the associated Lagrange basis polynomials. Suppose that $u \in P_N$ is an approximation to the underlying solution, and we write

$$u(x) = \sum_{j=0}^{N} u(x_j) h_j(x).$$

As shown in Sect. 3.1.6, the differentiation of u can be done through a matrix–vector multiplication:

$$\mathbf{u}^{(m)} = D^m \mathbf{u}, \qquad m \ge 1, \tag{3.158}$$

where $\mathbf{u}^{(k)} = (u^{(k)}(x_0), u^{(k)}(x_1), \ldots, u^{(k)}(x_N))^T$, $\mathbf{u} = \mathbf{u}^{(0)}$, and the first-order differentiation matrix:

$$D = \big(d_{kj} = h_j'(x_k)\big)_{k,j=0,1,\ldots,N}.$$

Hence, it suffices to compute the entries of the first-order differentiation matrix D, whose explicit formulas can be derived from Theorem 3.11.

3.2.5.1 Jacobi-Gauss-Lobatto Differentiation Matrix

In this case, the quadrature polynomial defined in (3.74) reads

$$Q(x) = (1 - x^2) J_{N-1}^{\alpha+1,\beta+1}(x).$$

To simplify the notation, we write

$$J(x) := \partial_x J_{N-1}^{\alpha+1,\beta+1}(x). \tag{3.159}$$

One verifies readily that (note: $x_0 = -1$ and $x_N = 1$):

$$Q'(x_j) = \begin{cases} \dfrac{2(-1)^{N-1}\Gamma(N+\beta+1)}{\Gamma(N)\Gamma(\beta+2)}, & j=0, \\ (1-x_j^2)J(x_j), & 1 \le j \le N-1, \\ \dfrac{-2\Gamma(N+\alpha+1)}{\Gamma(N)\Gamma(\alpha+2)}, & j=N. \end{cases}$$

Differentiating $Q(x)$ yields

$$\begin{aligned} Q''(x) &= -2J_{N-1}^{\alpha+1,\beta+1}(x) - 4x\partial_x J_{N-1}^{\alpha+1,\beta+1}(x) + (1-x^2)\partial_x^2 J_{N-1}^{\alpha+1,\beta+1}(x) \\ &\overset{(3.90)}{=} [(\alpha-\beta)+(\alpha+\beta)x]J(x) - (\lambda_{N-1}^{\alpha+1,\beta+1}+2)J_{N-1}^{\alpha+1,\beta+1}(x). \end{aligned}$$

Recalling that $\{J_{N-1}^{\alpha+1,\beta+1}(x_j) = 0\}_{j=1}^{N-1}$, and using the formulas (3.94), (3.107) and (3.100) to work out the constants, we find

$$Q''(x_j) = \begin{cases} \dfrac{2\big[\alpha - N(N+\alpha+\beta+1)\big]\Gamma(N+\beta+1)}{(-1)^{N+1}\Gamma(N)\Gamma(\beta+3)}, & j=0, \\ \big[\alpha-\beta+(\alpha+\beta)x_j\big]J(x_j), & 1 \le j \le N-1, \\ \dfrac{2\big[\beta - N(N+\alpha+\beta+1)\big]\Gamma(N+\alpha+1)}{\Gamma(N)\Gamma(\alpha+3)}, & j=N. \end{cases}$$

Applying the general formulas in Theorem 3.11, the entries of the first-order JGL differentiation matrix D are expressed as follows.

(a). The first column ($j=0$):

$$d_{k0} = \begin{cases} \dfrac{\alpha - N(N+\alpha+\beta+1)}{2(\beta+2)}, & k=0, \\ \dfrac{(-1)^{N-1}\Gamma(N)\Gamma(\beta+2)}{2\Gamma(N+\beta+1)}(1-x_k)J(x_k), & 1 \le k \le N-1, \\ \dfrac{(-1)^N}{2}\dfrac{\Gamma(\beta+2)\Gamma(N+\alpha+1)}{\Gamma(\alpha+2)\Gamma(N+\beta+1)}, & k=N. \end{cases} \tag{3.160}$$

(b). The second to the N-th column ($1 \le j \le N-1$):

$$d_{kj} = \begin{cases} \dfrac{2(-1)^N\Gamma(N+\beta+1)}{\Gamma(N)\Gamma(\beta+2)(1-x_j)(1+x_j)^2 J(x_j)}, & k=0, \\ \dfrac{(1-x_k^2)J(x_k)}{(1-x_j^2)J(x_j)}\dfrac{1}{x_k-x_j}, & k \ne j,\ 1 \le k \le N-1, \\ \dfrac{\alpha-\beta+(\alpha+\beta)x_k}{2(1-x_k^2)}, & 1 \le k=j \le N-1, \\ \dfrac{-2\Gamma(N+\alpha+1)}{\Gamma(N)\Gamma(\alpha+2)(1-x_j)^2(1+x_j)J(x_j)}, & k=N. \end{cases} \tag{3.161}$$

(c). The last column ($j = N$):

$$d_{kN} = \begin{cases} \dfrac{(-1)^{N+1}}{2} \dfrac{\Gamma(\alpha+2)\Gamma(N+\beta+1)}{\Gamma(\beta+2)\Gamma(N+\alpha+1)}, & k = 0, \\[2ex] \dfrac{\Gamma(N)\Gamma(\alpha+2)}{2\Gamma(N+\alpha+1)}(1+x_k)J(x_k), & 1 \le k \le N-1, \\[2ex] \dfrac{N(N+\alpha+\beta+1)-\beta}{2(\alpha+2)}, & k = N. \end{cases} \tag{3.162}$$

3.2.5.2 Jacobi-Gauss-Radau Differentiation Matrix

In this case, the quadrature polynomial in (3.74) is $Q(x) = (1+x)J_N^{\alpha,\beta+1}(x)$. Denoting $J(x) = \partial_x J_N^{\alpha,\beta+1}(x)$, one verifies that

$$Q'(x_j) = \begin{cases} \dfrac{(-1)^N \Gamma(N+\beta+2)}{N!\Gamma(\beta+2)}, & j = 0, \\[2ex] (1+x_j)J(x_j), & 1 \le j \le N. \end{cases}$$

We obtain from (3.100) and (3.107) that

$$Q''(x_0) = 2\partial_x J_N^{\alpha,\beta+1}(-1) = \frac{(-1)^{N-1}(N+\alpha+\beta+2)\Gamma(N+\beta+2)}{\Gamma(N)\Gamma(\beta+3)}.$$

Moreover, by (3.90),

$$\begin{aligned} Q''(x) &= 2\partial_x J_N^{\alpha,\beta+1}(x) + (1+x)\partial_x^2 J_N^{\alpha,\beta+1}(x) = 2\partial_x J_N^{\alpha,\beta+1}(x) \\ &+ \frac{1}{1-x}\Big[(\alpha-\beta-1+(\alpha+\beta+3)x)\partial_x J_N^{\alpha,\beta+1}(x) - \lambda_N^{\alpha,\beta+1} J_N^{\alpha,\beta+1}(x)\Big]. \end{aligned}$$

In view of $\{J_N^{\alpha,\beta+1}(x_j) = 0\}_{j=1}^N$, we derive that

$$Q''(x_j) = \frac{1}{1-x_j}(\alpha-\beta+1+(\alpha+\beta+1)x_j)J(x_j), \quad 1 \le j \le N.$$

Applying the general results in Theorem 3.11 leads to

$$
d_{kj} = \begin{cases}
\dfrac{-N(N+\alpha+\beta+2)}{2(\beta+2)}, & k=j=0, \\[2ex]
\dfrac{N!\Gamma(\beta+2)}{(-1)^N\Gamma(N+\beta+2)} J(x_k), & 1 \le k \le N,\ j=0, \\[2ex]
\dfrac{(-1)^{N+1}\Gamma(N+\beta+2)}{N!\Gamma(\beta+2)} \dfrac{1}{(1+x_j)^2 J(x_j)}, & k=0,\ 1 \le j \le N, \\[2ex]
\dfrac{(1+x_k)J(x_k)}{(1+x_j)J(x_j)} \dfrac{1}{x_k - x_j}, & 1 \le k \ne j \le N, \\[2ex]
\dfrac{\alpha-\beta+1+(\alpha+\beta+1)x_k}{2(1-x_k^2)}, & 1 \le k = j \le N.
\end{cases} \tag{3.163}
$$

3.2.5.3 Jacobi-Gauss Differentiation Matrix

In this case, the quadrature polynomial in (3.74) is $Q(x) = J_{N+1}^{\alpha,\beta}(x)$. One verifies by using (3.90) that

$$
\partial_x^2 J_{N+1}^{\alpha,\beta}(x_j) = \frac{1}{1-x_j^2}\big(\alpha-\beta+(\alpha+\beta+2)x_j\big)\partial_x J_{N+1}^{\alpha,\beta}(x_j), \quad 0 \le j \le N.
$$

Once again, we derive from Theorem 3.11 that

$$
d_{kj} = \begin{cases}
\dfrac{\partial_x J_{N+1}^{\alpha,\beta}(x_k)}{\partial_x J_{N+1}^{\alpha,\beta}(x_j)} \dfrac{1}{x_k - x_j}, & 0 \le k \ne j \le N, \\[2ex]
\dfrac{\alpha-\beta+(\alpha+\beta+2)x_k}{2(1-x_k^2)}, & 1 \le k = j \le N.
\end{cases} \tag{3.164}
$$

As a numerical illustration, we consider the approximation of the derivatives of $u(x) = \sin(4\pi x)$, $x \in [-1,1]$ by the Jacobi-Gauss-Lobatto interpolation associated with $\{x_j\}_{j=0}^N$ with $\alpha = \beta = 1$. More precisely, let

$$
u(x) \approx u_N(x) = I_N^{1,1} u(x) = \sum_{j=0}^{N} u(x_j) h_j(x) \in P_N. \tag{3.165}
$$

In Fig. 3.3a, we plot u' (solid line) versus $u_N'(x)$ ("$\cdot$") and u'' (solid line) versus $u_N''(x)$ ("$\star$") at $\{x_j\}_{j=0}^N$ with $N = 38$. In Fig. 3.3b, we depict the errors $\log_{10}\big(\|u' - u_N'\|_{N,\omega^{1,1}}\big)$ ("$\circ$") and $\log_{10}\big(\|u'' - u_N''\|_{N,\omega^{1,1}}\big)$ ("$\diamond$") against various N. We observe that the errors decay exponentially.

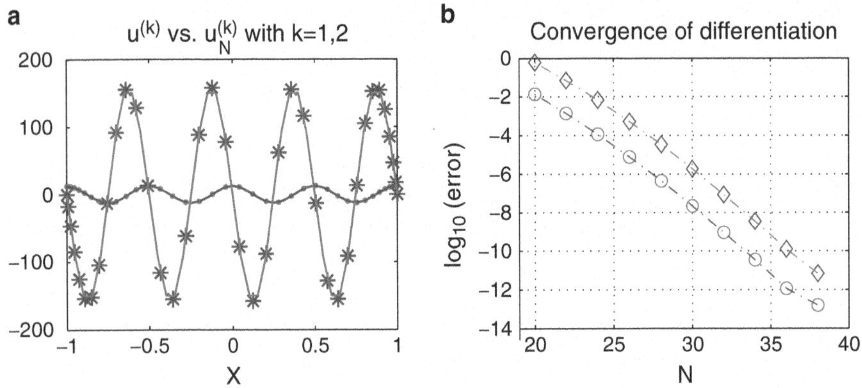

Fig. 3.3 Convergence of Jacobi differentiation in the physical space

3.2.6 Differentiation in the Frequency Space

We now describe the spectral differentiation by manipulating the expansion coefficients as in Sect. 3.1.7. For any $u \in P_N$, we write

$$u(x) = \sum_{n=0}^{N} \hat{u}_n J_n^{\alpha,\beta}(x) \in P_N, \quad u'(x) = \sum_{n=0}^{N-1} \hat{u}_n^{(1)} J_n^{\alpha,\beta}(x) \in P_{N-1}.$$

The process of differentiation in the frequency space is to express $\{\hat{u}_n^{(1)}\}$ in terms of $\{\hat{u}_n\}$.

Thanks to the recurrence formula (3.123), the corresponding coefficients in the relation (3.80) are

$$\tilde{a}_n = \widehat{A}_n^{\alpha,\beta}, \quad \tilde{b}_n = \widehat{B}_n^{\alpha,\beta}, \quad \tilde{c}_n = \widehat{C}_n^{\alpha,\beta},$$

where $\widehat{A}_n^{\alpha,\beta}$, $\widehat{B}_n^{\alpha,\beta}$ and $\widehat{C}_n^{\alpha,\beta}$ are given in (3.124a)–(3.124c), respectively.

Hence, by Theorem 3.12, the coefficients $\{\hat{u}_n^{(1)}\}_{n=0}^{N}$ can be exactly evaluated by the backward recurrence formula

$$\begin{cases} \hat{u}_N^{(1)} = 0, \quad \hat{u}_{N-1}^{(1)} = \dfrac{\hat{u}_N}{\widehat{C}_{N-1}^{\alpha,\beta}}, \\ \hat{u}_{n-1}^{(1)} = \dfrac{1}{\widehat{C}_{n-1}^{\alpha,\beta}} \left\{ \hat{u}_n - \widehat{B}_n^{\alpha,\beta} \hat{u}_n^{(1)} - \widehat{A}_{n+1}^{\alpha,\beta} \hat{u}_{n+1}^{(1)} \right\}, \\ \qquad n = N-1, N-2, \ldots, 2, 1. \end{cases} \tag{3.166}$$

In summary, given the physical values $\{u(x_j)\}_{j=0}^{N}$ at a set of Jacobi-Gauss-type points $\{x_j\}_{j=0}^{N}$, the evaluation of $\{u'(x_j)\}_{j=0}^{N}$ can be carried out in the following three steps:

- Find the coefficients $\{\hat{u}_n\}_{n=0}^N$ by using the *forward discrete Jacobi transform* (3.156).
- Compute the coefficients $\{\hat{u}_n^{(1)}\}_{n=0}^{N-1}$ by using (3.166).
- Find the derivative values $\{u'(x_j)\}_{j=0}^N$ by using the *backward discrete Jacobi transform* (3.157).

Higher-order derivatives can be computed by repeating the above procedure.

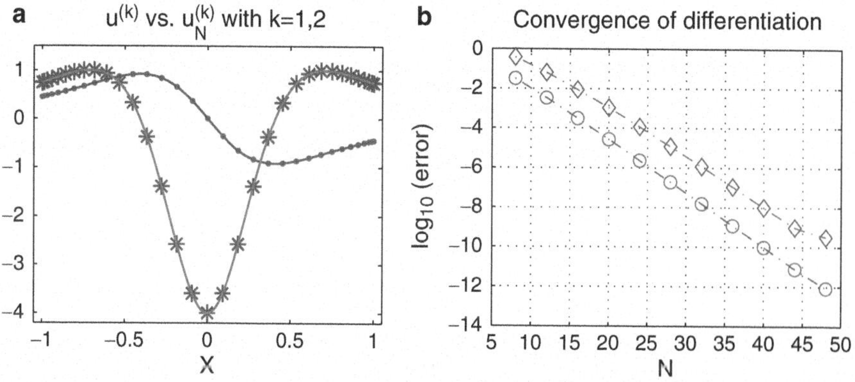

Fig. 3.4 Convergence of Jacobi differentiation in the frequency space

As an illustrative example, we fix the Jacobi index to be $(1,1)$, consider $u(x) = 1/(1+2x^2)$, $x \in [-1,1]$, and approximate its derivatives by taking the derivatives, in the frequency space, of its interpolation polynomial:

$$u(x) \approx u_N(x) = I_N^{1,1}u(x) = \sum_{n=0}^{N} \tilde{u}_n J_n^{1,1}(x) \in P_N. \tag{3.167}$$

We observe from Fig. 3.4 that the errors decay exponentially, similar to the differentiation in the physical space as shown in Fig. 3.3.

3.3 Legendre Polynomials

We discuss in this section an important special case of the Jacobi polynomials – then Legendre polynomials

$$L_n(x) = J_n^{0,0}(x), \quad n \geq 0, \; x \in I = (-1,1).$$

The distinct feature of the Legendre polynomials is that they are mutually orthogonal with respect to the uniform weight function $\omega(x) \equiv 1$. The first six Legendre polynomials and their derivatives are plotted in Fig. 3.5.

Since most of them can be derived directly from the corresponding properties of the Jacobi polynomials by taking $\alpha = \beta = 0$, we merely collect some relevant formulas without proof.

- Three-term recurrence relation:

$$(n+1)L_{n+1}(x) = (2n+1)xL_n(x) - nL_{n-1}(x), \quad n \geq 1, \tag{3.168}$$

and the first few Legendre polynomials are

$$L_0(x) = 1, \quad L_1(x) = x,$$
$$L_2(x) = \frac{1}{2}(3x^2 - 1), \quad L_3(x) = \frac{1}{2}(5x^3 - 3x).$$

- The Legendre polynomial has the expansion

$$L_n(x) = \frac{1}{2^n}\sum_{l=0}^{[n/2]}(-1)^l \frac{(2n-2l)!}{2^n l!(n-l)!(n-2l)!}x^{n-2l}, \tag{3.169}$$

and the leading coefficient is

$$k_n = \frac{(2n)!}{2^n(n!)^2}. \tag{3.170}$$

- Sturm-Liouville problem:

$$\big((1-x^2)L_n'(x)\big)' + \lambda_n L_n(x) = 0, \quad \lambda_n = n(n+1). \tag{3.171}$$

Equivalently,

$$(1-x^2)L_n''(x) - 2xL_n'(x) + n(n+1)L_n(x) = 0. \tag{3.172}$$

- Rodrigues' formula:

$$L_n(x) = \frac{1}{2^n n!}\frac{d^n}{dx^n}\Big[(x^2-1)^n\Big], \quad n \geq 0. \tag{3.173}$$

- Orthogonality:

$$\int_{-1}^{1} L_n(x)L_m(x)dx = \gamma_n \delta_{mn}, \quad \gamma_n = \frac{2}{2n+1}, \tag{3.174a}$$

$$\int_{-1}^{1} L_n'(x)L_m'(x)(1-x^2)dx = \gamma_n \lambda_n \delta_{mn}. \tag{3.174b}$$

- Symmetric property:

$$L_n(-x) = (-1)^n L_n(x), \quad L_n(\pm 1) = (\pm 1)^n. \tag{3.175}$$

Hence, $L_n(x)$ is an odd (resp. even) function, if n is odd (resp. even). Moreover, we have the uniform bound

$$|L_n(x)| \leq 1, \quad \forall x \in [-1,1], \, n \geq 0.$$

- Derivative recurrence relations:

$$(2n+1)L_n(x) = L'_{n+1}(x) - L'_{n-1}(x), \quad n \geq 1, \tag{3.176a}$$

$$L'_n(x) = \sum_{\substack{k=0 \\ k+n \text{ odd}}}^{n-1} (2k+1)L_k(x), \tag{3.176b}$$

$$L''_n(x) = \sum_{\substack{k=0 \\ k+n \text{ even}}}^{n-2} \left(k+\frac{1}{2}\right)\big(n(n+1) - k(k+1)\big)L_k(x), \tag{3.176c}$$

$$(1-x^2)L'_n(x) = \frac{n(n+1)}{2n+1}\big(L_{n-1}(x) - L_{n+1}(x)\big). \tag{3.176d}$$

- The boundary values of the derivatives:

$$L'_n(\pm 1) = \frac{1}{2}(\pm 1)^{n-1} n(n+1), \tag{3.177a}$$

$$L''_n(\pm 1) = (\pm 1)^n (n-1)n(n+1)(n+2)/8. \tag{3.177b}$$

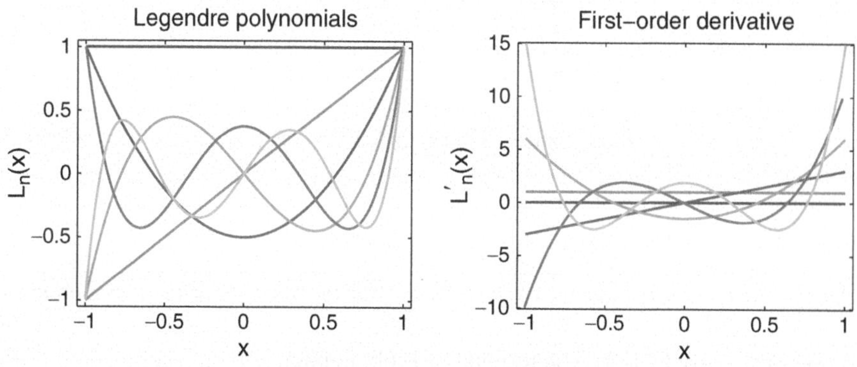

Fig. 3.5 The first six Legendre polynomials and their first-order derivatives

3.3.1 Legendre-Gauss-Type Quadratures

The Legendre-Gauss-type quadrature formulas can be derived from the Jacobi ones in the previous section.

Theorem 3.29. *Let* $\{x_j, \omega_j\}_{j=0}^N$ *be a set of Legendre-Gauss-type nodes and weights.*

- *For the Legendre-Gauss* (LG) *quadrature,*

$$\begin{aligned} &\{x_j\}_{j=0}^N \text{ are the zeros of } L_{N+1}(x); \\ &\omega_j = \frac{2}{(1-x_j^2)[L'_{N+1}(x_j)]^2}, \quad 0 \le j \le N. \end{aligned} \tag{3.178}$$

- *For the Legendre-Gauss-Radau* (LGR) *quadrature,*

$$\begin{aligned} &\{x_j\}_{j=0}^N \text{ are the zeros of } L_N(x) + L_{N+1}(x); \\ &\omega_j = \frac{1}{(N+1)^2} \frac{1-x_j}{[L_N(x_j)]^2}, \quad 0 \le j \le N. \end{aligned} \tag{3.179}$$

- *For the Legendre-Gauss-Lobatto* (LGL) *quadrature,*

$$\begin{aligned} &\{x_j\}_{j=0}^N \text{ are the zeros of } (1-x^2)L'_N(x); \\ &\omega_j = \frac{2}{N(N+1)} \frac{1}{[L_N(x_j)]^2}, \quad 0 \le j \le N. \end{aligned} \tag{3.180}$$

With the above quadrature nodes and weights, there holds

$$\int_{-1}^{1} p(x)dx = \sum_{j=0}^{N} p(x_j)\omega_j, \quad \forall p \in P_{2N+\delta}, \tag{3.181}$$

where $\delta = 1, 0, -1$ *for LG, LGR and LGL, respectively.*

Proof. The rule (3.181) with (3.178) follows directly from Theorem 3.25 with $\alpha = \beta = 0$.

We now prove (3.179). The formula (3.116b) implies

$$(1+x)J_N^{0,1}(x) = L_N(x) + L_{N+1}(x). \tag{3.182}$$

Hence, we infer from Theorem 3.26 that the nodes $\{x_j\}_{j=0}^N$ are the zeros of $L_N(x) + L_{N+1}(x)$, and the formulas of the weights are

$$\omega_0 = \frac{2}{(N+1)^2} = \frac{1}{(N+1)^2} \frac{1-x_0}{[L_N(x_0)]^2},$$

$$\omega_j = \frac{2(2N+1)}{N(N+1)} \frac{1}{(1+x_j)\left[J_{N-1}^{0,1}(x)\partial_x J_N^{0,1}(x)\right]\Big|_{x=x_j}}, \ 1 \le j \le N.$$

To derive the equivalent expression in (3.179), we deduce from the fact $\{J_N^{0,1}(x_j) = 0\}_{j=1}^N$ that

$$\partial_x J_N^{0,1}(x_j) \overset{(3.133)}{=} \frac{2N(N+1)}{2N+1} \frac{J_{N-1}^{0,1}(x_j)}{1-x_j^2}$$
$$\overset{(3.182)}{=} \frac{2N(N+1)}{2N+1} \frac{L_{N-1}(x_j)+L_N(x_j)}{(1+x_j)(1-x_j^2)},$$

which, together with (3.182), leads to

$$(1+x_j)J_{N-1}^{0,1}(x_j)\partial_x J_N^{0,1}(x_j) = \frac{2N(N+1)}{2N+1}\left[\frac{L_{N-1}(x_j)+L_N(x_j)}{1+x_j}\right]^2 \frac{1}{1-x_j}.$$

Due to $L_N(x_j)+L_{N+1}(x_j)=0$ for $1 \le j \le N$, using the three-term recurrence relation (3.168) gives

$$\begin{aligned} L_{N-1}(x_j) &= \frac{2N+1}{N} x_j L_N(x_j) - \frac{N+1}{N} L_{N+1}(x_j) \\ &= \frac{2N+1}{N} x_j L_N(x_j) + \frac{N+1}{N} L_N(x_j) \\ &= \frac{2N+1}{N}(1+x_j)L_N(x_j) - L_N(x_j). \end{aligned}$$

A combination of the above facts leads to (3.179).

We now turn to the derivation of (3.180). By Theorem 3.27 with $\alpha = \beta = 0$,

$$\begin{aligned} \omega_0 &= \omega_N = \frac{2}{N(N+1)}, \\ \omega_j &= \frac{8}{N+1} \frac{1}{(1-x_j^2)J_{N-2}^{1,1}(x_j)\partial_x J_{N-1}^{1,1}(x_j)}, \quad 1 \le j \le N-1. \end{aligned} \tag{3.183}$$

In view of $\{J_{N-1}^{1,1}(x_j) = 0\}_{j=1}^N$, we derive from (3.133) that

$$(1-x_j^2)\partial_x J_{N-1}^{1,1}(x_j) = N J_{N-2}^{1,1}(x_j), \quad 1 \le j \le N-1.$$

As a consequence of (3.98) and the above equality, we find that

$$(1-x_j^2)J_{N-2}^{1,1}(x_j)\partial_x J_{N-1}^{1,1}(x_j) = N\big[J_{N-2}^{1,1}(x_j)\big]^2 = \frac{4}{N}\big[L'_{N-1}(x_j)\big]^2.$$

Differentiating (3.168) and using (3.176a) and the fact $\{L'_N(x_j)=0\}_{j=1}^N$, yields

$$\begin{aligned} L'_{N-1}(x_j) &= \frac{2N+1}{N}L_N(x_j) - \frac{N+1}{N}L'_{N+1}(x_j) \\ &= \frac{2N+1}{N}L_N(x_j) - \frac{N+1}{N}\big(L'_{N-1}(x_j)+(2N+1)L_N(x_j)\big) \\ &= -(2N+1)L_N(x_j) - \frac{N+1}{N}L'_{N-1}(x_j), \end{aligned}$$

which leads to

$$L'_{N-1}(x_j) = -NL_N(x_j), \quad 1 \le j \le N-1.$$

Consequently,

$$(1-x_j^2)J^{1,1}_{N-2}(x_j)\partial_x J^{1,1}_{N-1}(x_j) = 4NL_N^2(x_j).$$

Plugging it into the second formula of (3.183) gives the desired result. □

3.3.2 Computation of Nodes and Weights

As a special case of (3.142), the interior Legendre-Gauss-type nodes are the eigenvalues of the following Jacobian matrix:

$$A_{M+1} = \begin{bmatrix} a_0 & \sqrt{b_1} & & & \\ \sqrt{b_1} & a_1 & \sqrt{b_2} & & \\ & \ddots & \ddots & \ddots & \\ & & \sqrt{b_{M-1}} & a_{M-1} & \sqrt{b_M} \\ & & & \sqrt{b_M} & a_M \end{bmatrix}, \tag{3.184}$$

where

- For LG: $a_j = 0, \quad b_j = \dfrac{j^2}{4j^2-1}, \quad M = N.$
- For LGR: $a_j = \dfrac{1}{(2j+1)(2j+3)}, \quad b_j = \dfrac{j(j+1)}{(2j+1)^2}, \quad M = N-1.$
- For LGL: $a_j = 0, \quad b_j = \dfrac{j(j+2)}{(2j+1)(2j+3)}, \quad M = N-2.$

The quadrature weights can be evaluated by using the formulas in Theorem 3.29. Alternatively, as a consequence of (3.144), the quadrature weights can be computed from the first component of the orthonormal eigenvectors of A_{M+1}.

The eigenvalue method is well-suited for the Gauss-quadratures of low or moderate order. However, for high-order quadratures, the eigenvalue method may suffer from round-off errors, so it is advisable to use a root-finding iterative approach. To fix the idea, we restrict our attention to the commonly used Legendre-Gauss-Lobatto case and compute the zeros of $L_N'(x)$. In this case, the *Newton method* (3.31) reads

$$\begin{cases} x_j^{k+1} = x_j^k - \dfrac{L_N'(x_j^k)}{L_N''(x_j^k)}, & k \ge 0, \\ \text{given } x_j^0, & 1 \le j \le N-1. \end{cases} \tag{3.185}$$

To avoid evaluating the values of L_N'', we use (3.171) to derive that

$$\frac{L_N'(x)}{L_N''(x)} = \frac{(1-x^2)L_N'(x)}{2xL_N'(x) - N(N+1)L_N(x)}.$$

For an iterative method, it is essential to start with a good initial approximation. In Lether (1978), an approximation of the zeros of $L_N(x)$ is given by

$$\sigma_k = \left[1 - \frac{N-1}{8N^3} - \frac{1}{384N^4}\left(39 - \frac{28}{\sin^2\theta_k}\right)\right]\cos\theta_k + O(N^{-5}), \tag{3.186}$$

where

$$\theta_k = \frac{4k-1}{4N+2}\pi, \quad 1 \le k \le N.$$

Notice from Corollary 3.4 (the interlacing property) that there exists exactly one zero of $L_N'(x)$ between two consecutive zeros of $L_N(x)$. Therefore, we can take the initial guess as

$$x_j^0 = \frac{\sigma_j + \sigma_{j+1}}{2}, \quad 1 \le j \le N-1. \tag{3.187}$$

We point out that due to $L_N'(-x) = (-1)^{N+1}L_N'(x)$, the computational cost of (3.185) can be halved.

After finding the nodes $\{x_j\}_{j=0}^N$, we can compute the corresponding weights by the formula (3.180):

$$w(x) = \frac{2}{N(N+1)}\frac{1}{L_N^2(x)}. \tag{3.188}$$

It is clear that $w'(x_j) = 0$ for $1 \le j \le N-1$. In other words, the interior nodes are the extremes of $w(x)$. We plot the graph of $w(x)$ with $N = 8$ in Fig. 3.6a. As a consequence, for a small perturbation of the nodes, we can obtain very accurate values $\omega_j = w(x_j)$ even for very large N.

In Fig. 3.6b, we depict the locations of the Legendre-Gauss-Lobatto nodes $\{x_j\}_{j=0}^8$, and $\{\theta_j = \arccos x_j\}_{j=0}^8$. We see that $\{\theta_j\}$ distribute nearly equidistantly along the upper half unit circle (i.e., in $[0,\pi]$). The projection of $\{\theta_j\}$ onto $[-1,1]$ yields the clustering of points $\{x_j\}$ near the endpoints $x=\pm 1$ with spacing $O(N^{-2})$.

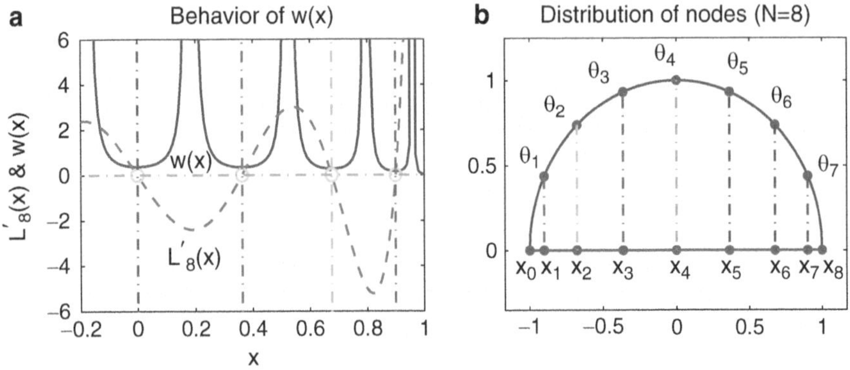

Fig. 3.6 (**a**) Behavior of $w(x)$ in (3.188) with $N=8$; (**b**) Distribution of the Legendre-Gauss-Lobatto nodes with $N=8$

We tabulate in Table 3.2 some samples of the LGL nodes and weights with $N=8,16$ (note that $x_{N-j}=-x_j$ and $\omega_{N-j}=\omega_j$) computed by the aforementioned method.

Table 3.2 LGL nodes and weights

Nodes x_j	Weights ω_j
1.000000000000000e+00	2.777777777777778e-02
8.997579954114601e-01	1.654953615608056e-01
6.771862795107377e-01	2.745387125001617e-01
3.631174638261782e-01	3.464285109730462e-01
0.000000000000000e+00	3.715192743764172e-01
1.000000000000000e+00	7.352941176470588e-03
9.731321766314184e-01	4.492194054325414e-02
9.108799959155736e-01	7.919827050368709e-02
8.156962512217703e-01	1.105929090070281e-01
6.910289806276847e-01	1.379877462019266e-01
5.413853993301015e-01	1.603946619976215e-01
3.721744335654770e-01	1.770042535156577e-01
1.895119735183174e-01	1.872163396776192e-01
0.000000000000000e+00	1.906618747534694e-01

3.3.3 Interpolation and Discrete Legendre Transforms

Given a set of Legendre-Gauss-type quadrature nodes and weights $\{x_j,\omega_j\}_{j=0}^N$, we define the associated interpolation operator I_N, discrete inner product $\langle\cdot,\cdot\rangle_N$ and discrete norm $\|\cdot\|_N$, as in Sect. 3.1.5.

Thanks to the exactness of the Legendre-Gauss-type quadrature (cf. (3.181)), we have

$$\langle u,v\rangle_N = (u,v), \quad \forall u\cdot v \in P_{2N+\delta}, \tag{3.189}$$

where $\delta = 1,0,-1$ for LG, LGR and LGL, respectively. Consequently,

$$\|u\|_N = \|u\|, \quad \forall u \in P_N \text{ for LG and LGR.} \tag{3.190}$$

Although the above formula does not hold for LGL, we derive from Lemma 3.3 with $\alpha = \beta = 0$ the following equivalence:

$$\|u\| \le \|u\|_N \le \sqrt{2+N^{-1}}\|u\|, \quad \forall u \in P_N. \tag{3.191}$$

Moreover, as a direct consequence of (3.154), we have

$$\langle L_N, L_N\rangle_N = \frac{2}{N}. \tag{3.192}$$

We now turn to the discrete Legendre transforms. The Lagrange interpolation polynomial $I_N u \in P_N$, so we write

$$(I_N u)(x) = \sum_{n=0}^{N} \tilde{u}_n L_n(x),$$

where the (discrete) Legendre coefficients $\{\tilde{u}_n\}$ are determined by the *forward discrete Legendre transform*:

$$\tilde{u}_n = \frac{1}{\gamma_n}\sum_{j=0}^{N} u(x_j)L_n(x_j)\omega_j = \frac{\langle u, L_n\rangle_N}{\|L_n\|_N^2}, \quad 0 \le n \le N, \tag{3.193}$$

where $\gamma_n = \frac{2}{2n+1}$ for $0 \le n \le N$, except for LGL case, $\gamma_N = \frac{2}{N}$. On the other hand, given the expansion coefficients $\{\tilde{u}_n\}$, the physical values $\{u(x_j)\}$ can be computed by the *backward discrete Legendre transform*:

$$u(x_j) = (I_N u)(x_j) = \sum_{n=0}^{N} \tilde{u}_n L_n(x_j), \quad 0 \le j \le N. \tag{3.194}$$

Assuming that $\big(L_n(x_j)\big)_{j,n=0,1,\ldots,N}$ have been precomputed, the discrete Legendre transforms (3.194) and (3.193) can be carried out by a standard matrix–vector multiplication routine in about N^2 flops. The cost of the discrete Legendre transforms can be halved, due to the symmetry: $L_n(x_j) = (-1)^n L_n(x_{N-j})$.

To illustrate the convergence of Legendre interpolation approximations, we consider the test function: $u(x) = \sin(k\pi x)$. Writing

$$\sin(k\pi x) = \sum_{n=0}^{\infty} \hat{u}_n L_n(x), \quad x \in [-1,1], \tag{3.195}$$

we can derive from the property of the Bessel functions (cf. Watson (1966)) that

$$\hat{u}_n = \frac{1}{\sqrt{2k}}(2n+1)J_{n+1/2}(k\pi)\sin(n\pi/2), \quad n \geq 0, \tag{3.196}$$

where $J_{n+1/2}(\cdot)$ is the Bessel function of the first kind. Using the asymptotic formula

$$J_\nu(x) \sim \frac{1}{2\pi\nu}\left(\frac{ex}{2\nu}\right)^\nu, \quad \nu \gg 1,\ \nu \in \mathbb{R}, \tag{3.197}$$

we find that the exponential decay of the expansion coefficients occurs when the mode

$$n > \frac{ek\pi}{2} - \frac{1}{2}. \tag{3.198}$$

We now approximate u by $I_N u = \sum_{n=0}^{N} \tilde{u}_n L_n(x)$, and consider the error in the coefficients $|\hat{u}_n - \tilde{u}_n|$. We observe from Fig. 3.7a that the errors between the exact and discrete expansion coefficients decay exponentially when $N > ek\pi/2$, and it verifies the estimate

$$\max_{0\leq n\leq N} |\hat{u}_n - \tilde{u}_n| \sim \hat{u}_{N+1} \quad \text{for } N \gg 1. \tag{3.199}$$

In Fig. 3.7b, we depict the exact expansion coefficients $\hat{u}_n$ (marked by "o") and the discrete expansion coefficients $\tilde{u}_n$ (marked by "•") against the subscript n, and in Fig. 3.7c, we plot the exact solution versus its interpolation. Observe that $I_N u$ provides an accurate approximation to u as long as $N > ek\pi/2$.

As with the Fourier case, when a discontinuous function is expanded in Legendre series, the Gibbs phenomena occur in the neighborhood of a discontinuity. For example, the Legendre series expansion of the sign function $\mathrm{sgn}(x)$ is

$$\mathrm{sgn}(x) = \sum_{n=0}^{\infty} \frac{(-1)^n(4n+3)(2n)!}{2^{2n+1}(n+1)!n!} L_{2n+1}(x). \tag{3.200}$$

One verifies readily that the expansion coefficients behave like

$$|\hat{u}_{2n+1}| = \frac{(4n+3)(2n)!}{2^{2n+1}(n+1)!n!} \simeq \frac{1}{\sqrt{n}}, \quad n \gg 1. \tag{3.201}$$

In Fig. 3.7d, we plot the numerical approximation $I_N u(x)$ and $\mathrm{sgn}(x)$ in the interval $[-0.3, 0.3]$, which indicates a Gibbs phenomenon near $x = 0$.

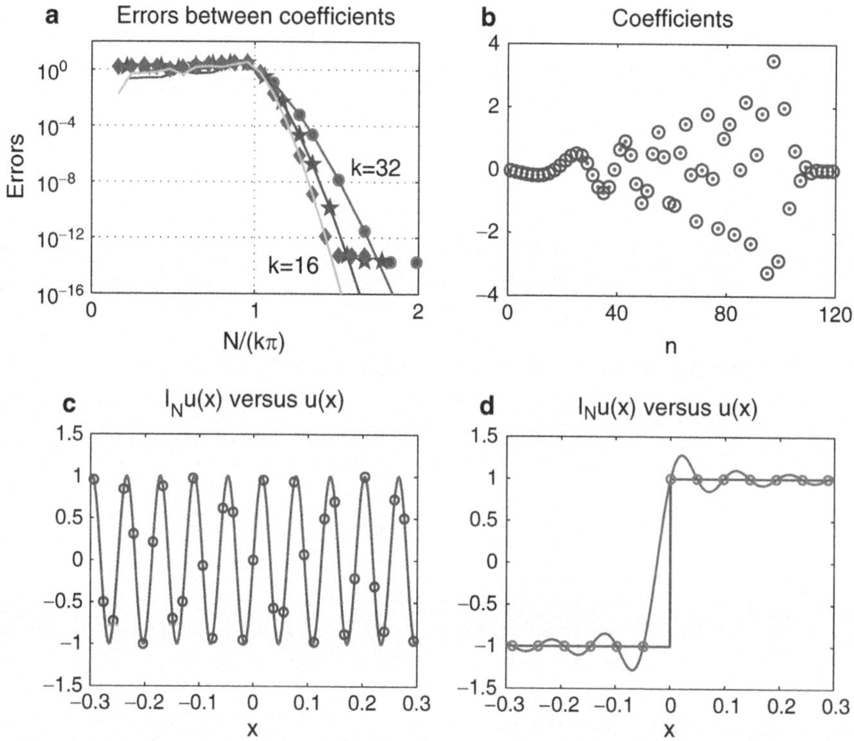

Fig. 3.7 (**a**) Error $\max_{0\le n\le N}|\hat{u}_n-\tilde{u}_n|$ & $\hat{u}_{N+1}$ (*solid line*) vs. $N/(k\pi)$ with $k=16,24,32$; (**b**) $\hat{u}_n$ vs. $\tilde{u}_n$ with $k=32$ and $N=128$; (**c**) $u(x)$ vs. $I_Nu(x)$, $x\in[-0.3,0.3]$ with $k=32$ and $N=128$; (**d**) $u(x)=\text{sgn}(x)$ vs. $I_Nu(x)$, $x\in[-0.3,0.3]$ with $k=32$ and $N=64$

3.3.4 Differentiation in the Physical Space

Given $u\in P_N$ and its values at a set of Legendre-Gauss-type points $\{x_j\}_{j=0}^N$, let $\{h_j\}_{j=0}^N$ be the associated Lagrange basis polynomials. According to the general approach described in Sect. 3.1.6, we have

$$\mathbf{u}^{(m)}=D^m\mathbf{u},\quad m\ge 1, \tag{3.202}$$

where

$$D=\big(d_{kj}=h_j'(x_k)\big)_{0\le k,j\le N},\ \mathbf{u}^{(m)}=\big(u^{(m)}(x_0),\ldots,u^{(m)}(x_N)\big)^T,\ \mathbf{u}=\mathbf{u}^{(0)}.$$

We derive below explicit representations of the entries of D for the three different cases by using the general formulas of the Jacobi polynomials in Sect. 3.2.5.

- For the Legendre-Gauss-Lobatto case ($x_0=-1$ and $x_N=1$): The general formulas (3.160)–(3.162) for JGL in Sect. 3.2.5 with $\alpha=\beta=0$ lead to the reduced formulas involving $(1-x^2)\partial_x J_{N-1}^{1,1}$ and $(1\pm x)\partial_x J_{N-1}^{1,1}$. Using (3.100) and (3.172) leads to

$$\begin{aligned}(1-x_j^2)\partial_x J_{N-1}^{1,1}(x_j) &= \frac{2}{N+1}(1-x_j^2)\partial_x^2 L_N(x_j)\\ &= -2NL_N(x_j), \quad 1 \le j \le N-1,\end{aligned}$$

and

$$(1\pm x_j)\partial_x J_{N-1}^{1,1}(x_j) = -2N\frac{L_N(x_j)}{1\mp x_j}, \quad 1 \le j \le N-1.$$

Plugging the above in (3.160)–(3.162) with $\alpha = \beta = 0$, we derive

$$d_{kj} = \begin{cases} -\dfrac{N(N+1)}{4}, & k=j=0,\\ \dfrac{L_N(x_k)}{L_N(x_j)}\dfrac{1}{x_k-x_j}, & k \ne j,\ 0 \le k,j \le N,\\ 0, & 1 \le k=j \le N-1,\\ \dfrac{N(N+1)}{4} & k=j=N. \end{cases} \tag{3.203}$$

- For the Legendre-Gauss-Radau case ($x_0=-1$): The general formula (3.163) in the case of $\alpha=\beta=0$ can be simplified to

$$d_{kj} = \begin{cases} -\dfrac{N(N+2)}{4}, & k=j=0,\\ \dfrac{x_k}{1-x_k^2} + \dfrac{(N+1)L_N(x_k)}{(1-x_k^2)Q'(x_k)}, & 1 \le k=j \le N,\\ \dfrac{Q'(x_k)}{Q'(x_j)}\dfrac{1}{x_k-x_j}, & k \ne j, \end{cases} \tag{3.204}$$

where $Q(x) = L_N(x) + L_{N+1}(x)$ (which is proportional to $(1+x)J_N^{0,1}(x)$). For $k=j$, we derive from Theorem 3.11 that

$$d_{kk} = \frac{Q''(x_k)}{2Q'(x_k)}, \quad 0 \le k \le N.$$

To avoid computing the second-order derivatives, we obtain from (3.172) that

$$Q''(x_k) = \frac{2x_kQ'(x_k) + 2(N+1)L_N(x_k)}{1-x_k^2}, \quad 1 \le k \le N.$$

For $k=j=0$, we can work out the constants by using (3.177).

- For the Legendre-Gauss case: The general formula (3.164) in the case of $\alpha = \beta = 0$ reduces to

$$d_{kj} = \begin{cases} \dfrac{L'_{N+1}(x_k)}{L'_{N+1}(x_j)} \dfrac{1}{x_k - x_j}, & k \neq j, \\ \dfrac{x_k}{1-x_k^2}, & k = j. \end{cases} \tag{3.205}$$

In all cases, the differentiation matrix D is a full matrix, so $O(N^2)$ flops are needed to compute $\{u'(x_j)\}_{j=0}^N$ from $\{u(x_j)\}_{j=0}^N$. Also note that since $u^{(N+1)}(x) \equiv 0$ for any $u \in P_N$, we have $D^{N+1}\mathbf{u} = 0$ for any $\mathbf{u} \in \mathbb{R}^{N+1}$. Hence, the only eigenvalue of D is zero which has a multiplicity $N+1$.

3.3.5 Differentiation in the Frequency Space

Given $u \in P_N$, we write

$$u(x) = \sum_{k=0}^{N} \hat{u}_k L_k(x) \in P_N,$$

and

$$u'(x) = \sum_{k=1}^{N} \hat{u}_k L'_k(x) = \sum_{k=0}^{N} \hat{u}_k^{(1)} L_k(x) \quad \text{with} \quad \hat{u}_N^{(1)} = 0.$$

Thanks to (3.176a), we find

$$\begin{aligned} u' &= \sum_{k=0}^{N} \hat{u}_k^{(1)} L_k = \hat{u}_0^{(1)} + \sum_{k=1}^{N-1} \hat{u}_k^{(1)} \frac{1}{2k+1}(L'_{k+1} - L'_{k-1}) \\ &= \frac{\hat{u}_{N-1}^{(1)}}{2N-1} L'_N + \sum_{k=1}^{N-1} \left\{ \frac{\hat{u}_{k-1}^{(1)}}{2k-1} - \frac{\hat{u}_{k+1}^{(1)}}{2k+3} \right\} L'_k. \end{aligned}$$

Since $\{L'_k\}$ are orthogonal polynomials (cf. (3.174b)), comparing the coefficients of L'_k leads to the backward recursive relation:

$$\begin{aligned} &\hat{u}_{k-1}^{(1)} = (2k-1)\Big(\hat{u}_k + \frac{\hat{u}_{k+1}^{(1)}}{2k+3}\Big), \quad k = N-1, N-2, \ldots, 1, \\ &\hat{u}_N^{(1)} = 0, \quad \hat{u}_{N-1}^{(1)} = (2N-1)\hat{u}_N. \end{aligned} \tag{3.206}$$

Higher-order differentiations can be performed by the above formula recursively.

3.4 Chebyshev Polynomials

In this section, we consider another important special case of the Jacobi polynomials – Chebyshev polynomials (of the first kind), which are proportional to Jacobi polynomials $\{J_n^{-1/2,-1/2}\}$ and are orthogonal with respect to the weight function $\omega(x) = (1-x^2)^{-1/2}$.

The three-term recurrence relation for the Chebyshev polynomials reads:

$$T_{n+1}(x) = 2xT_n(x) - T_{n-1}(x), \quad n \geq 1, \tag{3.207}$$

with $T_0(x) = 1$ and $T_1(x) = x$.

The Chebyshev polynomials are eigenfunctions of the Sturm-Liouville problem:

$$\sqrt{1-x^2}\big(\sqrt{1-x^2}T_n'(x)\big)' + n^2T_n(x) = 0, \tag{3.208}$$

or equivalently,

$$(1-x^2)T_n''(x) - xT_n'(x) + n^2T_n(x) = 0. \tag{3.209}$$

While we can derive the properties of Chebyshev polynomials from the general properties of Jacobi polynomials with $(\alpha,\beta) = (-1/2,-1/2)$, it is more convenient to explore the relation between Chebyshev polynomials and trigonometric functions. Indeed, using the trigonometric relation

$$\cos((n+1)\theta) + \cos((n-1)\theta) = 2\cos\theta\cos(n\theta),$$

and taking $\theta = \arccos x$, we find that $\cos(n \arccos x)$ satisfies the three-term recurrence relation (3.207), and it is $1, x$ for $n = 0, 1$, respectively. Thus, by an induction argument, $\cos(n \arccos x)$ is also a polynomial of degree n with the leading coefficient 2^{n-1} (Fig. 3.8). We infer from Theorem 3.1 of the uniqueness that

$$T_n(x) = \cos n\theta, \quad \theta = \arccos x, \quad n \geq 0, \ x \in I. \tag{3.210}$$

This explicit representation enables us to derive many useful properties.

An immediate consequence is the recurrence relation

$$2T_n(x) = \frac{1}{n+1}T_{n+1}'(x) - \frac{1}{n-1}T_{n-1}'(x), \quad n \geq 2. \tag{3.211}$$

One can also derive from (3.210) that

$$T_n(-x) = (-1)^n T(x), \quad T_n(\pm 1) = (\pm 1)^n, \tag{3.212a}$$

$$|T_n(x)| \leq 1, \quad |T_n'(x)| \leq n^2, \tag{3.212b}$$

$$(1-x^2)T_n'(x) = \frac{n}{2}T_{n-1}(x) - \frac{n}{2}T_{n+1}(x), \tag{3.212c}$$

$$2T_m(x)T_n(x) = T_{m+n}(x) + T_{m-n}(x), \quad m \geq n \geq 0, \tag{3.212d}$$

and

$$T_n'(\pm 1) = (\pm 1)^{n-1} n^2, \tag{3.213a}$$

$$T_n''(\pm 1) = \frac{1}{3}(\pm 1)^n n^2 (n^2 - 1). \tag{3.213b}$$

It is also easy to show that

$$\int_{-1}^{1} T_n(x) T_m(x) \frac{1}{\sqrt{1-x^2}} dx = \frac{c_n \pi}{2} \delta_{mn}, \tag{3.214}$$

where $c_0 = 2$ and $c_n = 1$ for $n \geq 1$. Hence, we find from (3.208) that

$$\int_{-1}^{1} T_n'(x) T_m'(x) \sqrt{1-x^2} dx = \frac{n^2 c_n \pi}{2} \delta_{mn}, \tag{3.215}$$

i.e., $\{T_n'(x)\}$ are mutually orthogonal with respect to the weight function $\sqrt{1-x^2}$.
We can obtain from (3.211) that

$$T_n'(x) = 2n \sum_{\substack{k=0 \\ k+n \text{ odd}}}^{n-1} \frac{1}{c_k} T_k(x), \tag{3.216a}$$

$$T_n''(x) = \sum_{\substack{k=0 \\ k+n \text{ even}}}^{n-2} \frac{1}{c_k} n(n^2 - k^2) T_k(x). \tag{3.216b}$$

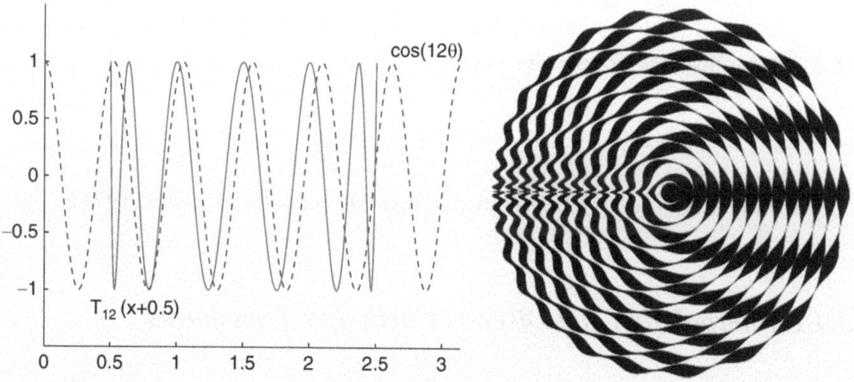

Fig. 3.8 *Left*: curves of $T_{12}(x+1.5)$ and $\cos(12\theta)$; *Right*: we plot $T_n(x)$ radially, increase the radius for each value of n, and fill in the areas between the curves (Trott (1999), pp. 10 and 84)

Another remarkable consequence of (3.210) is that the Gauss-type quadrature nodes and weights can be derived explicitly.

Theorem 3.30. *Let* $\{x_j, \omega_j\}_{j=0}^N$ *be a set of Chebyshev-Gauss-type quadrature nodes and weights.*

- *For Chebyshev-Gauss* (CG) *quadrature,*

$$x_j = -\cos\frac{(2j+1)\pi}{2N+2}, \quad \omega_j = \frac{\pi}{N+1}, \quad 0 \le j \le N.$$

- *For Chebyshev-Gauss-Radau* (CGR) *quadrature,*

$$x_j = -\cos\frac{2\pi j}{2N+1}, \quad 0 \le j \le N,$$
$$\omega_0 = \frac{\pi}{2N+1}, \quad \omega_j = \frac{2\pi}{2N+1}, \quad 1 \le j \le N.$$

- *For Chebyshev-Gauss-Lobatto* (CGL) *quadrature,*

$$x_j = -\cos\frac{\pi j}{N}, \quad \omega_j = \frac{\pi}{\tilde{c}_j N}, \quad 0 \le j \le N.$$

where $\tilde{c}_0 = \tilde{c}_N = 2$ *and* $\tilde{c}_j = 1$ *for* $j = 1, 2, \ldots, N-1$.

With the above choices, there holds

$$\int_{-1}^{1} p(x)\frac{1}{\sqrt{1-x^2}}dx = \sum_{j=0}^{N} p(x_j)\omega_j, \quad \forall p \in P_{2N+\delta}, \tag{3.217}$$

where $\delta = 1, 0, -1$ *for the* CG, CGR *and* CGL, *respectively.*

In the Chebyshev case, the nodes $\{\theta_j = \arccos(x_j)\}$ are equally distributed on $[0, \pi]$, whereas $\{x_j\}$ are clustered in the neighborhood of $x = \pm 1$ with density $O(N^{-2})$, for instance, for the CGL points

$$1 - x_1 = 1 - \cos\frac{\pi}{N} = 2\sin^2\frac{\pi}{2N} \simeq \frac{\pi^2}{2N^2} \quad \text{for } N \gg 1.$$

For more properties of Chebyshev polynomials, we refer to Rivlin (1974).

3.4.1 Interpolation and Discrete Chebyshev Transforms

Given a set of Chebyshev-Gauss-type quadrature nodes and weights $\{x_j, \omega_j\}_{j=0}^N$, we define the associated interpolation operator I_N, discrete inner product $\langle \cdot, \cdot \rangle_{N,\omega}$ and discrete norm $\|\cdot\|_{N,\omega}$, as in Sect. 3.1.5.

Thanks to the exactness of the Chebyshev-Gauss-type quadrature, we have

$$\langle u, v \rangle_{N,\omega} = (u, v)_\omega, \quad \forall uv \in P_{2N+\delta}, \tag{3.218}$$

where $\delta = 1, 0, -1$ for CG, CGR and CGL, respectively. Consequently,

$$\|u\|_{N,\omega} = \|u\|_{\omega}, \quad \forall u \in P_N, \text{ for CG and CGR.} \tag{3.219}$$

Although the above identity does not hold for the CGL, the following equivalence follows from Lemma 3.3:

$$\|u\|_{\omega} \le \|u\|_{N,\omega} \le \sqrt{2}\|u\|_{\omega}, \quad \forall u \in P_N. \tag{3.220}$$

Moreover, a direct computation leads to

$$\langle T_N, T_N \rangle_{N,\omega} = \frac{\pi}{N} \sum_{j=0}^{N} \frac{\cos^2 j\pi}{\tilde{c}_j} = \pi. \tag{3.221}$$

We now turn to the discrete Chebyshev transforms. To fix the idea, we only consider the Chebyshev-Gauss-Lobatto case. As a special family of Jacobi polynomials, the transforms can be performed via a matrix–vector multiplication with $O(N^2)$ operations as usual. However, thanks to (3.210), they can be carried out with $O(N \log_2 N)$ operations via FFT.

Given $u \in C[-1,1]$, let $I_N u$ be its Lagrange interpolation polynomial relative to the CGL points, and we write

$$(I_N u)(x) = \sum_{n=0}^{N} \tilde{u}_n T_n(x) \in P_N,$$

where $\{\tilde{u}_n\}$ are determined by the *forward discrete Chebyshev transform* (cf. Theorem 3.9):

$$\tilde{u}_n = \frac{2}{\tilde{c}_n N} \sum_{j=0}^{N} \frac{1}{\tilde{c}_j} u(x_j) \cos \frac{nj\pi}{N}, \quad 0 \le n \le N. \tag{3.222}$$

On the other hand, given the expansion coefficients $\{\tilde{u}_n\}$, the physical values $\{u(x_j)\}$ are evaluated by the *backward discrete Chebyshev transform*:

$$u(x_j) = (I_N u)(x_j) = \sum_{n=0}^{N} \tilde{u}_n T_n(x_j) = \sum_{n=0}^{N} \tilde{u}_n \cos \frac{nj\pi}{N}, \quad 0 \le j \le N. \tag{3.223}$$

Hence, it is clear that both the forward transform (3.222) and backward transform (3.223) can be computed by using FFT in $O(N \log_2 N)$ operations.

Let us conclude this part with a discussion of point-per-wavelength required for the approximation using Chebyshev polynomials. We have

$$\sin(k\pi x) = \sum_{n=0}^{\infty} \hat{u}_n T_n(x), \quad x \in [-1,1], \tag{3.224}$$

with

$$\hat{u}_n := \hat{u}_n(k) = \frac{2}{c_n} J_n(k\pi) \sin(n\pi/2), \quad n \ge 0, \tag{3.225}$$

where $J_n(\cdot)$ is again the Bessel function of the first kind. Hence, using the asymptotic formula (3.197), we find that the exponential decay of the expansion coefficients occurs when

$$n > \frac{ek\pi}{2}, \tag{3.226}$$

which is similar to (3.198) for the Legendre expansion.

3.4.2 Differentiation in the Physical Space

Given $u \in P_N$ and its values at a set of Chebyshev-Gauss-type collocation points $\{x_j\}_{j=0}^N$, let $\{h_j(x)\}_{j=0}^N$ be the associated Lagrange basis polynomials. According to the general results stated in Sect. 3.1.6, we have

$$\mathbf{u}^{(m)} = D^m\mathbf{u}, \quad m \ge 1, \tag{3.227}$$

where

$$D = \big(d_{kj} = h_j'(x_k)\big)_{0\le k,j\le N}, \; \mathbf{u}^{(m)} = \big(u^{(m)}(x_0),\ldots,u^{(m)}(x_N)\big)^T, \; \mathbf{u} = \mathbf{u}^{(0)}.$$

The entries of the first-order differentiation matrix D can be determined by the explicit formulas below.

- For the Chebyshev-Gauss-Lobatto case ($x_0 = -1$ and $x_N = 1$):

$$d_{kj} = \begin{cases} -\dfrac{2N^2+1}{6}, & k = j = 0, \\ \dfrac{\tilde{c}_k}{\tilde{c}_j}\dfrac{(-1)^{k+j}}{x_k - x_j}, & k \ne j, 0 \le k,j \le N, \\ -\dfrac{x_k}{2(1-x_k^2)}, & 1 \le k = j \le N-1, \\ \dfrac{2N^2+1}{6}, & k = j = N, \end{cases} \tag{3.228}$$

  where $\tilde{c}_0 = \tilde{c}_N = 2$ and $\tilde{c}_j = 1$ for $1 \le j \le N-1$.
- For the Chebyshev-Gauss-Radau case ($x_0 = -1$):

$$d_{kj} = \begin{cases} -\dfrac{N(N+1)}{3}, & k = j = 0, \\ \dfrac{x_k}{2(1-x_k^2)} + \dfrac{(2N+1)T_N(x_k)}{2(1-x_k^2)Q'(x_k)}, & 1 \le k = j \le N, \\ \dfrac{Q'(x_k)}{Q'(x_j)}\dfrac{1}{x_k - x_j}, & k \ne j, \end{cases} \tag{3.229}$$

where $Q(x) = T_N(x) + T_{N+1}(x)$. To derive (3.229), we find from Theorem 3.26 that $\{x_j\}_{j=0}^N$ are the zeros of $(1+x)J_N^{-1/2,1/2}(x)$. In view of the correspondence:

$$J_N^{-1/2,-1/2}(x) = J_N^{-1/2,-1/2}(1)T_N(x), \tag{3.230}$$

one verifies by using (3.116b) that

$$(1+x)J_N^{-1/2,1/2}(x) = J_N^{-1/2,-1/2}(1)\big(T_N(x) + T_{N+1}(x)\big).$$

Hence, for $k = j$, we find from Theorem 3.11 that

$$d_{kk} = \frac{Q''(x_k)}{2Q'(x_k)}, \quad 0 \le k \le N.$$

To avoid evaluating the second-order derivatives, we derive from (3.209) and the fact $Q(x_k) = 0$ that

$$Q''(x_k) = \frac{x_k Q'(x_k) + (2N+1)T_N(x_k)}{1 - x_k^2}, \quad 1 \le k \le N.$$

Hence,

$$d_{kk} = \frac{x_k}{2(1-x_k^2)} + \frac{(2N+1)T_N(x_k)}{2(1-x_k^2)Q'(x_k)}, \quad 1 \le k \le N.$$

The formula for the entry d_{00} follows directly from the Jacobi-Gauss-Radau case with $\alpha = \beta = 0$.

- For the Chebyshev-Gauss case:

$$d_{kj} = \begin{cases} \dfrac{T'_{N+1}(x_k)}{T'_{N+1}(x_j)} \dfrac{1}{x_k - x_j}, & k \ne j, \\ \dfrac{x_k}{2(1-x_k^2)}, & k = j. \end{cases} \tag{3.231}$$

3.4.3 Differentiation in the Frequency Space

Now, we describe the FFT algorithm for Chebyshev spectral differentiation. Let us start with the conventional approach. Given

$$u(x) = \sum_{k=0}^{N} \hat{u}_k T_k(x) \in P_N, \tag{3.232}$$

we derive from (3.211) that

$$\begin{aligned} u' &= \sum_{k=1}^{N} \hat{u}_k T_k' = \sum_{k=0}^{N} \hat{u}_k^{(1)} T_k \quad (\text{with } \hat{u}_N^{(1)} = 0) \\ &= \hat{u}_0^{(1)} + \hat{u}_1^{(1)} T_1 + \sum_{k=2}^{N-1} \hat{u}_k^{(1)} \Big(\frac{T_{k+1}'}{2(k+1)} - \frac{T_{k-1}'}{2(k-1)} \Big) \\ &= \frac{\hat{u}_{N-1}^{(1)}}{2N} T_N' + \sum_{k=1}^{N-1} \frac{1}{2k} \big(c_{k-1} \hat{u}_{k-1}^{(1)} - \hat{u}_{k+1}^{(1)} \big) T_k', \end{aligned} \tag{3.233}$$

where $c_0 = 2$ and $c_k = 1$ for $k \ge 1$. Since $\{T_k'\}$ are mutually orthogonal, we compare the expansion coefficients in terms of $\{T_k'\}$ and find that $\{\hat{u}_k^{(1)}\}$ can be computed from $\{\hat{u}_k\}$ via the backward recurrence relation:

$$\begin{aligned} &\hat{u}_N^{(1)} = 0, \; \hat{u}_{N-1}^{(1)} = 2N\hat{u}_N, \\ &\hat{u}_{k-1}^{(1)} = \big(2k\hat{u}_k + \hat{u}_{k+1}^{(1)}\big)/c_{k-1}, \quad k = N-1, \ldots, 1. \end{aligned} \tag{3.234}$$

Higher-order derivatives can be evaluated recursively by this relation.

Notice that given $\{u(x_j)\}_{j=0}^N$ at the Chebyshev-Gauss-Lobatto points $\{x_j\}_{j=0}^N$, the computation of $\{u'(x_j)\}_{j=0}^N$ through the process of differentiation in the physical space requires $O(N^2)$ operations due to the fact that the differentiation matrix (see the previous section) is full. However, thanks to the fast discrete Chebyshev transforms between the physical values and expansion coefficients, one can compute $\{u'(x_j)\}_{j=0}^N$ from $\{u(x_j)\}_{j=0}^N$ in $O(N\log_2 N)$ operations as follows:

- Compute the discrete Chebyshev coefficients $\{\hat{u}_k\}$ from $\{u(x_j)\}$ using (3.222) in $O(N\log_2 N)$ operations.
- Compute the Chebyshev coefficients $\{\hat{u}_k^{(1)}\}$ of u' using (3.234) in $O(N)$ operations.
- Compute $\{u'(x_j)\}$ from $\{\hat{u}_k^{(1)}\}$ using (3.223) (with $\{\hat{u}_k^{(1)}, u'(x_j)\}$ in place of $\{\hat{u}_k, u(x_j)\}$) in $O(N\log_2 N)$ operations.

To summarize, thanks to its relation with Fourier series (cf. (3.210)), the Chebyshev polynomials enjoy several distinct advantages over other Jacobi polynomials:

- The nodes and weights of Gauss-type quadratures are given explicitly, avoiding the potential loss of accuracy at large N when computing them through a numerical procedure.
- The discrete Chebyshev transforms can be carried out using FFT in $O(N\log_2 N)$ operations.
- Thanks to the fast discrete transforms, the derivatives as well as nonlinear terms can also be evaluated in $O(N\log_2 N)$ operations.

However, the fact that the Chebyshev polynomials are mutually orthogonal with respect to a weighted inner product may induce complications in analysis and/or implementations of a Chebyshev spectral method.

3.5 Error Estimates for Polynomial Approximations

The aim of this section is to perform error analysis, in anisotropic Jacobi-weighted Sobolev spaces, for approximating functions by Jacobi polynomials. These results play a fundamental role in analysis of spectral methods for PDEs. More specifically, we shall consider:

- Inverse inequalities for Jacobi polynomials
- Estimates for the best approximation by series of Jacobi polynomials
- Error analysis of Jacobi-Gauss-type polynomial interpolations

Many results presented in this section with estimates in anisotropic Jacobi-weighted Sobolev spaces are mainly based on the papers by Guo and Wang (2001, 2004) (also see Funaro (1992)). Similar estimates in standard Sobolev spaces can be found in the books by Bernardi and Maday (1992a, 1997) and Canuto et al. (2006).

3.5.1 Inverse Inequalities for Jacobi Polynomials

Since all norms of a function in any finite dimensional space are equivalent, we have

$$\|\partial_x \phi\| \le C_N \|\phi\|, \quad \forall \phi \in P_N,$$

which is an example of inverse inequalities. The inverse inequalities are very useful for analyzing spectral approximations of nonlinear problems. In this context, an important issue is to derive the optimal constant C_N. Recall that the notation $A \lesssim B$ means that there exists a generic positive constant c, independent of N and any function, such that $A \le cB$.

The first inverse inequality relates two norms weighted with different Jacobi weight functions.

Theorem 3.31. *For $\alpha, \beta > -1$ and any $\phi \in P_N$, we have*

$$\|\partial_x \phi\|_{\omega^{\alpha+1,\beta+1}} \le \sqrt{\lambda_N^{\alpha,\beta}} \|\phi\|_{\omega^{\alpha,\beta}}, \tag{3.235}$$

and

$$\|\partial_x^m \phi\|_{\omega^{\alpha+m,\beta+m}} \lesssim N^m \|\phi\|_{\omega^{\alpha,\beta}}, \quad m \ge 1, \tag{3.236}$$

where $\lambda_N^{\alpha,\beta} = N(N+\alpha+\beta+1)$.

Proof. For any $\phi \in P_N$, we write

$$\phi(x) = \sum_{n=0}^{N} \hat{\phi}_n^{\alpha,\beta} J_n^{\alpha,\beta}(x) \text{ with } \hat{\phi}_n^{\alpha,\beta} = \frac{1}{\gamma_n^{\alpha,\beta}} \int_{-1}^{1} \phi J_n^{\alpha,\beta} \omega^{\alpha,\beta} dx. \tag{3.237}$$

Hence, by the orthogonality of Jacobi polynomials,

$$\|\phi\|^2_{\omega^{\alpha,\beta}} = \sum_{n=0}^{N} \gamma_n^{\alpha,\beta} |\hat{\phi}_n^{\alpha,\beta}|^2.$$

Differentiating (3.237) and using the orthogonality (3.97), we obtain

$$\begin{aligned} \|\phi'\|^2_{\omega^{\alpha+1,\beta+1}} &= \sum_{n=1}^{N} \lambda_n^{\alpha,\beta} \gamma_n^{\alpha,\beta} |\hat{\phi}_n^{\alpha,\beta}|^2 \\ &\le \lambda_N^{\alpha,\beta} \sum_{n=1}^{N} \gamma_n^{\alpha,\beta} |\hat{\phi}_n^{\alpha,\beta}|^2 \le \lambda_N^{\alpha,\beta} \|\phi\|^2_{\omega^{\alpha,\beta}}, \end{aligned} \tag{3.238}$$

which yields (3.235).

Using the above inequality recursively leads to

$$\|\partial_x^m \phi\|_{\omega^{\alpha+m,\beta+m}} \le \Big(\prod_{k=0}^{m-1} \lambda_{N-k}^{\alpha+k,\beta+k} \Big)^{1/2} \|\phi\|_{\omega^{\alpha,\beta}}. \tag{3.239}$$

Hence, we obtain (3.236) by using (3.91). □

If the polynomial ϕ vanishes at the endpoints $x = \pm 1$, i.e.,

$$\phi \in P_N^0 := \{u \in P_N \,:\, u(\pm 1) = 0\}, \tag{3.240}$$

the following inverse inequality holds.

Theorem 3.32. *For $\alpha, \beta > -1$ and any $\phi \in P_N^0$,*

$$\|\partial_x \phi\|_{\omega^{\alpha,\beta}} \lesssim N \|\phi\|_{\omega^{\alpha-1,\beta-1}}. \tag{3.241}$$

Proof. We refer to Bernardi and Maday (1992b) for the proof of $\alpha = \beta$, and Guo and Wang (2004) for the derivation of the general case. Here, we merely sketch the proof of $\alpha = \beta = 0$. Since $\phi/(1-x^2) \in P_{N-2}$, we write

$$\phi(x)/(1-x^2) = \sum_{n=1}^{N-1} \tilde{\phi}_n L_n'(x).$$

Thus, by (3.174b),

$$\|\phi\|^2_{\omega^{-1,-1}} = \sum_{n=1}^{N-1} n(n+1)\gamma_n |\tilde{\phi}_n|^2,$$

where $\gamma_n = 2/(2n+1)$. In view of (3.171), we have

$$\phi'(x) = \sum_{n=1}^{N-1} \tilde{\phi}_n \big((1-x^2)L_n'(x)\big)' = -\sum_{n=1}^{N-1} n(n+1)\tilde{\phi}_n L_n(x),$$

and by (3.174a),

$$\|\partial_x \phi\|^2 = \sum_{n=1}^{N-1} n^2(n+1)^2 \gamma_n |\tilde{\phi}_n|^2.$$

Thus, we have

$$\|\partial_x \phi\|^2 \le N(N-1)\|\phi\|^2_{\omega^{-1,-1}}. \tag{3.242}$$

This gives (3.241) with $\alpha = \beta = 0$. □

The inverse inequality (3.236) is an algebraic analogy to the trigonometric inverse inequality (2.44), and both of them involve "optimal" constant $C_N = O(N)$. However, the norms in (3.236) are weighted with different weight functions. In most applications, we need to use inverse inequalities involving the same weighted norms. For this purpose, we present an inverse inequality with respect to the Legendre weight function $\omega(x) \equiv 1$ (cf. Canuto and Quarteroni (1982)).

Theorem 3.33. *For any* $\phi \in P_N$,

$$\|\partial_x \phi\| \le \frac{1}{2}(N+1)(N+2)\|\phi\|. \tag{3.243}$$

Proof. Using integration by parts, (3.174a), (3.175) and (3.177a), we obtain

$$\int_{-1}^{1} \left[L_n'(x)\right]^2 dx = L_n(x)L_n'(x)\Big|_{-1}^{1} - \int_{-1}^{1} L_n''(x)L_n(x)dx = n(n+1). \tag{3.244}$$

Hence, by (3.174a),

$$\|L_n'\| = \sqrt{\frac{n(n+1)(2n+1)}{2}}\|L_n\| \le (n+1)^{3/2}\|L_n\|, \quad n \ge 0. \tag{3.245}$$

Next, for any $\phi \in P_N$, we write

$$\phi(x) = \sum_{n=0}^{N} \hat{\phi}_n L_n(x) \text{ with } \hat{\phi}_n = \left(n+\frac{1}{2}\right)\int_{-1}^{1} \phi(x)L_n(x)dx,$$

so we have

$$\|\phi\|^2 = \sum_{n=0}^{N} \frac{2}{2n+1}|\hat{\phi}_n|^2.$$

On the other hand, we obtain from (3.244) and the Cauchy–Schwarz inequality that

$$\begin{aligned}
\|\partial_x \phi\| &\le \sum_{n=0}^{N} |\hat{\phi}_n|\|L_n'\| \le \sum_{n=0}^{N} |\hat{\phi}_n|\sqrt{n(n+1)} \\
&\le \Big(\sum_{n=0}^{N} \frac{2}{2n+1}|\hat{\phi}_n|^2\Big)^{1/2}\Big(\sum_{n=0}^{N} n(n+1)(n+1/2)\Big)^{1/2} \\
&\le \Big(\sum_{n=0}^{N}(n+1)^3\Big)^{1/2}\|\phi\| \le \frac{(N+1)(N+2)}{2}\|\psi\|.
\end{aligned}$$

This ends the proof. □

Remark 3.6. *The factor N^2 in* (3.243) *is sharp in the sense that for any positive integer N, there exists a polynomial $\psi \in P_N$ and a positive constant c independent of N such that*

$$\|\partial_x \psi\| \geq cN^2 \|\psi\|. \tag{3.246}$$

Indeed, taking $\psi(x) = L_N'(x)$, one verifies readily by using integration by parts, (3.174a), (3.177), (3.100), (3.94) *and* (3.107) *that*

$$\begin{aligned}\int_{-1}^{1} \left[L_N''(x)\right]^2 dx &= \left[L_N''(x)L_N'(x) - L_N'''(x)L_N(x)\right]\Big|_{-1}^{1} \\ &= \frac{1}{12}(N-1)N(N+1)(N+2)(N^2+N+3),\end{aligned} \tag{3.247}$$

which, together with (3.244), *implies*

$$\|L_N''\| = \frac{1}{2\sqrt{3}}\sqrt{(N-1)(N+2)(N^2+N+3)}\|L_N'\|.$$

This justifies the claim.

We now consider the extension of (3.243) to the Jacobi polynomials. We observe from the proof of Theorem 3.33 that the use of (3.244) allows for a simple derivation of (3.243). However, the explicit formula for $\int_{-1}^{1}\left(\partial_x J_n^{\alpha,\beta}\right)^2 \omega^{\alpha,\beta} dx$ for general (α,β) is much more involved, although one can derive them by using (3.119)–(3.120) (and (3.216a) for the Chebyshev case). We refer to Guo (1998a) for the following result, and leave the proof of the Chebyshev case as an exercise (see Problem 3.21).

Theorem 3.34. *For $\alpha,\beta > -1$ and any $\phi \in P_N$,*

$$\|\partial_x \phi\|_{\omega^{\alpha,\beta}} \lesssim N^2 \|\phi\|_{\omega^{\alpha,\beta}}.$$

3.5.2 Orthogonal Projections

A common procedure in error analysis is to compare the numerical solution u_N with a suitable orthogonal projection $\pi_N u$ (or interpolation $I_N u$) of the exact solution u in some appropriate Sobolev space with the norm $\|\cdot\|_S$ (cf. Remark 1.7), and use the triangle inequality,

$$\|u - u_N\|_S \leq \|u - \pi_N u\|_S + \|\pi_N u - u_N\|_S.$$

Hence, one needs to estimate the errors $\|u - \pi_N u\|_S$ and $\|I_N u - u\|_S$. Such estimates involving Jacobi polynomials will be the main concern of this section.

Let $I=(-1,1)$, and let $\omega^{\alpha,\beta}(x)=(1-x)^{\alpha}(1+x)^{\beta}$ with $\alpha,\beta>-1$, be the Jacobi weight function as before. For any $u\in L^2_{\omega^{\alpha,\beta}}(I)$, we write

$$u(x)=\sum_{n=0}^{\infty}\hat{u}_n^{\alpha,\beta}J_n^{\alpha,\beta}(x)\quad\text{with}\quad \hat{u}_n^{\alpha,\beta}=\frac{\big(u,J_n^{\alpha,\beta}\big)_{\omega^{\alpha,\beta}}}{\gamma_n^{\alpha,\beta}},\tag{3.248}$$

where $\gamma_n^{\alpha,\beta}=\|J_n^{\alpha,\beta}\|^2_{\omega^{\alpha,\beta}}$.

Define the $L^2_{\omega^{\alpha,\beta}}$-orthogonal projection $\pi_N^{\alpha,\beta}:L^2_{\omega^{\alpha,\beta}}(I)\to P_N$ such that

$$\big(\pi_N^{\alpha,\beta}u-u,\,v\big)_{\omega^{\alpha,\beta}}=0,\quad\forall v\in P_N,\tag{3.249}$$

or equivalently,

$$(\pi_N^{\alpha,\beta}u)(x)=\sum_{n=0}^{N}\hat{u}_n^{\alpha,\beta}J_n^{\alpha,\beta}(x).\tag{3.250}$$

We find from Theorem 3.14 that $\pi_N^{\alpha,\beta}u$ is the best polynomial approximation of u in $L^2_{\omega^{\alpha,\beta}}(I)$.

To measure the truncation error $\pi_N^{\alpha,\beta}u-u$, we introduce the non-uniformly (or anisotropic) Jacobi-weighted Sobolev space:

$$B^m_{\alpha,\beta}(I):=\Big\{u:\partial_x^k u\in L^2_{\omega^{\alpha+k,\beta+k}}(I),\ 0\le k\le m\Big\},\quad m\in\mathbb{N},\tag{3.251}$$

equipped with the inner product, norm and semi-norm

$$\begin{aligned}&\big(u,v\big)_{B^m_{\alpha,\beta}}=\sum_{k=0}^{m}\big(\partial_x^k u,\partial_x^k v\big)_{\omega^{\alpha+k,\beta+k}},\\&\|u\|_{B^m_{\alpha,\beta}}=\big(u,u\big)^{1/2}_{B^m_{\alpha,\beta}},\quad |u|_{B^m_{\alpha,\beta}}=\|\partial_x^m u\|_{\omega^{\alpha+m,\beta+m}}.\end{aligned}\tag{3.252}$$

The space $B^m_{\alpha,\beta}(I)$ distinguishes itself from the usual weighted Sobolev space $H^m_{\omega^{\alpha,\beta}}(I)$ (cf. Appendix B) by involving different weight functions for derivatives of different orders. It is obvious that $H^m_{\omega^{\alpha,\beta}}(I)$ is a subspace of $B^m_{\alpha,\beta}(I)$, that is, for any $m\ge 0$ and $\alpha,\beta>-1$,

$$\|u\|_{B^m_{\alpha,\beta}}\le c\|u\|_{H^m_{\omega^{\alpha,\beta}}}.$$

Before presenting the main result, we first derive from (3.101) to (3.102) and the orthogonality (3.109) that

$$\int_{-1}^{1}\partial_x^k J_n^{\alpha,\beta}(x)\partial_x^k J_l^{\alpha,\beta}(x)\omega^{\alpha+k,\beta+k}(x)dx=h_{n,k}^{\alpha,\beta}\delta_{nl},\tag{3.253}$$

where for $n \geq k$,

$$\begin{aligned} h_{n,k}^{\alpha,\beta} &= (d_{n,k}^{\alpha,\beta})^2 \gamma_{n-k}^{\alpha+k,\beta+k} \\ &= \frac{2^{\alpha+\beta+1}\Gamma(n+\alpha+1)\Gamma(n+\beta+1)\Gamma(n+k+\alpha+\beta+1)}{(2n+\alpha+\beta+1)(n-k)!\Gamma^2(n+\alpha+\beta+1)}. \end{aligned} \tag{3.254}$$

Summing (3.253) for all $0 \leq k \leq m$, we find that the Jacobi polynomials are orthogonal in the Sobolev space $B^m_{\alpha,\beta}(I)$, namely,

$$\left(J_n^{\alpha,\beta}, J_l^{\alpha,\beta}\right)_{B^m_{\alpha,\beta}} = 0, \quad \text{if} \quad n \neq l. \tag{3.255}$$

Now, we are ready to state the first fundamental result.

Theorem 3.35. *Let $\alpha, \beta > -1$. For any $u \in B^m_{\alpha,\beta}(I)$,*

- *if $0 \leq l \leq m \leq N+1$, we have*

$$\begin{aligned} &\|\partial_x^l(\pi_N^{\alpha,\beta} u - u)\|_{\omega^{\alpha+l,\beta+l}} \\ &\qquad \leq c\sqrt{\frac{(N-m+1)!}{(N-l+1)!}}(N+m)^{(l-m)/2}\|\partial_x^m u\|_{\omega^{\alpha+m,\beta+m}}, \end{aligned} \tag{3.256}$$

- *if $m > N+1$, we have*

$$\begin{aligned} &\|\partial_x^l(\pi_N^{\alpha,\beta} u - u)\|_{\omega^{\alpha+l,\beta+l}} \\ &\qquad \leq c(2\pi N)^{-1/4}\Big(\frac{\sqrt{e/2}}{N}\Big)^{N-l+1}\|\partial_x^{N+1} u\|_{\omega^{\alpha+N+1,\beta+N+1}}, \end{aligned} \tag{3.257}$$

where $c \approx 1$ for $N \gg 1$.

Proof. Denote $\tilde{m} = \min\{m, N+1\}$. Thanks to the orthogonality (3.253)–(3.254),

$$\|\partial_x^k u\|^2_{\omega^{\alpha+k,\beta+k}} = \sum_{n=k}^{\infty} h_{n,k}^{\alpha,\beta}|\hat{u}_n^{\alpha,\beta}|^2, \quad k \geq 0, \tag{3.258}$$

so we have

$$\begin{aligned} \|\partial_x^l(\pi_N^{\alpha,\beta} u - u)\|^2_{\omega^{\alpha+l,\beta+l}} &= \sum_{n=N+1}^{\infty} h_{n,l}^{\alpha,\beta}|\hat{u}_n^{\alpha,\beta}|^2 \\ &\leq \max_{n \geq N+1}\left\{\frac{h_{n,l}^{\alpha,\beta}}{h_{n,\tilde{m}}^{\alpha,\beta}}\right\}\sum_{n=N+1}^{\infty} h_{n,\tilde{m}}^{\alpha,\beta}|\hat{u}_n^{\alpha,\beta}|^2 \\ &\leq \frac{h_{N+1,l}^{\alpha,\beta}}{h_{N+1,\tilde{m}}^{\alpha,\beta}}\|\partial_x^{\tilde{m}} u\|^2_{\omega^{\alpha+\tilde{m},\beta+\tilde{m}}}. \end{aligned} \tag{3.259}$$

By (3.254),

$$\frac{h_{N+1,l}^{\alpha,\beta}}{h_{N+1,\tilde{m}}^{\alpha,\beta}} = \frac{\Gamma(N+l+\alpha+\beta+2)(N-\tilde{m}+1)!}{\Gamma(N+\tilde{m}+\alpha+\beta+2)(N-l+1)!}. \tag{3.260}$$

Using the Stirling's formula (A.7) yields

$$\frac{\Gamma(N+l+\alpha+\beta+2)}{\Gamma(N+\tilde{m}+\alpha+\beta+2)} \cong \frac{1}{(N+\tilde{m}+\alpha+\beta+2)^{\tilde{m}-l}} \cong (N+\tilde{m})^{l-\tilde{m}}. \tag{3.261}$$

Correspondingly,

$$\frac{h_{N+1,l}^{\alpha,\beta}}{h_{N+1,\tilde{m}}^{\alpha,\beta}} \le c^2 \frac{(N-\tilde{m}+1)!}{(N-l+1)!}(N+\tilde{m})^{l-\tilde{m}}, \tag{3.262}$$

where $c \approx 1$. A combination of the above estimates leads to

$$\|\partial_x^l(\pi_N^{\alpha,\beta}u-u)\|_{\omega^{\alpha+l,\beta+l}}^2 \le c^2 \frac{(N-\tilde{m}+1)!}{(N-l+1)!}(N+\tilde{m})^{l-\tilde{m}} \|\partial_x^{\tilde{m}}u\|_{\omega^{\alpha+\tilde{m},\beta+\tilde{m}}}^2. \tag{3.263}$$

Finally, if $0 \le l \le m \le N+1$, then $\tilde{m} = m$, so (3.256) follows. On the other hand, if $m > N+1$, then $\tilde{m} = N+1$, and the estimate (3.257) follows from (3.263) and Stirling's formula (A.8). □

Remark 3.7. *In contrast with error estimates for finite elements or finite differences, the convergence rate of spectral approximations is only limited by the regularity of the underlying function. Therefore, we made a special effort to characterize the explicit dependence of the errors on the regularity index m. For any fixed m, the estimate* (3.256) *becomes*

$$\|\partial_x^l(\pi_N^{\alpha,\beta}u-u)\|_{\omega^{\alpha+l,\beta+l}} \lesssim N^{l-m}\|\partial_x^m u\|_{\omega^{\alpha+m,\beta+m}}, \tag{3.264}$$

which is the typical convergence rate found in the literature.

Hereafter, the factor

$$\sqrt{\frac{(N-m+1)!}{N!}}, \quad 0 \le m \le N+1,$$

frequently appears in the characterization of the approximation errors. For a quick reference,

$$\begin{aligned} N^{(1-m)/2} \le \sqrt{\frac{(N-m+1)!}{N!}} &= \frac{1}{\sqrt{N(N-1)\dots(N-(m-2))}} \\ &\le (N-m+2)^{(1-m)/2}, \end{aligned} \tag{3.265}$$

so for $m = o(N)$ (in particular, for fixed m), we have

$$\sqrt{\frac{(N-m+1)!}{N!}} \cong N^{(1-m)/2}. \tag{3.266}$$

Some other remarks are also in order.

- Theorem 3.35 indicates that the truncated Jacobi series $\pi_N^{\alpha,\beta} u$ is the best polynomial approximation of u in both $L^2_{\omega^{\alpha,\beta}}(I)$ and the anisotropic Jacobi-weighted Sobolev space $B^l_{\alpha,\beta}(I)$.
- It must be pointed out that the truncation error $\pi_N^{\alpha,\beta} u - u$ measured in the usual weighted Sobolev space $H^l_{\omega^{\alpha,\beta}}(I)$ (with $l \geq 1$) does not have an optimal order of convergence. Indeed, one can always find a function such that its truncated Jacobi series converges in $L^2_{\omega^{\alpha,\beta}}(I)$, but diverges in $H^1_{\omega^{\alpha,\beta}}(I)$. For instance, we take $u = L_{N+1} - L_{N-1}$, and notice that $\pi_N^{0,0} u = -L_{N-1}$ and $\partial_x u = (2N+1)L_N$. It is clear that

$$\|\partial_x(\pi_N^{0,0} u - u)\| = \|L'_{N+1}\| \overset{(3.244)}{=} \sqrt{(N+1)(N+2)} \geq \frac{\sqrt{N}}{2} \|\partial_x u\|.$$

In general, we have the following estimates: for $\alpha > -1$ and $0 \leq l \leq m$,

$$\left\|\pi_N^{\alpha,\alpha} u - u\right\|_{l,\omega^{\alpha,\alpha}} \lesssim N^{2l-m-1/2} \|\partial_x^m u\|_{\omega^{\alpha+m,\alpha+m}}. \tag{3.267}$$

This estimate for the Legendre and Chebyshev cases was derived in Canuto and Quarteroni (1982), and in Guo (2000) for the general case with $\alpha, \beta > -1$.

Since $H^l_{\omega^{\alpha,\beta}}(I)$ is a Hilbert space, the best approximation polynomial for u is the orthogonal projection of u upon P_N under the inner product

$$\left(u,v\right)_{l,\omega^{\alpha,\beta}} = \sum_{k=0}^{l} \left(\partial_x^k u, \partial_x^k v\right)_{\omega^{\alpha,\beta}}, \tag{3.268}$$

which induces the norm $\|\cdot\|_{l,\omega^{\alpha,\beta}}$ of $H^l_{\omega^{\alpha,\beta}}(I)$. In fact, this type of approximation results are often needed in analysis of spectral methods for second-order elliptic PDEs. Therefore, we consider below the $H^1_{\omega^{\alpha,\beta}}$-orthogonal projection. Denote the inner product in $H^1_{\omega^{\alpha,\beta}}(I)$ by

$$a_{\alpha,\beta}(u,v) := \left(u',v'\right)_{\omega^{\alpha,\beta}} + \left(u,v\right)_{\omega^{\alpha,\beta}}, \quad \forall u,v \in H^1_{\omega^{\alpha,\beta}}(I),$$

and define the orthogonal projection $\pi^1_{N,\alpha,\beta} : H^1_{\omega^{\alpha,\beta}}(I) \to P_N$ by

$$a_{\alpha,\beta}(\pi^1_{N,\alpha,\beta} u - u, v) = 0, \quad \forall v \in P_N. \tag{3.269}$$

By definition, $\pi^1_{N,\alpha,\beta}u$ is the best approximation of u in the sense that

$$\|\pi^1_{N,\alpha,\beta}u - u\|_{1,\omega^{\alpha,\beta}} = \inf_{\phi\in P_N} \|\phi - u\|_{1,\omega^{\alpha,\beta}}. \tag{3.270}$$

By using the fundamental Theorem 3.35, we can derive the following estimate.

Theorem 3.36. *Let $\alpha,\beta > -1$. If $\partial_x u \in B^{m-1}_{\alpha,\beta}(I)$, then for $1 \le m \le N+1$,*

$$\|\pi^1_{N,\alpha,\beta}u - u\|_{1,\omega^{\alpha,\beta}} \le c\sqrt{\frac{(N-m+1)!}{N!}}(N+m)^{(1-m)/2}\|\partial_x^m u\|_{\omega^{\alpha+m-1,\beta+m-1}}, \tag{3.271}$$

where c is a positive constant independent of m, N and u.

Proof. Let $\pi^{\alpha,\beta}_{N-1}$ be the $L^2_{\omega^{\alpha,\beta}}$-orthogonal projection upon P_{N-1} as defined in (3.249). Set

$$\phi(x) = \int_{-1}^{x} \pi^{\alpha,\beta}_{N-1}u'(y)dy + \xi, \tag{3.272}$$

where the constant ξ is chosen such that $\phi(0) = u(0)$. In view of (3.270), we derive from the inequality (B.43) and Theorem 3.35 that

$$\begin{aligned}\|\pi^1_{N,\alpha,\beta}u - u\|_{1,\omega^{\alpha,\beta}} &\le \|\phi - u\|_{1,\omega^{\alpha,\beta}} \le c\|(\phi - u)'\|_{\omega^{\alpha,\beta}}\\ &\le c\|\pi^{\alpha,\beta}_{N-1}u' - u'\|_{\omega^{\alpha,\beta}} \le c\sqrt{\frac{(N-m+1)!}{N!}}(N+m)^{(1-m)/2}\|\partial_x^m u\|_{\omega^{\alpha+m-1,\beta+m-1}}.\end{aligned}$$

This completes the proof. □

While the estimate (3.271) is optimal in the $H^1_{\omega^{\alpha,\beta}}$-norm, it does not imply an optimal order in the $L^2_{\omega^{\alpha,\beta}}$-norm. An optimal estimate in the $L^2_{\omega^{\alpha,\beta}}$-norm can be obtained by using a duality argument, which is also known as the Aubin-Nitsche technique (see, e.g., Ciarlet (1978)).

The first step is to show the regularity of the solution for an auxiliary problem.

Lemma 3.4. *Let $\alpha,\beta > -1$. For each $g \in L^2_{\omega^{\alpha,\beta}}(I)$, there exists a unique $\psi \in H^1_{\omega^{\alpha,\beta}}(I)$ such that*

$$a_{\alpha,\beta}(\psi,v) = (g,v)_{\omega^{\alpha,\beta}}, \quad \forall v \in H^1_{\omega^{\alpha,\beta}}(I). \tag{3.273}$$

Moreover, the solution $\psi \in H^2_{\omega^{\alpha,\beta}}(I)$ and satisfies

$$\|\psi\|_{2,\omega^{\alpha,\beta}} \lesssim \|g\|_{\omega^{\alpha,\beta}}. \tag{3.274}$$

Proof. The bilinear form $a_{\alpha,\beta}(\cdot,\cdot)$ is the inner product of the Hilbert space $H^1_{\omega^{\alpha,\beta}}(I)$, so the existence and uniqueness of the solution ψ of (3.273) follows from the Riesz representation theorem (see Appendix B).Taking $v = \psi$ in (3.273) and using the Cauchy–Schwarz inequality leads to

$$\|\psi\|_{1,\omega^{\alpha,\beta}} \lesssim \|g\|_{\omega^{\alpha,\beta}}. \tag{3.275}$$

By taking $v \in \mathscr{D}(I)$ in (3.273) (where $\mathscr{D}(I)$ is the set of all infinitely differentiable functions with compact support in I, see Appendix B) and integrating by parts, we find that, in the sense of distributions,

$$-(\psi'\omega^{\alpha,\beta})' = (g-\psi)\omega^{\alpha,\beta}. \tag{3.276}$$

Next, we show that $\psi'\omega^{\alpha,\beta}$ is continuous on $[-1,1]$ with $(\psi'\omega^{\alpha,\beta})(\pm 1) = 0$. Indeed, integrating (3.276) over any interval $(x_1,x_2) \subseteq [-1,1]$, we obtain from the Cauchy–Schwarz inequality and (3.275) that

$$\begin{aligned} \big|(\psi'\omega^{\alpha,\beta})(x_1) - (\psi'\omega^{\alpha,\beta})(x_2)\big| &\le \int_{x_1}^{x_2} |(g-\psi)\omega^{\alpha,\beta}|dx \\ \le \Big(\int_{x_1}^{x_2} \omega^{\alpha,\beta}(x)dx\Big)^{1/2} &\|g-\psi\|_{\omega^{\alpha,\beta}} \lesssim \Big(\int_{x_1}^{x_2} \omega^{\alpha,\beta}(x)dx\Big)^{1/2} \|g\|_{\omega^{\alpha,\beta}}. \end{aligned}$$

Hence, $\psi'\omega^{\alpha,\beta} \in C[-1,1]$ and $(\psi'\omega^{\alpha,\beta})(\pm 1)$ are well-defined. Multiplying (3.276) by any function $v \in H^1_{\omega^{\alpha,\beta}}(I)$ and integrating the resulting equality by parts, we derive from (3.273) that

$$[\psi'\omega^{\alpha,\beta}v]\Big|_{-1}^{1} = a_{\alpha,\beta}(\psi,v) - (g,v)_{\omega^{\alpha,\beta}} = 0, \quad \forall v \in H^1_{\omega^{\alpha,\beta}}(I).$$

Hence, $(\psi'\omega^{\alpha,\beta})(\pm 1) = 0$.

We are now ready to prove (3.274). A direct computation from (3.276) leads to

$$-\psi'' = -\big((\alpha+\beta)x + (\alpha-\beta)\big)(1-x^2)^{-1}\psi' + (g-\psi). \tag{3.277}$$

One verifies readily that

$$\|\psi''\|^2_{\omega^{\alpha,\beta}} \le D_1 + D_2, \tag{3.278}$$

where $D_1 = D_1(I_1) + D_1(I_2)$ with $I_1 = (-1,0)$ and $I_2 = (0,1)$, and

$$\begin{aligned} D_1(I_j) &= 8(\alpha^2+\beta^2)\int_{I_j} |\psi'|^2\omega^{\alpha-2,\beta-2}dx, \quad j = 1,2, \\ D_2 &= 2\left|\int_{-1}^{1} (g-\psi)^2\omega^{\alpha,\beta}dx\right|. \end{aligned}$$

By (3.275),

$$D_2 \lesssim \|g-\psi\|^2_{\omega^{\alpha,\beta}} \lesssim \|g\|^2_{\omega^{\alpha,\beta}}.$$

Thus, it remains to estimate D_1. Due to $(\psi'\omega^{\alpha,\beta})(1) = 0$, integrating (3.276) over $(x,1)$ yields

$$\psi' = (1-x)^{-\alpha}(1+x)^{-\beta}\int_x^1 (g-\psi)\omega^{\alpha,\beta}dy.$$

Plugging it into $D_1(I_2)$ gives

$$\begin{aligned} D_1(I_2) &\lesssim \int_0^1 (1-x)^{-\alpha-2}(1+x)^{-\beta-2}\Big[\int_x^1 (g-\psi)\omega^{\alpha,\beta}dy\Big]^2 dx \\ &\lesssim \int_0^1 (1-x)^{-\alpha-2}\Big[\int_x^1 (g-\psi)\omega^{\alpha,\beta}dy\Big]^2 dx \\ &\lesssim \int_0^1 (1-x)^{-\alpha}\Big[\frac{1}{1-x}\int_x^1 (g-\psi)\omega^{\alpha,\beta}dy\Big]^2 dx. \end{aligned}$$

Since $-\alpha < 1$, using the Hardy inequality (B.39) leads to

$$D_1(I_2) \lesssim \int_0^1 (g-\psi)^2\omega^{\alpha,2\beta}dx \lesssim \int_0^1 (g-\psi)^2\omega^{\alpha,\beta}dx.$$

A similar inequality holds for $D_1(I_1)$. Therefore, a combination of the above estimates leads to

$$\|\psi''\|_{\omega^{\alpha,\beta}} \lesssim \|g\|_{\omega^{\alpha,\beta}},$$

which, together with (3.275), implies (3.274). □

We are now in a position to derive the optimal estimate in $L^2_{\omega^{\alpha,\beta}}$-norm via the duality argument.

Theorem 3.37. *Let $\alpha,\beta > -1$. If $\partial_x u \in B^{m-1}_{\alpha,\beta}(I)$, then for $1 \le m \le N+1$,*

$$\begin{aligned} &\|\pi^1_{N,\alpha,\beta}u - u\|_{\omega^{\alpha,\beta}} \\ &\quad \le c\sqrt{\frac{(N-m+1)!}{N!}}(N+m)^{-(m+1)/2}\|\partial_x^m u\|_{\omega^{\alpha+m-1,\beta+m-1}}, \end{aligned} \tag{3.279}$$

where c is a positive constant independent of m, N and u.

Proof. We have

$$\|\pi^1_{N,\alpha,\beta}u - u\|_{\omega^{\alpha,\beta}} = \sup_{0\neq g\in L^2_{\omega^{\alpha,\beta}}(I)} \frac{|(\pi^1_{N,\alpha,\beta}u - u, g)_{\omega^{\alpha,\beta}}|}{\|g\|_{\omega^{\alpha,\beta}}}. \tag{3.280}$$

Let ψ be the solution to the auxiliary problem (3.273) for given $g \in L^2_{\omega^{\alpha,\beta}}(I)$. Taking $v = \pi^1_{N,\alpha,\beta}u - u$ in (3.273), we obtain from (3.269) that

$$\begin{aligned} (\pi^1_{N,\alpha,\beta}u - u, g)_{\omega^{\alpha,\beta}} &= a_{\alpha,\beta}(\pi^1_{N,\alpha,\beta}u - u, \psi) \\ &= a_{\alpha,\beta}(\pi^1_{N,\alpha,\beta}u - u, \psi - \pi^1_{N,\alpha,\beta}\psi). \end{aligned}$$

Hence, by the Cauchy–Schwarz inequality, Theorem 3.36 and the regularity estimate (3.274), we have

$$\begin{aligned}|(\pi^1_{N,\alpha,\beta}u-u,g)_{\omega^{\alpha,\beta}}| &\le \|\pi^1_{N,\alpha,\beta}u-u\|_{1,\omega^{\alpha,\beta}}\|\pi^1_{N,\alpha,\beta}\psi-\psi\|_{1,\omega^{\alpha,\beta}}\\ &\le cN^{-1}\|\pi^1_{N,\alpha,\beta}u-u\|_{1,\omega^{\alpha,\beta}}\|\psi''\|_{\omega^{\alpha+1,\beta+1}}\\ &\le cN^{-1}\|\pi^1_{N,\alpha,\beta}u-u\|_{1,\omega^{\alpha,\beta}}\|g\|_{\omega^{\alpha,\beta}}.\end{aligned}$$

Consequently, by (3.280),

$$\|\pi^1_{N,\alpha,\beta}u-u\|_{\omega^{\alpha,\beta}} \le cN^{-1}\|\pi^1_{N,\alpha,\beta}u-u\|_{1,\omega^{\alpha,\beta}}.$$

Finally, the desired result follows from Theorem 3.36. □

The approximation results in the Sobolev norms are of great importance for spectral approximation of boundary value problems. Oftentimes, it is necessary to take the boundary conditions into account and consider the projection operators onto the space of polynomials built in homogeneous boundary data.

To this end, we assume that $-1<\alpha,\beta<1$, and denote

$$H^1_{0,\omega^{\alpha,\beta}}(I)=\{u\in H^1_{\omega^{\alpha,\beta}}(I): u(\pm1)=0\},\quad P^0_N=\{u\in P_N:\ u(\pm1)=0\}.$$

If $-1<\alpha,\beta<1$, then any function in $H^1_{\omega^{\alpha,\beta}}(I)$ is continuous on $[-1,1]$, and there holds

$$\max_{|x|\le1}|u(x)|\lesssim\|u\|_{1,\omega^{\alpha,\beta}},\quad \forall u\in H^1_{\omega^{\alpha,\beta}}(I). \tag{3.281}$$

We leave the proof of this statement as an exercise (see Problem 3.22). Define

$$\hat{a}_{\alpha,\beta}(u,v)=\int_{-1}^{1}u'(x)v'(x)\omega^{\alpha,\beta}(x)dx,$$

which is the inner product of $H^1_{0,\omega^{\alpha,\beta}}(I)$, and induces the semi-norm, equivalent to the norm of $H^1_{0,\omega^{\alpha,\beta}}(I)$ (see Lemma B.7).

Consider the orthogonal projection $\hat{\pi}^{1,0}_{N,\alpha,\beta}: H^1_{0,\omega^{\alpha,\beta}}(I)\to P^0_N$, defined by

$$\hat{a}_{\alpha,\beta}(\hat{\pi}^{1,0}_{N,\alpha,\beta}u-u,v)=0,\quad \forall v\in P^0_N. \tag{3.282}$$

The basic approximation result is stated as follows.

Theorem 3.38. *Let $-1<\alpha,\beta<1$. If $u\in H^1_{0,\omega^{\alpha,\beta}}(I)$ and $\partial_x u\in B^{m-1}_{\alpha,\beta}(I)$, then for $1\le m\le N+1$,*

$$\begin{aligned}&\|\hat{\pi}^{1,0}_{N,\alpha,\beta}u-u\|_{1,\omega^{\alpha,\beta}}\\ &\quad\le c\sqrt{\frac{(N-m+1)!}{N!}}(N+m)^{(1-m)/2}\|\partial_x^m u\|_{\omega^{\alpha+m-1,\beta+m-1}},\end{aligned} \tag{3.283}$$

where c is a positive constant independent of m,N and u.

Proof. Let $\pi_{N-1}^{\alpha,\beta}u$ be the $L^2_{\omega^{\alpha,\beta}}$- orthogonal projection as defined in (3.249). Setting

$$\phi(x) = \int_{-1}^{x}\left\{\pi_{N-1}^{\alpha,\beta}u' - \frac{1}{2}\int_{-1}^{1}\pi_{N-1}^{\alpha,\beta}u'd\eta\right\}d\xi, \tag{3.284}$$

we have $\phi \in P_N^0$, and

$$\phi' = \pi_{N-1}^{\alpha,\beta}u' - \frac{1}{2}\int_{-1}^{1}\pi_{N-1}^{\alpha,\beta}u'd\eta.$$

Hence, by the triangle inequality,

$$\begin{aligned}\|u'-\phi'\|_{\omega^{\alpha,\beta}} &\le \|u'-\pi_{N-1}^{\alpha,\beta}u'\|_{\omega^{\alpha,\beta}} + \frac{1}{2}\left\|\int_{-1}^{1}\pi_{N-1}^{\alpha,\beta}u'd\eta\right\|_{\omega^{\alpha,\beta}}\\ &\le \|u'-\pi_{N-1}^{\alpha,\beta}u'\|_{\omega^{\alpha,\beta}} + \frac{\sqrt{\gamma_0^{\alpha,\beta}}}{2}\left|\int_{-1}^{1}\pi_{N-1}^{\alpha,\beta}u'd\eta\right|,\end{aligned} \tag{3.285}$$

where $\gamma_0^{\alpha,\beta}$ is given in (3.109). Due to $u(\pm 1) = 0$, we derive from the Cauchy–Schwarz inequality that for $-1 < \alpha, \beta < 1$,

$$\left|\int_{-1}^{1}\pi_{N-1}^{\alpha,\beta}u'dx\right| = \left|\int_{-1}^{1}(\pi_{N-1}^{\alpha,\beta}u'-u')dx\right| \le \sqrt{\gamma_0^{-\alpha,-\beta}}\|\pi_{N-1}^{\alpha,\beta}u'-u'\|_{\omega^{\alpha,\beta}}. \tag{3.286}$$

Hence, by definition and Theorem 3.35,

$$\begin{aligned}\|(\hat{\pi}_{N,\alpha,\beta}^{1,0}u-u)'\|_{\omega^{\alpha,\beta}} &\le \|\phi'-u'\|_{\omega^{\alpha,\beta}} \le c\|\pi_{N-1}^{\alpha,\beta}u'-u'\|_{\omega^{\alpha,\beta}}\\ &\le c\sqrt{\frac{(N-m+1)!}{N!}}(N+m)^{(1-m)/2}\|\partial_x^m u\|_{\omega^{\alpha+m-1,\beta+m-1}}.\end{aligned} \tag{3.287}$$

Finally, using the Poincaré inequality (B.41) and (3.287) leads to

$$\begin{aligned}\|\hat{\pi}_{N,\alpha,\beta}^{1,0}u-u\|_{\omega^{\alpha,\beta}} &\le c\|(\hat{\pi}_{N,\alpha,\beta}^{1,0}u-u)'\|_{\omega^{\alpha,\beta}}\\ &\le c\sqrt{\frac{(N-m+1)!}{N!}}(N+m)^{(1-m)/2}\|\partial_x^m u\|_{\omega^{\alpha+m-1,\beta+m-1}}.\end{aligned} \tag{3.288}$$

This completes the proof. □

As in the proof of Theorem 3.37, we can derive an optimal estimate for $\hat{\pi}_{N,\alpha,\beta}^{1,0}u - u$ in the $L^2_{\omega^{\alpha,\beta}}$-norm by using a duality argument. One may refer to Canuto et al. (2006) for the Legendre and Chebyshev cases, and to Guo and Wang (2004) for the general cases. Moreover, we shall introduce in Chap. 5 a family of generalized Jacobi polynomials, and a concise analysis based on this notion will automatically lead to the desired results.

When we apply the Jacobi approximation (e.g., the Chebyshev approximation) to boundary-value problems, it is often required to use the projection operator associated with the bilinear form

$$a_{\alpha,\beta}(u,v) = \int_{-1}^{1} \partial_x u(x) \partial_x \big(v(x)\omega^{\alpha,\beta}(x)\big)dx, \tag{3.289}$$

which is closely related to the weighted Galerkin formulation for the model equation

$$-u''(x) + \mu u(x) = f(x), \quad \mu \geq 0; \quad u(\pm 1) = 0.$$

In contrast with (3.282), we define the orthogonal projection $\pi_{N,\alpha,\beta}^{1,0} : H_{0,\omega^{\alpha,\beta}}^{1}(I) \to P_N^0$, such that

$$a_{\alpha,\beta}(u - \pi_{N,\alpha,\beta}^{1,0} u, v) = 0, \quad \forall v \in P_N^0. \tag{3.290}$$

The bilinear form is continuous and coercive as stated in the following lemma.

Lemma 3.5. *If $-1 < \alpha, \beta < 1$, then for any $u, v \in H_{0,\omega^{\alpha,\beta}}^{1}(I)$,*

$$|a_{\alpha,\beta}(u,v)| \leq C_1 |u|_{1,\omega^{\alpha,\beta}} |v|_{1,\omega^{\alpha,\beta}}, \tag{3.291}$$

and

$$a_{\alpha,\beta}(v,v) \geq C_2 |v|_{1,\omega^{\alpha,\beta}}^2, \tag{3.292}$$

where C_1 and C_2 are two positive constants independent of u and v.

Proof. Since $-1 < \alpha, \beta < 1$, we have from (B.40) that

$$\begin{aligned} |a_{\alpha,\beta}(u,v)| &\leq |(u',v')_{\omega^{\alpha,\beta}} + (u', v(\omega^{\alpha,\beta})')| \\ &\leq |u|_{1,\omega^{\alpha,\beta}} |v|_{1,\omega^{\alpha,\beta}} + 2|u|_{1,\omega^{\alpha,\beta}} \|v\|_{\omega^{\alpha-2,\beta-2}} \\ &\leq C_1 |u|_{1,\omega^{\alpha,\beta}} |v|_{1,\omega^{\alpha,\beta}}. \end{aligned}$$

We now prove the coercivity. A direct calculation gives

$$a_{\alpha,\beta}(v,v) = |v|_{1,\omega^{\alpha,\beta}}^2 + \frac{1}{2}\big(v^2, W_{\alpha,\beta}\big)_{\omega^{\alpha-2,\beta-2}},$$

where

$$\begin{aligned} W_{\alpha,\beta}(x) &= (\alpha+\beta)(1-\alpha-\beta)x^2 \\ &\quad + 2(\alpha-\beta)(1-\alpha-\beta)x + \alpha + \beta - (\alpha-\beta)^2. \end{aligned}$$

By the property of quadratic polynomials, one verifies readily that $W_{\alpha,\beta}(x) \geq 0$, provided that

$$\begin{cases} (\alpha+\beta)(\alpha+\beta-1) \geq 0, \\ W_{\alpha,\beta}(-1) = -4\beta(\beta-1) \geq 0, \\ W_{\alpha,\beta}(1) = -4\alpha(\alpha-1) \geq 0, \end{cases}$$

or

$$\begin{cases} (\alpha+\beta)(\alpha+\beta-1) \le 0, \\ 4(\alpha-\beta)^2(\alpha+\beta-1)^2+4(\alpha+\beta)(\alpha+\beta-1)(\alpha+\beta-(\alpha-\beta)^2) \le 0. \end{cases}$$

If $0 \le \alpha,\beta \le 1$, then both of them are valid, which implies (3.292) with $0 \le \alpha$, $\beta \le 1$.

Next, let $-1 < \alpha,\beta < 0$ and $u(x) = \omega^{\alpha,\beta}(x)v(x)$. As $0 < -\alpha,-\beta < 1$, it follows from the above shown case that

$$a_{\alpha,\beta}(v,v) = a_{-\alpha,-\beta}(u,u) \ge |u|^2_{1,\omega^{-\alpha,-\beta}}. \tag{3.293}$$

On the other hand, by (B.40),

$$|v|^2_{1,\omega^{\alpha,\beta}} \le 2|u|^2_{1,\omega^{-\alpha,-\beta}} + 8(\alpha^2+\beta^2)\|u\|^2_{\omega^{-\alpha-2,-\beta-2}} \le c|u|^2_{1,\omega^{-\alpha,-\beta}}. \tag{3.294}$$

A combination of (3.293) and (3.294) leads to (3.292) with $-1 < \alpha,\beta < 0$.

Now, let $-1 < \alpha \le 0 \le \beta < 1$ and $u(x) = (1-x)^{\alpha}v(x)$. We deduce from Corollary B.1 that $u \in H^1_{0,\omega^{-\alpha,0}}(I)$, so by (B.40),

$$\begin{aligned} |v|^2_{1,\omega^{\alpha,\beta}} &= |(1-x)^{-\alpha}u|^2_{1,\omega^{\alpha,\beta}} \le 2|u|^2_{1,\omega^{-\alpha,\beta}} + 2\alpha^2\|u\|^2_{\omega^{-\alpha-2,\beta}} \\ &\le 2|u|^2_{1,\omega^{-\alpha,\beta}} + 8\alpha^2\|u\|^2_{\omega^{-\alpha-2,\beta-2}} \le c|u|^2_{1,\omega^{-\alpha,\beta}}. \end{aligned}$$

In view of $-1 < \alpha \le 0 \le \beta < 1$, we have

$$\begin{aligned} |u|^2_{1,\omega^{-\alpha,\beta}} &\le |u|^2_{1,\omega^{-\alpha,\beta}} - 2\alpha(\alpha+1)\|u\|^2_{\omega^{-\alpha-2,\beta}} + 2\beta(1-\beta)\|u\|^2_{\omega^{-\alpha,\beta-2}} \\ &= \big(\partial_x((1-x)^{-\alpha}u), \partial_x((1+x)^{\beta}u)\big) = a_{\alpha,\beta}(v,v). \end{aligned}$$

This leads to (3.292) with $-1 < \alpha \le 0 \le \beta < 1$.

We can treat the remaining case $-1 < \beta \le 0 \le \alpha < 1$ in the same fashion as above. □

Theorem 3.39. *Let $-1 < \alpha,\beta < 1$. If $u \in H^1_{0,\omega^{\alpha,\beta}}(I)$ and $\partial_x u \in B^{m-1}_{\alpha,\beta}(I)$, then for $1 \le m \le N+1$ and $\mu = 0,1$,*

$$\begin{aligned} &\|u - \pi^{1,0}_{N,\alpha,\beta}u\|_{\mu,\omega^{\alpha,\beta}} \\ &\le c\sqrt{\frac{(N-m+1)!}{N!}}(N+m)^{\mu-(m+1)/2}\|\partial_x^m u\|_{\omega^{\alpha+m-1,\beta+m-1}}, \end{aligned} \tag{3.295}$$

where c is a positive constant independent of m, N and u.

Proof. We first prove the case $\mu = 1$. Let $\hat{\pi}^{1,0}_{N,\alpha,\beta}$ be the projection operator defined in (3.282). By the definition (3.290),

$$a_{\alpha,\beta}(\pi^{1,0}_{N,\alpha,\beta}u - u, \pi^{1,0}_{N,\alpha,\beta}u - u) = a_{\alpha,\beta}(\pi^{1,0}_{N,\alpha,\beta}u - u, \hat{\pi}^{1,0}_{N,\alpha,\beta}u - u),$$

which, together with Lemma 3.5, gives

$$\begin{aligned}|\pi^{1,0}_{N,\alpha,\beta}u-u|^2_{1,\omega^{\alpha,\beta}} &\le c|a_{\alpha,\beta}(\pi^{1,0}_{N,\alpha,\beta}u-u,\hat{\pi}^{1,0}_{N,\alpha,\beta}u-u)|\\ &\le c|\pi^{1,0}_{N,\alpha,\beta}u-u|_{1,\omega^{\alpha,\beta}}|\hat{\pi}^{1,0}_{N,\alpha,\beta}u-u|_{1,\omega^{\alpha,\beta}}.\end{aligned}$$

Hence, the estimate (3.295) with $\mu=1$ follows from Theorem 3.38 and the inequality (B.41).

To prove the case $\mu=0$, we resort to the duality argument. Given $g\in L^2_{\omega^{\alpha-1,\beta-1}}(I)$, we consider a auxiliary problem. It is to find $v\in H^1_{0,\omega^{\alpha,\beta}}(I)$ such that

$$a_{\alpha,\beta}(v,z)=(g,z)_{\omega^{\alpha-1,\beta-1}},\quad \forall z\in H^1_{0,\omega^{\alpha,\beta}}(I). \tag{3.296}$$

Since by (B.40),

$$\begin{aligned}|(g,z)_{\omega^{\alpha-1,\beta-1}}| &\le c\|g\|_{\omega^{\alpha-1,\beta-1}}\|z\|_{\omega^{\alpha-2,\beta-2}}\\ &\le c\|g\|_{\omega^{\alpha-1,\beta-1}}|z|_{1,\omega^{\alpha,\beta}},\end{aligned}$$

we deduce from Lemma 3.5 and the Lax-Milgram lemma (see Chap. 1 or Appendix B) that the problem (3.296) has a unique solution in $H^1_{0,\omega^{\alpha,\beta}}(I)$. Moreover, in the sense of distributions, we have $v''(x)=-(1-x^2)^{-1}g(x)$. Therefore,

$$|v|_{2,\omega^{\alpha+1,\beta+1}}=\|g\|_{\omega^{\alpha-1,,\beta-1}}.$$

Taking $z=\pi^{1,0}_{N,\alpha,\beta}u-u$ in (3.296), we obtain from Lemma 3.5 and Theorem 3.39 that

$$\begin{aligned}&|(g,\pi^{1,0}_{N,\alpha,\beta}u-u)_{\omega^{\alpha-1,\beta-1}}|=|a_{\alpha,\beta}(v,\pi^{1,0}_{N,\omega}u-u)|\\ &=|a_{\alpha,\beta}(\pi^{1,0}_{N,\alpha,\beta}v-v,\pi^{1,0}_{N,\alpha,\beta}u-u)|\\ &\le c|\pi^{1,0}_{N,\alpha,\beta}v-v|_{1,\omega^{\alpha,\beta}}|\pi^{1,0}_{N,\alpha,\beta}u-u|_{1,\omega^{\alpha,\beta}}\\ &\le cN^{-1}|v|_{2,\omega^{\alpha+1,\beta+1}}|\pi^{1,0}_{N,\alpha,\beta}u-u|_{1,\omega^{\alpha,\beta}}\\ &\le c\sqrt{\frac{(N-m+1)!}{N!}}(N+m)^{-(m+1)/2}\|g\|_{\omega^{\alpha-1,,\beta-1}}\|\partial_x^m u\|_{\omega^{\alpha+m-1,\beta+m-1}}.\end{aligned}$$

Consequently,

$$\begin{aligned}\|\pi^{1,0}_{N,\alpha,\beta}u-u\|_{\omega^{\alpha-1,\beta-1}} &= \sup_{0\neq g\in L^2_{\omega^{\alpha-1,\beta-1}}(I)}\frac{|(\pi^{1,0}_{N,\alpha,\beta}u-u,g)_{\omega^{\alpha-1,\beta-1}}|}{\|g\|_{\omega^{\alpha-1,,\beta-1}}}\\ &\le c\sqrt{\frac{(N-m+1)!}{N!}}(N+m)^{-(m+1)/2}\|\partial_x^m u\|_{\omega^{\alpha+m-1,\beta+m-1}}.\end{aligned}$$

It is clear that

$$\|\pi^{1,0}_{N,\alpha,\beta}u-u\|_{\omega^{\alpha,\beta}}\le c\|\pi^{1,0}_{N,\alpha,\beta}u-u\|_{\omega^{\alpha-1,\beta-1}}.$$

Thus, the desired result follows. □

3.5.3 Interpolations

This section is devoted to the analysis of polynomial interpolation on Jacobi-Gauss-type points. The analysis essentially relies on the polynomial approximation results derived in the previous section, and the asymptotic properties of the nodes and weights of the associated quadrature formulas.

For clarity of presentation, we start with the Chebyshev-Gauss interpolation. Recall the Chebyshev-Gauss nodes and weights (see Theorem 3.30):

$$x_j = \cos\frac{2j+1}{2(N+1)}\pi, \quad \omega_j = \frac{\pi}{N+1}, \quad 0 \le j \le N.$$

To this end, we denote the Chebyshev weight function by $\omega = (1-x^2)^{-1/2}$.

An essential step is to show the stability of the interpolation operator I_N^c.

Lemma 3.6. *For any $u \in B^1_{-1/2,-1/2}(I)$, we have*

$$\|I_N^c u\|_\omega \le \|u\|_\omega + \frac{\pi}{N+1}\|(1-x^2)^{1/2}u'\|_\omega. \tag{3.297}$$

Proof. Let $x = \cos\theta$ and $\hat{u}(\theta) = u(\cos\theta)$. Thanks to the exactness of the Chebyshev-Gauss quadrature (cf. (3.217)), we have

$$\|I_N^c u\|_\omega^2 = \|I_N^c u\|_{N,\omega}^2 = \frac{\pi}{N+1}\sum_{j=0}^N u^2(x_j) = \frac{\pi}{N+1}\sum_{j=0}^N \hat{u}^2(\theta_j),$$

where

$$\theta_j = \arccos(x_j) = \frac{2j+1}{2(N+1)}\pi, \quad 0 \le j \le N.$$

Denote

$$a_j = \frac{j\pi}{N+1}, \quad 0 \le j \le N+1.$$

It is clear that

$$\theta_j \in K_j := [a_j, a_{j+1}], \quad 0 \le j \le N,$$

and the length of the subinterval is $|K_j| = \pi/(N+1)$. Applying the embedding inequality (B.34) on K_j yields

$$|\hat{u}(\theta_j)| \le \max_{\theta\in K_j}|\hat{u}(\theta)| \le \sqrt{\frac{N+1}{\pi}}\|\hat{u}\|_{L^2(K_j)} + \sqrt{\frac{\pi}{N+1}}\|\partial_\theta\hat{u}\|_{L^2(K_j)}.$$

Hence,

$$\begin{aligned}\|I_N^c u\|_\omega &\le \sqrt{\frac{\pi}{N+1}}\sum_{j=0}^{N}|\hat u(\theta_j)|\\ &\le \sum_{j=0}^{N}\Big(\|\hat u\|_{L^2(K_j)}+\frac{\pi}{N+1}\|\partial_\theta \hat u\|_{L^2(K_j)}\Big)\\ &\le \|\hat u\|_{L^2(0,\pi)}+\frac{\pi}{N+1}\|\partial_\theta \hat u\|_{L^2(0,\pi)}.\end{aligned}$$

Finally, the inverse change of variable $\theta \to x$ leads to (3.297). □

Now, we are in a position to present the main result on the Chebyshev-Gauss interpolation error estimates.

Theorem 3.40. *For any $u \in B^m_{-1/2,-1/2}(I)$ with $m \ge 1$, we have that for any $0 \le l \le m \le N+1$,*

$$\begin{aligned}&\|\partial_x^l(I_N^c u-u)\|_{\omega^{l-1/2,l-1/2}}\\ &\quad\le c\sqrt{\frac{(N-m+1)!}{N!}}(N+m)^{l-(m+1)/2}\|\partial_x^m u\|_{\omega^{m-1/2,m-1/2}},\end{aligned} \tag{3.298}$$

where c is a positive constant independent m, N and u.

Proof. Let $\pi_N^c := \pi_N^{-1/2,-1/2}$ be the Chebyshev orthogonal projection operator defined in (3.249). Since $\pi_N^c u \in P_N$, we have $I_N^c(\pi_N^c u) = \pi_N^c u$. Using Lemma 3.6 and Theorem 3.35 with $\alpha = \beta = -1/2$ leads to

$$\begin{aligned}\|I_N^c u-\pi_N^c u\|_\omega &= \|I_N^c(u-\pi_N^c u)\|_\omega\\ &\le c\big(\|u-\pi_N^c u\|_\omega + N^{-1}\|\partial_x(u-\pi_N^c u)\|_{\omega^{-1}}\big)\\ &\le c\sqrt{\frac{(N-m+1)!}{N!}}(N+m)^{-(m+1)/2}\|\partial_x^m u\|_{\omega^{m-1/2,m-1/2}},\end{aligned}$$

which, together with the inverse inequality (3.236), leads to

$$\begin{aligned}\|\partial_x^l(I_N^c u-\pi_N^c u)\|_{\omega^{l-1/2,l-1/2}} &\le cN^l\|I_N^c u-\pi_N^c u\|_\omega\\ &\le c\sqrt{\frac{(N-m+1)!}{N!}}(N+m)^{l-(m+1)/2}\|\partial_x^m u\|_{\omega^{m-1/2,m-1/2}}.\end{aligned}$$

Finally, it follows from the triangle inequality and Theorem 3.35 that

$$\begin{aligned}\|\partial_x^l(I_N^c u-u)\|_{\omega^{l-1/2,l-1/2}} &\le \|\partial_x^l(I_N^c u-\pi_N^c u)\|_{\omega^{l-1/2,l-1/2}}\\ &\quad+\|\partial_x^l(\pi_N^c u-u)\|_{\omega^{l-1/2,l-1/2}}\\ &\le c\sqrt{\frac{(N-m+1)!}{N!}}(N+m)^{l-(m+1)/2}\|\partial_x^m u\|_{\omega^{m-1/2,m-1/2}}.\end{aligned}$$

This ends the proof. □

We observe that the Chebyshev-Gauss interpolation shares the same optimal order of convergence with the orthogonal projection $\pi_N^{-1/2,-1/2}$ (cf. Theorem 3.35).

Next, we extend the above argument to the general Jacobi-Gauss-type interpolations. An essential difference is that unlike the Chebyshev case, the explicit expressions of the nodes and weights are not available. Hence, we have to resort to their asymptotic expressions.

Let $\{x_j, \omega_j\}_{j=0}^N$ be the set of Jacobi-Gauss, Jacobi-Gauss-Radau, or Jacobi-Gauss-Lobatto nodes and weights relative to the Jacobi weight function $\omega^{\alpha,\beta}$ (cf. Sect. 3.2). Assume that $\{x_j\}_{j=0}^N$ are arranged in descending order, and set $\{\theta_j = \arccos(x_j)\}_{j=0}^N$. For the variable transformation $x = \cos\theta, \theta \in [0,\pi], x \in [-1,1]$, it is clear that

$$\frac{d\theta}{dx} = -\frac{1}{\sqrt{1-x^2}}, \quad 1-x = 2\Big(\sin\frac{\theta}{2}\Big)^2, \quad 1+x = 2\Big(\cos\frac{\theta}{2}\Big)^2. \tag{3.299}$$

3.5.3.1 Jacobi-Gauss Interpolation

Recall the asymptotic formulas of the Jacobi-Gauss nodes and weights given by Theorem 8.9.1 and Formula (15.3.10) of Szegö (1975).

Lemma 3.7. *For $\alpha, \beta > -1$, we have*

$$\theta_j = \cos^{-1} x_j = \frac{1}{N+1}\{(j+1)\pi + O(1)\}, \tag{3.300}$$

with $O(1)$ being uniformly bounded for all values $j = 0, 1, \ldots, N$, and

$$\omega_j \cong \frac{2^{\alpha+\beta+1}\pi}{N+1}\left(\sin\frac{\theta_j}{2}\right)^{2\alpha+1}\left(\cos\frac{\theta_j}{2}\right)^{2\beta+1}, \quad 0 \le j \le N. \tag{3.301}$$

As with Lemma 3.6, we first show the stability of the Jacobi-Gauss interpolation operator $I_N^{\alpha,\beta}$.

Lemma 3.8. *For any $\alpha, \beta > -1$, and any $u \in B^1_{\alpha,\beta}(I)$,*

$$\|I_N^{\alpha,\beta} u\|_{\omega^{\alpha,\beta}} \lesssim \|u\|_{\omega^{\alpha,\beta}} + N^{-1}\|u'\|_{\omega^{\alpha+1,\beta+1}}. \tag{3.302}$$

Proof. Let $x = \cos\theta$ and $\hat{u}(\theta) = u(x)$ with $\theta \in (0,\pi)$. By the exactness of the Jacobi-Gauss quadrature (cf. Theorem 3.25) and Lemma 3.7,

$$\begin{aligned}\|I_N^{\alpha,\beta} u\|^2_{\omega^{\alpha,\beta}} &= \|I_N^{\alpha,\beta} u\|^2_{N,\omega^{\alpha,\beta}} = \sum_{j=0}^N u^2(x_j)\omega_j \\ &\lesssim N^{-1}\sum_{j=0}^N \hat{u}^2(\theta_j)\left(\sin\frac{\theta_j}{2}\right)^{2\alpha+1}\left(\cos\frac{\theta_j}{2}\right)^{2\beta+1}.\end{aligned}$$

The asymptotic formula (3.300) implies that $\theta_j \in K_j \subset [a_0, a_1] \subset (0,\pi)$, where $a_0 = \frac{O(1)}{N+1}$, $a_1 = \frac{N\pi + O(1)}{N+1}$ and the length of each closed subinterval K_j is $\frac{c}{N+1}$. Hence,

$$\|I_N^{\alpha,\beta} u\|_{\omega^{\alpha,\beta}} \lesssim N^{-\frac{1}{2}} \sum_{j=0}^{N} \max_{\theta\in K_j} \left| \hat{u}(\theta) \Big(\sin\frac{\theta}{2}\Big)^{\alpha+\frac{1}{2}} \Big(\cos\frac{\theta}{2}\Big)^{\beta+\frac{1}{2}} \right|.$$

For notational simplicity, we denote

$$\chi^{\alpha,\beta}(\theta) = \Big(\sin\frac{\theta}{2}\Big)^{\alpha+\frac{1}{2}} \Big(\cos\frac{\theta}{2}\Big)^{\beta+\frac{1}{2}}.$$

Applying the embedding inequality (B.34) on K_j yields

$$\begin{aligned}
\|I_N^{\alpha,\beta} u\|_{\omega^{\alpha,\beta}} &\lesssim \sum_{j=0}^{N} \Big(\|\hat{u}\chi^{\alpha,\beta}\|_{L^2(K_j)} + N^{-1} \|\partial_\theta [\hat{u}\chi^{\alpha,\beta}]\|_{L^2(K_j)} \Big) \\
&\lesssim \|\hat{u}\chi^{\alpha,\beta}\|_{L^2(0,\pi)} + N^{-1} \|\partial_\theta [\hat{u}\chi^{\alpha,\beta}]\|_{L^2(a_0,a_1)} \\
&\lesssim \|\hat{u}\chi^{\alpha,\beta}\|_{L^2(0,\pi)} + N^{-1} \|\chi^{\alpha,\beta}\partial_\theta \hat{u}\|_{L^2(0,\pi)} \\
&\quad + N^{-1} \|\hat{u}\chi^{\alpha-1,\beta-1}\|_{L^2(a_0,a_1)}.
\end{aligned}$$

In view of (3.299), an inverse change of variable leads to

$$\begin{aligned}
\|\hat{u}\chi^{\alpha,\beta}\|^2_{L^2(0,\pi)} &= \int_0^{\pi} \hat{u}^2(\theta) \Big(\sin\frac{\theta}{2}\Big)^{2\alpha+1} \Big(\cos\frac{\theta}{2}\Big)^{2\beta+1} d\theta \\
&\lesssim \int_{-1}^{1} u^2(x)(1-x)^{\alpha+1/2}(1+x)^{\beta+1/2} \frac{1}{\sqrt{1-x^2}} dx \\
&\lesssim \|u\|^2_{\omega^{\alpha,\beta}},
\end{aligned}$$

and similarly,

$$\|\chi^{\alpha,\beta}\partial_\theta \hat{u}\|_{L^2(0,\pi)} \lesssim \|\partial_x u\|_{\omega^{\alpha+1,\beta+1}}.$$

We treat the last term as

$$\begin{aligned}
N^{-1}\|\hat{u}\chi^{\alpha-1,\beta-1}\|_{L^2(a_0,a_1)} &\lesssim \Big(\sup_{a_0\le\theta\le a_1} \frac{1}{N\sin\theta} \Big) \|\hat{u}\chi^{\alpha,\beta}\|_{L^2(a_0,a_1)} \\
&\lesssim \|\hat{u}\chi^{\alpha,\beta}\|_{L^2(0,\pi)} \lesssim \|u\|_{\omega^{\alpha,\beta}},
\end{aligned}$$

where due to the fact $a_0 = O(N^{-1})$ and $a_1 = \pi - O(N^{-1})$, we have

$$\sup_{a_0\le\theta\le a_1} \frac{1}{N\sin\theta} \le c.$$

A combination of the above estimates leads to the desired result. □

As a consequence of Lemma 3.8, we have the following inequality in the polynomial space.

Corollary 3.7. *For any $\phi \in P_M$ and $\psi \in P_L$,*

$$\|I_N^{\alpha,\beta}\phi\|_{\omega^{\alpha,\beta}} \lesssim \Big(1+\frac{M}{N}\Big)\|\phi\|_{\omega^{\alpha,\beta}}, \tag{3.303a}$$

$$|\langle\phi,\psi\rangle_{N,\omega^{\alpha,\beta}}| \lesssim \Big(1+\frac{M}{N}\Big)\Big(1+\frac{L}{N}\Big)\|\phi\|_{\omega^{\alpha,\beta}}\|\psi\|_{\omega^{\alpha,\beta}}. \tag{3.303b}$$

Proof. Using the inverse inequality (3.236) and (3.302) gives

$$\|I_N^{\alpha,\beta}\phi\|_{\omega^{\alpha,\beta}} \lesssim \|\phi\|_{\omega^{\alpha,\beta}} + N^{-1}\|\partial_x\phi\|_{\omega^{\alpha+1,\beta+1}} \lesssim \Big(1+\frac{M}{N}\Big)\|\phi\|_{\omega^{\alpha,\beta}}.$$

Therefore,

$$\begin{aligned}|\langle\phi,\psi\rangle_{N,\omega^{\alpha,\beta}}| &= |\langle I_N^{\alpha,\beta}\phi, I_N^{\alpha,\beta}\psi\rangle_{N,\omega^{\alpha,\beta}}| \overset{(3.150)}{=} |(I_N^{\alpha,\beta}\phi, I_N^{\alpha,\beta}\psi)_{\omega^{\alpha,\beta}}| \\ &\lesssim \Big(1+\frac{M}{N}\Big)\Big(1+\frac{L}{N}\Big)\|\phi\|_{\omega^{\alpha,\beta}}\|\psi\|_{\omega^{\alpha,\beta}}.\end{aligned}$$

This ends the proof. □

With the aid of the stability result (3.302), we can estimate the Jacobi-Gauss interpolation errors by using an argument similar to that for Theorem 3.40.

Theorem 3.41. *Let $\alpha,\beta > -1$. For any $u \in B^m_{\alpha,\beta}(I)$ with $m \ge 1$, we have that for $0 \le l \le m \le N+1$,*

$$\begin{aligned}&\big\|\partial_x^l(I_N^{\alpha,\beta}u - u)\big\|_{\omega^{\alpha+l,\beta+l}} \\ &\quad \le c\sqrt{\frac{(N-m+1)!}{N!}}(N+m)^{l-(m+1)/2}\|\partial_x^m u\|_{\omega^{\alpha+m,\beta+m}},\end{aligned} \tag{3.304}$$

where c is a positive constant independent of m, N and u.

Similar to (3.267), the Jacobi-Gauss interpolation errors measured in the norms of the usual Sobolev spaces $H^l_{\omega^{\alpha,\beta}}(I)(l \ge 1)$ are not optimal. For instance, a standard argument using (3.302), Theorem 3.34, and Theorem 3.36 leads to that *for any $u \in B^m_{\alpha,\beta}(I)$ with $1 \le m \le N+1$,*

$$\begin{aligned}&\|I_N^{\alpha,\beta}u - u\|_{1,\omega^{\alpha,\beta}} \\ &\quad \le c\sqrt{\frac{(N-m+1)!}{N!}}(N+m)^{(3-m)/2}\|\partial_x^m u\|_{\omega^{\alpha+m-1,\beta+m-1}}.\end{aligned} \tag{3.305}$$

Now, we consider the Jacobi-Gauss-Radau and Jacobi-Gauss-Lobatto interpolations.

3.5.3.2 Jacobi-Gauss-Radau Interpolation

In view of (3.140), the N interior Jacobi-Gauss-Radau nodes $\{x_j\}_{j=1}^N$ turn out to be the Jacobi-Gauss nodes with the parameter $(\alpha, \beta+1)$. Hence, by (3.300) and (3.301),

$$\theta_j = \arccos(x_j) = \frac{1}{N}\{j\pi + O(1)\}, \quad 1 \le j \le N, \tag{3.306}$$

and

$$\begin{aligned}\omega_j &\cong \frac{2^{\alpha+\beta+2}\pi}{N}\frac{1}{1+x_j}\left(\sin\frac{\theta_j}{2}\right)^{2\alpha+1}\left(\cos\frac{\theta_j}{2}\right)^{2\beta+3} \\ &\cong \frac{2^{\alpha+\beta+1}\pi}{N}\left(\sin\frac{\theta_j}{2}\right)^{2\alpha+1}\left(\cos\frac{\theta_j}{2}\right)^{2\beta+1}, \quad 1 \le j \le N.\end{aligned} \tag{3.307}$$

Moreover, applying the Stirling's formula (A.7) to (3.134a) yields

$$\omega_0 = O\big(N^{-2\beta-2}\big). \tag{3.308}$$

Similar to Lemma 3.8, we have the following stability of the Jacobi-Gauss-Radau interpolation operator.

Lemma 3.9. *For any $u \in B^1_{\alpha,\beta}(I)$,*

$$\|I_N^{\alpha,\beta}u\|_{\omega^{\alpha,\beta}} \lesssim N^{-\beta-1}|u(-1)| + \|u\|_{\omega^{\alpha,\beta}} + N^{-1}|u|_{1,\omega^{\alpha+1,\beta+1}}. \tag{3.309}$$

Proof. By the exactness of the Jacobi-Gauss-Radau quadrature (cf. Theorem 3.26),

$$\|I_N^{\alpha,\beta}u\|^2_{\omega^{\alpha,\beta}} = \|I_N^{\alpha,\beta}u\|^2_{N,\omega^{\alpha,\beta}} = u^2(-1)\omega_0 + \sum_{j=1}^N u^2(x_j)\omega_j.$$

Thanks to (3.306) and (3.307), using the same argument as for Lemma 3.8 leads to

$$\sum_{j=1}^N u^2(x_j)\omega_j \lesssim \|u\|^2_{\omega^{\alpha,\beta}} + N^{-2}|u|^2_{1,\omega^{\alpha+1,\beta+1}}.$$

Hence, a combination of the above two results and (3.308) yields (3.309). □

As a direct consequence of Lemma 3.9, we have the following results.

Corollary 3.8. *For any $\phi \in P_M$ and $\psi \in P_L$ with $\phi(-1) = \psi(-1) = 0$,*

$$\|I_N^{\alpha,\beta}\phi\|_{\omega^{\alpha,\beta}} \lesssim \left(1+\frac{M}{N}\right)\|\phi\|_{\omega^{\alpha,\beta}}, \tag{3.310a}$$

$$|\langle\phi,\psi\rangle_{N,\omega^{\alpha,\beta}}| \lesssim \left(1+\frac{M}{N}\right)\left(1+\frac{L}{N}\right)\|\phi\|_{\omega^{\alpha,\beta}}\|\psi\|_{\omega^{\alpha,\beta}}. \tag{3.310b}$$

In order to deal with the boundary term in Lemma 3.9, we need to estimate the projection errors at the endpoints.

Lemma 3.10. *Let $\alpha,\beta > -1$. For $u \in B^m_{\alpha,\beta}(I)$,*

- *if $\alpha+1 < m \le N+1$, we have*

$$\big|(\pi_N^{\alpha,\beta} u - u)(1)\big| \le c m^{-1/2} N^{1+\alpha-m} \|\partial_x^m u\|_{\omega^{\alpha+m,\beta+m}}, \tag{3.311}$$

- *if $\beta+1 < m \le N+1$, we have*

$$\big|(\pi_N^{\alpha,\beta} u - u)(-1)\big| \le c m^{-1/2} N^{1+\beta-m} \|\partial_x^m u\|_{\omega^{\alpha+m,\beta+m}}, \tag{3.312}$$

where c is a positive constant independent of m, N and u.

Proof. Let $h^{\alpha,\beta}_{n,k}$ be the constant defined in (3.254) and let $\tilde m = \min\{m, N+1\}$. By the Cauchy–Schwarz inequality and (3.258),

$$\begin{aligned}
\big|(\pi_N^{\alpha,\beta} u - u)(1)\big| &\le \sum_{n=N+1}^{\infty} |\hat u_n^{\alpha,\beta}||J_n^{\alpha,\beta}(1)| \\
&\le \bigg(\sum_{n=N+1}^{\infty} |J_n^{\alpha,\beta}(1)|^2 (h^{\alpha,\beta}_{n,\tilde m})^{-1}\bigg)^{1/2} \bigg(\sum_{n=N+1}^{\infty} |\hat u_n^{\alpha,\beta}|^2 h^{\alpha,\beta}_{n,\tilde m}\bigg)^{1/2} \\
&\le \bigg(\sum_{n=N+1}^{\infty} |J_n^{\alpha,\beta}(1)|^2 (h^{\alpha,\beta}_{n,\tilde m})^{-1}\bigg)^{1/2} \|\partial_x^{\tilde m} u\|_{\omega^{\alpha+\tilde m,\beta+\tilde m}}.
\end{aligned}$$

By (3.94), (3.254) and the Stirling's formula (A.7), we find

$$\frac{|J_n^{\alpha,\beta}(1)|^2}{h^{\alpha,\beta}_{n,\tilde m}} \le c e^{-\tilde m} \frac{(n-\tilde m)!}{n!} \frac{n^{1+2\alpha}}{n^{\tilde m}}, \quad \forall n \ge N+1 \gg 1.$$

Moreover, by (A.8) and the inequality: $1-x \le e^{-x}$ for $x \in [0,1]$,

$$e^{-\tilde m} \frac{(n-\tilde m)!}{n!} \le c \frac{e^{-\tilde m}}{n^{\tilde m}} \Big(1 - \frac{\tilde m}{n}\Big)^{n-\tilde m+1/2} \le c n^{-\tilde m}.$$

Hence, for $\tilde m > \alpha+1$,

$$\sum_{n=N+1}^{\infty} \frac{|J_n^{\alpha,\beta}(1)|^2}{h^{\alpha,\beta}_{n,\tilde m}} \le c \sum_{n=N+1}^{\infty} n^{2\alpha+1-2\tilde m} \le c \int_N^{\infty} x^{2\alpha+1-2\tilde m} dx \le \frac{c}{\tilde m} N^{2(1+\alpha-\tilde m)}.$$

A combination of the above estimates leads to

$$|(\pi_N^{\alpha,\beta} u - u)(1)| \le \frac{c}{\sqrt{\tilde m}} N^{1+\alpha-\tilde m} \|\partial_x^{\tilde m} u\|_{\omega^{\alpha+\tilde m,\beta+\tilde m}}, \quad \tilde m > \alpha+1. \tag{3.313}$$

This gives (3.311).

Thanks to (3.105), we derive (3.312) easily □

Thanks to Lemma 3.9, Lemma 3.10 and Theorem 3.35, we can derive the following result by an argument analogous to that for Theorem 3.40.

Theorem 3.42. *For $\alpha, \beta > -1$ and any $u \in B^m_{\alpha,\beta}(I)$, we have that for $0 \le l \le m$ and $\beta + 1 < m \le N+1$,*

$$\|\partial_x^l (I_N^{\alpha,\beta} u - u)\|_{\omega^{\alpha+l,\beta+l}} \le c\sqrt{\frac{(N-m+1)!}{N!}} N^{l-(m+1)/2} \|\partial_x^m u\|_{\omega^{\alpha+m,\beta+m}}, \tag{3.314}$$

where c is a positive constant independent m, N and u.

3.5.3.3 Jacobi-Gauss-Lobatto Interpolation

The relation (3.141) indicates that the $N-1$ interior JGL nodes $\{x_j\}_{j=1}^{N-1}$ are the JG nodes with the parameter $(\alpha+1, \beta+1)$. Hence, by (3.300),

$$\theta_j = \arccos(x_j) = \frac{1}{N-1}\{j\pi + O(1)\}, \quad 1 \le j \le N-1. \tag{3.315}$$

Moreover, we find from (3.141) and (3.301) that the associated weights have the asymptotic property:

$$\omega_j \cong \frac{2^{\alpha+\beta+1}\pi}{N-1}\left(\sin\frac{\theta_j}{2}\right)^{2\alpha+1}\left(\cos\frac{\theta_j}{2}\right)^{2\beta+1}, \quad 1 \le j \le N-1. \tag{3.316}$$

Furthermore, applying the Stirling's formula (A.7) to the boundary weights in (3.139a) and (3.139b) yields

$$\omega_0 = O(N^{-2\beta-2}), \quad \omega_N = O(N^{-2\alpha-2}).$$

Hence, similar to Lemmas 3.8 and 3.9, we can derive the following stability result.

Lemma 3.11. *For any $u \in B^1_{\alpha,\beta}(I)$,*

$$\begin{aligned}\|I_N^{\alpha,\beta} u\|_{\omega^{\alpha,\beta}} \lesssim\, & N^{-\alpha-1}|u(1)| + N^{-\beta-1}|u(-1)| \\ & + \|u\|_{\omega^{\alpha,\beta}} + N^{-1}|u|_{1,\omega^{\alpha+1,\beta+1}}.\end{aligned} \tag{3.317}$$

As with Corollaries 3.7 and 3.8, the following bounds can be obtained directly from Lemma 3.11.

Corollary 3.9. *For any $\phi \in P_M$ and $\psi \in P_L$ with $\phi(\pm 1) = \psi(\pm 1) = 0$,*

$$\|I_N^{\alpha,\beta}\phi\|_{\omega^{\alpha,\beta}} \lesssim \left(1 + \frac{M}{N}\right)\|\phi\|_{\omega^{\alpha,\beta}}, \tag{3.318a}$$

$$|\langle \phi, \psi \rangle_{N,\omega^{\alpha,\beta}}| \lesssim \left(1 + \frac{M}{N}\right)\left(1 + \frac{L}{N}\right)\|\phi\|_{\omega^{\alpha,\beta}}\|\psi\|_{\omega^{\alpha,\beta}}. \tag{3.318b}$$

Similar to the Jacobi-Gauss-Radau case, we can derive the following estimates by using Lemmas 3.11 and 3.10, and Theorem 3.35.

Theorem 3.43. *For $\alpha, \beta > -1$, and any $u \in B^m_{\alpha,\beta}(I)$, we have*

$$\|\partial_x^l (I_N^{\alpha,\beta} u - u)\|_{\omega^{\alpha+l,\beta+l}} \le c\sqrt{\frac{(N-m+1)!}{N!}} N^{l-(m+1)/2} \|\partial_x^m u\|_{\omega^{\alpha+m,\beta+m}}, \quad (3.319)$$

for $0 \le l \le m$ and $\max\{\alpha+1, \beta+1\} < m \le N+1$, where c is a positive constant independent of m, N and u.

Note that in the analysis of interpolation errors, we used the approximation results of the $L^2_{\omega^{\alpha,\beta}}$-projection operator $\pi_N^{\alpha,\beta}$. This led to the estimates in the norms of $B^l_{\alpha,\beta}(I)$, but it induced the constraints $m > \alpha+1$ and/or $m > \beta+1$ for the Radau and Lobatto interpolations. As a result, for the Legendre-Gauss-Lobatto interpolation, the estimate stated in Theorem 3.43 does not hold for $m = 1$.

In Chap. 5 (see Sect. 6.5), we shall take a different approach to derive the following estimate for the Legendre-Gauss-Lobatto interpolation.

Theorem 3.44. *For any $u \in B^m_{-1,-1}(I)$, we have that for $1 \le m \le N+1$,*

$$\begin{aligned} &\|\partial_x(I_N u - u)\| + N\|I_N u - u\|_{\omega^{-1,-1}} \\ &\quad \le c\sqrt{\frac{(N-m+1)!}{N!}} (N+m)^{(1-m)/2} \|\partial_x^m u\|_{\omega^{m-1,m-1}}, \end{aligned} \quad (3.320)$$

where c is a positive constant independent of m, N and u.

Problems

3.1. Derive the properties stated in Corollary 3.2.

3.2. Let $\{p_n\}$ be a sequence of orthogonal polynomials defined on a finite interval (a,b), and let $x_n^{(n)}$ be the largest zero of p_n. Show that $\lim_{n\to\infty} x_n^{(n)}$ exists.

3.3. Regardless of the choice of $\{x_j, \omega_j\}_{j=0}^N$, the quadrature formula (3.33) cannot have degree of precision greater than $2N+1$.

3.4. Let

$$T = \big(t_{nj} := p_n(x_j)\big)_{0\le n,j\le N}, \quad S = \big(s_{jn} := \gamma_n^{-1} p_n(x_j)\omega_j\big)_{0\le n,j\le N}$$

be the transform matrices associated with (3.62) and (3.64). Show that $T = S^{-1}$.

3.5. For $\alpha > \rho > -1$ and $\beta > -1$, show that

$$\int_{-1}^{1} J_n^{\alpha,\beta}(x)\omega^{\rho,\beta}(x)dx = \frac{2^{\beta+\rho+1}\Gamma(\rho+1)\Gamma(n+\beta+1)}{n!\Gamma(\alpha-\rho)\Gamma(\rho+\beta+n+2)}.$$

3.6. Prove the following Rodrigues-like formula:

$$\begin{aligned}(1-x)^{\alpha}(1+x)^{\beta}J_n^{\alpha,\beta}(x) &= \frac{(-1)^m(n-m)!}{2^m n!}\times \\ &\partial_x^m\left\{(1-x)^{\alpha+m}(1+x)^{\beta+m}J_{n-m}^{\alpha+m,\beta+m}(x)\right\}, \\ &\alpha,\beta > -1,\ n\ge m\ge 0.\end{aligned} \tag{3.321}$$

3.7. Derive the formulas in Theorem 3.20.

3.8. Derive the formulas in Theorem 3.27.

3.9. Prove that the following equation holds for integers $n > m$,

$$\frac{d^m}{dx^m}\left[(1-x^2)^m\frac{d^m L_n}{dx^m}\right] + (-1)^{m+1}\lambda_{m,n}L_n = 0, \tag{3.322}$$

where

$$\lambda_{m,n} = \frac{(n+m)!}{(n-m)!}. \tag{3.323}$$

3.10. Prove the orthogonality

$$\int_{-1}^{1} L_n^{(m)}(x)L_k^{(m)}(x)(1-x^2)^m dx = \lambda_{m,n}\|L_n\|^2\delta_{nk}.$$

3.11. Show that the Legendre polynomials satisfy

$$\partial_x^m L_n(\pm 1) = (\pm 1)^{n-m}\frac{(n+m)!}{2^m m!(n-m)!}. \tag{3.324}$$

3.12. Let

$$\mathscr{A}\phi = -\partial_x((1-x^2)\partial_x\phi)$$

be the Sturm-Liouville operator. Verify that

$$\partial_x^k(\mathscr{A}\phi) = -(1-x^2)\partial_x^{k+2}\phi + 2(k+1)x\partial_x^{k+1}\phi + k(k+1)\partial_x^k\phi.$$

Hence, we have

$$\|\partial_x^k(\mathscr{A}\phi)\| \lesssim \|\phi\|_{k+2}.$$

3.13. Given $u \in P_N$, we consider the expansions

$$\partial_x^k u(x) = \sum_{n=k}^{N} \hat{u}_n^{(k)} L_n(x), \quad 0\le k<N.$$

Prove the following relations:

$$\hat{u}_n^{(1)} = (2n+1) \sum_{\substack{p=n+1 \\ n+p \text{ odd}}}^{N} \hat{u}_p^{(0)};$$

$$\hat{u}_n^{(1)} = \frac{(2n+1)}{2} \sum_{\substack{p=n+2 \\ n+p \text{ even}}}^{N} (p(p+1) - n(n+1))\hat{u}_p^{(0)};$$

$$\frac{1}{2n-1}\hat{u}_{n-1}^{(k)} - \frac{1}{2n+3}\hat{u}_{n+1}^{(k)} = \hat{u}_n^{(k-1)}.$$

3.14. Prove that

$$L_n(0) = \begin{cases} 0, & \text{if } n \text{ odd}, \\ n!2^{-n}\Big((n/2)!\Big)^{-2}, & \text{if } n \text{ even}. \end{cases}$$

3.15. According to the formula (4.8.11) of Szegö (1975), we have that for any $n \in \mathbb{N}$ and $x \in [-1,1]$,

$$L_n(x) = \int_0^{\pi} (x + \mathrm{i}\sqrt{1-x^2}\cos\theta)^n d\theta,$$

where $\mathrm{i} = \sqrt{-1}$. Prove that the Legendre polynomials are uniformly bounded between the parabolas

$$-\frac{1+x^2}{2} \le L_n(x) \le \frac{1+x^2}{2}, \quad \forall x \in [-1,1].$$

3.16. Given $u \in P_N$, we consider the expansions

$$\partial_x^k u(x) = \sum_{n=k}^{N} \hat{u}_n^{(k)} T_n(x), \quad 0 \le k < N.$$

Prove the following relations:

$$\hat{u}_n^{(1)} = \frac{2}{c_n} \sum_{\substack{p=n+1 \\ n+p \text{ odd}}}^{N} p\hat{u}_p^{(0)};$$

$$\hat{u}_n^{(2)} = \frac{1}{c_n} \sum_{\substack{p=n+2 \\ n+p \text{ even}}}^{N} p(p^2 - n^2)\hat{u}_p^{(0)};$$

$$\hat{u}_n^{(3)} = \frac{1}{4c_n} \sum_{\substack{p=n+3 \\ n+p \text{ odd}}}^{N} p(p^2(p^2-2) - 2p^2n^2 + (n^2-1)^2)\hat{u}_p^{(0)};$$

$$\hat{u}_n^{(4)} = \frac{1}{24c_n} \sum_{\substack{p=n+4 \\ n+p \text{ even}}}^{N} p(p^2(p^2-4)^2 - 3p^4n^2 + 3p^2n^4 - n^2(n^2-4)^2)\hat{u}_p^{(0)},$$

and the recurrence formula

$$c_{n-1}\hat{u}_{n-1}^{(k)} - \hat{u}_{n+1}^{(k)} = 2n\hat{u}_n^{(k-1)}.$$

3.17. Show that

$$\partial_x^m T_n(\pm 1) = (\pm 1)^{n+m} \prod_{k=0}^{m} \frac{n^2 - k^2}{2k+1}.$$

3.18. Prove that

$$\int_{-1}^{1} \left[T_n(x)\right]^2 dx = 1 - (4n^2 - 1)^{-1}, \quad n \geq 0.$$

3.19. Show that:

(a) The constants α_j and β_j in (3.25) are the same as the coefficients in (3.7).
(b) The characteristic polynomial of the matrix A_{n+1} is the monic polynomial $\bar{p}_{n+1}(x)$, namely,

$$\bar{p}_{n+1}(x) = \det(xI_{n+1} - A_{n+1}), \quad n \geq -1, \tag{3.325}$$

3.20. Prove the inverse inequalities

$$\|\phi\| \lesssim N^{\alpha} \|\phi\|_{\omega^{\alpha,\alpha}}, \quad \forall \phi \in P_N, \ \alpha \geq 0,$$

and

$$\|\phi\|_{\omega^{-1,-1}} \lesssim N\|\phi\|, \quad \forall \phi \in P_N, \ \phi(\pm 1) = 0.$$

3.21. Prove Theorem 3.34 for the Chebyshev case, that is, for any $\phi \in P_N$,

$$\|\partial_x \phi\|_{\omega} \lesssim N^2 \|\phi\|_{\omega}, \quad \omega(x) = \frac{1}{\sqrt{1-x^2}}.$$

3.22. Show that for $-1 < \alpha, \beta < 1$, we have $H^1_{\omega^{\alpha,\beta}}(I) \subseteq C(\bar{I})$ and (3.281) holds.

3.23. Let I_N be the Legendre-Gauss interpolation operator $N+1$ Legendre-Gauss-Lobatto points. Verify that for $u = L_{N+1} - L_{N-1}$,

$$\|I_N u - u\|_{H^1} \geq cN^{1/2} \|u'\|.$$

3.24. Let I_N be the interpolation operator on $N+1$ Legendre-Gauss-Lobatto points. Show that for any $u \in H_0^1(I)$,

$$\|I_N u\|_{\omega^{-1,-1}} \leq c\big(\|u\|_{\omega^{-1,-1}} + N^{-1} \|\partial_x u\|\big). \tag{3.326}$$

Chapter 4
Spectral Methods for Second-Order Two-Point Boundary Value Problems

We consider in this chapter spectral algorithms for solving the two-point boundary value problem:

$$-\varepsilon U'' + p(x)U' + q(x)U = F, \quad \text{in } I := (-1,1), \tag{4.1}$$

(where $\varepsilon > 0$) with the general boundary conditions

$$a_- U(-1) + b_- U'(-1) = c_-, \quad a_+ U(1) + b_+ U'(1) = c_+, \tag{4.2}$$

which include in particular the Dirichlet boundary conditions ($a_\pm = 1$ and $b_\pm = 0$), Neumann boundary conditions ($a_\pm = 0$ and $b_\pm = \pm 1$), and Robin (or mixed) boundary conditions ($a_- = b_+ = 0$ or $a_+ = b_- = 0$). Whenever possible, we shall give a uniform treatment for all these boundary conditions. Without loss of generality, we assume that:

$$\begin{aligned}
&\text{(i)}\ \ a_\pm \ge 0;\\
&\text{(ii)}\ \ a_-^2 + b_-^2 \neq 0,\ a_- b_- \le 0;\ \ a_+^2 + b_+^2 \neq 0,\ a_+ b_+ \ge 0;\\
&\text{(iii)}\ \ q(x) - p'(x)/2 \ge 0,\ \ \forall x \in I;\\
&\text{(iv)}\ \ p(1) > 0 \text{ if } b_+ \neq 0;\ \ p(-1) < 0 \text{ if } b_- \neq 0.
\end{aligned} \tag{4.3}$$

The above conditions are necessary for the well-posedness of (4.1)–(4.2).

Let us first reduce the problem (4.1)–(4.2) to a problem with homogeneous boundary conditions.

- *Case I.* $a_\pm = 0$ and $b_\pm \neq 0$
  We set $\tilde{u} = \beta x^2 + \gamma x$, where β and γ are uniquely determined by asking $\tilde{u}$ to satisfy (4.2), namely,

$$\begin{aligned}
-2b_-\beta + b_-\gamma &= c_-,\\
2b_+\beta + b_+\gamma &= c_+.
\end{aligned}$$

J. Shen et al., *Spectral Methods: Algorithms, Analysis and Applications*, Springer Series in Computational Mathematics 41, DOI 10.1007/978-3-540-71041-7_4,

- *Case II.* $a_-^2 + a_+^2 \neq 0$
  We set $\tilde{u} = \beta x + \gamma$, where β and γ again can be uniquely determined by requiring $\tilde{u}$ to satisfy (4.2). Indeed, we have

$$\begin{aligned}(-a_- + b_-)\beta + a_-\gamma &= c_-,\\ (a_+ + b_+)\beta + a_+\gamma &= c_+.\end{aligned}$$

  The determinant of the coefficient matrix is

$$\mathrm{DET} = -2a_-a_+ + a_+b_- - a_-b_+.$$

  The assumption (4.3) implies that $b_- \leq 0$ and $b_+ \geq 0$, so we have DET < 0.

Now, we set

$$u = U - \tilde{u}, \quad f = F - (-\varepsilon\tilde{u}'' + p(x)\tilde{u}' + q(x)\tilde{u}).$$

Then u satisfies the following equation

$$-\varepsilon u'' + p(x)u' + q(x)u = f, \quad \text{in } I = (-1,1), \tag{4.4}$$

with the homogeneous boundary condition

$$a_-u(-1) + b_-u'(-1) = 0, \quad a_+u(1) + b_+u'(1) = 0. \tag{4.5}$$

Let us denote

$$H^1_\diamond(I) = \{u \in H^1(I) : u(\pm 1) = 0 \text{ if } b_\pm = 0\}, \tag{4.6}$$

and

$$h_- = \begin{cases} 0, & \text{if } a_-b_- = 0,\\ \frac{a_-}{b_-}, & \text{if } a_-b_- \neq 0,\end{cases} \qquad h_+ = \begin{cases} 0, & \text{if } a_+b_+ = 0,\\ \frac{a_+}{b_+}, & \text{if } a_+b_+ \neq 0.\end{cases} \tag{4.7}$$

Then, a standard weak formulation for (4.4)-(4.5) is:

$$\begin{cases} \text{Find } u \in H^1_\diamond(I) \text{ such that}\\ \mathscr{B}(u,v) = (f,v), \quad \forall v \in H^1_\diamond(I),\end{cases} \tag{4.8}$$

where

$$\begin{aligned}\mathscr{B}(u,v) := \varepsilon\,(u',v') &+ \varepsilon h_+u(1)v(1) - \varepsilon h_-u(-1)v(-1)\\ &+ (p(x)u',v) + (q(x)u,v).\end{aligned} \tag{4.9}$$

It is easy to see that the bilinear form $\mathscr{B}(\cdot,\cdot)$ defined above is continuous and coercive in $H^1_\diamond(I) \times H^1_\diamond(I)$ under the conditions (4.3) (see Problem 4.1). One derives immediately from the Lax-Milgram lemma (see Appendix B) that the problem (4.8) admits a unique solution. Note that only the Dirichlet boundary condition(s) is enforced *exactly* in $H^1_\diamond(I)$, but all other boundary conditions are treated *naturally*.

The rest of this chapter is organized as follows. In the first section, we consider the problem (4.1)–(4.2) with constant coefficients and present several *Galerkin* schemes based on weak formulations using *continuous* inner products. In the second section, we consider the *Galerkin method with numerical integration* which is based on the weak formulation (4.8) using *discrete* inner products. In the third section, we present the *collocation methods* which look for approximate solutions to satisfy (4.2) and (4.1) *exactly* at a set of collocation points. In Sect. 4.4, we introduce some preconditioned iterative methods for solving the linear systems arising from spectral approximations of two-point boundary value problems. In Sect. 4.5, we provide error analysis for two model cases and the one-dimensional Helmholtz equation.

For a thorough discussion on other numerical methods for more general two-point boundary value problems, we refer to Ascher et al. (1995).

4.1 Galerkin Methods

To simplify the presentation, we shall restrict ourselves in this section to a special case of (4.4), namely,

$$\begin{aligned} & -u'' + \alpha u = f, \quad \text{in } I = (-1,1), \\ & a_- u(-1) + b_- u'(-1) = 0, \quad a_+ u(1) + b_+ u'(1) = 0, \end{aligned} \tag{4.10}$$

where $\alpha \geq 0$ is a given constant. The general case (4.1)-(4.2) will be treated in Sects. 4.2 and 4.3.

As a special case of (4.8), the standard weak formulation for (4.10) is

$$\begin{cases} \text{Find } u \in H^1_\diamond(I) \text{ such that} \\ (u',v') + h_+ u(1)v(1) - h_- u(-1)v(-1) \\ \qquad\qquad + \alpha(u,v) = (f,v), \quad \forall v \in H^1_\diamond(I). \end{cases} \tag{4.11}$$

4.1.1 Weighted Galerkin Formulation

We consider the approximation of (4.10) by using a weighted Galerkin method in the polynomial space

$$\tilde{X}_N = \{\phi \in P_N : \phi(\pm 1) = 0 \text{ if } b_\pm = 0\}. \tag{4.12}$$

A straightforward extension of (4.11) using the weighted inner product leads to the following formulation:

$$\begin{cases} \text{Find } u_N \in \tilde{X}_N \text{ such that} \\ (u_N', \omega^{-1}(v_N\omega)')_\omega + \omega(1)h_+u_N(1)v_N(1) \\ \qquad - \omega(-1)h_-u_N(-1)v_N(-1) + \alpha(u_N,v_N)_\omega \\ \qquad = (f,v_N)_\omega, \ \forall v_N \in \tilde{X}_N. \end{cases} \tag{4.13}$$

However, there are several problems associated with this formulation. First, the above formulation does not make sense if $\lim_{x\to\pm1}\omega(x)$ does not exist, except in the case of Dirichlet boundary conditions. Hence, it can not be used for the Jacobi weight function with $\alpha<0$ or $\beta<0$, including in particular the Chebyshev weight (cf. Canuto and Quarteroni (1994) and pp. 194–196 in Funaro (1992) for some special weighted weak formulations of (4.10)). Secondly, as it will become clear later in this section, even in the case $\omega(x)\equiv 1$, this formulation will not lead to a sparse or special linear system that can be inverted efficiently. The cure is to use a new weighted weak formulation in which the general boundary conditions in (4.10) are enforced *exactly* rather than *approximately* in (4.13).

Let us denote

$$X_N = \{v \in P_N : \ a_\pm v(\pm1) + b_\pm v'(\pm1) = 0\}. \tag{4.14}$$

The new weighted Galerkin method for (4.10) is

$$\begin{cases} \text{Find } u_N \in X_N \text{ such that} \\ -(u_N'', v_N)_\omega + \alpha(u_N,v_N)_\omega = (f_N,v_N)_\omega, \quad \forall\, v_N \in X_N, \end{cases} \tag{4.15}$$

where f_N is an appropriate polynomial approximation of f, which is usually taken to be the interpolation of f associated with the Gauss-type quadrature points. The main difference with (4.13) is that the Robin boundary conditions are enforced *exactly* here. We shall see below that by choosing appropriate basis functions of X_N, we shall be able to reduce (4.15) to a linear system with a sparse or special coefficient matrix that can be solved efficiently.

Given a set of basis functions $\{\phi_j\}_{j=0}^{N-2}$ of X_N, we denote

$$\begin{aligned} f_k &= \int_I f_N\,\phi_k\omega dx, \quad \mathbf{f} = (f_0, f_1, \ldots, f_{N-2})^T; \\ u_N &= \sum_{j=0}^{N-2} \hat{u}_j\phi_j, \quad \mathbf{u} = (\hat{u}_0, \hat{u}_1, \ldots, \hat{u}_{N-2})^T; \\ s_{kj} &= -\int_I \phi_j''\phi_k\omega dx, \quad m_{kj} = \int_I \phi_j\,\phi_k\omega dx, \end{aligned} \tag{4.16}$$

and

$$S = (s_{kj})_{0\le k,j\le N-2}, \quad M = (m_{kj})_{0\le k,j\le N-2}.$$

Taking $v_N=\phi_k$, $0\le k\le N-2$ in (4.15), we find that (4.15) is equivalent to the following linear system:

$$(S+\alpha M)\mathbf{u} = \mathbf{f}. \tag{4.17}$$

Below, we determine the entries of S and M for two special cases: $\omega = 1, (1-x^2)^{-1/2}$.

4.1.2 Legendre-Galerkin Method

We set $\omega(x) \equiv 1$ and $f_N = I_N f$ (the Legendre interpolation polynomial of f relative to the Legendre-Gauss-Lobatto points (cf. Sect. 3.3)). Then (4.15) becomes

$$-\int_I u_N'' v_N \, dx + \alpha \int_I u_N v_N \, dx = \int_I I_N f \, v_N \, dx, \quad \forall v_N \in X_N, \tag{4.18}$$

which is referred to as the Legendre-Galerkin method for (4.10).

The actual linear system for (4.18) depends on the choice of basis functions of X_N. Just as in the finite-element methods, where neighboring points are used to form basis functions so as to minimize their interactions in the physical space, neighboring orthogonal polynomials should be used to form basis functions in a spectral-Galerkin method so as to minimize their interactions in the frequency space. Therefore, we look for basis functions as a *compact combination of Legendre polynomials* (cf. Shen (1994)), namely,

$$\phi_k(x) = L_k(x) + a_k L_{k+1}(x) + b_k L_{k+2}(x), \tag{4.19}$$

where the parameters $\{a_k, b_k\}$ are chosen to satisfy the boundary conditions in (4.10). Such basis functions are referred to as *modal* basis functions.

Lemma 4.1. *For all $k \geq 0$, there exists a unique set of $\{a_k, b_k\}$ such that $\phi_k(x) = L_k(x) + a_k L_{k+1}(x) + b_k L_{k+2}(x)$ verifies the boundary conditions in* (4.10).

Proof. Since $L_k(\pm 1) = (\pm 1)^k$ and $L_k'(\pm 1) = \frac{1}{2}(\pm 1)^{k-1} k(k+1)$ (see Sect. 3.3), the boundary conditions in (4.10) lead to the following system for $\{a_k, b_k\}$:

$$\begin{aligned}
&\Big(a_+ + \frac{b_+}{2}(k+1)(k+2)\Big)a_k + \Big(a_+ + \frac{b_+}{2}(k+2)(k+3)\Big)b_k \\
&\qquad\qquad = -a_+ - \frac{b_+}{2}k(k+1), \\
-&\Big(a_- - \frac{b_-}{2}(k+1)(k+2)\Big)a_k + \Big(a_- - \frac{b_-}{2}(k+2)(k+3)\Big)b_k \\
&\qquad\qquad = -a_- + \frac{b_-}{2}k(k+1).
\end{aligned} \tag{4.20}$$

The determinant of the coefficient matrix is

$$\begin{aligned}
\mathrm{DET}_k = {} & 2a_+a_- + a_-b_+(k+2)^2 - a_+b_-(k+2)^2 \\
& - b_-b_+(k+1)(k+2)^2(k+3)/2.
\end{aligned}$$

We then derive from (4.3) that the four terms (including the signs before them) of DET_k are all nonnegative, and at least one is positive for any k. Hence, $\{a_k, b_k\}$ can be uniquely determined by solving (4.20), namely,

$$\begin{aligned} a_k =&(2k+3)(a_+b_- + a_-b+)/\mathrm{DET}_k, \\ b_k =&\Big\{-2a_-a_+ + (k+1)^2(a_+b_- - a_-b_+) \\ &+ \frac{b_-b_+}{2}k(k+1)^2(k+2)\Big\}\Big/\mathrm{DET}_k. \end{aligned} \tag{4.21}$$

This completes the proof. □

Note that in particular:

- If $a_\pm = 1$ and $b_\pm = 0$ (Dirichlet boundary conditions), we have $a_k = 0$ and $b_k = -1$.
- If $a_\pm = 0$, $b_\pm = \pm 1$ (Neumann boundary conditions), we have $a_k = 0$ and $b_k = -k(k+1)/((k+2)(k+3))$.

It is obvious that $\{\phi_k\}$ are linearly independent. Therefore, by dimension argument, we have

$$X_N = \mathrm{span}\{\phi_k : k = 0, 1, \ldots, N-2\}.$$

Remark 4.1. *In the very special case*

$$-u_{xx} = f, \ x \in (-1,1); \quad u_x(\pm 1) = 0,$$

with the condition $\int_{-1}^{1} f dx = 0$*, since the solution is only determined up to a constant, we should use*

$$X_N = \mathrm{span}\{\phi_k : k = 1, 2, \ldots, N-2\}.$$

This remark also applies to the Chebyshev-Galerkin method presented below.

Lemma 4.2. *The stiffness matrix S is a diagonal matrix with*

$$s_{kk} = -(4k+6)b_k, \quad k = 0, 1, \ldots. \tag{4.22}$$

The mass matrix M is symmetric penta-diagonal whose nonzero elements are

$$m_{jk} = m_{kj} = \begin{cases} \dfrac{2}{2k+1} + a_k^2 \dfrac{2}{2k+3} + b_k^2 \dfrac{2}{2k+5}, & j = k, \\ a_k \dfrac{2}{2k+3} + a_{k+1} b_k \dfrac{2}{2k+5}, & j = k+1, \\ b_k \dfrac{2}{2k+5}, & j = k+2. \end{cases} \tag{4.23}$$

Proof. Integrating by parts and using the fact that $\{\phi_k\}$ satisfy the boundary conditions (4.5), we find that

$$\begin{aligned} s_{jk} &= -\int_I \phi_k''(x)\,\phi_j(x)dx \\ &= \int_I \phi_k'(x)\,\phi_j'(x)dx + h_+\phi_k(1)\phi_j(1) - h_-\phi_k(-1)\phi_j(-1) \\ &= -\int_I \phi_k(x)\,\phi_j''(x)dx = s_{kj}, \end{aligned} \tag{4.24}$$

where $h_\pm$ are defined in (4.7). It is then obvious from (4.24) and the definition of $\{\phi_k\}$ that S is a diagonal matrix. Thanks to (3.176c) and (3.174), we find

$$\begin{aligned} s_{kk} &= -b_k \int_I L''_{k+2}(x)L_k(x)dx \\ &= -b_k(k+1/2)(4k+6)\int_I L_k^2(x)dx = -b_k(4k+6). \end{aligned}$$

The nonzero entries for M in (4.23) can be easily obtained by using (3.174). □

Remark 4.2. *An immediate consequence is that* $\{\phi_k\}_{k=0}^{N-2}$ *forms an orthogonal basis of* X_N *with respect to the inner product* $-(u_N'', v_N)$. *Furthermore, an orthonormal basis of* X_N *with respect to this inner product is*

$$\tilde{\phi}_k(x) := \frac{1}{\sqrt{-b_k(4k+6)}}\phi_k(x).$$

Notice that under the assumption (4.3), $b_k < 0$ *for all* k.

We now provide a detailed implementation procedure. Given the values of f at the LGL points $\{x_j\}_{j=0}^N$, we determine the values of u_N (solution of (4.15)) at $\{x_j\}_{j=0}^N$ as follows:

1. (Pre-computation) Compute the LGL points, $\{a_k, b_k\}$ and nonzero elements of S and M.
2. Evaluate the Legendre coefficients of $I_N f$ from $\{f(x_j)\}_{j=0}^N$ (forward Legendre transform, see (3.193)) and evaluate $\mathbf{f}$.
3. Solve $\mathbf{u}$ from (4.17).
4. Evaluate $u_N(x_j) = \sum_{i=0}^{N-2} \hat{u}_i \phi_i(x_j)$, $j = 0, 1, \ldots, N$ (backward Legendre transform, see (3.194)).

Although the solution of the linear system (4.17) can be done in $O(N)$ flops, the two discrete Legendre transforms in the above procedure cost about $2N^2$ flops. To reduce the cost of the discrete transforms between the physical and frequency spaces, a natural choice is to use Chebyshev polynomials so that the discrete Chebyshev transforms can be accelerated by using FFT.

4.1.3 Chebyshev-Galerkin Method

We set $\omega = (1-x^2)^{-1/2}$ and $f_N = I_N^c f$ (the Chebyshev interpolation polynomial of f relative to the Chebyshev-Gauss-Lobatto points (see Sect. 3.4)). Then, (4.15) becomes

$$-\int_I u_N'' v_N \,\omega\, dx + \alpha \int_I u_N v_N \,\omega\, dx = \int_I I_N^c f v_N \omega\, dx, \quad \forall v_N \in X_N, \tag{4.25}$$

which is referred to as the Chebyshev-Galerkin method for (4.10).

As before, we would like to seek the basis functions of X_N in the form

$$\phi_k(x) = T_k(x) + a_k T_{k+1}(x) + b_k T_{k+2}(x). \tag{4.26}$$

Lemma 4.3. *For all $k \geq 0$, there exists a unique set of $\{a_k, b_k\}$ such that $\phi_k(x) = T_k(x) + a_k T_{k+1}(x) + b_k T_{k+2}(x)$ satisfies the boundary conditions in* (4.10).

Proof. Since $T_k(\pm 1) = (\pm 1)^k$ and $T_k'(\pm 1) = (\pm 1)^{k-1} k^2$, we find from (4.5) that $\{a_k, b_k\}$ must satisfy the system

$$\begin{aligned} &(a_+ + b_+(k+1)^2)a_k + (a_+ + b_+(k+2)^2)b_k = -a_+ - b_+k^2, \\ &-(a_- - b_-(k+1)^2)a_k + (a_- - b_-(k+2)^2)b_k = -a_- + b_-k^2, \end{aligned} \tag{4.27}$$

whose determinant is

$$\begin{aligned} \mathrm{DET}_k = {} & 2a_+a_- + \{(k+1)^2 + (k+2)^2\}(a_-b_+ - a_+b_-) \\ & - 2b_-b_+(k+1)^2(k+2)^2. \end{aligned}$$

As in the Legendre case, the conditions in (4.3) imply that $\mathrm{DET}_k > 0$. Hence, $\{a_k, b_k\}$ are uniquely determined by

$$\begin{aligned} a_k = {} & 4(k+1)(a_+b_- + a_-b_+)/\mathrm{DET}_k, \\ b_k = {} & \{(-2a_-a_+ + (k^2 + (k+1)^2)(a_+b_- - a_-b_+) \\ & + 2b_-b_+k^2(k+1)^2\}/\mathrm{DET}_k. \end{aligned} \tag{4.28}$$

This ends the proof. □

Therefore, we have from the dimension argument that

$$X_N = \mathrm{span}\{\phi_k : k = 0, 1, \ldots, N-2\}.$$

One easily derives from (3.214) that the mass matrix M is a symmetric positive definite penta-diagonal matrix whose nonzero elements are

$$m_{jk} = m_{kj} = \begin{cases} \frac{\pi}{2}(c_k + a_k^2 + b_k^2), & j = k, \\ \frac{\pi}{2}(a_k + a_{k+1}b_k), & j = k+1, \\ \frac{\pi}{2}b_k, & j = k+2, \end{cases} \tag{4.29}$$

where $c_0 = 2$ and $c_k = 1$ for $k \geq 1$. However, the computation of s_{kj} is much more involved. Below, we shall derive the explicit expression of s_{kj} for two special cases.

Lemma 4.4. *For the case $a_\pm = 1$ and $b_\pm = 0$ (Dirichlet boundary conditions), we have $a_k = 0$, $b_k = -1$ and*

$$s_{kj} = \begin{cases} 2\pi(k+1)(k+2), & j = k, \\ 4\pi(k+1), & j = k+2, k+4, k+6, \ldots, \\ 0, & j < k \text{ or } j+k \text{ odd}. \end{cases} \tag{4.30}$$

For the case $a_\pm = 0$, $b_+ = 1$ and $b_- = -1$ (Neumann boundary conditions), we have $a_k = 0$, $b_k = -\frac{k^2}{(k+2)^2}$ and

$$s_{kj} = \begin{cases} 2\pi(k+1)k^2/(k+2), & j = k, \\ 4\pi j^2(k+1)/(k+2)^2, & j = k+2, k+4, k+6, \ldots, \\ 0, & j < k \text{ or } j+k \text{ odd}. \end{cases} \tag{4.31}$$

Proof. One observes immediately that

$$s_{kj} = -\int_I \phi_j'' \phi_k \omega \, dx = 0, \quad \text{for } j < k.$$

Hence, S is an upper triangular matrix. By the odd-even parity of the Chebyshev polynomials, we have also $s_{kj} = 0$ for $j + k$ odd.

Thanks to (3.216b), we have

$$\begin{aligned} T_{k+2}''(x) =& \frac{1}{c_k}(k+2)\big((k+2)^2 - k^2\big)T_k(x) \\ &+ \frac{1}{c_{k-2}}(k+2)\big((k+2)^2 - (k-2)^2\big)T_{k-2}(x) + \ldots. \end{aligned} \tag{4.32}$$

We first consider the case $a_\pm = 1$ and $b_\pm = 0$. From (4.21), we find $\phi_k(x) = T_k(x) - T_{k+2}(x)$. It follows immediately from (4.32) and (3.214) that

$$\begin{aligned} -(\phi_k'', \phi_k)_\omega = (T_{k+2}'', T_k)_\omega &= \frac{1}{c_k}(k+2)\big((k+2)^2 - k^2\big)(T_k, T_k)_\omega \\ &= 2\pi(k+1)(k+2). \end{aligned}$$

Setting $\phi_j''(x) = \sum_{n=0}^{j} d_n T_n(x)$, by a simple computation using (4.32), we derive

$$d_n = \begin{cases} -\dfrac{4}{c_j}(j+1)(j+2), & n = j, \\ -\dfrac{1}{c_n}\{(j+2)^3 - j^3 - 2n^2\}, & n < j. \end{cases}$$

Hence for $j = k+2, k+4, \ldots$, we find

$$-(\phi_j'', \phi_k)_\omega = -d_k(T_k, T_k)_\omega + d_{k+2}(T_{k+2}, T_{k+2})_\omega = 4\pi(k+1).$$

The case with $a_\pm = 0$ and $b_\pm = \pm 1$ can be treated similarly as above. □

Similar to the Legendre-Galerkin method, the implementation of the Chebyshev-Galerkin method for (4.10) involves the following steps:

1. (pre-computation) Compute $\{a_k, b_k\}$ and nonzero elements of S and M.
2. Evaluate the Chebyshev coefficients of $I_N^c f$ from $\{f(x_j)\}_{j=0}^N$ (forward Chebyshev transform, see (3.222)) and evaluate $\mathbf{f}$.
3. Solve $\mathbf{u}$ from (4.17).
4. Evaluate $u_N(x_j) = \sum_{i=0}^{N-2} \hat{u}_i \phi_i(x_j)$, $j = 0, 1, \ldots, N$ (backward Chebyshev transform, see (3.223)).

Remark 4.3. *Note that the forward and backward Chebyshev transforms can be performed by using FFT in $O(N \log_2 N)$ operations. However, the cost of Step* 3 *depends on the boundary conditions in* (4.5). *For the special but important cases described in the above lemma, the special structures of S would allow us to solve the system* (4.17) *in $O(N)$ operations. More precisely, in* (4.30) *and* (4.31), *the nonzero elements of S take the form $s_{kj} = a(j) * b(k)$. Hence, a special Gaussian elimination procedure for* (4.17) (cf. Shen (1995)) *would only require $O(N)$ flops instead of $O(N^3)$ flops for a general full matrix.*

Therefore, thanks to FFT, the computational complexity of Chebyshev-Galerkin method for the above cases is $O(N \log_2 N)$ which is quasi-optimal (i.e., optimal up to a logarithmic term).

Remark 4.4. *In the case of Dirichlet boundary conditions, one can also use the basis functions $\psi_k(x) = (1 - x^2)T_k(x)$* (cf. Heinrichs (1989)), *which lead to a banded stiffness matrix.*

4.1.4 Chebyshev-Legendre Galerkin Method

The main advantage of using Chebyshev polynomials is that the discrete Chebyshev transforms can be performed in $O(N \log_2 N)$ operations by using FFT. However, the Chebyshev-Galerkin method leads to non-symmetric formulations which may cause difficulties in analysis and implementation. On the other hand, the Legendre-Galerkin method leads to symmetric formulation and sparse matrices for problems with constant coefficients, but the discrete Legendre transforms are expensive (with $O(N^2)$ operations). In order to take advantage of both the Legendre and Chebyshev methods (cf. Don and Gottlieb (1994)), one may use the so-called Chebyshev-Legendre Galerkin method (cf. Shen (1996)):

$$-\int_I u_N'' v_N \, dx + \alpha \int_I u_N v_N \, dx = \int_I I_N^c f v_N \, dx, \tag{4.33}$$

where I_N^c denotes the interpolation operator relative to the Chebyshev-Gauss-Lobatto points. So the only difference with (4.18) is that the Chebyshev interpolation operator I_N^c is used here to replace the Legendre interpolation operator in (4.18). Therefore, (4.33) leads to the linear system (4.17) with $\mathbf{u}$, S and M defined in (4.16) and (4.22)-(4.23), but with $\mathbf{f}$ defined by

$$f_k = \int_I I_N^c f \, \phi_k dx, \quad \mathbf{f} = (f_0, f_1, \ldots, f_{N-2})^T. \tag{4.34}$$

Hence, the solution procedure of (4.33) is essentially the same as that of (4.18) except that *Chebyshev-Legendre transforms* (between the value of a function at the *CGL* points and the coefficients of its *Legendre* expansion) are needed instead of the *Legendre transforms*. More precisely, given the values of f at the CGL points $\{x_i = \cos(\frac{i\pi}{N})\}_{0 \le i \le N}$, we determine the values of u_N (solution of (4.33)) at the CGL points as follows:

1. (Pre-computation) Compute $\{a_k, b_k\}$ and nonzero elements of S and M.
2. Evaluate the Legendre coefficients of $I_N^c f$ from $\{f(x_i)\}_{i=0}^N$ (forward Chebyshev-Legendre transform).
3. Evaluate $\mathbf{f}$ from (4.34) and solve $\mathbf{u}$ from (4.17).
4. Evaluate $u_N(x_j) = \sum_{i=0}^{N-2} \hat{u}_i \phi_i(x_j)$, $j = 0, 1, \ldots, N$ ("modified" backward Chebyshev-Legendre transform).

The forward and ("modified") backward Chebyshev-Legendre transforms can be implemented efficiently. Indeed, each Chebyshev-Legendre transform can be split into two steps:

1. The transform between its physical values at Chebyshev-Gauss-Lobatto points and the coefficients of its Chebyshev expansion. This can be done by using FFT in $O(N \log_2 N)$ operations.
2. The transform between the coefficients of the Chebyshev expansion and of the Legendre expansion. Alpert and Rokhlin (1991) developed an $O(N)$ algorithm for this transform given a prescribed precision.

Therefore, the total computational cost for (4.33) is of order $O(N \log_2 N)$.

The algorithm in Alpert and Rokhlin (1991) is based on the fast multipole method (cf. Greengard and Rokhlin (1987)). Hence, it is most attractive for very large N. For small to moderate N, a simple algorithm described in Shen (1996) appears to be more competitive.

4.2 Galerkin Method with Numerical Integration

The *Galerkin* methods presented in the previous section lead to very efficient algorithms for problems with constant coefficients. However, they are not feasible for problems with general variable coefficients for which the exact integration is often not possible. Therefore, for problems with variable coefficients, we need to replace the continuous inner product by a suitable discrete inner product, leading to the so-called *Galerkin method with numerical integration*. More precisely, the Legendre-Galerkin method with numerical integration for (4.8) is

$$\begin{cases} \text{Find } u_N \in \tilde{X}_N = P_N \cap H^1_\diamond(I) \text{ such that} \\ \mathscr{B}_N(u_N, v_N) = \langle f, v_N \rangle_N, \quad \forall v_N \in \tilde{X}_N, \end{cases} \tag{4.35}$$

where

$$\begin{aligned} \mathscr{B}_N(u_N, v_N) := & \, \varepsilon \langle u_N', v_N' \rangle_N + \varepsilon h_+ u_N(1) v_N(1) - \varepsilon h_- u_N(-1) v_N(-1) \\ & + \langle p(x) u_N', v_N \rangle_N + \langle q(x) u_N, v_N \rangle_N, \end{aligned}$$

with $\langle \cdot, \cdot \rangle_N$ being the discrete inner product relative to the Legendre-Gauss-Lobatto quadrature.

Let $\{h_j\}$ be the Lagrange basis polynomials (also referred to as *nodal basis*) associated with $\{x_j\}_{j=0}^N$. To fix the idea, we assume $b_\pm \neq 0$, so $\tilde{X}_N = P_N$ and we can write

$$u_N(x) = \sum_{j=0}^{N} u_N(x_j) h_j(x). \tag{4.36}$$

Plugging the above expression into (4.35) and taking $v_N = h_k$, we find that (4.35) reduces to the linear system

$$B\mathbf{w} = W\mathbf{f}, \tag{4.37}$$

where

$$\begin{aligned} & \mathbf{w} = \big(u_N(x_0), u_N(x_1), \ldots, u_N(x_N)\big)^T; \\ & b_{kj} = \mathscr{B}_N(h_j, h_k), \quad B = (b_{kj})_{k,j=0,1,\ldots,N}; \\ & \mathbf{f} = \big(f(x_0), f(x_1), \ldots, f(x_N)\big)^T; \\ & W = \mathrm{diag}(\omega_0, \omega_1, \ldots, \omega_N), \end{aligned} \tag{4.38}$$

with $\{\omega_k\}_{k=0}^N$ being the weights of the Legendre-Gauss-Lobatto quadrature (see Theorem 3.29).

The entries b_{kj} can be determined as follows. Let $\{x_j\}_{j=0}^N$ be arranged in ascending order[1] with $x_0 = -1$ and $x_N = 1$. Using (3.59) and integration by parts, we have

$$\begin{aligned} \langle h_j', h_k' \rangle_N &= (h_j', h_k') = -(h_j'', h_k) + h_j' h_k \big|_{-1}^{1} \\ &= -\left(D^2\right)_{kj} \omega_k + d_{Nj} \delta_{Nk} - d_{0j} \delta_{0k}. \end{aligned} \tag{4.39}$$

Consequently,

$$\begin{aligned} b_{kj} =& \left[-\varepsilon \left(D^2\right)_{kj} + p(x_k) d_{kj} + q(x_k) \delta_{kj} \right] \omega_k \\ &+ \varepsilon \left(d_{Nj} + h_+ \delta_{Nj}\right) \delta_{Nk} - \varepsilon \left(d_{0j} + h_- \delta_{0j}\right) \delta_{0k}. \end{aligned} \tag{4.40}$$

We can also reinterpret (4.35) as a collocation form. Observe that

$$\langle u_N', h_k' \rangle_N = -u_N''(x_k) \omega_k + u_N'(1) \delta_{Nk} - u_N'(-1) \delta_{0k}, \quad 0 \le k \le N.$$

Then, taking $v_N = h_j$ in (4.35) for $j = 0, 1, \ldots, N$, since $\omega_0 = \omega_N = \frac{2}{N(N+1)}$, we find

$$\begin{cases} -\varepsilon u_N''(x_j) + p(x_j) u_N'(x_j) + q(x_j) u_N(x_j) = f(x_j), \ 1 \le j \le N-1, \\ a_- u_N(-1) + b_- u_N'(-1) = -\dfrac{b_-}{\varepsilon} \dfrac{2}{N(N+1)} \\ \qquad \left[f(-1) - \left(-\varepsilon u_N''(-1) + p(-1) u_N'(-1) + q(-1) u_N(-1)\right) \right], \\ a_+ u_N(1) + b_+ u_N'(1) = \dfrac{b_+}{\varepsilon} \dfrac{2}{N(N+1)} \\ \qquad \left[f(1) - \left(-\varepsilon u_N''(1) + p(1) u_N'(1) + q(1) u_N(1)\right) \right]. \end{cases} \tag{4.41}$$

Remark 4.5. *Note that the solution of* (4.35) *satisfies* (4.4) *exactly at the interior collocation points* $\{x_j\}_{j=1}^{N-1}$, *but the boundary conditions* (4.5) *are only satisfied approximately with an error proportional to the residual of* (4.4), *with* u *replaced by the approximate solution* u_N, *at the boundary. Thus,* (4.35) *does not correspond exactly to a collocation method, so it is sometimes referred to as a* collocation method in the weak form. *However, it is clear from* (4.41) *that in the Dirichlet case* (*i.e.,* $b_\pm = 0$), (4.41) *becomes* a collocation method (*see the next section*). *In other words, the Galerkin method with numerical integration* (4.35), *in the case of Dirichlet boundary conditions, is equivalent to the collocation method.*

Remark 4.6. *The matrix* B *in the linear system* (4.37), *even for the simplest differential equation, is full and ill-conditioned, so it is in general not advisable to solve* (4.37) *using a direct method for large* N. *Instead, an iterative method using an appropriate preconditioner should be used, see Sect.* 4.4.

[1] Historically (cf. Gottlieb and Orszag (1977)), the Chebyshev-collocation points were defined as $x_j = \cos \frac{j\pi}{N}$ which were in descending order. For the sake of consistency, we choose to arrange the collocation points in ascending order in this book.

4.3 Collocation Methods

The collocation method, or more specifically the *collocation method in the strong form*, is fundamentally different from the Galerkin method, in the sense that it is not based on a weak formulation. Instead, it looks for an approximate solution which enforces the boundary conditions in (4.5) and collocates (4.4) at a set of interior collocation points. On the other hand, the *collocation method in the weak form* presented in the last section is based on a weak formulation in which the general boundary conditions are treated *naturally* and are only satisfied *asymptotically*, and the approximate solution verifies (4.4) at a set of interior collocation points.

We describe below the collocation method for the two-point boundary value problem (4.1) with the general boundary conditions (4.2). Notice that the non-homogeneous boundary conditions can be treated directly in a collocation method so there is no need to "homogenize" the boundary conditions as we did previously for the Galerkin methods.

Given any set of distinct collocation points $\{x_j\}_{j=0}^N$ on $[-1,1]$ in ascending order with $x_0 = -1$ and $x_N = 1$, the collocation method for (4.1) with (4.2) is

$$\begin{cases} \text{Find } u_N \in P_N \text{ such that} \\ -\varepsilon u_N''(x_i) + p(x_i)u_N'(x_i) + q(x_i)u_N(x_i) = F(x_i), \ \ 1 \le i \le N-1, \\ a_- u_N(-1) + b_- u_N'(-1) = c_-, \quad a_+ u_N(1) + b_+ u_N'(1) = c_+. \end{cases} \tag{4.42}$$

Let $\{h_j\}$ be the Lagrange basis polynomials associated with $\{x_j\}_{j=0}^N$, and let $D = \big(d_{kj} := h_j'(x_k)\big)_{k,j=0,1,\ldots,N}$. Writing $w_j = u_N(x_j)$ and $u_N(x) = \sum_{j=0}^N w_j h_j(x)$, we have

$$\begin{aligned} u_N(x_k) &= \sum_{j=0}^N w_j h_j(x_k) = w_k, \\ u_N'(x_k) &= \sum_{j=0}^N w_j h_j'(x_k) = \sum_{j=0}^N d_{kj} w_j \\ &= \sum_{j=1}^{N-1} d_{kj} w_j + d_{k0} w_0 + d_{kN} w_N, \\ u_N''(x_k) &= \sum_{j=0}^N w_j h_j''(x_k) = \sum_{j=0}^N \big(D^2\big)_{kj} w_j \\ &= \sum_{j=1}^{N-1} \big(D^2\big)_{kj} w_j + \big(D^2\big)_{k0} w_0 + \big(D^2\big)_{kN} w_N. \end{aligned}$$

Substituting the above into (4.42) leads to

$$\begin{cases} \sum_{j=0}^{N} \left[-\varepsilon \left(D^2\right)_{ij} + p(x_i)d_{ij} + q(x_i)\delta_{ij} \right] w_j = F(x_i), \quad i = 1,2,\ldots,N-1, \\ a_- w_0 + b_- \sum_{j=0}^{N} d_{0j} w_j = c_-, \quad a_+ w_N + b_+ \sum_{j=0}^{N} d_{Nj} w_j = c_+. \end{cases} \tag{4.43}$$

Let us denote

$$\begin{aligned} &a_{ij} = -\varepsilon \left(D^2\right)_{ij} + p(x_i)d_{ij} + q(x_i)\delta_{ij}, \quad 1 \le i \le N-1,\, 0 \le j \le N, \\ &a_{0j} = a_- \delta_{0j} + b_- d_{0j}, \quad a_{Nj} = a_+ \delta_{Nj} + b_+ d_{Nj}, \quad 0 \le j \le N, \\ &\mathbf{b} = \left(c_-, F(x_1), F(x_2), \ldots, F(x_{N-1}), c_+\right)^T, \\ &\mathbf{w} = \left(w_0, w_1, \ldots, w_N\right)^T, \quad A = (a_{ij})_{0 \le i,j \le N}. \end{aligned} \tag{4.44}$$

Then, the linear system (4.43) reduces to

$$A\mathbf{w} = \mathbf{b}. \tag{4.45}$$

Remark 4.7. *Notice that the above formulation is valid for any set of collocation points. However, the choice of collocation points is essential for the stability, convergence and efficiency of the collocation method. For two-point boundary value problems, the Gauss-Lobatto points are commonly used. Due to the global nature of the Lagrange basis polynomials, the system matrix A in* (4.45) *is always full and ill-conditioned, even for problems with constant coefficients.*

Remark 4.8. *For the case of homogeneous Dirichlet boundary conditions, i.e., $u(\pm 1) = 0$, the collocation method* (4.42) *with $\{x_j\}$ being the Legendre-Gauss-Lobatto points, as observed in Remark* 4.5, *is equivalent to the Galerkin method with numerical integration* (4.35).

It is interesting to note that in the case of Dirichlet boundary conditions, after eliminating w_0 and w_N from (4.37) *and* (4.45)*, the reduced $(N-1) \times (N-1)$ matrices B and A are related by $B = WA$, where W is the diagonal matrix $W = \mathrm{diag}(\omega_1, \omega_2, \ldots, \omega_{N-1})$. Furthermore, the condition number of A behaves like $O(N^4)$* (cf. Orszag (1980))*, while that of B behaves like $O(N^3)$* (cf. Bernardi and Maday (1992a)).

Remark 4.9. *If the bilinear form $\mathscr{B}(\cdot,\cdot)$ in* (4.9) *is self-adjoint, then the matrix B in* (4.37) *from the Galerkin method with numerical integration is symmetric. However, the matrix A in* (4.45) *from the collocation method is always non-symmetric.*

4.3.1 Galerkin Reformulation

We show below that in the case of homogeneous Dirichlet boundary conditions, the collocation method (4.42) with $\{x_j\}$ being the Jacobi-Gauss-Lobatto points, can be reformulated as a Galerkin method with numerical integration.

Lemma 4.5. *Let* $\omega = (1+x)^\alpha(1-x)^\beta$ *be the Jacobi weight function with* $\alpha, \beta > -1$, $\{x_j\}_{j=0}^N$ *be the Jacobi-Gauss-Lobatto points, and* $\langle\cdot,\cdot\rangle_{N,\omega}$ *be the discrete inner product associated with the Jacobi-Gauss-Lobatto quadrature (cf. Theorem* 3.27). *Then* (4.42) *with* $b_\pm = c_\pm = 0$ *is equivalent to*

$$\begin{cases} \text{Find } u_N \in P_N^0 = P_N \cap H_0^1(I) \text{ such that} \\ \varepsilon\,\langle u_N', \omega^{-1}(v_N\omega)'\rangle_{N,\omega} + \langle p(x)u_N', v_N\rangle_{N,\omega} \\ \qquad + \langle q(x)u_N, v_N\rangle_{N,\omega} = \langle F, v_N\rangle_{N,\omega}, \quad \forall v_N \in P_N^0. \end{cases} \tag{4.46}$$

Proof. By a direct computation, we find that

$$\omega^{-1}(v_N\omega)' = \omega^{-1}(v_N'\omega + v_N\omega') = v_N' - \big(\alpha(1+x) - \beta(1-x)\big)\frac{v_N}{1-x^2}.$$

Since $v_N(\pm 1) = 0$ and $v_N \in P_N$, we derive that $\omega^{-1}(v_N\omega)' \in P_{N-1}$. Therefore, thanks to (3.59), we find that

$$\begin{aligned} \langle u_N', \omega^{-1}(v_N\omega)'\rangle_{N,\omega} &= (u_N', \omega^{-1}(v_N\omega)')_\omega \\ &= -(u_N'', v_N)_\omega = -\langle u_N'', v_N\rangle_{N,\omega}. \end{aligned} \tag{4.47}$$

Therefore, the formulation (4.46) is equivalent to

$$\langle -\varepsilon u_N'' + p(x)u_N' + q(x)u_N, v_N\rangle_{N,\omega} = \langle F, v_N\rangle_{N,\omega}, \quad \forall v_N \in P_N^0. \tag{4.48}$$

Notice that

$$P_N^0 = \operatorname{span}\{h_1(x), h_2(x), \ldots, h_{N-1}(x)\},$$

Taking $v_N = h_i$ for $1 \le i \le N-1$ in (4.48) leads to (4.42) with $b_\pm = c_\pm = 0$.

On the other hand, taking the discrete inner product of (4.42) with $h_k(x)$ for $1 \le k \le N-1$, we find that the solution u_N of (4.42) with $b_\pm = c_\pm = 0$ verifies (4.46). □

This lemma indicates that for (4.4) with Dirichlet boundary conditions, the Jacobi-collocation method, including the Legendre- and Chebyshev-collocation methods, can be reformulated as a *Galerkin method with numerical integration*. An obvious advantage of this reformulation is that error estimates for the Jacobi-collocation method can be carried out in the same way as the Jacobi-Galerkin method.

4.3.2 Petrov-Galerkin Reformulation

Except for the Dirichlet case, the collocation method (4.42) can not be reformulated as a Galerkin method with numerical integration. However, it can be reformulated as a Petrov-Galerkin method for which the trial functions and test functions are taken from different spaces.

Lemma 4.6. *Let* $\omega = (1+x)^{\alpha}(1-x)^{\beta}$ *be the Jacobi weight function with* $\alpha, \beta > -1$, $\{x_j\}_{j=0}^{N}$ *be the set of Jacobi-Gauss-Lobatto points, and* $\langle \cdot, \cdot \rangle_{N,\omega}$ *be the discrete inner product associated with the Jacobi-Gauss-Lobatto quadrature (cf. Theorem* 3.27*). Then,* (4.42) *with* $c_{\pm} = 0$ *is equivalent to the following Petrov-Galerkin method:*

$$\begin{cases} \text{Find } u_N \in X_N \text{ such that} \\ \varepsilon \left\langle u_N', \omega^{-1}(v_N \omega)' \right\rangle_{N,\omega} + \left\langle p(x) u_N', v_N \right\rangle_{N,\omega} \\ \qquad + \left\langle q(x) u_N, v_N \right\rangle_{N,\omega} = \left\langle F, v_N \right\rangle_{N,\omega}, \quad \forall v_N \in P_N^0, \end{cases} \tag{4.49}$$

where X_N *is defined in* (4.14).

Proof. By definition, the solution u_N of (4.42) with $c_{\pm} = 0$ is in X_N. The property (4.47) still holds for $u_N \in X_N$ and $v_N \in P_N^0$, so does (4.48). Taking the discrete inner product of (4.42) with $h_k(x)$ for $k = 1, 2, \ldots, N-1$, we find that the solution u_N of (4.42) with $c_{\pm} = 0$ verifies (4.49). Conversely, taking $v_N = h_i$ for $1 \le i \le N-1$ in (4.48) gives (4.42) with $c_{\pm} = 0$. □

This reformulation will allow us to obtain error estimates for the collocation method (4.42) by using the standard techniques developed for Petrov-Galerkin methods.

4.4 Preconditioned Iterative Methods

As noted in the previous two sections, there is no suitable direct spectral solver for equations with general variable coefficients. Hence, an appropriate iterative method should be used. Since the bilinear form associated with (4.4)–(4.5) is generally not symmetric nor necessarily positive definite, it is in general not advisable to apply an iterative method directly, unless the equation is diffusion dominant, i.e., ε is sufficiently large, when compared with $p(x)$. Instead, it is preferable to transform (4.4)-(4.5) into an equivalent equation whose bilinear form becomes positive definite. Indeed, multiplying (4.4) by the function

$$a(x) = \exp\left(-\frac{1}{\varepsilon}\int p(x)dx\right)$$

and using $-\varepsilon a'(x) = a(x)p(x)$, we find that (4.4) is equivalent to

$$-(a(x)u'(x))' + b(x)u(x) = g(x), \tag{4.50}$$

where $b(x) = a(x)q(x)/\varepsilon$ and $g(x) = a(x)f(x)/\varepsilon$. Hereafter, we assume that there are three constants c_1, c_2 and c_3 such that

$$0 < c_1 \le a(x) \le c_2, \quad 0 \le b(x) \le c_3, \quad \forall x \in [-1,1]. \tag{4.51}$$

We denote

$$\begin{aligned} \mathscr{B}(u,v) := & \int_{-1}^{1} a(x)u'v'dx + a(1)h_+u(1)v(1) - a(-1)h_-u(-1)v(-1) \\ & + \int_{-1}^{1} b(x)uvdx, \quad \forall u,v \in H^1_\diamond(I), \end{aligned} \tag{4.52}$$

where $H^1_\diamond(I)$ and $h_\pm$ are defined in (4.6) and (4.7), respectively. The weak formulation associated with (4.50) with general boundary conditions (4.5) is

$$\begin{cases} \text{Find } u \in H^1_\diamond(I) \text{ such that} \\ \mathscr{B}(u,v) = (g,v), \quad \forall v \in H^1_\diamond(I). \end{cases} \tag{4.53}$$

Hence, under the conditions (4.3) and (4.51), we find that $\mathscr{B}(u,v)$ is self-adjoint, continuous and coercive in $H^1_\diamond(I)$ so that the problem (4.53) admits a unique solution. Instead of dealing with the original equation (4.4)–(4.5), we shall consider below the equivalent problem (4.53) whose bilinear form is symmetric and positive definite.

4.4.1 Preconditioning in the Modal Basis

Let p_k be the Legendre or Chebyshev polynomial of degree k, X_N be defined in (4.14), and $\{\phi_k = p_k + a_k p_{k+1} + b_k p_{k+2}\}_{k=0}^{N-2}$ be the basis functions of X_N constructed in Sect. 4.1. Let I_N be the interpolation operator based on the Legendre or Chebyshev Gauss-Lobatto points $\{x_j\}_{j=0}^{N}$, and $\langle\cdot,\cdot\rangle_{N,\omega}$ (with $\omega = 1, (1-x^2)^{-1/2}$) be the associated discrete inner product. We consider the following *Galerkin method with numerical integration* for (4.53):

$$\begin{cases} \text{Find } u_N = \sum_{k=0}^{N-2} \hat{u}_k \phi_k \in X_N \text{ such that} \\ \mathscr{B}_{N,\omega}(u_N, \phi_j) = \langle g, \phi_j \rangle_{N,\omega}, \quad j = 0,1,\ldots,N-2, \end{cases} \tag{4.54}$$

where

$$\mathscr{B}_{N,\omega}(u_N, v_N) := -\big\langle [I_N(au_N')]', v_N \big\rangle_{N,\omega} + \big\langle bu_N, v_N \big\rangle_{N,\omega}. \tag{4.55}$$

Let us denote

$$b_{jk} = \mathscr{B}_{N,\omega}(\phi_k,\phi_j), \quad B = \left(b_{jk}\right)_{j,k=0,1,\ldots,N-2};$$
$$g_j = \langle g,\phi_j\rangle_{N,\omega}, \quad \mathbf{g} = \left(g_0,g_1,\ldots,g_{N-2}\right)^T;$$
$$\mathbf{u} = \left(\hat{u}_0,\hat{u}_1,\ldots,\hat{u}_{N-2}\right)^T.$$

Then, (4.54) is equivalent to the following linear system:

$$B\mathbf{u} = \mathbf{g}. \tag{4.56}$$

We observe that for $u_N = \sum_{k=0}^{N-2} \hat{u}_k\phi_k \in X_N$ and $v_N = \sum_{k=0}^{N-2} \hat{v}_k\phi_k \in X_N$, we have

$$\langle B\mathbf{u},\mathbf{v}\rangle_{l^2} = \mathscr{B}_{N,\omega}(u_N,v_N), \tag{4.57}$$

where $\langle \mathbf{a},\mathbf{b}\rangle_{l^2} = \sum_{j=0}^{N-2} a_jb_j$ for any vectors $\mathbf{a},\mathbf{b} \in \mathbb{R}^{N-1}$ with components $\{a_j,b_j\}$. It is easy to see that in general B is a full matrix, so we shall resort to an iterative method for which an efficient evaluation of the matrix–vector product $B\mathbf{u}$ is essential.

We now describe how to evaluate

$$(B\mathbf{u})_j = -\left\langle [I_N(au_N')]',\phi_j\right\rangle_{N,\omega} + \left\langle bu_N,\phi_j\right\rangle_{N,\omega}, \quad j=0,1,\ldots,N-2$$

without explicitly forming the matrix B. Given $u_N = \sum_{k=0}^{N-2} \hat{u}_k\phi_k$, we compute "$-\left\langle [I_N(au_N')]',\phi_j\right\rangle_{N,\omega}$" as follows:

1. Using (3.206) or (3.234) to determine $\{\tilde{u}_k^{(1)}\}$ from

$$u_N'(x) = \sum_{k=0}^{N-2} \hat{u}_k\phi_k'(x) = \sum_{k=0}^{N} \tilde{u}_k^{(1)} p_k(x);$$

2. (Forward discrete transform) Compute

$$u_N'(x_j) = \sum_{k=0}^{N} \tilde{u}_k^{(1)} p_k(x_j), \quad j=0,1,\ldots,N;$$

3. (Backward discrete transform) Determine $\{\tilde{w}_k\}$ from

$$I_N(au_N')(x_j) = \sum_{k=0}^{N} \tilde{w}_k p_k(x_j), \quad j=0,1,\ldots,N;$$

4. Using (3.206) or (3.234) to determine $\{\tilde{w}_k^{(1)}\}$ from

$$[I_N(au_N')]'(x) = \sum_{k=0}^{N} \tilde{w}_k p_k'(x) = \sum_{k=0}^{N} \tilde{w}_k^{(1)} p_k(x);$$

5. For $j=0,1,\dots,N-2$, compute

$$-\langle[I_N(au_N')]',\phi_j\rangle_{N,\omega}=-\sum_{k=0}^{N}\tilde{w}_k^{(1)}\langle p_k,\phi_j\rangle_{N,\omega}.$$

Note that the main cost in the above procedure is the two discrete transforms in Steps 2 and 3. The cost for each of Steps 1, 4 and 5 is $O(N)$ flops. The term $\langle bu_N,\phi_j\rangle_{N,\omega}$ can also be computed similarly as follows:

1. Compute

$$u_N(x_j)=\sum_{k=0}^{N}\hat{u}_k\phi_k(x_j),\quad j=0,1,\dots,N;$$

2. Determine $\{\tilde{w}_k\}$ from

$$I_N(bu_N)(x_j)=\sum_{k=0}^{N}\tilde{w}_k p_k(x_j),\quad j=0,1,\dots,N;$$

3. Compute

$$-\langle bu_N,\phi_j\rangle_{N,\omega},\quad j=0,1,\dots,N-2.$$

Hence, if b is not a constant, two additional discrete transforms are needed. In summary, the total cost for evaluate $B\mathbf{u}$ is dominated by four (only two if b is a constant) discrete transforms, and is $O(N^2)$ (resp. $O(N\log_2 N)$) flops in the Legendre (resp. Chebyshev) case.

4.4.1.1 Legendre Case

Thanks to (3.59), we have for any $u_N,v_N\in X_N$,

$$\begin{aligned}-\langle[I_N(au_N')]',v_N\rangle_N=&\langle au_N',v_N'\rangle_N+a(1)h_+u_N(1)v_N(1)\\&-a(-1)h_-u_N(-1)v_N(-1),\end{aligned}\tag{4.58}$$

where $h_\pm$ are defined in (4.7). Hence,

$$\mathscr{B}_N(u_N,v_N)=\mathscr{B}_N(v_N,u_N),\quad\forall u_N,v_N\in X_N.$$

Consequently, B is symmetric.

To simplify the presentation, we shall assume that $b_+b_-=0$ so that the Poincaré inequality is applicable to u_N.

Under the conditions (4.3) and (4.51), we have

$$\begin{aligned}\mathscr{B}_N(u_N,u_N)=&\langle au_N',u_N'\rangle_N+a(1)h_+u_N^2(1)-a(-1)h_-u_N^2(-1)\\&+\langle bu_N,u_N\rangle_N\ge c_1\langle u_N',u_N'\rangle_N=c_1(u_N',u_N').\end{aligned}\tag{4.59}$$

On the other hand, using the Poincaré inequality (B.21) and the Sobolev inequality (B.33), it is easy to show that there exists $c_4 > 0$ such that

$$\mathscr{B}_N(u_N, u_N) \leq c_4(u_N', u_N').$$

Hence, let $s_{ij} = (\phi_j', \phi_i')$ and $S = (s_{ij})_{i,j=0,1,\ldots,N-2}$. We have

$$0 < c_1 \leq \frac{\langle \mathbf{Bu}, \mathbf{u} \rangle_{l^2}}{\langle \mathbf{Su}, \mathbf{u} \rangle_{l^2}} = \frac{\mathscr{B}_N(u_N, u_N)}{(u_N', u_N')} \leq c_4. \tag{4.60}$$

Since $S^{-1}B$ is symmetric with respect to the inner product $\langle \mathbf{u}, \mathbf{v} \rangle_S := \langle \mathbf{Su}, \mathbf{v} \rangle_{l^2}$, (4.60) implies immediately

$$\operatorname{cond}(S^{-1}B) \leq \frac{c_4}{c_1}. \tag{4.61}$$

In other words, S^{-1} is an optimal preconditioner for B in the sense that the convergence rate of the conjugate gradient method applied to the preconditioned system

$$S^{-1}B\mathbf{u} = S^{-1}\mathbf{g} \tag{4.62}$$

will be independent of N. We recall from Sect. 4.1 that S is a diagonal matrix so the cost of applying S^{-1} is negligible. Hence, the main cost in each iteration is the evaluation of $B\mathbf{u}$ for given $\mathbf{u}$.

Remark 4.10. *In the case of Dirichlet boundary conditions, we have* $\phi_k = L_k - L_{k+2}$ *which, together with* (3.176a)*, implies that* $\phi_k' = -(2k+3)L_{k+1}$. *Therefore, from* $u = \sum_{k=0}^{N-2} \hat{u}_k \phi_k$, *we can obtain the derivative* $u' = -\sum_{k=0}^{N-2}(2k+3)\hat{u}_k L_{k+1}$ *in the modal basis without using* (3.206).

Remark 4.11. *If we use the normalized basis functions*

$$\tilde{\phi}_k := \big(-b_k(4k+6)\big)^{-1/2}\phi_k \ \text{ with } (\tilde{\phi}_j', \tilde{\phi}_i') = \delta_{ij},$$

the condition number of the corresponding matrix B with $b_{ij} = \mathscr{B}_N(\tilde{\phi}_j, \tilde{\phi}_i)$ is uniformly bounded. Hence, we can apply the conjugate gradient method directly to this system without preconditioning.

Remark 4.12. *If c_3 in* (4.51) *is large, the condition number in* (4.61) *will be large even though independent of N. In this case, one may improve the situation by replacing the bilinear form (u_N', v_N') with $\hat{a}(u_N', v_N') + \hat{b}(u_N, v_N)$ where*

$$\hat{a} = \frac{1}{2}\Big(\max_{|x|\leq 1} a(x) + \min_{|x|\leq 1} a(x)\Big), \quad \hat{b} = \frac{1}{2}\Big(\max_{|x|\leq 1} b(x) + \min_{|x|\leq 1} b(x)\Big).$$

The matrix corresponding to this new bilinear form is $\hat{a}S + \hat{b}M$ which is positive definite and penta-diagonal (cf. Sect. 4.1).

4.4.1.2 Chebyshev Case

In the Chebyshev case, an appropriate preconditioner for the inner product $\mathscr{B}_{N,\omega}(u_N, v_N)$ in $X_N \times X_N$ is $\left(u_N', \omega^{-1}(v_N\omega)'\right)_\omega$ for which the associated linear system can be solved in $O(N)$ flops as shown in Sect. 4.1. Ample numerical results indicate that the convergence rate of a conjugate gradient type method for non-symmetric systems such as Conjugate Gradient Square (CGS) or BICGStab methods (see Appendix C) is similar to that in the Legendre case.

The advantage of using the Chebyshev polynomials is of course that the evaluation of $B\mathbf{u}$ can be accelerated by FFT in $O(N\log_2 N)$ operations, instead of $O(N^2)$ in the Legendre case.

A few remarks on the use of modal basis functions are in order.

- For problems with constant coefficients, using appropriate modal basis functions leads to sparse matrices.
- For problems with variable coefficients, one can use a suitable problem with constant coefficients as an effective preconditioner.
- With the modal basis, the choice of collocation points (as long as they are Gauss-type quadrature points) is not important, as it is merely used to define an approximation $I_N f$ to f. Therefore, we can use the same set of Gauss-Lobatto points for almost any problem. On the other hand, with the nodal basis, the choice of quadrature rules/collocation points plays an important role and should be made in accordance with the underlying differential equations and boundary conditions (see Sect. 6.4), particularly for high-order equations and mixed type boundary conditions.

We emphasize that the preconditioning in the modal basis will be less effective if the coefficients $a(x)$ and $b(x)$ have large variations, since the variation of the coefficients is not taken into account in the construction of the preconditioner. However, preconditioners which are robust to the variation in coefficients can be constructed in the nodal basis as shown below.

4.4.2 Preconditioning in the Nodal Basis

For problems with large variations in coefficients $a(x)$ and $b(x)$, it is preferable to construct preconditioners in the physical space, i.e., in the nodal basis. We shall consider two approaches: (a) a finite difference preconditioner (cf. Orszag (1980)) for the collocation method for (4.50) with general boundary conditions (4.5); and (b) a finite element preconditioner (cf. Canuto and Quarteroni (1985), Deville and Mund (1985)) for the Galerkin method with numerical integration for (4.53).

4.4.2.1 Finite Difference Preconditioning

The collocation method in the strong form for (4.50) with (4.5) is

$$\begin{cases} \text{Find } u_N \in P_N \text{ such that} \\ -(au_N')'(x_j) + b(x_j)u_N(x_j) = g(x_j), \quad 1 \le j \le N-1, \\ a_-u_N(-1) + b_-u_N'(-1) = 0, \quad a_+u_N(1) + b_+u_N'(1) = 0. \end{cases} \tag{4.63}$$

As in Sect. 4.3, (4.63) can be rewritten as an $(N+1)\times(N+1)$ linear system

$$A\mathbf{w} = \mathbf{b}, \tag{4.64}$$

where the unknowns are $\{w_j := u_N(x_j)\}_{j=0}^N$, and

$$\mathbf{w} = \big(w_0, w_1, \ldots, w_N\big)^T, \quad \mathbf{b} = \big(0, g(x_1), g(x_2), \ldots, g(x_{N-1}), 0\big)^T. \tag{4.65}$$

As suggested by Orszag (1980), we can build a preconditioner for A by using a finite difference approximation to (4.50) with (4.5). Let us denote

$$h_k = x_k - x_{k-1}, \quad \tilde{h}_k = (x_{k+1} - x_{k-1})/2, \quad a_{k+1/2} = a\big((x_{k+1} + x_k)/2\big). \tag{4.66}$$

Then, the second-order finite difference scheme for (4.50) with (4.5) with first-order one-sided difference at the boundaries reads:

$$\begin{cases} -\dfrac{a_{i-1/2}}{\tilde{h}_i h_i} w_{i-1} + \left(\dfrac{a_{i-1/2}}{\tilde{h}_i h_i} + \dfrac{a_{i+1/2}}{\tilde{h}_i h_{i+1}}\right) w_i - \dfrac{a_{i+1/2}}{\tilde{h}_i h_{i+1}} w_{i+1} \\ \qquad\qquad + b(x_i) w_i = g(x_i), \quad 1 \le i \le N-1, \\ a_- w_0 + b_- \dfrac{w_1 - w_0}{h_1} = 0, \quad a_+ w_N + b_+ \dfrac{w_N - w_{N-1}}{h_N} = 0. \end{cases} \tag{4.67}$$

We can rewrite (4.67) in the linear system:

$$A_{fd}\mathbf{w} = \mathbf{b}, \tag{4.68}$$

where A_{fd} is a non-symmetric tridiagonal matrix.

It has been shown (cf. Orszag (1980), Canuto et al. (1987), Kim and Parter (1997)) that in the Dirichlet case, A_{fd}^{-1} is an optimal preconditioner for A, but $\mathrm{cond}(A_{fd}^{-1}A)$ deteriorates with other types of boundary conditions.

Remark 4.13. *The above discussion is valid for both the Legendre and Chebyshev collocation methods.*

4.4.2.2 Finite Element Preconditioning

A more robust preconditioner can be constructed by using a finite element approximation to (4.53).

Let us denote

$$X_h = \{u \in H^1_\diamond(I) : u|_{[x_{i+1},x_i]} \in P_1,\ i = 0,1,\ldots,N-1\}. \tag{4.69}$$

Then, the piecewise linear finite element approximation to (4.53) is

$$\begin{cases} \text{Find } u_h \in X_h \text{ such that} \\ \mathscr{B}_h(u_h,v_h) = \langle g, v_h\rangle_h, \quad \forall v_h \in X_h, \end{cases} \tag{4.70}$$

where

$$\begin{aligned} \mathscr{B}_h(u_h,v_h) :=& \langle a u_h', v_h'\rangle_h + a(1)h_+ u_h(1)v_h(1) \\ & - a(-1)h_- u_h(-1)v_h(-1) + \langle b u_h, v_h\rangle_h, \end{aligned}$$

and $\langle\cdot,\cdot\rangle_h$ is an appropriate discrete inner product associated with the piecewise linear finite element approximation.

To fix the idea, we assume $b_\pm \neq 0$. Let us denote for $k = 1,2,\ldots,N-1$,

$$\hat{h}_k(x) = \begin{cases} \dfrac{x - x_{k+1}}{x_k - x_{k+1}}, & x \in [x_k, x_{k+1}], \\ \dfrac{x_{k-1} - x}{x_{k-1} - x_k}, & x \in [x_{k-1}, x_k], \\ 0, & \text{otherwise}, \end{cases} \tag{4.71}$$

and

$$\begin{aligned} \hat{h}_0(x) &= \begin{cases} \dfrac{x - x_1}{x_0 - x_1}, & x \in [x_0, x_1], \\ 0, & \text{otherwise}, \end{cases} \\ \hat{h}_N(x) &= \begin{cases} \dfrac{x_{N-1} - x}{x_{N-1} - x_N}, & x \in [x_{N-1}, x_N], \\ 0, & \text{otherwise}. \end{cases} \end{aligned} \tag{4.72}$$

Then

$$X_h = \text{span}\{\hat{h}_0, \hat{h}_1, \ldots, \hat{h}_N\}. \tag{4.73}$$

Setting

$$
\begin{aligned}
& u_h(x)=\sum_{j=0}^N u_h(x_j)\hat{h}_j(x), \quad \mathbf{w}=(u_h(x_0),u_h(x_1),\ldots,u_h(x_N))^T; \\
& b_{kj}=\mathscr{B}_h(\hat{h}_j,\hat{h}_k), \quad B_{fe}=(b_{kj})_{k,j=0,1,\ldots,N}; \\
& m_{kj}=\langle \hat{h}_j,\hat{h}_k\rangle_h, \quad M_{fe}=(m_{kj})_{k,j=0,1,\ldots,N}; \\
& \mathbf{g}=(g(x_0),g(x_1),\ldots,g(x_N))^T,
\end{aligned} \tag{4.74}
$$

we can reduce (4.70) to the following linear system

$$
B_{fe}\mathbf{w}=M_{fe}\mathbf{g} \quad \text{or} \quad M_{fe}^{-1}B_{fe}\mathbf{w}=\mathbf{g}. \tag{4.75}
$$

Since both (4.37) and (4.75) provide approximate solutions to (4.53), it is expected that $(M_{fe}^{-1}B_{fe})^{-1}$ (resp. B_{fe}^{-1}) is a good preconditioner for $W^{-1}B$ (resp. B). The optimality of $(M_{fe}^{-1}B_{fe})^{-1}$ as a preconditioner for $W^{-1}B$ has been shown in Franken et al. (1990), while the optimality of B_{fe}^{-1} as a preconditioner for B has been shown in Parter and Rothman (1995).

4.5 Error Estimates

In this section, we perform error analysis for several typical spectral approximation schemes proposed in the previous sections and a spectral-Galerkin method for the 1-D Helmholtz equation.

4.5.1 Legendre-Galerkin Method

We first consider the Legendre-Galerkin method (4.18) (with $f_N = I_N f$ and $\omega \equiv 1$) for (4.10) with homogeneous Dirichlet boundary conditions, i.e., $b_\pm = 0$. In this case, the error analysis is standard. Indeed, applying Theorem 1.3 with $X = H_0^1(I)$, we find immediately

$$
\|u-u_N\|_1 \lesssim \inf_{v_N\in X_N} \|u-v_N\|_1 + \|f-I_N f\|.
$$

Applying Theorem 3.38 with $\alpha=\beta=0$ and Theorem 3.44 to the above leads to the following estimate.

Theorem 4.1. *Let u and u_N be the solutions of* (4.10) *with $b_\pm = 0$ and* (4.18), *respectively. If $u\in H_0^1(I), \partial_x u \in B_{0,0}^{m-1}(I)$ and $f\in B_{-1,-1}^k(I)$ with $1\le m\le N+1$ and $1\le k\le N+1$, we have*

$$\begin{aligned}\|u-u_N\|_1 \le\, & c\sqrt{\frac{(N-m+1)!}{N!}}(N+m)^{(1-m)/2}\|\partial_x^m u\|_{\omega^{m-1,m-1}} \\ & + c\sqrt{\frac{(N-k+1)!}{N!}}(N+k)^{-(k+1)/2}\|\partial_x^k f\|_{\omega^{k-1,k-1}},\end{aligned} \tag{4.76}$$

where c is a positive constant independent of m,k,N,f and u.

Remark 4.14. *Recall from Remark 3.7 that the factor*

$$N^{(1-m)/2} \le \sqrt{\frac{(N-m+1)!}{N!}} \le (N-m+2)^{(1-m)/2}, \tag{4.77}$$

and it is of order $O(N^{(1-m)/2})$ for fixed m.

We now consider the Legendre-Galerkin method (4.18) (with $f_N = I_N f$ and $\omega = 1$) with the general boundary conditions (4.5). To handle the boundary conditions involving derivatives, we need to make use of the H_0^2-orthogonal projection: $\Pi_N^{2,0} : H_0^2(I) \to P_N \cap H_0^2(I)$, defined by

$$\big(\partial_x^2(\Pi_N^{2,0}u-u), \partial_x^2 v_N\big) = 0, \quad \forall v_N \in P_N \cap H_0^2(I), \tag{4.78}$$

whose approximation property is stated in the following lemma.

Lemma 4.7. *If $u \in H_0^2(I)$ and $\partial_x^2 u \in B_{0,0}^{m-2}(I)$ with $2 \le m \le N+1$, then we have*

$$\|\Pi_N^{2,0}u-u\|_\mu \le c\sqrt{\frac{(N-m+1)!}{N!}}(N+m)^{\mu-(1+m)/2}\|\partial_x^m u\|_{\omega^{m-2,m-2}}, \tag{4.79}$$

for $0 \le \mu \le 2$, where c is a positive constant independent of m,N and u.

Proof. We first prove the case: $\mu = 2$. Let $\Pi_N^{1,0}$ be the H_0^1-orthogonal projection operator defined by (3.290) with $\alpha=\beta=0$, and set

$$\phi(x) = \int_{-1}^{x}\Big(\Pi_{N-1}^{1,0}\partial_y u(y) - \frac{3}{4}(1-y^2)\phi^*\Big)dy,$$

where the constant

$$\phi^* = \int_{-1}^{1}\Pi_{N-1}^{1,0}\partial_x u(x)dx.$$

One verifies readily that $\phi \in P_N$ and $\phi(\pm 1) = \phi'(\pm 1) = 0$. Moreover, thanks to the fact $u(\pm 1) = 0$, we derive from Theorem 3.39 with $\alpha=\beta=0$ that

$$\begin{aligned}|\phi^*| &\le \int_{-1}^{1}|\Pi_{N-1}^{1,0}\partial_x u(x) - \partial_x u(x)|dx \le \sqrt{2}\|\Pi_{N-1}^{1,0}\partial_x u - \partial_x u\| \\ &\le c\sqrt{\frac{(N-m+1)!}{N!}}(N+m)^{(1-m)/2}\|\partial_x^m u\|_{\omega^{m-2,m-2}},\end{aligned}$$

and

$$\|\partial_x^2(\Pi_N^{2,0}u-u)\| \overset{(4.78)}{\leq} \|\partial_x^2(\phi-u)\| \leq \|\partial_x(\Pi_{N-1}^{1,0}\partial_x u-\partial_x u)\|+c|\phi^*|$$
$$\leq c\sqrt{\frac{(N-m+1)!}{N!}}(N+m)^{(3-m)/2}\|\partial_x^m u\|_{\omega^{m-2,m-2}},$$

which, together with the Poincaré inequality (B.21), yields the desired result with $\mu=2$.

We now use a duality argument to prove (4.79) with $\mu=0$. Given $f\in L^2(I)$, we consider the following auxiliary problem:

$$\begin{cases} \text{Find } w\in H_0^2(I) \text{ such that} \\ \mathscr{B}(w,z):=(\partial_x^2 w,\partial_x^2 z)=(f,z), \quad \forall z\in H_0^2(I), \end{cases} \tag{4.80}$$

which admits a unique solution in $H_0^2(I)$ satisfying

$$\|w\|_4\leq c\|f\|.$$

Hence, taking $z=\Pi_N^{2,0}u-u$ in (4.80), we have from the shown case (i.e., (4.79) with $\mu=2$) that

$$|(f,\Pi_N^{2,0}u-u)| = |\mathscr{B}(\Pi_N^{2,0}u-u,\Pi_N^{2,0}w-w)|$$
$$\leq \|\partial_x^2(\Pi_N^{2,0}u-u)\|\|\partial_x^2(\Pi_N^{2,0}w-w)\|$$
$$\leq c\sqrt{\frac{(N-m+1)!}{N!}}(N+m)^{-(1+m)/2}\|\partial_x^m u\|_{\omega^{m-2,m-2}}\|\partial_x^4 w\|_{\omega^{2,2}}$$
$$\leq c\sqrt{\frac{(N-m+1)!}{N!}}(N+m)^{-(1+m)/2}\|\partial_x^m u\|_{\omega^{m-2,m-2}}\|f\|.$$

Consequently,

$$\|\Pi_N^{2,0}u-u\| = \sup_{0\neq f\in L^2(I)}\frac{|(f,\Pi_N^{2,0}u-u)|}{\|f\|}$$
$$\leq c\sqrt{\frac{(N-m+1)!}{N!}}(N+m)^{-(1+m)/2}\|\partial_x^m u\|_{\omega^{m-2,m-2}}.$$

Finally, we prove the cases $0<\mu<2$ by using space interpolation. Let $\theta=1-\mu/2$. Since $H^\mu(I)=[H^2(I),L^2(I)]_\theta$, we have from the Gagliardo-Nirenberg inequality (see Theorem B.7) and (4.79) with $\mu=0,2$ that

$$\|\Pi_N^{2,0}u-u\|_\mu \leq \|\Pi_N^{2,0}u-u\|_2^{1-\theta}\|\Pi_N^{2,0}u-u\|^\theta$$
$$\leq c\sqrt{\frac{(N-m+1)!}{N!}}(N+m)^{\mu-(1+m)/2}\|\partial_x^m u\|_{\omega^{m-2,m-2}}.$$

This ends the proof. □

Remark 4.15. *We shall provide in Chap. 5 a much simpler proof of the above estimates by using the notion of generalized Jacobi polynomials.*

With the aid of the above lemma, we can derive the following result, which will be useful for the convergence analysis.

Theorem 4.2. *There exists a mapping $\Pi_N^2 : H^2(I) \to P_N$ such that*

$$(\Pi_N^2 u)(\pm 1) = u(\pm 1), \quad (\Pi_N^2 u)'(\pm 1) = u'(\pm 1). \tag{4.81}$$

Moreover, if $u \in H^2(I)$ and $\partial_x^2 u \in B_{0,0}^{m-2}(I)$ with $2 \le m \le N+1$, then for $0 \le \mu \le 2$, we have

$$\begin{aligned} &\|\Pi_N^2 u - u\|_\mu \\ &\le c\sqrt{\frac{(N-m+1)!}{N!}}(N+m)^{\mu-(1+m)/2}\big(\|u\|_2 + \|\partial_x^m u\|_{\omega^{m-2,m-2}}\big), \end{aligned} \tag{4.82}$$

where c is a positive constant independent of m, N and u.

Proof. Recall the Hermite interpolation basis polynomials associated with two points $x_0 = -1$ and $x_1 = 1$:

$$\begin{aligned} H_0(x) &= \frac{(2+x)(1-x)^2}{4}, \quad H_1(x) = H_0(-x), \\ \hat{H}_0(x) &= \frac{(1+x)(1-x)^2}{4}, \quad \hat{H}_1(x) = -\hat{H}_0(-x). \end{aligned}$$

Setting

$$\Phi(x) = u(-1)H_0(x) + u(1)H_1(x) + u'(-1)\hat{H}_0(x) + u'(1)\hat{H}_1(x) \in P_3,$$

we find that $\Phi(\pm 1) = u(\pm 1)$ and $\Phi'(\pm 1) = u'(\pm 1)$. For any $u \in H^2(I)$, we have $u_* := u - \Phi \in H_0^2(I)$. Defining

$$\Pi_N^2 u = \Pi_N^{2,0} u_* + \Phi,$$

we find that $\Pi_N^2 u$ satisfies (4.81), and

$$u - \Pi_N^2 u = u_* - \Pi_N^{2,0} u_*.$$

Therefore, by Lemma 4.7,

$$\begin{aligned} \|u - \Pi_N^2 u\|_\mu &= \|u_* - \Pi_N^{2,0} u_*\|_\mu \\ &\le c\sqrt{\frac{(N-m+1)!}{N!}}(N+m)^{\mu-(1+m)/2}\|\partial_x^m u_*\|_{\omega^{m-2,m-2}}. \end{aligned} \tag{4.83}$$

It is clear that for $m \geq 4$, we have $\partial_x^m u_* = \partial_x^m u$. For $m = 2,3$, we obtain from the Sobolev inequality (B.33) that

$$\max_{|x|\leq 1} |\partial_x^m \Phi(x)| \leq c\|u\|_2 \;\Rightarrow\; \|\partial_x^m u_*\|_{\omega^{m-2,m-2}} \leq c\big(\|u\|_2 + \|\partial_x^m u\|_{\omega^{m-2,m-2}}\big).$$

The estimate (4.82) follows. □

With the above preparations, we are ready to carry out error analysis of the Legendre-Galerkin approximation of (4.10) with general boundary conditions (4.5).

Theorem 4.3. *Let u and u_N be the solutions of* (4.10) *and* (4.18), *respectively. If $u \in H^2(I)$, $\partial_x^2 u \in B_{0,0}^{m-2}(I)$ and $f \in B_{-1,-1}^k(I)$ with $2 \leq m \leq N+1$ and $1 \leq k \leq N+1$, we have*

$$\begin{aligned} \|u-u_N\|_1 \leq\, & c\sqrt{\frac{(N-m+1)!}{N!}}(N+m)^{(1-m)/2}\big(\|u\|_2 + \|\partial_x^m u\|_{\omega^{m-2,m-2}}\big) \\ & + c\sqrt{\frac{(N-k+1)!}{N!}}(N+k)^{-(k+1)/2}\|\partial_x^k f\|_{\omega^{k-1,k-1}}, \end{aligned} \tag{4.84}$$

where c is a positive constant independent of m, k, N, f and u.

Proof. We derive from (4.10) and (4.18) that

$$A(u-u_N, v_N) := \alpha(u-u_N, v_N) - ((u-u_N)'', v_N) = (f - I_N f, v_N), \quad \forall v_N \in X_N.$$

Under the assumption (4.3), one verifies the continuity and coercivity:

$$\begin{aligned} & A(v,w) \leq c_1\|v\|_1\|w\|_1, \quad \forall v,w \in H^2(I) \cap H_\diamond^1(I), \\ & A(v,v) \geq c_2|v|_1^2, \quad \forall v \in H^2(I) \cap H_\diamond^1(I). \end{aligned} \tag{4.85}$$

Applying Theorem 1.3 with $X = H^2(I) \cap H_\diamond^1(I)$, and using Theorems 3.44 and 4.2, we find

$$\begin{aligned} \|u-u_N\|_1 & \leq c\big(\|u - \Pi_N^2 u\|_1 + \|I_N f - f\|\big) \\ & \leq c\sqrt{\frac{(N-m+1)!}{N!}}(N+m)^{(1-m)/2}\big(\|u\|_2 + \|\partial_x^m u\|_{\omega^{m-2,m-2}}\big) \\ & \quad + c\sqrt{\frac{(N-k+1)!}{N!}}(N+k)^{-(k+1)/2}\|\partial_x^k f\|_{\omega^{k-1,k-1}}. \end{aligned}$$

This completes the proof. □

4.5.2 Chebyshev-Collocation Method

We consider in this section the Chebyshev-collocation method for the model equation

$$\gamma u - u_{xx} = f, \quad \text{in } (-1,1),\ \gamma > 0; \quad u(\pm 1) = 0. \tag{4.86}$$

Let $\{x_j\}_{j=0}^N$ be the Chebyshev-Gauss-Lobatto points. As shown previously, the collocation approximation is

$$\begin{cases} \text{Find } u_N \in P_N^0 \text{ such that} \\ \gamma u_N(x_j) - u_N''(x_j) = f(x_j), \quad 1 \le j \le N-1. \end{cases} \tag{4.87}$$

Let $\omega = (1-x^2)^{-1/2}$ be the Chebyshev weight function, and define the bilinear form as in (3.289):

$$a_\omega(u,v) := \big(u_x, \omega^{-1}(v\omega)_x\big)_\omega = \int_{-1}^{1} u_x (v\omega)_x \, dx. \tag{4.88}$$

We find from Lemma 3.5 (with $\alpha = \beta = -1/2$) that $a_\omega(\cdot,\cdot)$ is continuous and coercive in $H_{0,\omega}^1(I) \times H_{0,\omega}^1(I)$. As a special case of Lemma 4.5, we can reformulate the Chebyshev-collocation scheme (4.87) as

$$\begin{cases} \text{Find } u_N \in P_N^0 \text{ such that} \\ \gamma \langle u_N, v_N \rangle_{N,\omega} + a_\omega(u_N, v_N) = \langle f, v_N \rangle_{N,\omega}, \quad \forall v_N \in P_N^0. \end{cases} \tag{4.89}$$

Then its convergence can be analyzed by using Theorem 1.3 and a standard argument.

Theorem 4.4. *If $u \in H_{0,\omega}^1(I), \partial_x u \in B_{-1/2,-1/2}^{m-1}(I)$ and $f \in B_{-1/2,-1/2}^k(I)$ with $1 \le m,k \le N+1$, then we have*

$$\begin{aligned} \|u - u_N\|_{1,\omega} \le\ & c\sqrt{\frac{(N-m+1)!}{N!}}(N+m)^{(1-m)/2}\|\partial_x^m u\|_{\omega^{m-3/2,m-3/2}} \\ & + c\sqrt{\frac{(N-k+1)!}{N!}}N^{-(1+k)/2}\|\partial_x^k f\|_{\omega^{k-1/2,k-1/2}}, \end{aligned} \tag{4.90}$$

where c is a positive constant independent of m,k,N,f and u.

Proof. Let $\Pi_{N,\omega}^{1,0}$ be the orthogonal projection operator defined in (3.290) with $\alpha = \beta = -1/2$. Applying Theorem 1.3 with $X = H_{0,\omega}^1(I)$ leads to

$$\begin{aligned} \|u - u_N\|_{1,\omega} \le c\Bigg(& \|u - \Pi_{N,\omega}^{1,0} u\|_{1,\omega} + \sup_{0 \ne v_N \in P_N^0} \frac{|(\Pi_{N,\omega}^{1,0} u, v_N)_\omega - \langle \Pi_{N,\omega}^{1,0} u, v_N \rangle_{N,\omega}|}{\|v_N\|_{1,\omega}} \\ & + \sup_{0 \ne v_N \in P_N^0} \frac{|(f, v_N)_\omega - \langle f, v_N \rangle_{N,\omega}|}{\|v_N\|_{1,\omega}} \Bigg). \end{aligned}$$

Therefore, it is necessary to estimate the error between the discrete and inner products. For this purpose, let π_N^c be the L^2_ω-orthogonal projection as defined in (3.249), and I_N^c be the Chebyshev-Gauss-Lobatto interpolation operator. Then we derive from (3.218) and Theorems 3.35 and 3.43 with $\alpha = \beta = -1/2$ that

$$\begin{aligned} |(f,v_N)_\omega - \langle f,v_N\rangle_{N,\omega}| &\le |(f-\pi_{N-1}^c f, v_N)_\omega - \langle I_N^c f - \pi_{N-1}^c f, v_N\rangle_{N,\omega}| \\ &\overset{(3.220)}{\le} c\big(\|f-\pi_{N-1}^c f\|_\omega + \|f - I_N^c f\|_\omega\big)\|v_N\|_\omega \\ &\le c\sqrt{\frac{(N-k+1)!}{N!}}N^{-(1+k)/2}\|\partial_x^k f\|_{\omega^{k-1/2,k-1/2}}\|v_N\|_\omega, \end{aligned} \tag{4.91}$$

and similarly,

$$|(\Pi_{N,\omega}^{1,0}u, v_N)_\omega - \langle \Pi_{N,\omega}^{1,0}u, v_N\rangle_{N,\omega}| \le c\big(\|\Pi_{N,\omega}^{1,0}u - u\|_\omega + \|\pi_{N-1}^c u - u\|_\omega\big)\|v_N\|_\omega.$$

Hence, the estimate (4.90) follows from Theorems 3.35 and 3.39. □

Remark 4.16. *As shown in (4.91), we have the following error estimate between the continuous and discrete inner products relative to the Chebyshev-Gauss-Lobatto setting: If $u \in B^m_{-1/2,-1/2}(I)$ with $1 \le m \le N+1$, then for any $\phi \in P_N$, we have*

$$\begin{aligned} &|(u,\phi)_\omega - \langle u,\phi\rangle_{N,\omega}| \\ &\quad \le c\sqrt{\frac{(N-m+1)!}{N!}}N^{-(1+m)/2}\|\partial_x^m u\|_{\omega^{m-1/2,m-1/2}}\|\phi\|_\omega, \end{aligned} \tag{4.92}$$

where c is a positive constant independent of m, N, ϕ and u. This result is quite useful for error analysis of Chebyshev spectral methods.

4.5.3 Galerkin Method with Numerical Integration

We considered in previous two sections error analysis of problems with constant coefficients. We now discuss the general variable coefficient problem (4.50) with general boundary conditions (4.5), whose variational formulation is given by (4.52)–(4.53). Correspondingly, the Legendre Galerkin method with numerical integration is given by (4.54)–(4.55) with $\omega \equiv 1$. For clarity of presentation, we recall the formulation. Let

$$X_N = \{v \in P_N : a_\pm v(\pm 1) + b_\pm v'(\pm 1) = 0\}. \tag{4.93}$$

We look for $u_N \in X_N$ such that

$$\mathscr{B}_N(u_N, v_N) = \langle g, v_N\rangle_N, \quad \forall v_N \in X_N, \tag{4.94}$$

where

$$\begin{aligned}\mathscr{B}_N(u_N,v_N) =& \langle a u_N', v_N'\rangle_N + \langle b u_N, v_N\rangle_N + a(1)h_+ u_N(1)v_N(1)\\ & - a(-1)h_- u_N(-1)v_N(-1),\end{aligned} \tag{4.95}$$

with $h_\pm$ being defined in (4.7). Observe from (4.59) that for any $u_N, v_N \in X_N$,

$$\mathscr{B}_N(u_N,u_N) \geq c\|u_N'\|^2 + \langle b u_N, u_N\rangle_N, \tag{4.96}$$

and

$$\big|\mathscr{B}_N(u_N,v_N)\big| \leq c\|u_N\|_1\|v_N\|_1. \tag{4.97}$$

For simplicity, we assume $b(x) \geq b_0 > 0$, if $b_\pm \neq 0$, so we have the coercivity:

$$\mathscr{B}_N(u_N,u_N) \geq c\|u_N\|_1^2. \tag{4.98}$$

As a preparation, we first obtain the following result. As its proof is very similar to that of (4.92), we leave it as an exercise (see Problem 4.3).

Lemma 4.8. *If $u \in B^m_{-1,-1}(I)$ with $1 \leq m \leq N+1$, then for any $\phi \in P_N$,*

$$|(u,\phi) - \langle u,\phi\rangle_N| \leq c\sqrt{\frac{(N-m+1)!}{N!}}(N+m)^{-(m+1)/2}\|\partial_x^m u\|_{\omega^{m-1,m-1}}\|\phi\|, \tag{4.99}$$

where c is a positive constant independent of m, N, ϕ and u.

The convergence of the scheme (4.94), under the aforementioned assumptions on $a_\pm, b_\pm$ and the variable coefficients a, b, is presented below.

Theorem 4.5. *Let u and u_N be the solutions of (4.52)–(4.53) and (4.94), respectively. If*

$$\begin{aligned}& a,b,a',b' \in L^\infty(I),\ \ u \in H^2(I),\ \ \partial_x^2 u \in B^{m-2}_{0,0}(I),\\ & \partial_x(au') \in B^{m-2}_{0,0}(I),\ \ \partial_x(bu) \in B^{m-1}_{0,0}(I),\ \ g \in B^k_{-1,-1}(I),\end{aligned} \tag{4.100}$$

with $2 \leq m \leq N+1$ and $1 \leq k \leq N+1$, then

$$\begin{aligned}&\|u-u_N\|_1\\ &\leq c\sqrt{\frac{(N-m+1)!}{N!}}(N+m)^{(1-m)/2}\Big(\|u\|_2 + \|\partial_x^m u\|_{\omega^{m-2,m-2}}\\ &+ \|\partial_x^{m-1}(au')\|_{\omega^{m-2,m-2}} + \|\partial_x^m(bu)\|_{\omega^{m-1,m-1}}\Big)\\ &+ c\sqrt{\frac{(N-k+1)!}{N!}}(N+k)^{-(k+1)/2}\|\partial_x^k g\|_{\omega^{k-1,k-1}},\end{aligned} \tag{4.101}$$

where c is a positive constant only depending on the L^∞-norms of a, b, a' and b'.

Proof. Let Π_N^2 be the same as in Theorem 4.2, and set $\phi = \Pi_N^2 u$ and $e_N = u_N - \phi$. Then by (4.53) and (4.94),

$$\begin{aligned}\mathscr{B}_N(e_N, e_N) &= \mathscr{B}_N(u_N, e_N) - \mathscr{B}_N(\phi, e_N) \\ &= \langle g, e_N \rangle_N - (g, e_N) + \mathscr{B}(u, e_N) - \mathscr{B}_N(\phi, e_N).\end{aligned}$$

Using Lemma 4.8 yields

$$|\langle g, e_N \rangle_N - (g, e_N)| \le c\sqrt{\frac{(N-k+1)!}{N!}}(N+k)^{-(k+1)/2}\|\partial_x^k g\|_{\omega^{k-1,k-1}}\|e_N\|. \quad (4.102)$$

By the definitions (4.52) and (4.95),

$$\begin{aligned}\left|\mathscr{B}(u, e_N) - \mathscr{B}_N(\phi, e_N)\right| &\le \left|\left(au' - I_N(a\phi'), e_N'\right)\right| + \left|(bu, e_N) - \langle b\phi, e_N \rangle_N\right| \\ &:= T_a + T_b,\end{aligned}$$

where we used the exactness (3.189) and the property (4.81) to eliminate the boundary values. Using (3.191) and Theorem 3.44, we find that

$$\begin{aligned}T_a &\le \left|\left(au' - I_N(au'), e_N'\right)\right| + \left|\left(I_N(au' - a\phi'), e_N'\right)\right| \\ &\le \left(\|au' - I_N(uu')\| + \|I_N(au' - a\phi')\|\right)\|e_N'\| \\ &\le c\Big(\sqrt{\frac{(N-m+1)!}{N!}}(N+m)^{(1-m)/2}\|\partial_x^{m-1}(au')\|_{\omega^{m-2,m-2}} \\ &\quad + \|I_N(au' - a\phi')\|\Big)\|e_N'\|.\end{aligned}$$

Moreover, by (3.191) and Lemma 4.8,

$$\begin{aligned}T_b &\le \left|(bu, e_N) - \langle bu, e_N \rangle_N\right| + \left|\langle bu - b\phi, e_N \rangle_N\right| \\ &\le c\Big(\sqrt{\frac{(N-m+1)!}{N!}}(N+m)^{-(m+1)/2}\|\partial_x^m(bu)\|_{\omega^{m-1,m-1}}\|e_N\| \\ &\quad + \|I_N(bu - b\phi)\|_N\|e_N\|_N\Big) \\ &\le c\Big(\sqrt{\frac{(N-m+1)!}{N!}}(N+m)^{-(m+1)/2}\|\partial_x^m(bu)\|_{\omega^{m-1,m-1}} \\ &\quad + \|I_N(bu - b\phi)\|\Big)\|e_N\|.\end{aligned}$$

Thanks to (4.81), we have

$$(u-\phi)(\pm 1) = 0, \quad (u-\phi)'(\pm 1) = 0.$$

Thus, we obtain from Lemma 3.11 and Theorem 4.2 that

$$\begin{aligned}\|I_N(au'-a\phi')\| &\le c(\|au'-a\phi'\|+N^{-1}\|(au'-a\phi')'\|_{\omega^{1,1}})\\ &\le c\Big((\|a\|_\infty+N^{-1}\|a'\|_\infty)\|(u-\phi)'\|+N^{-1}\|a\|_\infty\|(u-\phi)''\|\Big)\\ &\le c\sqrt{\frac{(N-m+1)!}{N!}}(N+m)^{(1-m)/2}\big(\|u\|_2+\|\partial_x^m u\|_{\omega^{m-2,m-2}}\big),\end{aligned}$$

and

$$\begin{aligned}\|I_N(bu-b\phi)\| &\le c(\|bu-b\phi\|+N^{-1}\|(bu-b\phi)'\|_{\omega^{1,1}})\\ &\le c\sqrt{\frac{(N-m+1)!}{N!}}(N+m)^{-(1+m)/2}\big(\|u\|_2+\|\partial_x^m u\|_{\omega^{m-2,m-2}}\big).\end{aligned}$$

Consequently, we derive from (4.98) and the above estimates that

$$\begin{aligned}\|e_N\|_1 \le\, & c\sqrt{\frac{(N-m+1)!}{N!}}(N+m)^{(1-m)/2}\big(\|u\|_2+\|\partial_x^m u\|_{\omega^{m-2,m-2}}+\\ &+\|\partial_x^{m-1}(au')\|_{\omega^{m-2,m-2}}+\|\partial_x^m(bu)\|_{\omega^{m-1,m-1}}\big)\\ &+c\sqrt{\frac{(N-k+1)!}{N!}}(N+k)^{-(k+1)/2}\|\partial_x^k g\|_{\omega^{k-1,k-1}}.\end{aligned}$$

We complete the proof by using the triangle inequality and Theorem 4.2. □

4.5.4 Helmholtz Equation

As the last example of this chapter, we consider the 1-D Helmholtz equation with complex-valued solution:

$$\begin{aligned}&-u''-k^2u=f, \quad r\in I:=(0,1),\\ &u(0)=0, \quad u'(1)-iku(1)=h,\end{aligned} \tag{4.103}$$

where k is called the wave number. We refer to Sect. 9.1 for more details on the background of the Helmholtz equation as well as its spectral approximation in multi-dimensional settings.

Note that this problem does not fit the general framework that we used for previous examples, since the problem is indefinite due to the negative sign in front of k^2.

The solution of (4.103) is increasingly oscillatory as k increases, so the number of unknowns in a numerical approximation should increase properly with k and it is thus important to derive error estimates with explicit dependence on k. The first step is to derive *a priori* estimates for the exact solution and characterize the dependence on k explicitly. To this end, we consider the following weak formulation of (4.103):

$$\begin{cases} \text{Find } u \in X = \{u \in H^1(I) \ : \ u(0) = 0\} \text{ such that} \\ \mathscr{B}(u,v) = \int_0^1 \big(u'\bar{v}' - k^2 u\bar{v}\big)dr - iku(1)\bar{v}(1) \\ \qquad = \int_0^1 f\bar{v}dr + h\bar{v}(1), \quad \forall v \in X, \end{cases} \tag{4.104}$$

where $\bar{v}$ is the complex conjugate of v. With a slight abuse of notation, we still use $X, H^1(I), X_N$ etc. to denote the spaces of complex-valued functions.

One can show, using a standard "Fredholm alternative" argument, that the problem (4.104) admits a unique solution (see, e.g., Douglas et al. (1993), Ihlenburg and Babuška (1995)). One may also refer to Theorem 2.1 in Shen and Wang (2005) for a direct proof.

4.5.4.1 *A Priori* Estimates

Theorem 4.6. *Let u be the solution of* (4.104). *If $f \in L^2(I)$, then we have*

$$\|u'\| + k\|u\| \le c(\|f\| + |h|), \tag{4.105}$$

and

$$|u|_2 \le ck\big(\|f\| + |h|\big) + \|f\|, \tag{4.106}$$

where c is a positive constant independent of k, u, f and h.

Proof. We shall use the argument in Melenk (1995) (see also Cummings and Feng (2006)). The key step is to choose a suitable second test function which enables us to obtain *a priori* estimates without using the Green's functions as in Douglas et al. (1993) and Ihlenburg and Babuška (1995, 1997). In the following proof, $\varepsilon_j > 0$, $1 \le j \le 3$, are adjustable real numbers.

We first take $v = u$ in (4.104). The imaginary and real parts of the resulting equation are

$$\begin{aligned} -k|u(1)|^2 &= \mathrm{Im}(h\bar{u}(1)) + \mathrm{Im}(f,u), \\ \|u'\|^2 - k^2\|u\|^2 &= \mathrm{Re}(h\bar{u}(1)) + \mathrm{Re}(f,u). \end{aligned} \tag{4.107}$$

Applying the Cauchy–Schwarz inequality to the imaginary part leads to

$$k|u(1)|^2 \le \frac{k}{2}|u(1)|^2 + \frac{1}{2k}|h|^2 + \frac{\varepsilon_1 k}{2}\|u\|^2 + \frac{1}{2\varepsilon_1 k}\|f\|^2, \tag{4.108}$$

and likewise, we obtain from the real part that

$$\|u'\|^2 \le k^2\|u\|^2 + \varepsilon_2 k^2|u(1)|^2 + \frac{1}{4\varepsilon_2 k^2}|h|^2 + \frac{\varepsilon_3 k^2}{2}\|u\|^2 + \frac{1}{2\varepsilon_3 k^2}\|f\|^2. \tag{4.109}$$

As a consequence of (4.108)-(4.109) with $\varepsilon_2 = \frac{\varepsilon_3}{2\varepsilon_1}$, we have

$$\begin{aligned} |u(1)|^2 &\le \varepsilon_1 \|u\|^2 + \frac{1}{k^2}|h|^2 + \frac{1}{\varepsilon_1 k^2}\|f\|^2, \\ \|u'\|^2 &\le (1+\varepsilon_3)k^2\|u\|^2 + \Big(\frac{\varepsilon_3}{2\varepsilon_1} + \frac{\varepsilon_1}{2\varepsilon_3 k^2}\Big)|h|^2 \\ &\quad + \Big(\frac{\varepsilon_3}{2\varepsilon_1^2} + \frac{1}{2\varepsilon_3 k^2}\Big)\|f\|^2. \end{aligned} \tag{4.110}$$

Hence, it remains to bound $k^2\|u\|^2$.

We now take $v = 2ru'$ in (4.104), which belongs to X via a usual regularity argument. By integration by parts,

$$\begin{aligned} 2\mathrm{Re}\big(u', (ru')'\big) &= |u'(1)|^2 + \|u'\|^2, \\ -2k^2\mathrm{Re}\big(u, ru'\big) &= -k^2|u(1)|^2 + k^2\|u\|^2. \end{aligned} \tag{4.111}$$

Therefore, the real part of (4.104) with $v = 2ru'$ is

$$\begin{aligned} \|u'\|^2 + k^2\|u\|^2 + |u'(1)|^2 &= k^2|u(1)|^2 \\ &\quad + 2\mathrm{Re}\big((\mathrm{i}ku(1)+h)\bar{u}'(1)\big) + 2\mathrm{Re}(f, ru'). \end{aligned} \tag{4.112}$$

Using Cauchy–Schwarz inequality leads to

$$\begin{aligned} \|u'\|^2 + k^2\|u\|^2 + |u'(1)|^2 &\le k^2|u(1)|^2 + \frac{1}{2}|u'(1)|^2 \\ &\quad + 2k^2|u(1)|^2 + 2|h|^2 + \frac{1}{2}\|u'\|^2 + 2\|f\|^2. \end{aligned} \tag{4.113}$$

Then we obtain from (4.110) that

$$\begin{aligned} \frac{1}{2}\|u'\|^2 + k^2\|u\|^2 + \frac{1}{2}|u'(1)|^2 &\le 3\varepsilon_1 k^2\|u\|^2 + c\big(|h|^2 + (\varepsilon_1^{-1}+2)\|f\|^2\big) \\ &\le \frac{k^2}{2}\|u\|^2 + c\big(|h|^2 + \|f\|^2\big), \end{aligned}$$

where we took $\varepsilon_1 = 1/6$ to derive the last inequality. Hence, (4.105) follows.

Taking L^2-norm on both sides of the equation: $-u'' - k^2 u = f$, and using (4.105), we obtain (4.106). □

4.5.4.2 Convergence Analysis

The Legendre-Galerkin approximation to (4.104) is

$$\begin{cases} \text{Find } u_N \in X_N = \big\{u \in P_N \,:\, u(0) = 0\big\} \text{ such that} \\ \mathscr{B}(u_N, v_N) = (f, v_N) + h\bar{v}_N(1), \quad \forall v_N \in X_N. \end{cases} \tag{4.114}$$

Since for $u_N \in X_N$, we have $ru_N' \in X_N$, the proof of Theorem 4.6 is also valid for the discrete system (4.114).

Theorem 4.7. *Let u_N be a solution of* (4.114)*. Then Theorem* 4.6 *holds with u_N in the place of u.*

An immediate consequence is the following:

Corollary 4.1. *The problem* (4.114) *admits a unique solution.*

Proof. Since (4.114) is a finite dimensional linear system, it suffices to prove the uniqueness. Now, let u_N be a solution of (4.114) with $f \equiv 0$ and $h = 0$, we derive from Theorem 4.7 that $u_N \equiv 0$ which implies the uniqueness. □

The following approximation results play an important role in the error analysis.

Lemma 4.9. *Let $I = (0,1)$. There exists a mapping ${}_0\Pi_N^1 : X \to X_N$ such that*

$$\big(({}_0\Pi_N^1 u - u)', v_N'\big) = 0, \quad \forall v_N \in X_N. \tag{4.115}$$

Moreover, for any

$$u \in X \cap \widehat{H}^m(I) := \{u : (r - r^2)^{(l-1)/2} \partial_r^l u \in L^2(I),\ 1 \le l \le m\}$$

with $1 \le m \le N+1$,

$$\begin{aligned} &\|{}_0\Pi_N^1 u - u\|_\mu \\ &\le c\sqrt{\frac{(N-m+1)!}{N!}}(N+m)^{\mu-(m+1)/2}\|(r-r^2)^{(m-1)/2}\partial_r^m u\|, \end{aligned} \tag{4.116}$$

where $\mu = 0, 1$ and c is a positive constant independent of m, N and u.

Proof. It suffices to show the result for real-valued function $v(x)$ in $\Lambda = (-1,1)$. Let $\Pi_N^{1,0}$ be the H_0^1-orthogonal projection operator defined in (3.290) with $\alpha = \beta = 0$. For any $v \in H^1(\Lambda)$ with $v(-1) = 0$, let

$$v_*(x) = v(x) - \frac{1+x}{2}v(1) \quad \Rightarrow \quad v_* \in H_0^1(\Lambda),$$

and likewise for ϕ_* with $\phi \in P_N$ and $\phi(-1) = 0$. Define

$${}_0\widehat{\Pi}_N^1 v(x) = \Pi_N^{1,0} v_*(x) + \frac{1+x}{2}v(1).$$

Observe from (3.282) that

$$\begin{aligned} \big(\partial_x({}_0\widehat{\Pi}_N^1 v - v), \partial_x\phi\big) &= \big(\partial_x(\Pi_N^{1,0} v_* - v_*), \partial_x\phi_*\big) + \frac{v(1)}{2}\int_{-1}^1 \partial_x(\Pi_N^{1,0} v_* - v_*)(x)dx \\ &= \big(\partial_x(\Pi_N^{1,0} v_* - v_*), \partial_x\phi_*\big) = 0, \end{aligned}$$

for all $\phi \in P_N$ with $\phi(-1)=0$. Hence, by Theorem 3.39,

$$\begin{aligned}\|{}_0\widehat{\Pi}^1_N v - v\|_\mu &= \|\Pi^{1,0}_N v_* - v_*\|_\mu \\ &\le c\sqrt{\frac{(N-m+1)!}{N!}}(N+m)^{\mu-(m+1)/2}\|\partial_x^m v_*\|_{\omega^{m-1,m-1}}.\end{aligned}$$

Obviously, $\partial_x^m v_* = \partial_x^m v$ for $m \ge 2$, and for $m=1$, we obtain from (B.44) that

$$\|\partial_x v_*\| \le \|\partial_x v\| + c|v(1)| \le c\|\partial_x v\|.$$

Thus, setting

$$x = 2r-1, \quad r \in (0,1), \quad u(r) = v(x), \quad {}_0\Pi^1_N u = {}_0\widehat{\Pi}^1_N v,$$

we obtain (4.115) and (4.116). □

Now, we are ready to prove the following convergence result.

Theorem 4.8. *Let u and u_N be the solutions of* (4.104) *and* (4.114), *respectively. If $u \in X \cap \widehat{H}^m(I)$ with $1 \le m \le N+1$, we have*

$$\begin{aligned}|u-u_N|_1 + k\|u-u_N\| \le c\big(1+k^2N^{-1}+kN^{-1/2}\big)\times \\ \sqrt{\frac{(N-m+1)!}{N!}}(N+m)^{(1-m)/2}\|(r-r^2)^{(m-1)/2}\partial_r^m u\|,\end{aligned} \tag{4.117}$$

where c is a positive constant independent of k, m, N and u.

Proof. Let $e_N = u_N - {}_0\Pi^1_N u$ and $\tilde{e}_N = u - {}_0\Pi^1_N u$. By (4.104) and (4.114), we have

$$\mathscr{B}(u-u_N, v_N) = 0, \quad \forall v_N \in X_N.$$

Hence, we derive from (4.115) and the definition of $\mathscr{B}(\cdot,\cdot)$ that for any $v_N \in X_N$,

$$\begin{aligned}\mathscr{B}(e_N, v_N) =& \mathscr{B}(u - {}_0\Pi^1_N u, v_N) \\ =& -k^2(\tilde{e}_N, v_N) - ik\tilde{e}_N(1)\bar{v}_N(1).\end{aligned} \tag{4.118}$$

We can view (4.118) in the form of (4.104) with $u = e_N$, $h = -ik\tilde{e}_N(1)$ and $f = -k^2\tilde{e}_N$. Hence, as a direct consequence of Theorem 4.6, we have

$$|e_N|_1^2 + k^2\|e_N\|^2 \le ck^2\big(k^2\|\tilde{e}_N\|^2 + |\tilde{e}_N(1)|^2\big). \tag{4.119}$$

Furthermore, using the Sobolev inequality (B.33) and Lemma 4.9 leads to

$$\begin{aligned}|\tilde{e}_N(1)| &\le c\|\tilde{e}_N\|^{1/2}\|\tilde{e}_N\|_1^{1/2} \\ &\le c\sqrt{\frac{(N-m+1)!}{N!}}(N+m)^{-m/2}\|(r-r^2)^{(m-1)/2}\partial_r^m u\|.\end{aligned}$$

Hence, using Lemma 4.9 again yields

$$|u-u_N|_1+k\|u-u_N\| \le c\big(|e_N|_1+k\|e_N\|+|\tilde{e}_N|_1+k\|\tilde{e}_N\|\big)$$
$$\le c\big(1+k^2N^{-1}+kN^{-1/2}\big)\sqrt{\frac{(N-m+1)!}{N!}}(N+m)^{(1-m)/2}\|(r-r^2)^{(m-1)/2}\partial_r^m u\|.$$

This ends the proof. □

Problems

4.1. Show that under the assumption (4.3), the bilinear form $\mathscr{B}(\cdot,\cdot)$ defined by (4.9) is continuous and coercive in $H^1_\diamond(I)\times H^1_\diamond(I)$.

4.2. Let $\{h_j\}_{j=0}^N$ be the Lagrange basis polynomials relative to the Jacobi-Gauss-Radau points $\{x_j\}_{j=0}^N$ with $x_0=-1$ (see Theorem 3.26). Let $\widetilde{D}=(d_{kj}:=h_j'(x_k))_{1\le k,j\le N}$ be the differentiation matrix corresponding to the interior collocation point (see (3.163)). Write down the matrix form of the Jacobi-Gauss-Radau collocation method for

$$u'(x)=f(x),\quad x\in(-1,1);\quad u(-1)=c_-,$$

where $f\in C[-1,1]$ and c_- is a given value. Use the uniqueness of the approximate solution to show that the matrix $\tilde{D}$ is nonsingular.

4.3. Prove Lemma 4.8.

4.4. Consider the Burgers' equation:

$$\frac{\partial u}{\partial t}=\varepsilon\frac{\partial^2 u}{\partial x^2}-u\frac{\partial u}{\partial x},\quad \varepsilon>0. \tag{4.120}$$

(i) Verify that it has the soliton solution

$$u(x,t)=\kappa\Big[1-\tanh\Big(\frac{\kappa(x-\kappa t-x_c)}{2\varepsilon}\Big)\Big], \tag{4.121}$$

where the parameter $\kappa>0$ and the center $x_c\in\mathbb{R}$.

(ii) Take $\varepsilon=0.1,\kappa=0.5,x_c=-3,x\in[-5,5]$, and impose the initial value $u(x,0)$ and the boundary conditions $u(\pm5,t)$ by using the exact solution. Use the Crank-Nicolson leap-frog scheme to in time (see (1.2)–(1.3)), and the Chebyshev collocation method in space to solve the equation. Output the discrete maximum errors for $\tau=10^{-k}$ (time step size) with $k=2,3,4$ and $N=32,64,128$ at $t=12$. Refer to Table 1 in Wu et al. (2003) for the behavior of the errors (obtained by other means).

(iii) Replace the Chebyshev-collocation method in (ii) by the Chebyshev-Galerkin method. Do the same test and compare two methods. Refer to Sect. 3.4.3 for the

Chebyshev differentiation process using FFT and to Trefethen (2000) for a handy MATLAB code for this process.

(iv) Consider the Burgers' equation (4.120) in $(-1,1)$ with the given data

$$u(\pm 1,t)=0, \quad u(x,0)=-\sin(\pi x), \quad x\in[-1,1]. \tag{4.122}$$

Solve this problem by the methods in (ii) and (iii) by taking $\varepsilon=0.02, \tau=10^{-4}$ and $N=128$ and plot the numerical solution at $t=1$. Refer to Shen and Wang (2007b) for some profiles of the numerical solution (obtained by other means).

4.5. Consider the Fisher equation

$$\frac{\partial u}{\partial t}=\frac{\partial^2 u}{\partial x^2}+u(1-u). \tag{4.123}$$

(i) Verify that it has the traveling solution

$$u(x,t)=\Big[1+\exp\Big(\frac{x}{\sqrt{6}}-\frac{5}{6}t\Big)\Big]^{-2}. \tag{4.124}$$

(ii) Since $u(x,t)\to 0$ (resp. 1) as $x\to+\infty$ (resp. $-\infty$), we can approximate (4.123) in $(-L,L)$, where L is large enough so that the wave front does not reach the boundary $x=L$, by imposing the boundary conditions

$$u(-L,t)=1, \quad u(L,t)=0,$$

and taking the initial value as $u(x,0)$. Use the second-order splitting scheme (D.30) with $Au=\partial_x^2 u$ and $Bu=u(1-u)$ in time, and the Legendre-Galerkin method in space to solve this problem with $\tau=10^{-3}, N=128, L=100$ up to $t=6$. Output the discrete maximum errors between the exact and approximate solutions at $t=1,2,\ldots,6$. An advantage of the splitting scheme is that the subproblem (a Bernoulli's equation for t):

$$\frac{\partial u}{\partial t}=u(1-u)$$

can be solved exactly, so it suffices to solve a linear equation in each step. Refer to Wang and Shen (2005) for this numerical study by a mapping technique.

Chapter 5
Volterra Integral Equations

This chapter is devoted to spectral approximations of the Volterra integral equation (VIE):

$$y(t) + \int_0^t R(t,\tau)y(\tau)d\tau = f(t), \quad t \in [0,T], \tag{5.1}$$

where the source function f and the kernel function R are given, and $y(t)$ is the unknown function. We shall also implement and analyze spectral algorithms for solving the VIE with weakly singular kernel:

$$y(t) + \int_0^t (t-\tau)^{-\mu} R(t,\tau)y(\tau)d\tau = f(t), \quad t \in [0,T], \ \ 0 < \mu < 1, \tag{5.2}$$

where $R(t,t) \neq 0$ for $t \in [0,T]$.

While there have been many existing numerical methods for solving VIEs (see, e.g., Brunner (2004) and the references therein), very few are based on spectral approximations. In Elnagar and Kazemi (1996), a Chebyshev spectral method was developed to solve nonlinear Volterra-Hammerstein integral equations, and in Fujiwara (2006), it was applied to the Fredholm integral equations of the first kind under multiple-precision arithmetic. However, no theoretical analysis was provided to justify the high accuracy of the proposed methods.

It is known that the Fredholm type equations behave more or less like a boundary value problem (see, e.g., Delves and Mohanmed (1985)). As a result, some efficient numerical methods useful for boundary values problems (such as spectral methods) can be used directly to handle the Fredholm type equations (cf. Delves and Mohanmed (1985)). However, the Volterra equation (5.1) behaves like an initial value problem. Therefore, it is not straightforward to apply spectral methods to the Volterra type equations. On the other hand, an essential difference between (5.1) and a standard initial value problem is that numerical methods for the former require storage of values at all the grid points, while they only requires information at a fixed number of previous grid points for the latter.

This chapter is organized as follows. We devote the first two sections to describing spectral algorithms, including one with Legendre-collocation method

J. Shen et al., *Spectral Methods: Algorithms, Analysis and Applications*,
Springer Series in Computational Mathematics 41, DOI 10.1007/978-3-540-71041-7_5,

and one with Jacobi-Galerkin method, for VIEs with regular kernels. We then propose an efficient Jacobi-collocation method for VIEs with weakly singular kernel in Sect. 5.3. Finally, we discuss applications of these spectral methods to delay differential equations.

5.1 Legendre-Collocation Method for VIEs

For ease of implementation and analysis, we make the change of variable

$$t = T(1+x)/2, \quad x = 2t/T - 1, \quad x \in I := [-1,1], \; t \in [0,T], \tag{5.3}$$

under which (5.1) is transformed into

$$u(x) + \int_0^{T(1+x)/2} R\big(T(1+x)/2, \tau\big) y(\tau) d\tau = g(x), \quad x \in I, \tag{5.4}$$

where we have set

$$u(x) = y\big(T(1+x)/2\big), \quad g(x) = f\big(T(1+x)/2\big). \tag{5.5}$$

We further convert the interval $[0, T(1+x)/2]$ to $[-1,x]$ by using the linear transformation: $\tau = T(1+s)/2, s \in [-1,x]$. Then, (5.4) becomes

$$u(x) + \int_{-1}^{x} K(x,s) u(s) ds = g(x), \quad x \in I, \tag{5.6}$$

where

$$K(x,s) = \frac{T}{2} R\big(T(1+x)/2, T(1+s)/2\big), \quad x \in I, \; s \in [-1,x]. \tag{5.7}$$

5.1.1 Numerical Algorithm

Let $\{x_i\}_{i=0}^N$ be a set of Legendre-Gauss, or Legendre-Gauss-Radau or Legendre-Gauss-Lobatto collocation points (see Theorem 3.29). A first approximation to (5.6) using a Legendre collocation approach is

$$\begin{cases} \text{Find } u_N \in P_N \text{ such that} \\ u_N(x_i) + \displaystyle\int_{-1}^{x_i} K(x_i,s) u_N(s) ds = g(x_i), \quad 0 \le i \le N. \end{cases} \tag{5.8}$$

However, the integral term in (5.8) can not be evaluated exactly. So we transform the integral interval $[-1,x_i]$ to $[-1,1]$ and use a Gaussian type quadrature rule to approximate the integral. More precisely, under the linear transformation

$$
\begin{aligned}
s &:= s^{(i)} = \frac{1+x_i}{2}\theta + \frac{x_i - 1}{2}, \\
\theta &:= \theta^{(i)} = \frac{2}{1+x_i}s + \frac{1-x_i}{1+x_i}, \qquad \theta \in I, \; s \in [-1, x_i],
\end{aligned}
\tag{5.9}
$$

the scheme (5.8) becomes

$$
u_N(x_i) + \frac{1+x_i}{2}\int_{-1}^{1} K(x_i, s(x_i,\theta))u_N(s(x_i,\theta))d\theta = g(x_i), \quad 0 \le i \le N. \tag{5.10}
$$

We then approximate the integral term by a Legendre-Gauss type quadrature formula with the notes and weights denoted by $\{\theta_j, \omega_j\}_{j=0}^{M}$, leading to the Legendre collocation scheme (with numerical integration) for (5.6):

$$
\begin{cases}
\text{Find } u_N \in P_N \text{ such that} \\
u_N(x_i) + \dfrac{1+x_i}{2}\displaystyle\sum_{j=0}^{M} K(x_i, s(x_i,\theta_j))u_N(s(x_i,\theta_j))\omega_j = g(x_i), \; 0 \le i \le N.
\end{cases}
\tag{5.11}
$$

It is worthwhile to point out that the collocation points $\{x_i\}_{i=0}^{N}$ and quadrature points $\{\theta_j\}_{j=0}^{M}$ could be chosen differently in type and number. As a result, we can also use Legendre-Gauss-Radau or Legendre-Gauss-Lobatto for the integral term.

Next, we discuss the implementation of (5.11). Let $\{h_j\}_{j=0}^{N}$ be the Lagrange basis polynomials associated with the Legendre-Gauss-type points $\{x_j\}_{j=0}^{N}$. We expand the approximate solution u_N as

$$
u_N(x) = \sum_{k=0}^{N} u_N(x_k)h_k(x). \tag{5.12}
$$

Inserting it into (5.11) leads to

$$
u_N(x_i) + \frac{1+x_i}{2}\sum_{k=0}^{N}\Big(\sum_{j=0}^{M} K(x_i, s(x_i,\theta_j))h_k(s(x_i,\theta_j))\omega_j\Big)u_N(x_k) = g(x_i), \tag{5.13}
$$

for all $0 \le i \le N$. Setting

$$
\begin{aligned}
a_{ik} &= \frac{1+x_i}{2}\sum_{j=0}^{M} K(x_i, s(x_i,\theta_j))h_k(s(x_i,\theta_j))\omega_j, \quad \mathbf{A} = (a_{ik})_{0\le i,k\le N}, \\
\mathbf{u} &= (u_N(x_0), u_N(x_1), \ldots, u_N(x_N))^T, \quad \mathbf{g} = (g(x_0), g(x_1), \ldots, g(x_N))^T,
\end{aligned}
$$

the system (5.13) reduces to

$$
(\mathbf{A} + \mathbf{I})\mathbf{u} = \mathbf{g}. \tag{5.14}
$$

We observe that, as with a typical collocation scheme, the coefficient matrix of (5.14) is full. Moreover, all unknowns $\{u_N(x_i)\}_{i=0}^{N}$ are coupled together and the scheme (5.13) requires the semi-local information $\{K(x_i, s(x_i,\theta_j))\}_{j=0}^{i}$

(note that $-1 \le s(x_i, \theta_j) \le x_i$). As a comparison, to compute $u_N(x_i)$, piecewise-polynomial collocation methods or product integration methods only use the semi-local information of both the approximate solution u_N and the kernel K, namely, $\{u_N(x_j)\}_{j=0}^{i-1}$ and $\{K(x_i, \beta_j)\}$ where $\{-1 \le \beta_j \le x_i\}$ are some collocation points. Indeed, this allows us to obtain, as to be demonstrated below, a spectral accuracy instead of an algebraic order of accuracy for the proposed scheme (5.13).

We see that the entries of $\mathbf{A}$ involve the computations of the Lagrange basis polynomials at the non-interpolation points, i.e., $\{h_k(s(x_i, \theta_j))\}$. The idea for their efficient computation is to express h_k in terms of the Legendre polynomials:

$$h_k(s) = \sum_{p=0}^{N} \alpha_p^k L_p(s) \in P_N, \tag{5.15}$$

and by (3.193),

$$\alpha_p^k = L_p(x_k)\omega_k/\gamma_p \quad \text{where} \quad \gamma_p = \frac{2}{2p+1}, \quad 0 \le p < N, \tag{5.16}$$

and $\gamma_N = 2/(2N+1)$ for the Legendre-Gauss and Legendre-Gauss-Radau formulas, and $\gamma_N = 2/N$ for the Legendre-Gauss-Lobatto case. Consequently,

$$h_k(s) = \omega_k \sum_{p=0}^{N} \frac{L_p(x_k)}{\gamma_p} L_p(s), \quad 0 \le k \le N. \tag{5.17}$$

5.1.2 Convergence Analysis

We now analyze the convergence of the scheme (5.11). For clarity of presentation, we assume that the collocation and quadrature points in (5.11) are of the Legendre-Gauss-Lobatto type with $M = N$. The other cases can be treated in a similar fashion.

In what follows, we need to use the asymptotic estimate of the Lebesgue constant (see, e.g., Qu and Wong (1988)):

$$\Lambda_N := \max_{|x| \le 1} \sum_{j=0}^{N} |h_j(x)| \simeq \sqrt{N}, \quad N \gg 1. \tag{5.18}$$

The notation and Sobolev spaces used below are the same as those in Chap. 3.

Theorem 5.1. *Let u and u_N be the solutions of* (5.6) *and* (5.11) *with $M = N$, respectively. Assume that*

$$K \in L^\infty(D) \cap L^\infty(I; B^k_{-1,-1}(I)), \ \partial_x K \in L^\infty(D), \ u \in B^m_{-1,-1}(I), \tag{5.19}$$

where $D = \{(x,s) : -1 \le s \le x \le 1\}$ *and* $1 \le k,m \le N+1$. *Then we have*

$$\begin{aligned}\|u-u_N\| \le\, & c\sqrt{\frac{(N-k+1)!}{N!}}(N+k)^{-k/2}\max_{|x|\le 1}\|\partial_s^k K(x,\cdot)\|_{\omega^{k-1,k-1}}\|u\| \\ & + c\sqrt{\frac{(N-m+1)!}{N!}}(N+m)^{-(m+1)/2}\|\partial_x^m u\|_{\omega^{m-1,m-1}},\end{aligned} \tag{5.20}$$

and

$$\begin{aligned}\|u-u_N\|_\infty \le\, & c\sqrt{\frac{(N-k+1)!}{N!}}(N+k)^{-k/2}\max_{|x|\le 1}\|\partial_s^k K(x,\cdot)\|_{\omega^{k-1,k-1}}\|u\| \\ & + c\sqrt{\frac{(N-m+1)!}{N!}}(N+m)^{-m/2}\|\partial_x^m u\|_{\omega^{m-1,m-1}},\end{aligned} \tag{5.21}$$

where c is a positive constant independent of k, m, N and u.

Proof. We first prove (5.20). Rewrite (5.11) as

$$\begin{aligned}u_N(x_i) + & \frac{1+x_i}{2}\int_{-1}^{1} K(x_i, s(x_i,\theta))u_N(s(x_i,\theta))d\theta \\ & = g(x_i) + J_1(x_i), \quad 0 \le i \le N,\end{aligned} \tag{5.22}$$

where

$$\begin{aligned}J_1(x) = & \frac{1+x}{2}\int_{-1}^{1} K(x, s(x,\theta))u_N(s(x,\theta))d\theta \\ & - \frac{1+x}{2}\sum_{j=0}^{N} K(x, s(x,\theta_j))u_N(s(x,\theta_j))\omega_j.\end{aligned} \tag{5.23}$$

Let I_N be the Legendre-Gauss-Lobatto interpolation operator. Transforming the integral term in (5.22) back to $[-1,x]$ by using (5.9), we reformulate (5.22) as

$$u_N(x) + I_N\int_{-1}^{x} K(x,s)u_N(s)ds = (I_N g)(x) + (I_N J_1)(x), \quad x \in I. \tag{5.24}$$

Clearly, by (5.6),

$$(I_N g)(x) = (I_N u)(x) + I_N\int_{-1}^{x} K(x,s)u(s)ds, \quad x \in I. \tag{5.25}$$

Denote $e = u_N - u$. Inserting (5.25) into (5.24) leads to the error equation:

$$e(x) + \int_{-1}^{x} K(x,s)e(s)ds = (I_N J_1)(x) + J_2(x) + J_3(x), \tag{5.26}$$

where

$$\begin{aligned}J_2(x) &= (I_N u - u)(x), \\ J_3(x) &= \int_{-1}^{x} K(x,s)e(s)ds - I_N\Big(\int_{-1}^{x} K(x,s)e(s)ds\Big).\end{aligned}$$

Thus, we have

$$|e(x)| \le G(x) + K_{\max} \int_{-1}^{x} |e(s)|ds, \tag{5.27}$$

where

$$K_{\max} := \max_{D} |K(x,s)|, \quad G := |I_N J_1| + |J_2| + |J_3|.$$

Using the Gronwall inequality (B.9) leads to

$$|e(x)| \le G(x) + K_{\max} e^{2K_{\max}} \int_{-1}^{x} G(s)ds, \quad \forall x \in I. \tag{5.28}$$

This implies

$$\|e\| \le c\|G\| \le c\big(\|I_N J_1\| + \|J_2\| + \|J_3\|\big), \tag{5.29}$$

where c depends on $K_{\max}$.

It remains to estimate the three terms on the right hand side of (5.29). By Lemma 4.8,

$$\begin{aligned} |J_1(x)| &= \frac{1+x}{2}\big\{\big(K(x,s(x,\cdot)),u_N(s(x,\cdot))\big) - \big\langle K(x,s(x,\cdot)),u_N(s(x,\cdot))\big\rangle_N\big\} \\ &\le c\sqrt{\frac{(N-k+1)!}{N!}}(N+k)^{-(k+1)/2}\times \\ &\quad \frac{1+x}{2}\|\partial_\theta^k K(x,s(x,\cdot))\|_{\omega^{k-1,k-1}}\|u_N(s(x,\cdot))\|. \end{aligned}$$

A direct calculation using (5.9) yields

$$\begin{aligned} \|\partial_\theta^k K(x,s(x,\cdot))\|^2_{\omega^{k-1,k-1}} &= \int_{-1}^{1} |\partial_\theta^k K(x,s(x,\theta))|^2(1-\theta^2)^{k-1}d\theta \\ &= \frac{1+x}{2}\int_{-1}^{x} |\partial_s^k K(x,s)|^2(x-s)^{k-1}(1+s)^{k-1}ds \\ &\le \|\partial_s^k K(x,\cdot)\|^2_{\omega^{k-1,k-1}}, \end{aligned}$$

and

$$\frac{1+x}{2}\|u_N(s(x,\cdot))\|^2 = \int_{-1}^{x} |u_N(s)|^2 ds \le \|u_N\|^2.$$

Hence, we obtain the estimate of $|J_1|$:

$$|J_1(x)| \le c\sqrt{\frac{(N-k+1)!}{N!}}(N+k)^{-(k+1)/2}\|\partial_s^k K(x,\cdot)\|_{\omega^{k-1,k-1}}\|u_N\|,$$

which, together with (5.18), implies

$$\begin{aligned}\|I_N J_1\| &\le \sqrt{2}\|I_N J_1\|_\infty \le c\|J_1\|_\infty \max_{|x|\le 1}\sum_{j=0}^{N}|h_j(x)|\\ &\le c\sqrt{\frac{(N-k+1)!}{N!}}(N+k)^{-k/2}\max_{|x|\le 1}\|\partial_s^k K(x,\cdot)\|_{\omega^{k-1,k-1}}\|u_N\|\\ &\le c\sqrt{\frac{(N-k+1)!}{N!}}(N+k)^{-k/2}\max_{|x|\le 1}\|\partial_s^k K(x,\cdot)\|_{\omega^{k-1,k-1}}\big(\|e\|+\|u\|\big).\end{aligned} \tag{5.30}$$

Next, by Theorem 3.44,

$$\|J_2\| \le c\sqrt{\frac{(N-m+1)!}{N!}}(N+m)^{-(m+1)/2}\|\partial_x^m u\|_{\omega^{m-1,m-1}}. \tag{5.31}$$

Moreover, using Theorem 3.44 with $m=1$ yields

$$\begin{aligned}\|J_3\| &\le cN^{-1}\left\|K(x,x)e(x)+\int_{-1}^{x}\partial_x K(x,s)e(s)ds\right\|\\ &\le cN^{-1}\Big(\max_{|x|\le 1}|K(x,x)|+\max_{D}\|\partial_x K\|_\infty\Big)\|e\|.\end{aligned} \tag{5.32}$$

The estimate (5.20) follows from (5.29)–(5.32), provided that N is large enough. We now turn to the proof of (5.21). Clearly, it follows from (5.27) that

$$\|e\|_\infty \le c\big(\|I_N J_1\|_\infty + \|J_2\|_\infty + \|J_3\|_\infty\big). \tag{5.33}$$

Using the inequalities (B.33) and (B.44), we obtain from Theorem 3.44 that

$$\begin{aligned}\|J_2\|_\infty &\le c\|u-I_N u\|^{1/2}\|\partial_x(u-I_N u)\|^{1/2}\\ &\le c\sqrt{\frac{(N-m+1)!}{N!}}(N+m)^{-m/2}\|\partial_x^m u\|_{\omega^{m-1,m-1}},\end{aligned} \tag{5.34}$$

and

$$\begin{aligned}\|J_3\|_\infty &\le c\|J_3\|^{1/2}\|\partial_x J_3\|^{1/2} \overset{(5.32)}{\le} cN^{-1/2}\|e\|^{1/2}\left\|\partial_x\int_{-1}^{x}K(x,s)e(s)ds\right\|\\ &\le cN^{-1/2}\|e\| \le cN^{-1/2}\|e\|_\infty.\end{aligned} \tag{5.35}$$

Finally, a combination of (5.30) (with $\|e\|_\infty$ in place of $\|e\|$) and (5.33)–(5.35) leads to the estimate (5.21). □

Remark 5.1. *As pointed out in Remark* 3.7, *if the regularity index k (resp. m) is fixed, the order of convergence in* (5.20) *is $O(N^{-k})$ (resp. $O(N^{-m})$).*

5.1.3 Numerical Results and Discussions

We present below some numerical results and discuss the extension of the proposed methods to nonlinear VIEs.

Without lose of generality, we only consider the Legendre-Gauss-Lobatto quadrature rule in (5.11), and numerical evidences show that the other two types of rules produce similar results. Consider the VIE (5.6) with

$$K(x,s) = e^{xs}, \quad g(x) = e^{4x} + \frac{1}{x+4}\left(e^{x(x+4)} - e^{-(x+4)}\right), \tag{5.36}$$

which has the exact solution $u(x) = e^{4x}$. In Table 5.1, we tabulate the maximum point-wise errors obtained by (5.11) with various N, which indicate that the desired spectral accuracy is obtained.

Table 5.1 The maximum point-wise errors

N	6	8	10	12	14
Error	3.66e-01	1.88e-02	6.57e-04	1.65e-05	3.11e-07
N	16	18	20	22	24
Error	4.57e-09	5.37e-11	5.19e-13	5.68e-14	4.26e-14

In practice, many VIEs are usually nonlinear. For instance, the nonlinear version of (5.6) may take the form

$$u(x) + \int_{-1}^{x} K(x,s,u(s))\,ds = g(x), \quad x \in [-1,1]. \tag{5.37}$$

However, the nonlinearity adds rather little to the difficulty of obtaining accurate numerical solutions. The methods described earlier remain applicable. Although our convergence theory does not cover the nonlinear case, it should be quite straightforward to establish a convergence result similar to Theorem 5.1 provided that the kernel K is Lipschitz continuous with respect to its third argument. A similar technique for the piecewise-polynomial collocation methods was used by Brunner and Tang (1989) for solving nonlinear Volterra equations. Here, we just show the basic idea and provide a numerical example to illustrate the spectral accuracy.

Let $\{x_i, \omega_i\}_{i=0}^{N}$ be the Legendre-Gauss-type quadrature nodes and weights as before. We can design a collocation method for the nonlinear VIE (5.37) similar to the linear case. More precisely, we seek $u_N \in P_N$ such that

$$u_N(x_i) + \frac{1+x_i}{2}\int_{-1}^{1} K(x_i, s(x_i,\theta), u_N(s(x_i,\theta)))d\theta = g(x_i), \quad 0 \le i \le N, \tag{5.38}$$

where $s(x,\theta)$ is given by (5.9). We further approximate the integral by the quadrature rule:

$$u_N(x_i) + \frac{1+x_i}{2}\sum_{j=0}^{N} K\big(x_i, s(x_i,\theta_j), u_N(s(x_i,\theta_j))\big)\,\omega_j = g(x_i), \quad 0 \le i \le N. \tag{5.39}$$

Notice that inserting (5.12) into the numerical scheme (5.39) leads to a nonlinear system for $\{u_N(x_i)\}_{i=0}^N$, so a suitable iterative solver for the nonlinear system (e.g., Newton's method) should be used. In the following computations, we just use a simple Jacobi-type iteration method to solve the nonlinear system, which takes about 5 to 6 iterations. More detailed discussions on solving nonlinear VIEs with iteration methods can be found in Tang and Xu (2009).

Consider (5.37) with $K(x,s,u(s)) = e^{x-3s}u^2(s)$, and

$$g(x) = -\frac{1}{2(1+36\pi^2)}(e^{-x}+36\pi^2 e^{-x} - e^{-x}\cos 6\pi x + 6\pi e^{-x}\sin 6\pi x \\ - 36e\pi^2)e^x + e^x \sin 3\pi x, \tag{5.40}$$

so that the nonlinear VIE (5.37) has the exact solution $u(x) = e^x \sin 3\pi x$.

The maximum point-wise errors are displayed in Table 5.2, and once again, the exponential convergence is observed.

Table 5.2 The maximum point-wise errors

N	6	8	10	12	14
Error	2.33e-02	7.22e-04	1.82e-05	3.15e-07	4.06e-09
N	16	18	20	22	24
Error	3.98e-11	3.05e-13	3.86e-15	3.33e-15	3.98e-15

5.2 Jacobi-Galerkin Method for VIEs

As an alternative to the Legendre collocation method, we introduce and analyze in this section a Jacobi-Galerkin method for (5.6).

Rewrite (5.6) as

$$u(x) + Su(x) = g(x) \quad \text{with} \quad Su(x) = \int_{-1}^{x} K(x,s)u(s)ds. \tag{5.41}$$

The Jacobi-Galerkin approximation to (5.41) is

$$\begin{cases} \text{Find } u_N \in P_N \text{ such that} \\ (u_N, v_N)_{\omega^{\alpha,\beta}} + (Su_N, v_N)_{\omega^{\alpha,\beta}} = (g, v_N)_{\omega^{\alpha,\beta}}, \quad \forall v_N \in P_N, \end{cases} \tag{5.42}$$

where $\omega^{\alpha,\beta}(x) = (1-x)^\alpha(1+x)^\beta$ with $\alpha, \beta > -1$, is the Jacobi weight function. Let $\pi_N^{\alpha,\beta}$ be the $L^2_{\omega^{\alpha,\beta}}$-orthogonal projection operator. We find from (3.249) that (5.42) is equivalent to

$$u_N + \pi_N^{\alpha,\beta} Su_N = \pi_N^{\alpha,\beta} g. \tag{5.43}$$

Theorem 5.2. *Let u and u_N be the solutions of* (5.41) *and* (5.42), *respectively. If*

$$K,\ \partial_x K \in L^\infty(D), \quad u \in B^m_{\alpha,\beta}(I), \tag{5.44}$$

where $D = \{(x,s) : -1 \le s \le x \le 1\}$ *and* $1 \le m \le N+1$, *then for* $-1 < \alpha, \beta < 1$,

$$\|u - u_N\|_{\omega^{\alpha,\beta}} \le c\sqrt{\frac{(N-m+1)!}{N!}}(N+m)^{-(1+m)/2}\|\partial_x^m u\|_{\omega^{\alpha+m,\beta+m}}, \tag{5.45}$$

where c *is a positive constant independent of* m, N *and* u.

Proof. Subtracting (5.41) from (5.43) yields

$$u - u_N + Su - \pi_N^{\alpha,\beta} Su_N = g - \pi_N^{\alpha,\beta} g. \tag{5.46}$$

Set $e = u - u_N$. One verifies that

$$\begin{aligned} Su - \pi_N^{\alpha,\beta} Su_N &= Su - \pi_N^{\alpha,\beta} Su + \pi_N^{\alpha,\beta} S(u - u_N) \\ &= Su - \pi_N^{\alpha,\beta} Su + S(u-u_N) - \big(S(u-u_N) - \pi_N^{\alpha,\beta} S(u-u_N)\big) \\ &= (g-u) - \pi_N^{\alpha,\beta}(g-u) + S(u-u_N) - \big(S(u-u_N) - \pi_N^{\alpha,\beta} S(u-u_N)\big) \\ &= (g - \pi_N^{\alpha,\beta} g) - (u - \pi_N^{\alpha,\beta} u) + Se - (Se - \pi_N^{\alpha,\beta} Se). \end{aligned} \tag{5.47}$$

It follows from (5.46)-(5.47) that

$$e(x) = -\int_{-1}^{x} K(x,s)e(s)ds + (u - \pi_N^{\alpha,\beta} u) + (Se - \pi_N^{\alpha,\beta} Se).$$

Consequently,

$$|e(x)| \le K_{\max} \int_{-1}^{x} |e(s)|ds + |J_1| + |J_2|,$$

where $K_{\max} = \|K\|_{L^\infty(D)}$, and

$$J_1 = u - \pi_N^{\alpha,\beta} u, \quad J_2 = Se - \pi_N^{\alpha,\beta} Se.$$

By the Gronwall inequality (B.9),

$$\|e\|_{\omega^{\alpha,\beta}} \le c\big(\|J_1\|_{\omega^{\alpha,\beta}} + \|J_2\|_{\omega^{\alpha,\beta}}\big),$$

where c depends on $K_{\max}$. Using Theorem 3.35 yields

$$\|J_1\|_{\omega^{\alpha,\beta}} \le c\sqrt{\frac{(N-m+1)!}{N!}}(N+m)^{-(1+m)/2}\|\partial_x^m u\|_{\omega^{\alpha+m,\beta+m}}.$$

Moreover, using Theorem 3.35 with $l = 0$ and $m = 1$ gives

$$\|J_2\|_{\omega^{\alpha,\beta}} \le cN^{-1}\left\|K(x,x)e(x) + \int_{-1}^{x} \partial_x K(x,s)e(s)ds\right\|_{\omega^{\alpha,\beta}}.$$

Using the Cauchy-Schwartz inequality, we find that for $-1 < \alpha, \beta < 1$,

$$\begin{aligned}
\Big\| \int_{-1}^{x} \partial_x K(x,s) e(s) ds \Big\|^2_{\omega^{\alpha,\beta}} &\le \int_{-1}^{1} \Big(\int_{-1}^{x} |\partial_x K(x,s)||e(s)|ds \Big)^2 \omega^{\alpha,\beta}(x)dx \\
&\le \int_{-1}^{1} \Big(\int_{-1}^{x} (\partial_x K(x,s))^2 \omega^{-\alpha,-\beta}(s)ds \int_{-1}^{x} e^2(s)\omega^{\alpha,\beta}(s)ds \Big) \omega^{\alpha,\beta}(x)dx \\
&\le \|\partial_x K\|^2_{L^\infty(D)} \|e\|^2_{\omega^{\alpha,\beta}} \Big(\int_{-1}^{1} \Big(\int_{-1}^{x} \omega^{-\alpha,-\beta}(s)ds \Big) \omega^{\alpha,\beta}(x)dx \Big) \\
&\overset{(A.6)}{\le} \|\partial_x K\|^2_{L^\infty(D)} \|e\|^2_{\omega^{\alpha,\beta}} \gamma_0^{-\alpha,-\beta} \gamma_0^{\alpha,\beta}.
\end{aligned}$$

This implies

$$\|J_2\|_{\omega^{\alpha,\beta}} \le cN^{-1} \|e\|_{\omega^{\alpha,\beta}}.$$

Finally, a combination of the above estimates leads to the desired result. □

Remark 5.2. *The scheme* (5.41) *does not incorporate numerical integrations for both the kernel and source terms. In practice, we need to use the Galerkin method with numerical integration by replacing the continuous inner products by the discrete ones, namely,*

$$\begin{cases} \text{Find } u_N \in P_N \text{ such that} \\ \langle u_N, v_N \rangle_{N,\omega^{\alpha,\beta}} + \langle Su_N, v_N \rangle_{N,\omega^{\alpha,\beta}} = \langle g, v_N \rangle_{N,\omega^{\alpha,\beta}}, \quad \forall v_N \in P_N, \end{cases} \tag{5.48}$$

where $\langle \cdot,\cdot \rangle_{N,\omega^{\alpha,\beta}}$ is the discrete inner product associated with a Jacobi-Gauss-type quadrature rule (see Chap. 3*). Convergence results similar to Theorem* 5.2 *can be established for* (5.48)*. We leave the convergence analysis of the Legendre-Gauss-Lobatto case as an exercise (see Problem* 5.2*).*

5.3 Jacobi-Collocation Method for VIEs with Weakly Singular Kernels

In this section, we consider spectral approximation of the VIE (5.2) with singular kernels. As before, our starting point is to use (5.9) to reformulate (5.2) as:

$$\begin{aligned}
u(x) &= f(x) + \int_{-1}^{x} (x-s)^{-\mu} K(x,s) u(s) ds \\
&\overset{(5.9)}{=} f(x) + \Big(\frac{1+x}{2} \Big)^{1-\mu} \int_{-1}^{1} (1-\theta)^{-\mu} K(x, s(x,\theta)) u(s(x,\theta)) d\theta.
\end{aligned} \tag{5.49}$$

Let $\{x_j\}_{j=0}^N$ be any set of Jacobi-Gauss-Lobatto points, and $\{\theta_j, \omega_j\}_{j=0}^M$ be a set of Jacobi-Gauss-Lobatto points and weights with $\alpha = -\mu$ and $\beta = 0$ (see Theorem 3.27). The corresponding Jacobi-collocation method for (5.49) is:

$$\begin{cases} \text{Find } u_N \in P_N \text{ such that for } 0 \le j \le N, \\ u_N(x_j) = f(x_j) + \left(\frac{1+x_j}{2}\right)^{1-\mu} \sum_{k=0}^{M} K(x_j, s(x_j,\theta_k)) u_N(s(x_j,\theta_k))\omega_k. \end{cases} \tag{5.50}$$

As with the scheme (5.11), the points $\{x_j\}$ and $\{\theta_j\}$ can be chosen differently in type and in number. For simplicity, we assume that they are the same below.

Let $\{h_j\}_{j=0}^N$ be the Lagrange basis polynomials associated with $\{x_j\}_{j=0}^N$. We expand the approximate solution u_N as

$$u_N(x) = \sum_{j=0}^{N} u_N(x_j)h_j(x) \Rightarrow u_N(s(x_j,\theta_k)) = \sum_{i=0}^{N} u_N(x_i)h_i(s(x_j,\theta_k)). \tag{5.51}$$

Then, the scheme (5.50) becomes

$$u_N(x_i) = f(x_i) + \left(\frac{1+x_i}{2}\right)^{1-\mu} \sum_{j=0}^{N} \Big(\sum_{k=0}^{N} K(x_i, s(x_i,\theta_k)) h_j(s(x_i,\theta_k))\omega_k \Big) u_N(x_j), \tag{5.52}$$

for $0 \le i \le N$.

Typically, there is a weak singularity of the solution of (5.49) even if the given functions in (5.49) are sufficiently smooth (see, e.g., Brunner (2004)). We only consider here the case that the underlying unknown solution u is sufficiently smooth. Our attention in this case is to handle the weakly singular kernel occurred in (5.49). The details of the numerical implementation can be found in Chen and Tang (2010).

We now turn to the convergence analysis of the scheme (5.50). Compared with the regular kernel case, the analysis for (5.52) is much more involved.

We first make some necessary preparations. Let $I = [-1,1]$. For $r \ge 0$ and $0 \le \kappa \le 1$, we denote by $C^{r,\kappa}(I)$ the space of functions whose r-th derivatives are Hölder continuous with exponent κ, endowed with the usual norm

$$\|v\|_{C^{r,\kappa}} = \max_{0 \le l \le r} \max_{x \in I} |\partial_x^l v(x)| + \max_{0 \le l \le r} \sup_{x \ne y} \frac{|\partial_x^l v(x) - \partial_x^l v(y)|}{|x-y|^\kappa}.$$

If $\kappa = 0$, $C^{r,0}(I)$ turns out to be the space of functions with continuous derivatives up to r-th order on I, which is also commonly denoted by $C^r(I)$ with the norm $\|\cdot\|_{C^r}$.

Lemma 5.1. (cf. Ragozin (1970, 1971)). *For any non-negative integer r and $0 < \kappa < 1$, there exists a linear transform $T_N : C^{r,\kappa}(I) \to P_N$ such that*

$$\|v - T_N v\|_{L^\infty} \le c_{r,\kappa} N^{-(r+\kappa)} \|v\|_{C^{r,\kappa}}, \quad \forall v \in C^{r,\kappa}(I), \tag{5.53}$$

where $c_{r,\kappa}$ is a positive constant.

Another useful result is on the stability of the linear operator:

$$\mathcal{M}v(x) = \int_{-1}^{x} (x-s)^{-\mu} K(x,s) v(s) ds. \tag{5.54}$$

Below we prove that M is a compact operator from $C(I)$ to $C^{0,\kappa}(I)$, provided that the index $0<\kappa<1-\mu$. This result will play a crucial role in the convergence analysis of this section.

Lemma 5.2. *Let $0<\mu<1$. If $0<\kappa<1-\mu$, then for any function $v\in C(I)$ and any $x_1,\ x_2\in I=[-1,1]$ with $x_1\neq x_2$, there exists a positive constant c (may depend on $\|K\|_{C^{0,\kappa}}$ and $\|K\|_{L^\infty(D)}$ with $D=[-1,1]^2$), such that*

$$\frac{|Mv(x_1)-Mv(x_2)|}{|x_1-x_2|^\kappa}\le c\|v\|_\infty, \tag{5.55}$$

which implies

$$\|Mv\|_{C^{0,\kappa}}\le c\|v\|_\infty. \tag{5.56}$$

Proof. Without loss of generality, we assume that $x_1<x_2$. We first show that

$$\int_{-1}^{x_1}\left[(x_1-\tau)^{-\mu}-(x_2-\tau)^{-\mu}\right]d\tau\le c|x_2-x_1|^{1-\mu}. \tag{5.57}$$

As $x_1<x_2$, we have from the linear transformation (5.9) that

$$\begin{aligned}
&\int_{-1}^{x_1}\left[(x_1-\tau)^{-\mu}-(x_2-\tau)^{-\mu}\right]d\tau\\
&\le\left|\int_{-1}^{x_1}(x_1-\tau)^{-\mu}d\tau-\int_{-1}^{x_2}(x_2-\tau)^{-\mu}d\tau\right|+\left|\int_{x_1}^{x_2}(x_2-\tau)^{-\mu}d\tau\right|\\
&\le\left[\left(\frac{x_2+1}{2}\right)^{1-\mu}-\left(\frac{x_1+1}{2}\right)^{1-\mu}\right]\int_{-1}^{1}(1-\theta)^{-\mu}d\theta+\frac{|x_2-x_1|^{1-\mu}}{1-\mu}.
\end{aligned}$$

Observe that

$$\begin{aligned}
\left(\frac{x_2+1}{2}\right)^{1-\mu}-\left(\frac{x_1+1}{2}\right)^{1-\mu}&=\frac{1-\mu}{2^{1-\mu}}\int_{x_1}^{x_2}(y+1)^{-\mu}dy\\
&\le\frac{1-\mu}{2^{1-\mu}}\int_{x_1}^{x_2}(y-x_1)^{-\mu}dy=2^{\mu-1}|x_2-x_1|^{1-\mu},
\end{aligned}$$

where we used the fact that $y+1\ge y-x_1$ for $x_1\in[-1,1]$. Thus, (5.57) follows.

Next, we obtain from the triangle inequality that

$$\begin{aligned}
&|Mv(x_1)-Mv(x_2)|\\
&\le\left|\int_{-1}^{x_1}\left[(x_1-\tau)^{-\mu}K(x_1,\tau)-(x_2-\tau)^{-\mu}K(x_2,\tau)\right]v(\tau)d\tau\right|\\
&\quad+\left|\int_{x_1}^{x_2}(x_2-\tau)^{-\mu}K(x_2,\tau)v(\tau)d\tau\right|
\end{aligned}$$

$$
\begin{aligned}
&\le \int_{-1}^{x_1} \left|(x_1-\tau)^{-\mu}-(x_2-\tau)^{-\mu}\right|\cdot|K(x_1,\tau)|\cdot|v(\tau)|d\tau \\
&+\int_{-1}^{x_1}(x_2-\tau)^{-\mu}\,|K(x_1,\tau)-K(x_2,\tau)|\cdot|v(\tau)|d\tau \\
&+\int_{x_1}^{x_2}(x_2-\tau)^{-\mu}\,|K(x_2,\tau)|\cdot|v(\tau)|d\tau \\
&:= E_1+E_2+E_3.
\end{aligned}
$$

We now estimate the three terms one by one. By (5.57),

$$
\begin{aligned}
E_1 &\le \|v\|_\infty\|K\|_{L^\infty(D)}\int_{-1}^{x_1}\left|(x_1-\tau)^{-\mu}-(x_2-\tau)^{-\mu}\right|d\tau \\
&\le c\|v\|_\infty|x_2-x_1|^{1-\mu}.
\end{aligned}
$$

Moreover, we have

$$
\begin{aligned}
E_2 &\le \|v\|_\infty|x_2-x_1|^\kappa\int_{-1}^{x_1}(x_2-\tau)^{-\mu}\frac{|K(x_2,\tau)-K(x_1,\tau)|}{|x_2-x_1|^\kappa}d\tau \\
&\le \|v\|_\infty\|K\|_{C^{0,\kappa}}|x_2-x_1|^\kappa\frac{1}{1-\mu}\left[(x_2+1)^{1-\mu}-(x_2-x_1)^{1-\mu}\right] \\
&\le c\|v\|_\infty|x_2-x_1|^\kappa,
\end{aligned}
$$

where c depends on $\|K\|_{0,\kappa}$. Finally, we have

$$
E_3 \le \|K\|_{L^\infty(D)}\|v\|_\infty\int_{x_1}^{x_2}(x_2-\tau)^{-\mu}d\tau \le c\|v\|_{L^\infty}|x_2-x_1|^{1-\mu}.
$$

Using the above estimates and the assumption $0<\kappa<1-\mu$ completes the proof of the lemma. □

The following lemma on the Lebesgue constant of the Jacobi-Gauss-Lobatto interpolation (see Theorem 3.1 of Mastroianni and Occorsio (2001b)) also plays an important role in the convergence analysis.

Lemma 5.3. *Let $\{h_i\}_{i=0}^N$ be the Lagrange basis polynomials associated with the Jacobi-Gauss-Lobatto interpolations with the parameter pair $\{-\mu,0\}$. Then, for $-1/2\le\mu<3/2$, we have*

$$
\Lambda_N := \max_{|x|\le 1}\sum_{i=0}^N |h_i(x)| \sim \ln N. \tag{5.58}
$$

Theorem 5.3. *Let u and u_N be the solutions to the VIE* (5.49) *and* (5.50) *with $0<\mu<1$, respectively. Assume $u\in L^\infty(I)\cap B^r_{-1,-1}(I)$ with integer $1\le r\le N+1$, and*

$$
K_m^* := \max_{0\le i\le N}\left(\int_{-1}^{x_i}|\partial_s^m K(x_i,s)|^2(x_i-s)^{m-1-\mu}(1+s)^{m-1}ds\right)^{1/2} < \infty \tag{5.59}
$$

for certain integer $1 \le m \le N+1$. *Then we have the estimate:*

$$\begin{aligned} \|u - u_N\|_\infty \le\, & c\sqrt{\frac{(N-r+1)!}{N!}}(N+r)^{-r/2}(\ln N)\|\partial_x^r u\|_{\omega^{r-1,r-1}} \\ & + c\sqrt{\frac{(N-m+1)!}{N!}}(N+m)^{-(m+1)/2}(\ln N)\, K_m^*\|u\|_\infty, \end{aligned} \tag{5.60}$$

where c *is a positive constant independent of* N, r, m *and* u.

Proof. In what follows, let $(\cdot,\cdot)_{\omega^{-\mu,0}}$ and $\langle\cdot,\cdot\rangle_{N,\omega^{-\mu,0}}$ be the weighted continuous and discrete inner products, respectively, as defined in Chap. 3. Furthermore, let $I_N^{-\mu,0}$ be the corresponding interpolation operator. Firstly, we rewrite (5.49) as

$$u(x_i) = f(x_i) + \Big(\frac{1+x_i}{2}\Big)^{1-\mu}\big(K(x_i,s(x_i,\cdot)),u(s(x_i,\cdot))\big)_{\omega^{-\mu,0}}, \quad 0 \le i \le N, \tag{5.61}$$

and reformulate (5.50) into

$$u_N(x_i) = f(x_i) + \Big(\frac{1+x_i}{2}\Big)^{1-\mu}\big\langle K(x_i,s(x_i,\cdot)),u_N(s(x_i,\cdot))\big\rangle_{N,\omega^{-\mu,0}}, \quad 0 \le i \le N. \tag{5.62}$$

Denoting $e = u - u_N$, we have the error equation:

$$\begin{aligned} e(x_i) &= \Big(\frac{1+x_i}{2}\Big)^{1-\mu}\big(K(x_i,s(x_i,\cdot)),e(s(x_i,\cdot))\big)_{\omega^{-\mu,0}} + G(x_i) \\ &= \int_{-1}^{x_i}(x_i - s)^{-\mu}K(x_i,s)e(s)ds + G(x_i), \end{aligned} \tag{5.63}$$

where

$$\begin{aligned} G(x) = \Big(\frac{1+x}{2}\Big)^{1-\mu}\Big\{&\big(K(x,s(x,\cdot)),u_N(s(x,\cdot))\big)_{\omega^{-\mu,0}} \\ &- \big\langle K(x,s(x,\cdot)),u_N(s(x,\cdot))\big\rangle_{N,\omega^{-\mu,0}}\Big\}. \end{aligned} \tag{5.64}$$

Equivalently, we write (5.63) as

$$I_N^{-\mu,0}u - u_N = I_N^{-\mu,0}\Big(\int_{-1}^{x}(x-s)^{-\mu}K(x,s)e(s)ds\Big) + I_N^{-\mu,0}G. \tag{5.65}$$

Consequently,

$$e = \int_{-1}^{x}(x-s)^{-\mu}K(x,s)e(s)ds + G_1 + G_2 + I_N^{-\mu,0}G, \tag{5.66}$$

where

$$\begin{aligned} G_1 &= u - I_N^{-\mu,0}u, \\ G_2 &= I_N^{-\mu,0}\Big(\int_{-1}^{x}(x-s)^{-\mu}K(x,s)e(s)ds\Big) - \int_{-1}^{x}(x-s)^{-\mu}K(x,s)e(s)ds. \end{aligned} \tag{5.67}$$

It follows from the Gronwall inequality (see Lemma B.9) that

$$\|e\|_\infty \le c\big(\|G_1\|_\infty + \|G_2\|_\infty + \|I_N^{-\mu,0}G\|_\infty\big). \tag{5.68}$$

It remains to estimate the three terms on the right hand side of (5.68). Firstly, by Lemma 5.3 and an estimate similar to Lemma 4.8,

$$\begin{aligned}
\|I_N^{-\mu,0}G\|_\infty &\le \max_{0\le i\le N}|G(x_i)|\sum_{i=0}^{N}|h_i(x)| \le c\ln N \max_{0\le i\le N}|G(x_i)| \\
&\le c\sqrt{\frac{(N-m+1)!}{N!}}(N+m)^{-(m+1)/2}\ln N\times \\
&\max_{0\le i\le N}\Big\{\Big(\frac{1+x_i}{2}\Big)^{1-\mu}\|\partial_\theta^m(K(x_i,s(x_i,\cdot)))\|_{\omega^{m-1-\mu,m-1}}\|u_N(s(x_i,\cdot))\|_{\omega^{-\mu,0}}\Big\}.
\end{aligned} \tag{5.69}$$

A direct computation shows that

$$\begin{aligned}
&\|\partial_\theta^m(K(x_i,s(x_i,\cdot)))\|_{\omega^{m-1-\mu,m-1}} \\
&= \Big(\frac{1+x_i}{2}\Big)^{(1+\mu)/2}\Big(\int_{-1}^{x_i}|\partial_s^m K(x_i,s)|^2(x_i-s)^{m-1-\mu}(1+s)^{m-1}ds\Big)^{1/2},
\end{aligned} \tag{5.70}$$

and

$$\begin{aligned}
&\|u_N(s(x_i,\cdot))\|_{\omega^{-\mu,0}} \\
&= \Big(\frac{2}{1+x_i}\Big)^{(1-\mu)/2}\Big(\int_{-1}^{x_i}|u_N(s)|^2(x_i-s)^{-\mu}ds\Big)^{1/2} \\
&\le c\Big(\frac{2}{1+x_i}\Big)^{(1-\mu)/2}\|u_N\|_\infty.
\end{aligned} \tag{5.71}$$

Hence, we have

$$\begin{aligned}
\|I_N^{-\mu,0}G\|_\infty &\le c\sqrt{\frac{(N-m+1)!}{N!}}(N+m)^{-(m+1)/2}K_m^*\ln N\big(\|e\|_\infty + \|u\|_\infty\big) \\
&\le \frac{1}{3}\|e\|_\infty + c\sqrt{\frac{(N-m+1)!}{N!}}(N+m)^{-(m+1)/2}K_m^*\ln N\|u\|_\infty,
\end{aligned} \tag{5.72}$$

provided that N is large enough, where K_m^* is defined in (5.59).

We now turn to the estimation of G_1. Let I_N be the Legendre-Gauss-Lobatto polynomial interpolation operator. Using Lemma 5.3, the Sobolev inequality (B.33) and Theorem 3.44 gives

$$\begin{aligned}
\|G_1\|_\infty &= \|u - I_N^{-\mu,0}u\|_\infty = \|u - I_N u + I_N^{-\mu,0}(I_N u - u)\|_\infty \\
&\le (1+c\ln N)\|u - I_N u\|_\infty \le c\ln N\|u - I_N u\|^{1/2}\|u - I_N u\|_1^{1/2} \\
&\le c\sqrt{\frac{(N-r+1)!}{N!}}(N+r)^{-r/2}\ln N\|\partial_x^r u\|_{\omega^{r-1,r-1}}.
\end{aligned} \tag{5.73}$$

To estimate G_2, we obtain from Lemmas 5.1–5.3 that

$$\begin{aligned}\|G_2\|_\infty &= \|I_N^{-\mu,0}(Me) - Me\|_\infty \\ &\le \|I_N^{-\mu,0}(Me) - T_N(Me)\|_\infty + \|T_N(Me) - Me\|_\infty \\ &\le (1 + c\ln N)\|T_N(Me) - Me\|_\infty \\ &\le cN^{-\kappa}\ln N\|Me\|_{C^{0,\kappa}} \le cN^{-\kappa}\ln N\|e\|_\infty.\end{aligned} \tag{5.74}$$

Consequently, if $\kappa > 0$ and N is large enough, we have

$$\|G_2\|_\infty \le \frac{1}{3}\|e\|_\infty. \tag{5.75}$$

Finally, a combination of (5.68), (5.72), (5.73), and (5.75) leads to the desired estimate. □

5.4 Application to Delay Differential Equations

We discuss in this section numerical solutions of delay differential equations. To demonstrate the main idea, we consider the delay differential equation with proportional delay:

$$u'(x) = a(x)u(qx), \quad 0 < x \le T; \quad u(0) = y_0, \tag{5.76}$$

where $0 < q < 1$ is a given constant and a is a smooth function on $[0,T]$. This problem belongs to the class of the so-called pantograph delay differential equations (see Fox et al. (1971), Iserles (1993) for details on their theory and physical applications).

The existing numerical methods for solving (5.76) include Runge–Kutta type methods (see, e.g., Bellen and Zennaro (2003)) and the piecewise-polynomial collocation methods (see, for instance, Brunner (2004)). The main difficulty in the application of Runge–Kutta methods to (5.76) is the lack of information at the grid points for the function on the right hand side of (5.76), so these numerical data have to be generated by some local interpolation process. While the piecewise-polynomial collocation methods yield globally defined approximations, the corresponding numerical solutions are not globally smooth. Moreover, it has been shown in Brunner and Hu (2007) that for arbitrarily smooth solutions of (5.76) the optimal order at the grid points obtained using piecewise polynomials of degree m cannot exceed $p = m + 2$ when $m \ge 2$ (in contrast to their application to ordinary differential equations where collocation at the Gauss points leads to $O(h^{2m})$-convergence).

If the function a is in $C^d[0,T]$, then the corresponding solution of the initial-value problem (5.76) lies in $C^{d+1}[0,T]$. In this case, it is suitable to employ spectral-type methods since they produce approximate solutions that are defined globally on $[0,T]$ and globally smooth.

For ease of notation, we implement and analyze the spectral method on the reference interval $I := [-1,1]$. Hence, using the transformation

$$x = \frac{T}{2}(1+t), \quad t = \frac{2x}{T} - 1,$$

the problem (5.76) becomes

$$y'(t) = b(t)y(qt+q_1), \ -1 < t \leq 1; \quad y(-1) = y_0, \tag{5.77}$$

where

$$y(t) = u\big(T(1+t)/2\big), \quad b(t) = \frac{T}{2}a\big(T(1+t)/2\big), \quad q_1 = q-1. \tag{5.78}$$

To fix the idea, we only consider the Legendre-collocation method for solving (5.77). To this end, let $\{t_j, \omega_j\}_{j=0}^N$ be the set of Legendre-Gauss-Lobatto points and weights. Integrating (5.77) from -1 to t_j gives

$$y(t_j) = y_0 + \int_{-1}^{t_j} b(s)y(qs+q_1)ds, \quad 1 \leq j \leq N. \tag{5.79}$$

Using the linear transformation

$$s = \frac{t_j+1}{2}v + \frac{t_j-1}{2}, \quad v \in [-1,1],$$

yields

$$y(t_j) = y_0 + \int_{-1}^{1} \tilde{b}(v;t_j)y\Big(\frac{t_j+1}{2}qv + q_{1j}\Big)dv, \tag{5.80}$$

where

$$\tilde{b}(v;t_j) := \frac{1+t_j}{2}b\Big(\frac{t_j+1}{2}v + \frac{t_j-1}{2}\Big), \quad q_{1j} := \frac{t_j+1}{2}q - 1.$$

The Legendre-collocation scheme for (5.80) is to find $y_N \in P_N$ such that

$$y_N(t_j) = y_0 + \sum_{k=0}^{N} \tilde{b}(v_k;t_j)y_N\Big(\frac{t_j+1}{2}qv_k + q_{1j}\Big)\omega_k, \quad 0 \leq j \leq N, \tag{5.81}$$

where $\{v_k = t_k\}_{k=0}^N$ are the Legendre-Gauss-Lobatto points. We now describe in more detail how to efficiently implement (5.81).

Let $\{Y_j = y_N(t_j)\}_{j=0}^N$, and write

$$y_N(t) = \sum_{j=0}^{N} Y_j h_j(t), \tag{5.82}$$

where $\{h_j\}_{j=0}^N$ are the Lagrange basis polynomials relative to $\{t_j\}_{j=0}^N$. To evaluate y_N at non-interpolation points efficiently, we compute $h_j(t)$ by using (5.15)–(5.17). More precisely, we expand $h_k(v)$ in terms of the Legendre polynomials:

$$h_k(v) = \sum_{m=0}^{N} c_m^k L_m(v), \tag{5.83}$$

and find that

$$c_m^k = \frac{2m+1}{N(N+1)} \sum_{s=0}^{N} h_k(x_s) \frac{L_m(x_s)}{[L_N(x_s)]^2} = \frac{2m+1}{N(N+1)} \frac{L_m(x_k)}{[L_N(x_k)]^2}. \tag{5.84}$$

Hence, the scheme (5.81) becomes: find $y_N \in P_N$ such that

$$Y_j = y_0 + \sum_{i=0}^{N} a_{ji} Y_i, \quad 0 \le j \le N \tag{5.85}$$

with $a_{ji} = \sum_{k=0}^{N} \tilde{b}(v_k; t_j) h_i\left(\frac{t_j+1}{2} q v_k + q_{1j}\right) \omega_k$, which is a linear system (with a full matrix $A = (a_{ji})$) for the unknown vector $(Y_0, Y_1, \ldots, Y_N)^t$, and the entries of the matrix A can be computed by using (5.83)–(5.84).

Remark 5.3. *We may consider more general delay differential or integral equations with two or more vanishing delays:*

$$\begin{cases} y'(t) = a(t)y(t) + \sum_{\ell=1}^{r} b_\ell(t) y(q_\ell t), & t \in I := [a,b], \\ y(0) = y_0, \end{cases} \tag{5.86}$$

and the analogous multiple-delay Volterra integral equation

$$y(t) = g(t) + \sum_{\ell=1}^{r} \int_0^{q_\ell t} K_\ell(t,s) y(s)\, ds, \quad t \in I, \tag{5.87}$$

where $0 < q_1 < \ldots < q_r < 1$ $(r \ge 2)$. *It is demonstrated numerically in Ali et al. (2009) that for the pantograph-type functional equations the spectral methods proposed yield the exponential order of convergence.*

Next, we present some numerical results. Without lose of generality, we only consider the Legendre-Gauss-Lobatto quadrature rule in (5.11). We first consider (5.76) with $q = 0.7, y_0 = 1, T = 1$; the function $a(x)$ is chosen such that the exact solution of u is given by $u(x) = \cos(2x - 1)$.

In Table 5.3, we tabulate the maximum point-wise errors obtained by (5.85) with various N, which indicate that the desired spectral accuracy is obtained.

Table 5.3 The maximum point-wise errors

N	6	8	10	12	14
Error	6.41e-03	6.15e-05	3.06e-07	9.26e-10	1.79e-12

Below we consider the spectral methods for the case of two proportional delays; that is, for the functional equation

$$\begin{cases} y'(t) = a(t)y(t) + b_1(t)y(q_1t) + b_2(t)y(q_2t), & t \in I, \\ y(0) = y_0. \end{cases} \tag{5.88}$$

The numerical schemes proposed previously can be readily adapted to deal with (5.88). In the following, we use numerical examples to illustrate the accuracy and efficiency of the spectral methods. In (5.88), let $b_1(t) = \cos(t), b_2(t) = \sin(t)$ and $a(t) = 0$. We choose $g(t)$ such that the exact solution is given by $y(t) = \sin(tq_1^{-1}) + \cos(tq_2^{-1})$.

Table 5.4 The maximum point-wise errors with $q_1 = 0.05, q_2 = 0.95$

N	12	14	16	18	20
Error	1.14e-02	1.66e-03	2.07e-04	1.37e-5	7.22e-07

In Table 5.4, the maximum point-wise errors with $q_1 = 0.05, q_2 = 0.95$ are listed. This is a quite extreme case with very small value of the delay parameter q_1. For the piecewise-polynomial collocation methods, it will require few hundred collocation points to reach the errors of about 10^{-7}; while with the spectral approach only 20 points are needed.

Problems

5.1. Consider the numerical example for (5.6) with the given functions (5.36).
(i) Provide a maximum point-wise errors table similar to Table 5.1 using the Trapezoidal method.
(ii) Verify the results in Table 5.1.

5.2. Derive the L^2-estimate of the Legendre-Galerkin method with numerical integration for (5.48), where the discrete inner product is associated with the Legendre-Gauss-Lobatto quadrature.

5.3. Design a Legendre-collocation method for the delay Volterra integral equation

$$y(t) = g(t) + \int_0^{qt} K(t,s)y(s)\,ds,$$

with $0 < q < 1$. Try to provide a convergence analysis.

Chapter 6
Higher-Order Differential Equations

High-order differential equations often arise from mathematical modeling of a variety of physical phenomena. For example, higher even-order differential equations may appear in astrophysics, structural mechanics and geophysics, and higher odd-order differential equations, such as the Korteweg–de Vries (KdV) equation, are routinely used in modeling nonlinear waves and nonlinear optics.

In this chapter, we introduce a family of generalized Jacobi polynomials (GJPs) (cf. Guo et al. (2006a, 2009)), which serve as natural basis functions for spectral approximations of higher-order boundary value problems. The GJPs generalize the classical Jacobi polynomials with parameters $\alpha, \beta > -1$ to allow α and/or β being negative integers. The use of GJPs leads to much concise analysis and more precise error estimates for spectral approximations of PDEs, particularly for higher-order PDEs considered in this chapter.

6.1 Generalized Jacobi Polynomials

The definition of GJPs is motivated by the observation that the Legendre basis functions in (4.19) satisfy

$$\phi_k(x) = L_k(x) - L_{k+2}(x) \overset{(3.116)}{=} \frac{2k+3}{2(k+1)}(1-x^2)J_k^{1,1}(x).$$

Hence, $\{\phi_k\}$ are orthogonal with respect to the weight function $\omega^{-1,-1}(x) = (1-x^2)^{-1}$, which, apart from a multiplicative constant, can be referred to as the GJP with index-$(-1,-1)$.

J. Shen et al., *Spectral Methods: Algorithms, Analysis and Applications*, Springer Series in Computational Mathematics 41, DOI 10.1007/978-3-540-71041-7_6,

We now extend the definition of the classical Jacobi polynomials to the cases where one or both parameter(s) (k,l) being negative integer(s) by

$$J_n^{k,l}(x) = \begin{cases} (1-x)^{-k}(1+x)^{-l}J_{n-n_0}^{-k,-l}(x), & \text{if } k,l \le -1, \\ (1-x)^{-k}J_{n-n_0}^{-k,l}(x), & \text{if } k \le -1,\, l > -1, \\ (1+x)^{-l}J_{n-n_0}^{k,-l}(x), & \text{if } k > -1,\, l \le -1, \end{cases} \tag{6.1}$$

where $n \ge n_0$ with $n_0 := -(k+l), -k, -l$ for the above three cases, respectively. It is clear that the so-defined GJP $J_n^{k,l}$ is a polynomial of degree n.

Remark 6.1. *The definition* (6.1) *is also valid for $-1 < k \in \mathbb{R}$ and/or $-1 < l \in \mathbb{R}$. However, we shall restrict our attention below to $k,l \in \mathbb{Z}$ only. It is also possible to extend definition of the classical Jacobi polynomials $J_n^{\alpha,\beta}$ to all $\alpha,\beta \in \mathbb{R}$* (cf. Guo et al. (2009)). *However, this process is much more involved so it will not be discussed in this chapter.*

To simplify the notation, we introduce

$$\hat{k} := \begin{cases} -k, & k \le -1, \\ 0, & k > -1, \end{cases} \qquad \bar{k} := \begin{cases} -k, & k \le -1, \\ k, & k > -1. \end{cases} \tag{6.2}$$

It is clear that for any $k \in \mathbb{Z}$,

$$\hat{k} \ge 0, \quad \bar{k} > -1, \quad k = \bar{k} - 2\hat{k}. \tag{6.3}$$

With the above notation, it is easy to check that the GJPs can be expressed as

$$J_n^{k,l}(x) = \omega^{\hat{k},\hat{l}}(x) J_{n-n_0}^{\bar{k},\bar{l}}(x), \quad n \ge n_0,\ k,l \in \mathbb{Z}, \tag{6.4}$$

where $\omega^{a,b}(x) = (1-x)^a(1+x)^b$ and $n_0 = \hat{k} + \hat{l}$.

We now present some basic properties of the GJPs.

First of all, the GJPs are orthogonal with respect to the generalized Jacobi weight $\omega^{k,l}$ for all integers k and l, i.e.,

$$\int_{-1}^{1} J_n^{k,l}(x) J_m^{k,l}(x)\omega^{k,l}(x)dx = \gamma_{n-n_0}^{\bar{k},\bar{l}}\delta_{mn}, \quad n \ge n_0, \tag{6.5}$$

where the constant $\gamma_{n-n_0}^{\bar{k},\bar{l}}$ is given in (3.109).

As the classical Jacobi polynomials, the GJPs $\{J_n^{k,l}\}$ form a complete orthogonal system in $L^2_{\omega^{k,l}}(I)$. Indeed, we observe from (6.3) that for any $u \in L^2_{\omega^{k,l}}(I)$, we have $u\omega^{-\hat{k},-\hat{l}} \in L^2_{\omega^{\bar{k},\bar{l}}}(I)$. Since the classical Jacobi polynomials $\{J^{\bar{k},\bar{l}}_{n-n_0} : n \ge n_0\}$ are complete in $L^2_{\omega^{\bar{k},\bar{l}}}(I)$, we can write

$$(u\omega^{-\hat{k},-\hat{l}})(x) = \sum_{n=n_0}^{\infty} \hat{u}_n J^{\bar{k},\bar{l}}_{n-n_0}(x),$$

where

$$\hat{u}_n = \frac{1}{\gamma^{\bar{k},\bar{l}}_{n-n_0}} \int_{-1}^{1} u\omega^{-\hat{k},-\hat{l}} J^{\bar{k},\bar{l}}_{n-n_0} \omega^{\bar{k},\bar{l}} dx = \frac{1}{\gamma^{\bar{k},\bar{l}}_{n-n_0}} \int_{-1}^{1} u J^{k,l}_n \omega^{k,l} dx. \tag{6.6}$$

Therefore, any $u \in L^2_{\omega^{k,l}}(I)$ can be expanded as

$$u(x) = \sum_{n=n_0}^{\infty} \hat{u}_n J^{k,l}_n(x), \quad k,l \in \mathbb{Z}. \tag{6.7}$$

Thanks to the orthogonality (6.5), the expansion (6.7) must be unique. Hence, the system $\{J^{k,l}_n\}$ is complete in $L^2_{\omega^{k,l}}(I)$.

It is interesting that the GJPs with negative integer parameters can be expressed as *compact combinations* of Legendre polynomials. Indeed, one verifies by using (3.116) and the definition of the GJPs that

$$\begin{aligned}
J^{-1,-1}_n &= c^{-1,-1}_n \big(L_{n-2} - L_n\big),\\
J^{-2,-1}_n &= c^{-2,-1}_n \Big(L_{n-3} - \frac{2n-3}{2n-1} L_{n-2} - L_{n-1} + \frac{2n-3}{2n-1} L_n\Big),\\
J^{-1,-2}_n &= c^{-1,-2}_n \Big(L_{n-3} + \frac{2n-3}{2n-1} L_{n-2} - L_{n-1} - \frac{2n-3}{2n-1} L_n\Big),\\
J^{-2,-2}_n &= c^{-2,-2}_n \Big(L_{n-4} - \frac{2(2n-3)}{2n-1} L_{n-2} + \frac{2n-5}{2n-1} L_n\Big),
\end{aligned} \tag{6.8}$$

where

$$\begin{aligned}
c^{-1,-1}_n &= \frac{2(n-1)}{2n-1}, \quad & c^{-2,-1}_n &= \frac{2(n-2)}{2n-3},\\
c^{-1,-2}_n &= \frac{2(n-2)}{2n-3}, \quad & c^{-2,-2}_n &= \frac{4(n-2)(n-3)}{(2n-3)(2n-5)}.
\end{aligned}$$

In general, we have the following result.

Lemma 6.1. *Let $k,l \in \mathbb{Z}$ and $k,l \ge 1$. Then there exists a unique set of constants $\{a_j\}$ such that*

$$J^{-k,-l}_n(x) = \sum_{j=n-k-l}^{n} a_j L_j(x), \quad n \ge k+l. \tag{6.9}$$

Proof. In this proof, we denote by $\{c_j\}$ a set of generic constants. Then, using the properties (3.116a) and (3.116b) repeatedly leads to

$$
\begin{aligned}
J_n^{-k,-l}(x) &= (1-x)^k(1+x)^l J_{n-k-l}^{k,l}(x) \\
&= (1+x)^l(1-x)^{k-1}\Big(c_1 J_{n-k-l}^{k-1,l}(x) + c_2 J_{n-k-l+1}^{k-1,l}(x)\Big) \\
&= \ldots = (1+x)^l \sum_{j=n-k-l}^{n-l} c_j J_j^{0,l}(x) \\
&= \ldots = \sum_{j=n-k-l}^{n} a_j L_j(x).
\end{aligned}
$$

This completes the proof. □

Another attractive property of the GJPs is that for any $k,l \in \mathbb{Z}$ and $k,l \geq 1$,

$$
\begin{aligned}
&\partial_x^i J_n^{-k,-l}(1) = 0, \quad \text{for } i = 0,1,\ldots,k-1, \\
&\partial_x^j J_n^{-k,-l}(-1) = 0, \quad \text{for } j = 0,1,\ldots,l-1.
\end{aligned} \tag{6.10}
$$

Hence, $\{J_n^{-k,-l}\}$ are natural candidates as basis functions for PDEs with the following boundary conditions:

$$
\begin{aligned}
&\partial_x^i u(1) = a_i, \quad \text{for } i = 0,1,\ldots,k-1, \\
&\partial_x^j u(-1) = b_j, \quad \text{for } j = 0,1,\ldots,l-1.
\end{aligned} \tag{6.11}
$$

Remark 6.2. *Note that the GJPs can only be used as basis functions for problems with essential boundary conditions of the form* (6.11). *For other type of boundary conditions, one should still use the general approach to construct basis functions by using a compact combination of orthogonal polynomials* (*see Problem* 6.2).

As with the classical Jacobi polynomials (see (3.98)), the generalized Jacobi polynomials satisfy a similar derivative recurrence relation.

Lemma 6.2. *For* $k,l \in \mathbb{Z}$, *we have*

$$
\partial_x J_n^{k,l}(x) = C_n^{k,l} J_{n-1}^{k+1,l+1}(x), \tag{6.12}
$$

where

$$
C_n^{k,l} = \begin{cases} -2(n+k+l+1), & \text{if } k,l \leq -1, \\ -n, & \text{if } k \leq -1,\ l > -1, \\ -n, & \text{if } k > -1,\ l \leq -1, \\ \frac{1}{2}(n+k+l+1), & \text{if } k,l > -1. \end{cases} \tag{6.13}
$$

Proof. We prove (6.12) case by case.

(i) The formula (6.12)–(6.13) with $k,l > -1$ is a direct consequence of (3.100).

(ii) If $k,l \le -1$, then (3.321) implies

$$\begin{aligned} J_{n-1}^{k+1,l+1}(x) &\overset{(6.1)}{=} (1-x)^{-k-1}(1+x)^{-l-1}J_{n+k+l+1}^{-k-1,-l-1}(x) \\ &\overset{(3.321)}{=} -\frac{1}{2(n+k+l+1)}\partial_x\Big((1-x)^{-k}(1+x)^{-l}J_{n+k+l}^{-k,-l}(x)\Big) \\ &\overset{(6.1)}{=} -\frac{1}{2(n+k+l+1)}\partial_x J_n^{k,l}(x), \end{aligned}$$

which leads to (6.12) with $k,l \le -1$.

(iii) If $k \le -1$ and $l > -1$, then we have

$$\begin{aligned} \partial_x J_n^{k,l} &\overset{(6.1)}{=} \partial_x\big((1-x)^{-k}J_{n+k}^{-k,l}\big) \\ &= (1-x)^{-k-1}\Big(kJ_{n+k}^{-k,l} + (1-x)\partial_x J_{n+k}^{-k,l}\Big) \\ &\overset{(3.100)}{=} (1-x)^{-k-1}\Big(kJ_{n+k}^{-k,l} + \frac{1}{2}(n+l+1)(1-x)J_{n+k-1}^{-k+1,l+1}\Big) \\ &\overset{(3.116a)}{=} (1-x)^{-k-1}\Big\{kJ_{n+k}^{-k,l} + \frac{n+l+1}{2n+k+l+1}\Big(nJ_{n+k-1}^{-k,l+1} \\ &\quad - (n+k)J_{n+k}^{-k,l+1}\Big)\Big\}. \end{aligned}$$

The formula (3.118a) implies

$$J_{n+k-1}^{-k,l+1} = J_{n+k}^{-k,l} - J_{n+k}^{-k-1,l+1}.$$

Plugging it into the previous formula leads to

$$\begin{aligned} \partial_x J_n^{k,l} &= (1-x)^{-k-1}\Big\{kJ_{n+k}^{-k,l} + \frac{n+l+1}{2n+k+l+1}\Big(nJ_{n+k}^{-k,l} \\ &\quad - nJ_{n+k}^{-k-1,l+1} - (n+k)J_{n+k}^{-k,l+1}\Big)\Big\} \\ &= (1-x)^{-k-1}\Big\{\frac{n+k}{2n+k+l+1}\Big((n+k+l+1)J_{n+k}^{-k,l} \\ &\quad - (n+l+1)J_{n+k}^{-k,l+1}\Big) \\ &\quad - \frac{n(n+l+1)}{2n+k+l+1}J_{n+k}^{-k-1,l+1}\Big\}. \end{aligned}$$

Thanks to (3.118b), we have

$$(n+l+1)J_{n+k}^{-k,l+1} = (n+k+l+1)J_{n+k}^{-k,l} + nJ_{n+k}^{-k-1,l+1},$$

Consequently,

$$\begin{aligned}\partial_x J_n^{k,l} &= (1-x)^{-k-1}\Big(\frac{-(n+k)n}{2n+k+l+1}J_{n+k}^{-k-1,l+1}\\ &\quad -\frac{n(n+l+1)}{2n+k+l+1}J_{n+k}^{-k-1,l+1}\Big)\\ &= -n(1-x)^{-k-1}J_{n+k}^{-k-1,l+1}\\ &\overset{(6.1)}{=} -nJ_{n-1}^{k+1,l+1}.\end{aligned}$$

This gives (6.12) with $k \le -1$ and $l > -1$.

(iv) The case: $k > -1$ and $l \le -1$ can be proved in the same fashion as above.
□

In the rest of this chapter, we shall develop and analyze Galerkin and Petrov-Galerkin methods using GJPs as basis functions for high-order boundary value problems.

6.2 Galerkin Methods for Even-Order Equations

For the sake of clarity, we start with fourth-order equations followed by general even-order equations.

6.2.1 Fourth-Order Equations

Consider

$$\begin{aligned}&u^{(4)} - \alpha u'' + \beta u = f, \quad x \in I = (-1,1),\\ &u(\pm 1) = u'(\pm 1) = 0,\end{aligned} \tag{6.14}$$

where $\alpha \ge 0$ and $\beta > 0$ are given constants. This equation may arise from many scientific applications, such as the biharmonic equation, the Stokes equations in stream function form, and semi-implicit time discretizations of the Kuramoto–Sivashinsky equation and of the Cahn–Hilliard equation.

The weak formulation of (6.14) is

$$\begin{cases}\text{Find } u \in H_0^2(I) \text{ such that}\\ a(u,v) := (u'',v'') + \alpha(u',v') + \beta(u,v) = (f,v), \quad \forall v \in H_0^2(I).\end{cases} \tag{6.15}$$

Using the Poincaré inequality (cf. (B.44)),

$$\|u\|_2 \lesssim |u|_2, \quad \forall u \in H_0^2(I), \tag{6.16}$$

we find that the bilinear form $a(\cdot,\cdot)$ is continuous and coercive in $H_0^2(I) \times H_0^2(I)$, i.e.,

$$|a(u,v)| \lesssim |u|_2 |v|_2, \quad \|u\|_2^2 \lesssim a(u,u), \quad \forall u,v \in H_0^2(I). \tag{6.17}$$

Hence, by the Lax-Milgram lemma, the problem (6.15) admits a unique solution $u \in H_0^2(I)$, if $f \in H^{-2}(I)$.

The Legendre-Galerkin approximation of (6.14) is

$$\begin{cases} \text{Find } u_N \in V_N := P_N \cap H_0^2(I) \text{ such that} \\ a(u_N, v_N) = (I_N f, v_N), \quad \forall v_N \in V_N, \end{cases} \tag{6.18}$$

where I_N is the interpolation operator associated with, say, Legendre-Gauss-Lobatto points. Notice that an advantage of this approach is that the choice of interpolation points for I_N can be quite flexible. It is not necessary to construct special quadrature rules/collocation points corresponding to (6.14), as in the case of a collocation method (cf. Bernardi et al. (1992)).

In view of (6.17), the discrete problem (6.18) also admits a unique solution $u_N \in V_N$.

It is clear from (6.10) that

$$V_N = \text{span}\{J_k^{-2,-2} : k = 4,5\ldots,N\}.$$

Set

$$\phi_k(x) = \gamma_k J_{k+4}^{-2,-2}(x) \text{ with } \gamma_k \text{ such that } (\phi_k'', \phi_k'') = 1.$$

Thanks to (6.8), it is easy to verify that

$$\phi_k(x) = d_k\Big(L_k(x) - \frac{2(2k+5)}{2k+7}L_{k+2}(x) + \frac{2k+3}{2k+7}L_{k+4}(x)\Big), \tag{6.19}$$

with

$$d_k = \frac{1}{\sqrt{2(2k+3)^2(2k+5)}}.$$

Then, for all $0 \le k, j \le N-4$, we have

$$a_{kj} := (\phi_j'', \phi_k'') = (\phi_j'''', \phi_k) = (\phi_j, \phi_k'''') = \delta_{kj}. \tag{6.20}$$

Therefore, setting $q = N-4$ and

$$\begin{aligned} b_{kj} &= (\phi_j, \phi_k), \quad B = (b_{kj})_{0\le k,j\le q}; \\ c_{kj} &= (\phi_j', \phi_k'), \quad C = (c_{kj})_{0\le k,j\le q}; \\ f_k &= (I_N f, \phi_k), \quad \mathbf{f} = (f_0, f_1, \ldots, f_q)^T; \\ u_N &= \sum_{n=0}^{q} \hat{u}_n \phi_n(x), \quad \mathbf{u} = (\hat{u}_0, \hat{u}_1, \ldots, \hat{u}_q)^T, \end{aligned}$$

the system (6.18) reduces to

$$\big(I+\beta B+\alpha C\big)\mathbf{u}=\mathbf{f}, \tag{6.21}$$

where I is the q-by-q identity matrix.

It is obvious that B and C are symmetric positive definite matrices, and it is easy to shown that

$$b_{kj}=0 \text{ if } k\neq j,\ j\pm 2,\ j\pm 4; \quad c_{kj}=0 \text{ if } k\neq j,\ j\pm 2.$$

The nonzero entries of B and C can be easily determined from the properties of Legendre polynomials. Hence, the above system can be solved as efficiently as a second-order equation.

6.2.2 General Even-Order Equations

We now consider a more general $2m$th-order equation:

$$\begin{aligned}&b_0\partial_x^{2m}u(x)+\sum_{k=0}^{2m-1}b_{2m-k}\partial_x^k u(x)=f(x), \quad \text{in } I=(-1,1),\ m\geq 1,\\&\partial_x^k u(\pm 1)=0, \quad 0\leq k\leq m-1,\end{aligned} \tag{6.22}$$

where $\{b_j\}_{j=0}^{2m}$ and f are given functions.

We introduce the bilinear form associated with (6.22):

$$\begin{aligned}a_m(u,v):=&(-1)^m\big(\partial_x^m u,\partial_x^m(b_0v)\big)+(-1)^m\big(\partial_x^{m-1}u,\partial_x^m(b_1v)\big)\\&+(-1)^{m-1}\big(\partial_x^{m-1}u,\partial_x^{m-1}(b_2v)\big)+\ldots+\big(b_{2m}u,v\big).\end{aligned} \tag{6.23}$$

As usual, we assume that the coefficients $\{b_j\}$ are such that the bilinear form is continuous and coercive:

$$|a_m(u,v)|\leq C_0\|u\|_m\|v\|_m, \qquad \forall u,v\in H_0^m(I), \tag{6.24a}$$

$$a_m(u,u)\geq C_1\|u\|_m^2, \qquad \forall u\in H_0^m(I), \tag{6.24b}$$

where C_0 and C_1 are two positive constants depending on $\{b_j\}_{j=0}^{2m}$.

The weak formulation of (6.22) is

$$\begin{cases}\text{Find } u\in H_0^m(I) \text{ such that}\\ a_m(u,v)=(f,v), \quad \forall v\in H_0^m(I).\end{cases} \tag{6.25}$$

Thanks to (6.24a)-(6.24b), the above problem admits a unique solution in $H_0^m(I)$ if $f\in H^{-m}(I)$.

The Legendre-Galerkin approximation of (6.25) is

$$\begin{cases} \text{Find } u_N \in V_N := P_N \cap H_0^m(I) \text{ such that} \\ a_m(u_N, v_N) = (I_N f, v_N), \quad \forall v_N \in V_N, \end{cases} \tag{6.26}$$

where I_N is the interpolation operator associated with Legendre-Gauss-Lobatto points.

In view of the homogeneous boundary conditions built in $J_n^{-m,-m}$, we find

$$V_N = \text{span}\{J_{2m}^{-m,-m}, J_{2m+1}^{-m,-m}, \ldots, J_N^{-m,-m}\}.$$

Using the facts that $\omega^{m,m}\partial_x^{2m} J_l^{-m,-m} \in V_l$ and $J_k^{-m,-m}$ is orthogonal to V_l if $k > l$, we find

$$\begin{aligned} (\partial_x^m J_k^{-m,-m}, \partial_x^m J_l^{-m,-m}) &= (-1)^m (J_k^{-m,-m}, \partial_x^{2m} J_l^{-m,-m}) \\ &= (J_k^{-m,-m}, \omega^{m,m}\partial_x^{2m} J_l^{-m,-m})_{\omega^{-m,-m}} = 0, \end{aligned} \tag{6.27}$$

which is also true if $k < l$.

Define the basis functions

$$\phi_k(x) := c_{k,m} J_{k+2m}^{-m,-m}(x), \quad 0 \le k \le N-2m,$$

and choose a proper scaling factor $c_{k,m}$ such that

$$(\partial_x^m \phi_k, \partial_x^m \phi_l) = \delta_{kl}.$$

Hence, by setting $q = N - 2m$ and

$$\begin{aligned} u_N &= \sum_{l=0}^{q} \hat{u}_l \phi_l, \quad \mathbf{u} = (\hat{u}_0, \hat{u}_1, \ldots, \hat{u}_q)^T; \\ a_{kl} &= a_m(\phi_l, \phi_k), \quad A = (a_{kl})_{0 \le k,l \le q}; \\ f_k &= (I_N f, \phi_k), \quad \mathbf{f} = (f_0, f_1, \ldots, f_q)^T, \end{aligned}$$

the linear system associated with (6.26) becomes $A\mathbf{u} = \mathbf{f}$. Thanks to (6.24a)-(6.24b), we have

$$\begin{aligned} C_0 \|\mathbf{u}\|_{l^2}^2 = C_0 |u_N|_m^2 &\le a_m(u_N, u_N) = (A\mathbf{u}, \mathbf{u})_{l^2} \\ &\le C_1 |u_N|_m^2 = C_1 \|\mathbf{u}\|_{l^2}^2, \end{aligned} \tag{6.28}$$

which implies that $\text{cond}(A) \le C_1/C_0$ and is independent of N. It can be easily shown that A is a sparse matrix with bandwidth $2m+1$, if $\{b_j\}$ are constants. Hence, higher even-order equations in the form of (6.22) can be solved as efficiently as a second-order equation. Furthermore, the use of GJPs also leads to simplified error analysis, see Sect. 6.5.

6.3 Dual-Petrov-Galerkin Methods for Odd-Order Equations

In this section, we present a spectral dual-Petrov-Galerkin method using the GJPs as basis functions for odd-order differential equations (cf. Shen (2003)). For the sake of clarity, we shall start with a third-order equation, and then extend it to general odd-order differential equations.

6.3.1 Third-Order Equations

Consider the third-order equation

$$\begin{cases} \alpha u - \beta u_x - \gamma u_{xx} + u_{xxx} = f, \;\; x \in I = (-1,1), \\ u(\pm 1) = u_x(1) = 0, \end{cases} \tag{6.29}$$

where α, β, γ are given constants. Without loss of generality, we only consider homogeneous boundary conditions, since non-homogeneous boundary conditions $u(-1) = c_1$, $u(1) = c_2$ and $u_x(1) = c_3$ can be easily handled by considering $v = u - \hat{u}$, where $\hat{u}$ is the unique quadratic polynomial satisfying the non-homogeneous boundary conditions.

Since the leading third-order differential operator is not self-adjoint, it is natural to use a Petrov-Galerkin method, in which the trial and test functions are taken from different spaces.

Define

$$\begin{aligned} V &= \{u : \; u \in H_0^1(I), \; u_x \in L^2_{\omega^{-2,0}}(I)\}, \\ W &= \{u : \; u \in V, \; u_{xx} \in L^2_{\omega^{0,2}}(I)\}. \end{aligned} \tag{6.30}$$

A Petrov-Galerkin formulation for (6.29) is

$$\begin{cases} \text{Find } u \in V \text{ such that} \\ \alpha(u,v) - \beta(\partial_x u, v) + \gamma(\partial_x u, \partial_x v) + (\partial_x u, \partial_x^2 v) \\ \qquad\qquad = (f,v), \quad \forall v \in W. \end{cases} \tag{6.31}$$

We refer to Goubet and Shen (2007) for a rigorous mathematical analysis of (6.31) and related nonlinear problems.

We now construct a spectral approximation scheme for (6.31). Let us denote

$$\begin{aligned} V_N &= \{u \in P_N : \; u(\pm 1) = u_x(1) = 0\}, \\ V_N^* &= \{u \in P_N : \; u(\pm 1) = u_x(-1) = 0\}. \end{aligned} \tag{6.32}$$

It is clear that $V_N \subset V$ and $V_N^* \subset W$.

The Legendre dual-Petrov-Galerkin (LDPG) (cf. Shen (2003)) approximation of (6.31) is

$$\begin{cases} \text{Find } u_N \in V_N \text{ such that} \\ \alpha(u_N, v_N) - \beta(\partial_x u_N, v_N) + \gamma(\partial_x u_N, \partial_x v_N) + (\partial_x u_N, \partial_x^2 v_N) \\ \qquad = (I_N f, v_N), \ \forall v_N \in V_N^*, \end{cases} \tag{6.33}$$

where I_N is the interpolation operator associated with the Legendre-Gauss-Lobatto points.

Notice that for any $u_N \in V_N$, we have $\omega^{-1,1} u_N \in V_N^*$, so the above dual-Petrov-Galerkin formulation is equivalent to the following weighted Galerkin formulation:

$$\begin{cases} \text{Find } u_N \in V_N \text{ such that} \\ \alpha(u_N, v_N)_{\omega^{-1,1}} - \beta(\partial_x u_N, v_N)_{\omega^{-1,1}} + \gamma\big(\partial_x u_N, \omega^{1,-1}\partial_x(v_N \omega^{-1,1})\big)_{\omega^{-1,1}} \\ \quad + (\partial_x u_N, \omega^{1,-1}\partial_x^2(v_N \omega^{-1,1}))_{\omega^{-1,1}} = (f, v_N)_{\omega^{-1,1}}, \ \forall v_N \in V_N. \end{cases} \tag{6.34}$$

In fact, the dual-Petrov-Galerkin formulation (6.33) is most suitable for implementation while the weighted Galerkin formulation (6.34) is more convenient for error analysis.

At this point, one important issue is to show the well-posedness of the dual-Petrov-Galerkin scheme. For this purpose, we first prove the following Hardy-type inequalities.

Lemma 6.3.

$$\int_I \frac{u^2}{(1-x)^4} dx \le \frac{4}{9} \int_I \frac{(u_x)^2}{(1-x)^2} dx, \ \forall u \in V_N, \tag{6.35a}$$

$$\int_I \frac{u^2}{(1-x)^3} dx \le \int_I \frac{(u_x)^2}{1-x} dx, \ \forall u \in V_N. \tag{6.35b}$$

Proof. Let $u \in V_N$ and $h \le 2$. Then, for any constant q, we have

$$\begin{aligned} 0 &\le \int_I \Big(\frac{u}{1-x} + q u_x\Big)^2 \frac{1}{(1-x)^h} dx \\ &= \int_I \Big(\frac{u^2}{(1-x)^{2+h}} + q\frac{(u^2)_x}{(1-x)^{1+h}} + q^2 \frac{(u_x)^2}{(1-x)^h}\Big) dx \\ &= (1-(1+h)q) \int_I \frac{u^2}{(1-x)^{2+h}} dx + q^2 \int_I \frac{(u_x)^2}{(1-x)^h} dx. \end{aligned}$$

We obtain (6.35a) and (6.35b) by taking $h = 2, q = \frac{2}{3}$ and $h = q = 1$, respectively.

□

Remark 6.3. *Note that with a change of variable $x \to -x$ in the above lemma, we obtain the corresponding inequalities for $u \in V_N^*$.*

The leading third-order differential operator is coercive in the following sense.

Lemma 6.4.

$$\frac{1}{3}\|u_x\|^2_{\omega^{-2,0}} \le \left(u_x, (u\omega^{-1,1})_{xx}\right) \le 3\|u_x\|^2_{\omega^{-2,0}}, \quad \forall u \in V_N. \tag{6.36}$$

Proof. For any $u \in V_N$, we have $u\omega^{-1,1} \in V_N^*$. Thanks to the homogeneous boundary conditions built into the spaces V_N and V_N^*, all the boundary terms from the integration by parts of the third-order term would vanish. Therefore, using the identity

$$\partial_x^k \omega^{-1,1}(x) = \frac{2k!}{(1-x)^{k+1}}$$

and Lemma 6.3, we find

$$\begin{aligned}
\left(u_x, (u\omega^{-1,1})_{xx}\right) &= (u_x, u_{xx}\omega^{-1,1} + 2u_x\omega_x^{-1,1} + u\omega_{xx}^{-1,1}) \\
&= \frac{1}{2}\int_I \left((u_x^2)_x\omega^{-1,1} + (u^2)_x\omega_{xx}^{-1,1} + 4u_x^2\omega_x^{-1,1}\right) dx \\
&= \int_I \left(\frac{3}{2}u_x^2\omega_x^{-1,1} - \frac{1}{2}u^2\omega_{xxx}^{-1,1}\right) dx \\
&= 3\int_I \frac{u_x^2}{(1-x)^2} dx - 6\int_I \frac{u^2}{(1-x)^4} dx \ge \frac{1}{3}\int_I \frac{u_x^2}{(1-x)^2} dx.
\end{aligned}$$

The desired results follow immediately from the above. □

Set

$$\phi_k(x) = \gamma_k J_{n+3}^{-2,-1}(x), \quad \psi_k(x) = \gamma_k J_{k+3}^{-1,-2}(x),$$

where γ_k is chosen such that $(\phi_k', \psi_k'') = 1$. It is clear from (6.10) that

$$V_N = \text{span}\{\phi_0, \phi_1, \ldots, \phi_{N-3}\}, \quad V_N^* = \text{span}\{\psi_0, \psi_1, \ldots, \psi_{N-3}\}. \tag{6.37}$$

Therefore, denoting

$$\begin{aligned}
&u_N = \sum_{k=0}^{N-3} \hat{u}_k\phi_k, \quad \mathbf{u} = \left(\hat{u}_0, \hat{u}_1, \ldots, \hat{u}_{N-3}\right)^T; \\
&f_k = (I_N f, \psi_k), \quad \mathbf{f} = \left(f_0, f_1, \ldots, f_{N-3}\right)^T; \\
&m_{ij} = (\phi_j, \psi_i), \quad p_{ij} = -(\phi_j', \psi_i), \\
&q_{ij} = (\phi_j', \psi_i'), \quad s_{ij} = (\phi_j', \psi_i''),
\end{aligned} \tag{6.38}$$

the linear system (6.33) becomes

$$\left(\alpha M + \beta P + \gamma Q + S\right)\mathbf{u} = \mathbf{f}, \tag{6.39}$$

where M, P, Q and S are $(N-2)\times(N-2)$ matrices with entries m_{ij}, p_{ij} q_{ij} and s_{ij}, respectively.

By using (6.8) and the orthogonality of the Legendre polynomials, we can easily show that

$$\begin{aligned} m_{ij} &= 0 \text{ for } |i-j|>3; \quad p_{ij}=0 \text{ for } |i-j|>2; \\ q_{ij} &= 0 \text{ for } |i-j|>1; \quad s_{ij}=0 \text{ for } i\neq j. \end{aligned}$$

Non-zero elements of M, P, Q can be easily determined from the properties of Legendre polynomials.

In summary, the coefficient matrix in (6.39) is seven-diagonal, so it can be inverted efficiently as in the second-order case.

Remark 6.4. Ma and Sun (2000, 2001) *proposed an alternative Petrov-Galerkin method for solving third-order differential equation by using a different test function space*:

$$W_N := \{u \in P_{N-1} \, : \, u(\pm 1) = 0\}.$$

It is shown in Ma and Sun (2000, 2001) *that their Petrov-Galerkin method also leads to sparse matrices for problems with constant coefficients.*

Remark 6.5. *In* Shen and Wang (2007c), *the Legendre and Chebyshev dual-Petrov-Galerkin methods were implemented and analyzed for hyperbolic equations.*

6.3.2 General Odd-Order Equations

Consider the following problem:

$$\begin{aligned} &(-1)^{m+1}\partial_x^{2m+1}u(x) + S_m(u) + \gamma u(x) = f(x), \quad \text{in } I=(-1,1), \; m \geq 0, \\ &\partial_x^k u(\pm 1) = \partial_x^m u(1) = 0, \quad 0 \leq k \leq m-1, \end{aligned} \tag{6.40}$$

where $S_m(u)$ is a linear combination of $\{\partial_x^j u\}_{j=1}^{2m-1}$. Without loss of generality, we assume that

$$S_m(u) = (-1)^m \delta \partial_x^{2m-1} u(x), \quad \delta \geq 0, \tag{6.41}$$

since other linear terms with derivatives lower than $2m-1$ can be treated similarly.

Let us define

$$\begin{aligned} V_N &= \{u \in P_N : \, \partial_x^k u(\pm 1) = 0, \, 0 \leq k \leq m-1, \, \partial_x^m u(1) = 0\}, \\ V_N^* &= \{u \in P_N : \, \partial_x^k u(\pm 1) = 0, \, 0 \leq k \leq m-1, \, \partial_x^m u(-1) = 0\}. \end{aligned} \tag{6.42}$$

Then, the Legendre dual-Petrov-Galerkin (LDPG) approximation of (6.40) is

$$\begin{cases} \text{Find } u_N \in V_N \text{ such that} \\ -(\partial_x^{m+1}u_N, \partial_x^m v_N) - \delta(\partial_x^m u_N, \partial_x^{m-1} v_N) \\ \qquad + \gamma(u_N, v_N) = (I_N f, v_N), \quad \forall v_N \in V_N^*, \end{cases} \tag{6.43}$$

where I_N is the interpolation operator associated with the Legendre-Gauss-Lobatto points.

Note that for any $v_N \in V_N$, we have $\omega^{-1,1} v_N \in V_N^*$. Hence, we can rewrite (6.43) into the following weighted spectral-Galerkin formulation:

$$\begin{cases} \text{Find } u_N \in V_N \text{ such that} \\ b_m(u_N, v_N) := -\left(\partial_x^{m+1} u_N, \omega^{1,-1} \partial_x^m (\omega^{-1,1} v_N)\right)_{\omega^{-1,1}} \\ \qquad - \delta\left(\partial_x^m u_N, \omega^{1,-1} \partial_x^{m-1} (\omega^{-1,1} v_N)\right)_{\omega^{-1,1}} \\ \qquad + \gamma(u_N, v_N)_{\omega^{-1,1}} = (I_N f, v_N)_{\omega^{-1,1}}, \quad \forall v_N \in V_N. \end{cases} \tag{6.44}$$

Denote by

$$Q_N^{-k,-l} := \operatorname{span}\{J_{k+l}^{-k,-l}, \ldots, J_N^{-k,-l}\}, \quad \forall k, l \in \mathbb{Z}. \tag{6.45}$$

Thanks to the homogeneous boundary conditions built in V_N and V_N^*, we have

$$V_N = Q_N^{-m-1,-m}, \quad V_N^* = Q_N^{-m,-m-1}. \tag{6.46}$$

The following "coercivity" property (cf. Guo et al. (2006a)) is a direct extension of Lemma 6.4.

Lemma 6.5.

$$-\left(\partial_x^{m+1} u, \partial_x^m (u\omega^{-1,1})\right) = (2m+1) \int_I \left(\partial_x^m \left(\frac{u}{1-x}\right)\right)^2 dx, \quad \forall u \in V_N. \tag{6.47}$$

Proof. For any $u \in V_N$, we set $u = (1-x)\Phi$ with $\Phi \in Q_{N-1}^{-m,-m}$. Then using integration by parts yields

$$\begin{aligned} &-\left(\partial_x^{m+1} u, \partial_x^m (u\omega^{-1,1})\right) \\ &= -\left((1-x)\partial_x^{m+1}\Phi - (m+1)\partial_x^m \Phi, (1+x)\partial_x^m \Phi + m\partial_x^{m-1}\Phi\right) \\ &= -\frac{1}{2} \int_I \partial_x \{(\partial_x^m \Phi)^2\}(1-x^2)dx + (m+1) \int_I (\partial_x^m \Phi)^2 (1+x) dx \\ &\quad + \frac{m(m+1)}{2} \int_I \partial_x \{(\partial_x^{m-1}\Phi)^2\} dx + m \int_I \partial_x^m \Phi \partial_x ((1-x)\partial_x^{m-1}\Phi) dx \\ &= -\int_I (\partial_x^m \Phi)^2 x dx + (m+1) \int_I (\partial_x^m \Phi)^2 (1+x) dx \\ &\quad + m \int_I (\partial_x^m \Phi)^2 (1-x) dx - \frac{m}{2} \int_I \partial_x \{(\partial_x^{m-1}\Phi)^2\} dx \\ &= (2m+1) \int_I (\partial_x^m \Phi)^2 dx = (2m+1) \int_I \left(\partial_x^m \left(\frac{u}{1-x}\right)\right)^2 dx. \end{aligned}$$

This ends the proof. □

Notice that (6.47) is valid for all $m \ge 0$, and $u_N/(1-x) \in Q_N^{-m,-m}$. Applying the Poincaré inequality repeatedly, we derive that for $\gamma > 0$ and $\delta \ge 0$, there exists $C_2 > 0$ such that

$$\begin{aligned}(2m+1)\int_I \Big(\partial_x^m\Big(\frac{u_N}{1-x}\Big)\Big)^2 dx &\le b_m(u_N,u_N)\\ &\le C_2(2m+1)\int_I \Big(\partial_x^m\Big(\frac{u_N}{1-x}\Big)\Big)^2 dx, \quad \forall u_N \in V_N.\end{aligned} \tag{6.48}$$

Hence, we conclude from the Lax-Milgram lemma that (6.44) admits a unique solution.

Let us denote

$$\Phi_n := d_{m,n}J_{n+2m}^{-m-1,-m}, \quad \Psi_n := d_{m,n}J_{n+2m}^{-m,-m-1}. \tag{6.49}$$

It is easy to see that we can choose $d_{m,n}$ such that

$$a_{kl} := -(\partial_x^{m+1}\Phi_l, \partial_x^m\Psi_k) = \delta_{kl}. \tag{6.50}$$

We also have

$$b_{kl} := (\Phi_l, \Psi_k) = 0, \quad \text{if } |k-l| > 2m+1; \tag{6.51a}$$

$$c_{kl} := -(\partial_x^m\Phi_l, \partial_x^{m-1}\Psi_k) = 0, \quad \text{if } k+1 < l \text{ or } l < k-3. \tag{6.51b}$$

Hence, by setting

$$\begin{aligned}&u_N = \sum_{l=0}^{N-2m} \hat u_l \Phi_l, \quad \mathbf{u} = \big(\hat u_0, \hat u_1, \ldots, \hat u_{N-2m}\big)^T;\\ &B = (b_{kl})_{0\le k,l\le N-2m}, \quad C = (c_{kl})_{0\le k,l\le N-2m};\\ &f_k = (I_N f, \Psi_k), \quad \mathbf{f} = \big(f_0, f_1, \ldots, f_{N-2m}\big)^T,\end{aligned}$$

the system (6.43) reduces to

$$\big(I + \gamma B + \delta C\big)\mathbf{u} = \mathbf{f}. \tag{6.52}$$

Since B and C are sparse, the above system can be easily inverted.

The use of GJP basis functions leads to sparse linear systems for problems with constant or polynomial coefficients. Moreover, the linear system (6.52) is well-conditioned as we show below.

Indeed, since $\Phi_k\omega^{-1,1} \in V_N^*$, there exists a unique set of coefficients $\{h_{kj}\}$ such that

$$\Phi_k\omega^{-1,1} = \sum_{j=0}^{N+2m} h_{kj}\Psi_j, \quad 0 \le k \le N-2m.$$

Hence, let $H=(h_{kj})$, we derive from (6.50) and Lemma 6.5 that

$$\begin{aligned}\langle H\mathbf{u},\mathbf{u}\rangle_{l^2}&=-\Big(\sum_{j=0}^{N-2m}\hat{u}_j\partial_x^{m+1}\Phi_j,\sum_{k,j=0}^{N-2m}\hat{u}_k h_{kj}\partial_x^m\Psi_j\Big)\\&=-\Big(\sum_{j=0}^{N-2m}\hat{u}_j\partial_x^{m+1}\Phi_j,\sum_{k=0}^{N-2m}\hat{u}_k\partial_x^m(\Phi_k\omega^{-1,1})\Big)\\&=-\big(\partial_x^{m+1}u_N,\partial_x^m(u_N\omega^{-1,1})\big),\end{aligned}$$

where $\langle\mathbf{u},\mathbf{v}\rangle_{l^2}:=\sum_{j=0}^{N-2m}\hat{u}_j\hat{v}_j$ is the inner product in l^2. Similarly, we can verify that

$$\langle H(I+\gamma B+\delta C)\mathbf{u},\mathbf{u}\rangle_{l^2}=b_m(u_N,u_N).$$

We then derive from the above and (6.48) that

$$\langle H\mathbf{u},\mathbf{u}\rangle_{l^2}\le\langle H(I+\gamma B+\delta C)\mathbf{u},\mathbf{u}\rangle_{l^2}\le C_2\langle H\mathbf{u},\mathbf{u}\rangle_{l^2}.$$

Therefore, the condition number of $I+\gamma B+\delta C$, in the norm $\|\mathbf{v}\|_H:=\langle H\mathbf{v},\mathbf{v}\rangle_{l^2}^{1/2}$, is independent of N. It is clear that the above argument also applies to problems with variable coefficients as long as (6.48) holds.

6.3.3 Higher Odd-Order Equations with Variable Coefficients

As described in Chap. 4, an efficient approach for solving elliptic problems with variable coefficients is to use the Galerkin method with numerical integration coupled with an iterative technique. We show below that this procedure also works for higher odd-order equations.

To fix the idea, we consider the third-order equation:

$$\begin{aligned}&a(x)u-b(x)u_x+u_{xxx}=f,\;\; x\in I=(-1,1),\\&u(\pm1)=u_x(1)=0.\end{aligned}\tag{6.53}$$

The Legendre dual-Petrov-Galerkin method with numerical integration for (6.53) is

$$\begin{cases}\text{Find } u_N\in V_N \text{ such that}\\ \langle a(x)u_N,v_N\rangle_N-\langle b(x)u_N',v_N\rangle_N+\langle u_N',v_N''\rangle_N\\ \qquad\qquad=\langle f,v_N\rangle_N,\;\;\forall v_N\in V_N^*,\end{cases}\tag{6.54}$$

where $\langle\cdot,\cdot\rangle_N$ is the discrete inner product associated with the Legendre-Gauss-Lobatto quadrature. We recall

$$\langle u,v\rangle_N=(u,v),\;\;\forall u\cdot v\in P_{2N-1}.\tag{6.55}$$

Let $\{\phi_k; \psi_j\}$ be the basis functions defined in (6.37) with

$$\langle \phi_k', \psi_j'' \rangle_N = (\phi_k', \psi_j'') = \delta_{kj}, \quad 0 \le k, j \le N-3.$$

Hence, by setting

$$\begin{aligned} u_N &= \sum_{k=0}^{N-3} \tilde{u}_k \phi_k, \quad \mathbf{u} = \left(\tilde{u}_0, \tilde{u}_1, \ldots, \tilde{u}_{N-3}\right)^T; \\ m_{jk} &= \langle a(x)\phi_k, \psi_j \rangle_N, \quad p_{jk} = -\langle b(x)\phi_k', \psi_j \rangle_N; \\ \tilde{f}_j &= \langle f, \psi_j \rangle_N, \quad \mathbf{f} = \left(\tilde{f}_0, \tilde{f}_1, \ldots, \tilde{f}_{N-3}\right)^T, \end{aligned}$$

the linear system (6.54) becomes

$$\big(M + P + I\big)\mathbf{u} = \mathbf{f}. \tag{6.56}$$

It is clear that the matrices M and P are full and their formation involves $O(N^3)$ operations as well as the inversion of (6.56). Hence, a direct approach is advisable only if one uses a small or moderate number of modes. Otherwise, an iterative method can be efficiently implemented as follows:

- Note that a conjugate gradient type iterative method does not require the explicit formation of the matrix, but only the action of the matrix upon a given vector is needed at each iteration. Although the formation of M and P involves $O(N^3)$ operations, their action on a given vector $\mathbf{u}$, i.e. $M\mathbf{u}$ and $P\mathbf{u}$, can be computed in $O(N^2)$ operations.
- The number of operations can be further reduced to $O(N \log_2 N)$ if we use the following Chebyshev-Legendre dual-Petrov-Galerkin method:

$$\begin{cases} \text{Find } u_N \in V_N \text{ such that} \\ \langle I_N^c(a(x)u_N), v_N \rangle_N - \langle I_N^c(b(x)u_N'), v_N \rangle_N \\ \qquad + \langle u_N', v_N'' \rangle_N = \langle f, v_N \rangle_N, \ \ \forall v_N \in V_N^*, \end{cases} \tag{6.57}$$

  where I_N^c is the interpolation operator based on the Chebyshev-Gauss-Lobatto points, while $\langle \cdot, \cdot \rangle_N$ is still the discrete inner product of u and v associated with the Legendre-Gauss-Lobatto quadrature. Hence, the only difference between (6.57) and (6.54) is that $a(x)u_N$ and $b(x)u_N'$ in (6.54) are replaced by $I_N^c(a(x)u_N)$ and $I_N^c(b(x)u_N')$. Thanks to the Fast Fourier Transform (FFT) and the fast Chebyshev-Legendre transform (cf. Alpert and Rokhlin (1991)), the Legendre coefficients of $I_N^c(a(x)u_N)$ and $I_N^c(b(x)u_N')$ can be computed in $O(N \log_2 N)$ operations given the Legendre coefficients of u_N.
- Under reasonable assumptions on $a(x)$ and $b(x)$ the matrix $M + P + I$ is well-conditioned, that is, its condition number is independent of N. We may follow a similar procedure as for the system (6.52) to justify it. This statement is confirmed by our numerical results.

In Table 6.1, we list the condition numbers of $M+P+I$ in (6.56) with various $a(x)$ and $b(x)$. Notice that in all cases, the condition numbers are small, and more importantly, independent of N.

Table 6.1 Condition numbers of (6.56)

N	$a(x)=1$ $b(x)=0$	$a(x)=10$ $b(x)=0$	$a(x)=50$ $b(x)=0$	$a(x)=\sin x$ $b(x)=2x-1$	$a(x)=10\exp(x)$ $b(x)=\cos x$
16	1.119	2.218	7.219	1.188	2.393
64	1.119	2.218	7.219	1.188	2.393
128	1.119	2.218	7.219	1.188	2.393

Therefore, a conjugate gradient type iterative method like BICGSTAB or CGS for (6.56) will converge in a small and fixed number (i.e., independent of N) of iterations. In short, the Chebyshev-Legendre dual-Petrov-Galerkin method for (6.53) can be solved in a quasi-optimal $O(N\log_2 N)$ operations.

6.4 Collocation Methods

As an alternative to the Galerkin approach, one can also use a collocation method for higher-order equations. However, the choice of collocation points is more delicate, as it needs to take into account the underlying boundary conditions. More precisely, one needs to construct generalized quadratures involving derivatives at endpoints.

For $r,l\in\mathbb{N}$, let $\{x_j\}_{j=1}^{N-1}$ be the zeros of the Jacobi polynomial $J_{N-1}^{r,l}(x)$. The generalized Gauss-Lobatto quadrature formula takes the form (see Krylov (1962), Huang and Sloan (1992)):

$$\begin{aligned}\int_{-1}^{1} f(x)dx &= \sum_{j=1}^{N-1} f(x_j)\omega_j + \sum_{\nu=0}^{l-1} f^{(\nu)}(-1)\omega_-^{(\nu)}\\ &+\sum_{\mu=0}^{r-1} f^{(\mu)}(1)\omega_+^{(\mu)} + E_N[f],\end{aligned} \tag{6.58}$$

where $\{\omega_j\}$, $\{\omega_-^{(\nu)}\}$ and $\{\omega_+^{(\mu)}\}$ are quadrature weights, and $E_N[f]$ is the quadrature error. This formula is exact for all polynomials of degree $\le 2N+r+l-3$.

Given the values of the function $f(x)$ and its derivatives as follows

$$\begin{aligned}&f_j=f(x_j),\;\; 1\le j\le N-1;\quad f_-^{(\nu)}=f^{(\nu)}(-1),\;\; 0\le\nu\le l-1;\\ &f_+^{(\mu)}=f^{(\mu)}(1),\;\; 0\le\mu\le r-1,\end{aligned}$$

the polynomial of degree $N+l+r-2$ interpolating these data is given by

$$(I_N^{(G)}f)(x)=\sum_{j=1}^{N-1} f_j h_j(x)+\sum_{\nu=0}^{l-1} f_-^{(\nu)} h_0^{(\nu)}(x)+\sum_{\mu=0}^{r-1} f_+^{(\mu)} h_N^{(\mu)}(x). \tag{6.59}$$

Here, the interior (generalized) Lagrange basis polynomials are given by

$$h_j(x) = \frac{J_{N-1}^{r,l}(x)}{\partial_x J_{N-1}^{r,l}(x_j)(x-x_j)} \frac{(1+x)^l(1-x)^r}{(1+x_j)^l(1-x_j)^r}, \quad 1 \le j \le N-1, \tag{6.60}$$

and the expressions of $\{h_0^{(\nu)}\}$ and $\{h_N^{(\mu)}\}$ can be found in Huang and Sloan (1992). Correspondingly, the quadrature weights are expressed as

$$\{\omega_j, \omega_-^{(\nu)}, \omega_+^{(\mu)}\} = \int_{-1}^{1} \{h_j(x), h_0^{(\nu)}(x), h_N^{(\mu)}(x)\} dx. \tag{6.61}$$

We refer to Huang and Sloan (1992) for the explicit expressions of these weights. Some estimates for the interpolation errors were carried out in Bernardi and Maday (1997) and Wang et al. (2002).

As an illustrative example, we consider the implementation of the collocation method for the fifth-order equation:

$$\begin{aligned} &u^{(5)}(x) + a_1(x)u'(x) + a_0(x)u(x) = f(x), \quad \text{in } (-1,1), \\ &u(\pm 1) = u'(\pm 1) = u''(1) = 0, \end{aligned} \tag{6.62}$$

where a_0, a_1 and f are given functions.

Let us denote

$$X_N = \{u \in P_{N+3} \ : \ u(\pm 1) = u'(\pm 1) = u''(1) = 0\},$$

whose dimension is $N-1$. Let $\{x_j\}_{j=1}^{N-1}$ be the zeros of the Jacobi polynomial $J_{N-1}^{3,2}(x)$. The collocation scheme of (6.62) is

$$\begin{cases} \text{Find } u_N \in X_N \text{ such that for } 1 \le j \le N-1, \\ u_N^{(5)}(x_j) + a_1(x_j)u_N'(x_j) + a_0(x_j)u_N(x_j) = f(x_j). \end{cases} \tag{6.63}$$

It is clear that

$$X_N = \text{span}\{h_j \ : \ 1 \le j \le N-1\}.$$

Setting

$$\begin{aligned} &u_N(x) = \sum_{j=1}^{N-1} u_j h_j(x), \quad \mathbf{u} = (u_1, u_2, \ldots, u_{N-1})^T; \\ &d_{ij}^{(k)} = h_j^{(k)}(x_i), \quad D^{(k)} = (d_{ij}^{(k)})_{1 \le i,j \le N-1}; \\ &A_m = \text{diag}(a_m(x_1), a_m(x_2), \ldots, a_m(x_{N-1})), \ m = 0, 1; \\ &\mathbf{f} = (f(x_1), f(x_2), \ldots, f(x_{N-1}))^T, \end{aligned}$$

we find that the linear system (6.63) reduces to

$$\big(D^{(5)} + A_1 D^{(1)} + A_0\big)\mathbf{u} = \mathbf{f}. \tag{6.64}$$

Hence, the collocation method is easy to implement once the differentiation matrices have been pre-computed. However, it should be pointed out that

- Unlike the situation in Theorem 3.10, the high-order differentiation matrix can not be computed from the product of the first-order differentiation matrix, when the derivatives are involved in the boundary conditions (cf. Wang and Guo (2009)).
- The matrices $\{D^{(k)}\}$ are full with $\mathrm{cond}(D^{(k)}) \sim N^{2k}$ (see Table 6.2).
- When N is large, the accuracy of the nodes and of the entries of $\{D^{(k)}\}$ is subject to severe roundoff errors.

We note that although one can build effective preconditioners using finite difference or finite element approximations for the collocation matrices of second- and fourth-order derivatives (cf. Orszag (1980)), it is not clear how to construct effective preconditioners for collocation matrices of odd-order.

As a numerical comparison of the generalized Jacobi spectral-Galerkin method (GJS) with numerical integration (as described in the previous section) and the collocation method (COL), we tabulate in Table 6.2 the condition numbers of systems resulting from two methods. We see that for various $a_0(x)$ and $a_1(x)$, the condition numbers of the GJS systems are all small and independent of N, while those of the COL systems increase like $O(N^{10})$.

Table 6.2 Condition numbers of COL and GJS

N	Method	$a_0=0$, $a_1=0$	$a_0=10$, $a_1=0$	$a_0=50$, $a_1=1$	$a_0=100x$, $a_1=50$	$a_0=10e^{10x}$, $a_1=\sin(10x)$
16	COL	3.30E+05	3.77E+05	4.46E+05	2.49E+05	4.09E+05
16	GJS	1.00	1.07	1.42	1.62	33.05
32	COL	2.70E+08	2.78E+08	3.36E+08	1.37E+08	8.22E+08
32	GJS	1.00	1.07	1.42	1.62	33.05
64	COL	2.58E+11	2.64E+11	4.43E+11	8.11E+10	1.37E+11
64	GJS	1.00	1.07	1.42	1.62	33.05
128	COL	2.05E+14	2.10E+14	2.39E+14	1.86E+14	2.64E+14
128	GJS	1.00	1.07	1.42	1.62	33.05

Next, we examine the effect of roundoff errors. We take $a_0(x) = 10e^{10x}$ and $a_1(x) = \sin(10x)$, and let $u(x) = \sin^3(8\pi x)$ be the exact solution of (6.62). The L^2-errors of two methods against various N are depicted in Fig. 6.1. We observe that the effect of roundoff errors is much severer in the collocation method.

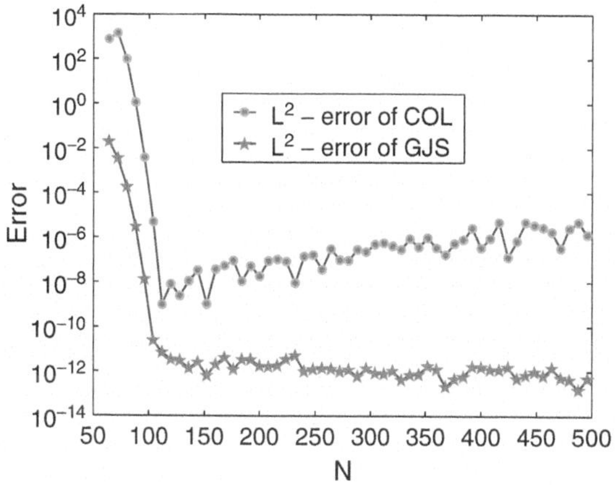

Fig. 6.1 L^2-errors of COL and GJS

6.5 Error Estimates

This section is devoted to error analysis of several spectral methods presented in the previous sections. We start by establishing some approximation results for the GJPs.

For $k,l \in \mathbb{Z}$, let $Q_N^{k,l}$ be the polynomial space defined in (6.45). We consider the orthogonal projection $\pi_N^{k,l} : L^2_{\omega^{k,l}}(I) \to Q_N^{k,l}$, defined by

$$\left(u - \pi_N^{k,l} u, v_N\right)_{\omega^{k,l}} = 0, \quad \forall v_N \in Q_N^{k,l}. \tag{6.65}$$

Thanks to the orthogonality (6.5) and the derivative relation (6.12), the following theorem is a direct extension of Theorem 3.35.

Theorem 6.1. *For any $k,l \in \mathbb{Z}$, and $u \in B^m_{k,l}(I)$, we have that for $0 \le \mu \le m \le N+1$,*

$$\begin{aligned} &\|\partial_x^\mu (\pi_N^{k,l} u - u)\|_{\omega^{k+\mu,l+\mu}} \\ &\le c\sqrt{\frac{(N-m+1)!}{(N-\mu+1)!}}(N+m)^{(\mu-m)/2}\|\partial_x^m u\|_{\omega^{k+m,l+m}}, \end{aligned} \tag{6.66}$$

where c is a positive constant independent of m, N and u.

Proof. The proof is essential the same as that of Theorem 3.35 as sketched below. Using (6.12) repeatedly yields

$$\partial_x^i J_n^{k,l}(x) = d_{n,i}^{k,l} J_{n-i}^{k+i,l+i}(x),$$

which implies

$$\int_{-1}^{1} \partial_x^i J_n^{k,l} \partial_x^i J_m^{k,l} \omega^{k+i,l+i} dx = h_{n,i}^{k,l} \delta_{mn}.$$

One can follow the same procedure as in the proof of Theorem 3.35 to work out the constants $d_{n,i}^{k,l}$ and $h_{n,i}^{k,l}$ to obtain the desired result. □

We would like to point out that the analysis based on the GJPs leads to a more concise and precise estimates than those based on the classical Jacobi polynomials in Chaps. 3 and 4.

To show this, we revisit the H_0^1-orthogonal projection (cf. (3.290)). The inequality (B.40) implies that $H_0^1(I) \subseteq L^2_{\omega^{-1,-1}}(I)$, so for any $u \in H_0^1(I)$, we can write

$$u = \sum_{n=2}^{\infty} \hat{u}_n J_n^{-1,-1}(x), \quad \hat{u}_n = \int_I u(x) J_n^{-1,-1}(x)(1-x^2)^{-1} dx. \tag{6.67}$$

Moreover, we find

$$\big(\partial_x(\pi_N^{-1,-1} u - u), \partial_x v_N\big) = 0, \quad \forall v_N \in P_N^0,$$

which implies that $\pi_N^{-1,-1}$ is identical with the H_0^1-orthogonal projection operator defined in (3.290) (with $\alpha = \beta = 0$). Hence, we obtain the optimal estimates (3.295) (with $\alpha = \beta = 0$) from (6.66) (with $k = l = -1$) directly, and the duality argument is not needed. This also applies to the estimates for the H_0^m-orthogonal projections with $m \geq 2$ (see Lemma 4.7).

As a second example, we provide below a proof for Theorem 3.44 by using Theorem 6.1 and the stability result in Problem 3.24.

Proof of Theorem 3.44:

For any $u \in H^1(I)$, we find from (B.33) that $u \in C(\bar{I})$. Define

$$u^*(x) = \frac{1-x}{2} u(-1) + \frac{1+x}{2} u(1).$$

It is clear that

$$\tilde{u} := u - u^* \in H_0^1(I) \text{ and } I_N u - u = I_N \tilde{u} - \tilde{u}.$$

Hence, by (3.326) and Theorem 6.1,

$$\begin{aligned}
\|I_N \tilde{u} - \pi_N^{-1,-1} \tilde{u}\|_{\omega^{-1,-1}} &= \|I_N\big(\tilde{u} - \pi_N^{-1,-1} \tilde{u}\big)\|_{\omega^{-1,-1}} \\
&\leq c\big(\|\tilde{u} - \pi_N^{-1,-1} \tilde{u}\|_{\omega^{-1,-1}} + N^{-1} \|\partial_x(\tilde{u} - \pi_N^{-1,-1} \tilde{u})\|\big) \\
&\leq c\sqrt{\frac{(N-m+1)!}{N!}} (N+m)^{-(1+m)/2} \|\partial_x^m \tilde{u}\|_{\omega^{m-1,m-1}},
\end{aligned}$$

which, together with the inverse inequality (3.241), implies

$$\begin{aligned}\|\partial_x(I_N\tilde{u}-\pi_N^{-1,-1}\tilde{u})\| &\le cN\|I_N\tilde{u}-\pi_N^{-1,-1}\tilde{u}\|_{\omega^{-1,-1}}\\ &\le c\sqrt{\frac{(N-m+1)!}{N!}}(N+m)^{(1-m)/2}\|\partial_x^m\tilde{u}\|_{\omega^{m-1,m-1}}.\end{aligned}$$

Therefore, using the triangle inequality and Theorem 6.1 yields

$$\begin{aligned}&\|\partial_x(I_N\tilde{u}-\tilde{u})\|+N\|I_N\tilde{u}-\tilde{u}\|_{\omega^{-1,-1}}\\ &\le c\sqrt{\frac{(N-m+1)!}{N!}}(N+m)^{(1-m)/2}\|\partial_x^m\tilde{u}\|_{\omega^{m-1,m-1}}.\end{aligned}\tag{6.68}$$

It is clear that if $m>1$, we have $\partial_x^m\tilde{u}=\partial_x^m u$, while for $m=1$, we verify from the inequalities (B.33) and (B.44) that

$$\|\partial_x\tilde{u}\|\le c\big(\|\partial_x u\|+|u(-1)|+|u(1)|\big)\le c\|u\|_1\le c\|\partial_x u\|.$$

This ends the proof of Theorem 3.44. □

6.5.1 Even-Order Equations

We consider here the error analysis of the Legendre-Galerkin scheme (6.18). Observe from (6.65) that

$$\begin{aligned}\big(\partial_x^2(\pi_N^{-2,-2}u-u),\partial_x^2 v_N\big)&=\big(\pi_N^{-2,-2}u-u,\partial_x^4 v_N\big)\\ &=\big(\pi_N^{-2,-2}u-u,\omega^{2,2}\partial_x^4 v_N\big)_{\omega^{-2,-2}}=0,\quad \forall v_N\in V_N.\end{aligned}\tag{6.69}$$

Hence, $\pi_N^{-2,-2}$ is also the orthogonal projector in $H_0^2(I)$ as defined in (4.78).

Theorem 6.2. *Let u and u_N be the solutions of (6.15) and (6.18), respectively. If $\alpha,\beta>0$ and $u\in H_0^2(I)\cap B^m_{-2,-2}(I)$ and $f\in B^k_{-1,-1}(I)$ with $2\le m\le N+1$ and $1\le k\le N+1$, then we have*

$$\begin{aligned}\|\partial_x^\mu(u-u_N)\|&\le c\sqrt{\frac{(N-m+1)!}{(N-\mu+1)!}}(N+m)^{(\mu-m)/2}\|\partial_x^m u\|_{\omega^{m-2,m-2}}\\ &\quad+c\sqrt{\frac{(N-k+1)!}{N!}}(N+k)^{-(k+1)/2}\|\partial_x^k f\|_{\omega^{k-1,k-1}},\end{aligned}$$

where $\mu=0,1,2$, and c is a positive constant independent of m,k,N,f and u.

Proof. Using (6.15) and (6.18) leads to the error equation

$$a(u-u_N,v_N)=(f-I_Nf,v_N),\quad \forall v_N\in V_N.$$

Let us denote $\hat{e}_N = \pi_N^{-2,-2}u - u_N$ and $\tilde{e}_N = \pi_N^{-2,-2}u - u$. Taking $v_N = \hat{e}_N$ in the above equality, we obtain from (6.69) that

$$\|\hat{e}_N''\|^2 + \alpha\|\hat{e}_N'\|^2 + \beta\|\hat{e}_N\|^2 = \alpha(\tilde{e}_N', \hat{e}_N') + \beta(\tilde{e}_N, \hat{e}_N) + (f - I_N f, \hat{e}_N).$$

Since $(\tilde{e}_N', \hat{e}_N') = -(\tilde{e}_N, \hat{e}_N'')$, using the Cauchy–Schwarz inequality and the Poincaré inequality (6.16) yields

$$\|\hat{e}_N\|_2^2 \le c\|\tilde{e}_N\|^2.$$

Thus, the estimate (6.69) follows from the triangle inequality and Theorems 6.1 and 3.44. □

A similar analysis can be performed for the spectral-Galerkin scheme (6.26), and the estimate is stated as follows.

Theorem 6.3. *Let u and u_N be the solutions of* (6.25) *and* (6.26)*, respectively. If $u \in H_0^m(I) \cap B^r_{-m,-m}(I)$ and $f \in B^k_{-1,-1}(I)$ with $m \le r \le N+1$ and $1 \le k \le N+1$, then we have*

$$\begin{aligned}\|u - u_N\|_\mu &\le c\sqrt{\frac{(N-r+1)!}{(N-\mu+1)!}}(N+m)^{(\mu-r)/2}\|\partial_x^r u\|_{\omega^{r-m,r-m}} \\ &+ c\sqrt{\frac{(N-k+1)!}{N!}}(N+k)^{-(k+1)/2}\|\partial_x^k f\|_{\omega^{k-1,k-1}},\end{aligned}$$

where $0 \le \mu \le m \le r \le N+1$ and c is a positive constant independent of r, k, N, f and u.

6.5.2 Odd-Order Equations

Next, we turn to the analysis of dual-Petrov-Galerkin methods for odd-order equations. We start with the following important observation.

Lemma 6.6. *Let $\pi_N^{-2,-1}$ be the orthogonal projector defined in* (6.65)*. Then,*

$$\left(\partial_x(u - \pi_N^{-2,-1}u), \partial_x^2 v_N\right) = 0, \quad \forall u \in V,\ v_N \in V_N^*, \tag{6.70}$$

where V and V_N^* are defined in (6.30) *and* (6.32)*, respectively.*

Proof. Using integration by parts yields

$$\left(\partial_x(u - \pi_N^{-2,-1}u), \partial_x^2 v_N\right) = -\left(u - \pi_N^{-2,-1}u, \omega^{2,1}\partial_x^3 v_N\right)_{\omega^{-2,-1}}.$$

In view of $\omega^{2,1}\partial_x^3 v_N \in V_N$, (6.70) follows from (6.65). □

This lemma indicates that $\pi_N^{-2,-1}$ is simultaneously an orthogonal projector with respect to two bilinear forms.

Theorem 6.4. *Let u and u_N be the solutions of* (6.29) *and* (6.34), *respectively. If* $\alpha, \beta \geq 0$ *and* $-\frac{1}{3} < \gamma < \frac{1}{6}$, $u \in V \cap B^m_{-2,-1}(I)$ *and* $f \in B^k_{-1,-1}(I)$ *with* $2 \leq m \leq N+1$ *and* $1 \leq k \leq N+1$, *then we have*

$$\begin{aligned}
&\alpha \|e_N\|_{\omega^{-1,1}} + N^{-1} \|\partial_x e_N\|_{\omega^{-1,0}} \\
&\qquad \leq c(1+|\gamma| N) \sqrt{\frac{(N-m+1)!}{N!}} (N+m)^{-(1+m)/2} \|\partial_x^m u\|_{\omega^{m-2,m-1}} \\
&\qquad\quad + c \sqrt{\frac{(N-k+1)!}{N!}} (N+k)^{-(k+1)/2} \|\partial_x^k f\|_{\omega^{k-1,k-1}},
\end{aligned}$$

where c is a positive constant independent of m, k, N, f and u.

Proof. Denote $\hat{e}_N = \pi_N^{-2,-1} u - u_N$ and $e_N = u - u_N = (u - \pi_N^{-2,-1} u) + \hat{e}_N$.
We derive from (6.29), (6.34) and Lemma 6.6 that

$$\begin{aligned}
&\alpha (e_N, v_N)_{\omega^{-1,1}} - \beta (\partial_x e_N, v_N)_{\omega^{-1,1}} + \gamma (\partial_x e_N, \omega^{1,-1} \partial_x (v_N \omega^{-1,1}))_{\omega^{-1,1}} \\
&\quad + (\partial_x \hat{e}_N, \omega^{1,-1} \partial_x^2 (v_N \omega^{-1,1}))_{\omega^{-1,1}} = (f - I_N f, v_N), \quad \forall v_N \in V_N.
\end{aligned}$$

Thus, by (6.70),

$$\begin{aligned}
&\alpha (\hat{e}_N, v_N)_{\omega^{-1,1}} - \beta (\partial_x \hat{e}_N, v_N)_{\omega^{-1,1}} + \gamma (\partial_x \hat{e}_N, \omega^{1,-1} \partial_x (v_N \omega^{-1,1}))_{\omega^{-1,1}} \\
&\quad = \alpha (\pi_N^{-2,-1} u - u, v_N)_{\omega^{-1,1}} - \beta (\partial_x (\pi_N^{-2,-1} u - u), v_N)_{\omega^{-1,1}} \\
&\qquad + \gamma (\partial_x (\pi_N^{-2,-1} u - u), \partial_x (v_N \omega^{-1,1})) \\
&\qquad - (\partial_x \hat{e}_N, \omega^{1,-1} \partial_x^2 (v_N \omega^{-1,1}))_{\omega^{-1,1}} + (f - I_N f, v_N).
\end{aligned} \tag{6.71}$$

One verifies readily that for any $v \in V_N$,

$$\begin{aligned}
&-(\partial_x v, v)_{\omega^{-1,1}} = -\frac{1}{2} \int_I \partial_x (v^2) \omega^{-1,1} dx = \|v\|^2_{\omega^{-2,0}}, \\
&(\partial_x v, \partial_x (v \omega^{-1,1})) = (\partial_x v, \partial_x v \omega^{-1,1} + 2 v \omega^{-2,0}) = \|\partial_x v\|^2_{\omega^{-1,1}} - 2\|v\|^2_{\omega^{-3,0}}.
\end{aligned}$$

In view of this, taking $v_N = \hat{e}_N$ in (6.71) leads to

$$\begin{aligned}
&\alpha \|\hat{e}_N\|^2_{\omega^{-1,1}} + \beta \|\hat{e}_N\|^2_{\omega^{-2,0}} + \gamma \|\partial_x \hat{e}_N\|^2_{\omega^{-1,1}} - 2\gamma \|\hat{e}_N\|^2_{\omega^{-3,0}} + \frac{1}{3} \|\partial_x \hat{e}_N\|^2_{\omega^{-2,0}} \\
&\quad \leq \alpha (\pi_N^{-2,-1} u - u, \hat{e}_N)_{\omega^{-1,1}} - \beta (\partial_x (\pi_N^{-2,-1} u - u), \hat{e}_N)_{\omega^{-1,1}} \\
&\qquad + \gamma (\partial_x (\pi_N^{-2,-1} u - u), \partial_x (\hat{e}_N \omega^{-1,1})) + (f - I_N f, \hat{e}_N).
\end{aligned}$$

The right-hand side can be bounded by using Lemma 6.3, the Cauchy–Schwarz inequality and the fact that $\omega^{-1,2} \le 2\omega^{-1,1} \le 2\omega^{-2,0}$:

$$
\begin{aligned}
\big(\pi_N^{-2,-1}u-u,\hat{e}_N\big)_{\omega^{-1,1}} &\le \|\hat{e}_N\|_{\omega^{-1,1}}\|\pi_N^{-2,-1}u-u\|_{\omega^{-1,1}} \\
&\le 2\|\hat{e}_N\|_{\omega^{-1,1}}\|\pi_N^{-2,-1}u-u\|_{\omega^{-2,-1}}, \\
\big(\partial_x(\pi_N^{-2,-1}u-u),\hat{e}_N\big)_{\omega^{-1,1}} &= \big(\pi_N^{-2,-1}u-u,\partial_x\hat{e}_N\omega^{-1,1}+2\hat{e}_N\omega^{-2,0}\big) \\
&\le c\|\pi_N^{-2,-1}u-u\|_{\omega^{-2,-1}}\|\partial_x\hat{e}_N\|_{\omega^{-2,0}}, \\
\big(\partial_x(\pi_N^{-2,-1}u-u),\partial_x(\hat{e}_N\omega^{-1,1})\big) &= \big(\partial_x(\pi_N^{-2,-1}u-u),\partial_x\hat{e}_N\omega^{-1,1}+2\hat{e}_N\omega^{-2,0}\big) \\
&\le \|\partial_x(\pi_N^{-2,-1}u-u)\|_{\omega^{-1,0}}\|\partial_x\hat{e}_N\|_{\omega^{-2,0}}.
\end{aligned}
$$

For $0 \le \gamma < \frac{1}{6}$, we choose δ sufficiently small such that $\frac{1}{3}-2\gamma-\delta>0$. Combining the above inequalities, using the inequality

$$
ab \le \varepsilon a^2 + \frac{1}{4\varepsilon}b^2, \quad \forall \varepsilon > 0, \tag{6.72}
$$

and dropping some unnecessary terms, we derive from Theorem 6.1 that

$$
\begin{aligned}
&\frac{\alpha}{2}\|\hat{e}_N\|^2_{\omega^{-1,1}} + \Big(\frac{1}{3}-2\gamma-\delta\Big)\|\partial_x\hat{e}_N\|^2_{\omega^{-2,0}} \\
&\qquad \le c\big(\|\pi_N^{-2,-1}u-u\|^2_{\omega^{-2,-1}} + \gamma\|\partial_x(\pi_N^{-2,-1}u-u)\|^2_{\omega^{-1,0}}\big) \\
&\qquad \le c(1+\gamma N^2)\frac{(N-m+1)!}{N!}(N+m)^{-(1+m)}\|\partial_x^m u\|^2_{\omega^{m-2,m-1}} \\
&\qquad\quad + (f-I_N f,\hat{e}_N).
\end{aligned}
$$

For $-\frac{1}{3} < \gamma < 0$, we choose δ sufficiently small such that $\frac{1}{3}+\gamma-\delta>0$, and we obtain

$$
\begin{aligned}
&\frac{\alpha}{2}\|\hat{e}_N\|^2_{\omega^{-1,1}} + \Big(\frac{1}{3}+\gamma-\delta\Big)\|\partial_x\hat{e}_N\|^2_{\omega^{-2,0}} \\
&\qquad \le c(1+|\gamma|N^2)\frac{(N-m+1)!}{N!}(N+m)^{-(1+m)}\|\partial_x^m u\|^2_{\omega^{m-2,m-1}} \\
&\qquad\quad + (f-I_N f,\hat{e}_N).
\end{aligned}
$$

Then, the desired results follow from the triangle inequality, Poincaré inequality, Theorems 6.1 and 3.44, and the fact that $\|u\|_{\omega^{-1,0}} \le 2\|u\|_{\omega^{-2,0}}$. □

Remark 6.6. *Note that the error estimate in the above theorem is optimal for $\gamma = 0$ but sub-optimal for $\gamma \ne 0$.*

The error analysis of the general $(2m+1)$th-order equation can be done in a very similar fashion as above. As a direct extension of Lemma 6.6, we have the following result.

Lemma 6.7. *Let V_N and V_N^* be defined in* (6.42). *Then,*

$$\left(\partial_x^{m+1}(\pi_N^{-m-1,-m}u-u),\partial_x^m v_N\right)=0,\quad \forall u\in V,\ v_N\in V_N^*, \tag{6.73a}$$

$$\begin{aligned}&\left(\partial_x^{m+1}(\pi_N^{-m-1,-m}u-u),\omega^{1,-1}\partial_x^m(\omega^{-1,1}v_N)\right)_{\omega^{-1,1}}\\&\quad=0,\quad \forall u\in V,\ v_N\in V_N.\end{aligned} \tag{6.73b}$$

Proof. By the definition (6.65), for any $v_N\in V_N^*$,

$$\begin{aligned}&\left(\partial_x^{m+1}(\pi_N^{-m-1,-m}u-u),\partial_x^m v_N\right)\\&=(-1)^{m+1}\left(\pi_N^{-m-1,-m}u-u,\omega^{m+1,m}\partial_x^{2m+1}v_N\right)_{\omega^{-m-1,-m}}=0.\end{aligned}$$

Here, we used the fact that for any $v_N\in V_N^*$, $\omega^{m+1,m}\partial_x^{2m+1}v_N\in V_N$. Since for any $v_N\in V_N$, we have $\omega^{-1,1}v_N\in V_N^*$. Hence, (6.73b) is a direct consequence of (6.73a). □

The convergence rate of the scheme (6.44) is given below.

Theorem 6.5. *Let u and u_N be the solutions of* (6.40) *and* (6.44), *respectively. Given $\gamma,\delta>0$. If $u\in V\cap B^r_{-m-1,-m}(I)$ and $f\in B^k_{-1,-1}(I)$ with $m+1\le r\le N+1$ and $1\le k\le N+1$, then we have*

$$\begin{aligned}&\|\partial_x^m((1-x)^{-1}(u-u_N))\|_{\omega^{1,0}}+N\|\partial_x^{m-1}((1-x)^{-1}(u-u_N))\|+N\|u-u_N\|_{\omega^{-1,1}}\\&\le c\sqrt{\frac{(N-r+1)!}{(N-m+1)!}}(N+m)^{(m-r)/2}\|\partial_x^r u\|_{\omega^{r-m-1,r-m}}\\&\quad+c\sqrt{\frac{(N-k+1)!}{N!}}(N+k)^{-(k+1)/2}\|\partial_x^k f\|_{\omega^{k-1,k-1}},\end{aligned}$$

where c is a positive constant independent of r,k,N,f and u.

The analysis is similar to the third-order case (cf. Guo et al. (2009) for the details). We leave the proof of this theorem as an exercise (see Problem 6.5).

6.6 Applications

In this section, we apply the spectral-Galerkin methods using generalized Jacobi polynomials as basis functions to time-dependent problems. These include the Cahn–Hilliard equation, and the third-order and fifth-order Korteweg–de Vries (KdV) equations.

6.6.1 Cahn–Hilliard Equation

Consider the following Cahn–Hilliard equation:

$$\begin{aligned} &u_t = -\gamma\big(u_{xx} - \varepsilon^{-2}(u^2-1)u\big)_{xx}, \quad x \in (-1,1),\, t>0,\ \gamma>0,\\ &u(\pm 1,t) = u'(\pm 1,t) = 0, \quad t \ge 0,\\ &u(x,0) = u_0(x), \quad x \in [-1,1]. \end{aligned} \tag{6.74}$$

The Cahn–Hilliard equation was originally introduced by Cahn and Hilliard (1958) to describe the complicated phase separation and coarsening phenomena in a solid. It has been widely used in materials science and fluid dynamics applications. We refer to Sect. 9.3 for a thorough discussion on the Cahn–Hilliard equation.

Let $V_N = \{u \in P_N : u(\pm 1) = u'(\pm 1) = 0\}$ and τ be the time step size. A fully discrete scheme with Crank-Nicolson leap-frog in time and Legendre-Galerkin method in space is

$$\begin{cases} \text{Find } u_N^{k+1} \in V_N \text{ such that}\\ \dfrac{1}{2\tau}\big(u_N^{k+1} - u_N^{k-1}, v_N\big) + \gamma\big(\partial_x^2 \hat{u}_N^{k+1}, \partial_x^2 v_N\big)\\ \qquad = \dfrac{1}{\varepsilon^2}\big(I_N[(u_N^k)^3 - u_N^k], \partial_x^2 v_N\big), \quad \forall v_N \in V_N, \end{cases} \tag{6.75}$$

where $\hat{u}_N^{k+1} = \frac{1}{2}(u_N^{k+1} + u_N^{k-1})$, and I_N is the Legendre-Gauss-Lobatto interpolation operator. Note that the above scheme requires u_N^0 and u_N^1 to start. We can set $u_N^0 = I_N u_0$ and compute u_N^1 using a first-order (in time) scheme.

At each time step, one only needs to solve the following problem:

$$\begin{cases} \text{Find } u_N^{k+1} \in V_N \text{ such that}\\ \tau\gamma\big(\partial_x^2 \hat{u}_N^{k+1}, \partial_x^2 v_N\big) + \big(\hat{u}_N^{k+1}, v_N\big) = \big(u_N^{k-1}, v_N\big)\\ \qquad + \dfrac{\tau}{\varepsilon^2}\big(I_N[(u_N^k)^3 - u_N^k], \partial_x^2 v_N\big), \quad \forall v_N \in V_N. \end{cases} \tag{6.76}$$

The above system is of the form (6.18) so it can be solved by using the method presented in Sect. 6.2.1.

We implemented the above scheme with $N = 64$, $\varepsilon = 0.02$, $\gamma = 0.01$ and $\tau = 0.000002$. Note that we need to take τ very small to ensure the stability due to the explicit treatment of the nonlinear term. Much stabler schemes will be presented in Sect. 9.3. We start with $u|_{t=0} = \sin^2(\pi x)$ and compute the numerical solution up to time $t = 0.2$. Note that due to the explicit treatment of the nonlinear term, we have to use a very small time step for the scheme to be stable. We refer to Sect. 9.3 for a discussion on the stability issue of the scheme (6.75) and some strategies to design more robust and efficient numerical schemes for Cahn–Hilliard equations. In Fig. 6.2, we plot, on the left, time evolution of the numerical solution, and on the right, the final steady state solution of (6.74).

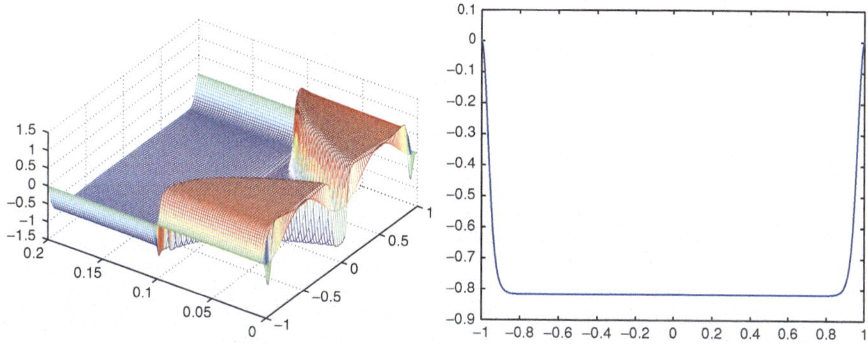

Fig. 6.2 Left: time evolution of the numerical solution for (6.74); Right: steady state solution for (6.74)

6.6.2 Korteweg–de Vries (KdV) Equation

Consider the KdV equation (cf. Korteweg and de Vries (1895)):

$$U_t + UU_x + U_{xxx} = 0, \quad U(x,0) = U_0(x), \tag{6.77}$$

with the exact soliton solution

$$U(x,t) = 12\kappa^2 \text{sech}^2\big(\kappa(x - 4\kappa^2 t - x_0)\big), \tag{6.78}$$

where κ and x_0 are given parameters. Since $U(x,t)$ approaches to 0 exponentially as $|x| \to \infty$, we can approximate the initial-value problem (6.77) by an initial boundary value problem for $x \in (-L, L)$ as long as the soliton does not reach the boundaries.

For computational convenience, we first rescale the problem into $(-1,1)$. Setting

$$y = x/L, \quad u(y,t) = U(x,t), \quad u_0(y) = U_0(x), \quad \alpha = 1/L^3, \quad \beta = 1/L,$$

the problem of interest becomes

$$\begin{aligned} &u_t + \beta u u_y + \alpha u_{yyy} = 0, \quad y \in I = (-1,1),\ t \in (0,T], \\ &u(y,0) = u_0(y), \quad u(\pm 1, t) = u'(1,t) = 0. \end{aligned} \tag{6.79}$$

Let τ be the time step size, a fully discrete scheme with Crank-Nicolson leap-frog in time and Legendre dual-Petrov-Galerkin method in space for (6.79) is

$$\begin{cases} \text{Find } u_N^{k+1} \in V_N \text{ such that} \\ \dfrac{1}{2\tau}\big(u_N^{k+1} - u_N^{k-1}, v_N\big) + \alpha\big(\partial_y \hat{u}_N^{k+1}, \partial_y^2 v_N\big) \\ \qquad = -\dfrac{\beta}{2}\big(\partial_y I_N (u_N^k)^2, v_N\big), \quad \forall v_N \in V_N^*, \end{cases} \tag{6.80}$$

where $\hat{u}_N^{k+1} = \frac{1}{2}(u_N^{k+1} + u_N^{k-1})$, the "dual" spaces V_N and V_N^* are defined in (6.32), and I_N is the Legendre-Gauss-Lobatto interpolation operator. Hence, at each time step, one only needs to solve the following problem:

$$\begin{cases} \text{Find } u_N^{k+1} \in V_N \text{ such that} \\ \alpha\tau\big(\partial_y \hat{u}_N^{k+1}, \partial_y^2 v_N\big) + (\hat{u}_N^{k+1}, v_N) = (u_N^{k-1}, v_N) \\ \qquad\qquad - \dfrac{\beta\tau}{2}\big(\partial_y I_N(u_N^k)^2, v_N\big), \ \ \forall v_N \in V_N^*. \end{cases} \tag{6.81}$$

The above system is of the form (6.33) so it can be solved by using the method presented in Sect. 6.3.1. We refer to Shen (2003) for a detailed stability and convergence analysis for the scheme (6.80).

Example 1. Single soliton solution. We take $\kappa = 0.3$, $x_0 = -20$, $L = 50$ and $\tau = 0.001$ so that for $N \lesssim 160$, the time discretization error is negligible compared with the spatial discretization error. In Fig. 6.3a, we plot the time evolution of the approximate solution, and in (b), we depict the maximum errors in the semi-log scale at $t = 1$ and $t = 50$. Note that the straight lines indicate that the errors decay like $\exp(-cN)$. The excellent accuracy for this known exact solution indicates that the KdV equation on a finite interval can be used to effectively simulate the KdV equation on a semi-infinite interval before the wave reaches the boundary.

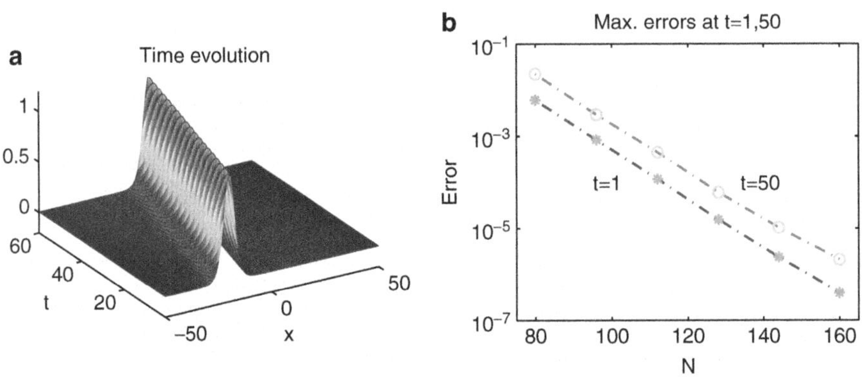

Fig. 6.3 (**a**) Time evolution of the numerical solution for the KdV equation, and (**b**) the maximum errors at $t = 1, 50$ with different N

Example 2. Interaction of five solitons. We start with the following initial condition

$$u_0(x) = \sum_{i=1}^{5} 12\kappa_i^2 \operatorname{sech}^2(\kappa_i(x - x_i)) \tag{6.82}$$

with

$$\begin{aligned}&\kappa_1 = .3,\ \kappa_2 = .25,\ \kappa_3 = .2,\ \kappa_4 = .15,\ \kappa_5 = .1,\\&x_1 = -120,\ x_1 = -90,\ x_3 = -60,\ x_4 = -30,\ x_5 = 0.\end{aligned}$$

In the following computations, we fix $L = 150$, $\tau = 0.02$ and $N = 256$. In Fig. 6.4a, we plot the time evolution of the solution in the (x,t) plane. We also plot the initial profile and the profile at the final step ($t = 600$) in Fig. 6.4b. We observe that the soliton with higher amplitude travels at a faster speed, and the amplitudes of the five solitary waves are well preserved at the final time. This indicates that our scheme has an excellent conservation property.

Example 3. Solitary waves generated by a Gaussian profile. We start with the initial condition $u_0(x) = e^{-1.5(7x)^2}$. We plot the time evolution of the solution in Fig. 6.5a, and the profiles at $t = 0, 100$ in Fig. 6.5b. The initial Gaussian profile has evolved into four separated solitary waves by the time $t = 100$.

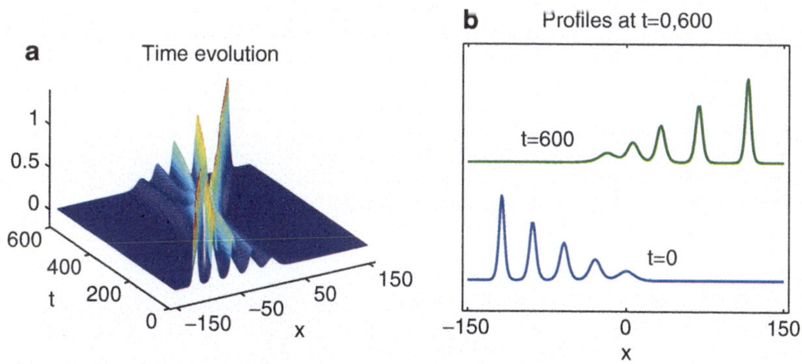

Fig. 6.4 Interaction of five solitons generated by (6.82)

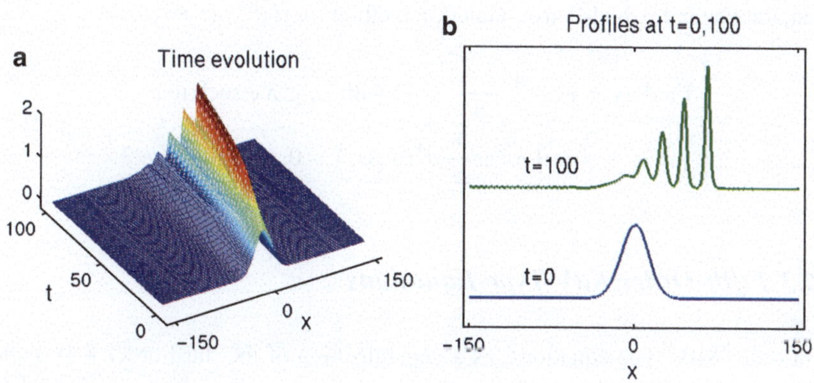

Fig. 6.5 Solitary waves generated by an initial Gaussian profile

It must be noted that our dual-Petrov-Galerkin approach is quite flexible and can be used for other unconventional boundary conditions. For instance, Colin and Ghidaglia (2001) studied the KdV equation

$$u_t + \frac{2}{L}(u_x + uu_x) + \frac{8}{L^3}u_{xxx} = 0, \quad x \in (-1,1),\ t > 0, \tag{6.83}$$

with the following boundary conditions

$$u(-1) = g(t), \quad u_x(1) = u_{xx}(1) = 0. \tag{6.84}$$

Let us denote

$$X_N = \{u \in P_N : u(-1) = u_x(1) = u_{xx}(1) = 0\}. \tag{6.85}$$

Then, the "dual" space is

$$X_N^* = \{v \in P_N : v(1) = v_x(-1) = v_{xx}(-1) = 0\}. \tag{6.86}$$

There exists a unique set of coefficients $\{a_j^{(k)},\ \tilde{a}_j^{(k)}\}$ such that

$$\begin{aligned} \phi_k(x) &= L_k(x) + \sum_{j=1}^{3} a_j^{(k)} L_{k+j}(x) \in X_N, \\ \psi_k(x) &= L_k(x) + \sum_{j=1}^{3} \tilde{a}_j^{(k)} L_{k+j}(x) \in X_N^*, \end{aligned} \tag{6.87}$$

and

$$\begin{aligned} X_N &= \text{span}\{\phi_0, \phi_1, \ldots, \phi_{N-3}\}, \\ X_N^* &= \text{span}\{\psi_0, \psi_1, \ldots, \psi_{N-3}\}. \end{aligned} \tag{6.88}$$

Then, the Legendre dual-Petrov-Galerkin method for (6.83)-(6.84) is

$$\begin{cases} \text{Find } u_N = v_N + \dfrac{(1-x)^3}{8} g(t) \text{ with } v_N \in X_N \text{ such that} \\ \left(\partial_t u_N + \dfrac{2}{L} u_N \partial_x u_N + \dfrac{8}{L^3}\partial_x^3 u_N, \psi_j\right) = 0, \ \ 0 \le j \le N-3. \end{cases} \tag{6.89}$$

6.6.3 Fifth-Order KdV Type Equations

Fifth-order KdV type equations, as a generalization of the third-order KdV equation, arise naturally in the modeling of many wave phenomena (see, e.g., Kawahara

(1972), Kichenassamy and Olver (1992)). As an example, we consider a typical fifth-order KdV type equation:

$$u_t + \gamma u u_x + \nu u_{xxx} - \mu u_{xxxxx} = 0, \quad u(x,0) = u_0(x). \tag{6.90}$$

For $\gamma \neq 0$ and $\mu\nu > 0$, it has the following exact solution (cf. Parkes et al. (1998))

$$u(x,t) = \eta_0 + A\,\mathrm{sech}^4\big(\kappa(x - ct - x_0)\big), \tag{6.91}$$

where x_0 and η_0 are arbitrary constants, and

$$A = \frac{105\nu^2}{169\mu\gamma}, \quad \kappa = \sqrt{\frac{\nu}{52\mu}}, \quad c = \gamma\eta_0 + \frac{36\nu^2}{169\mu}. \tag{6.92}$$

It should be pointed out that in contrast to the solution (6.78), the amplitude of (6.91) is fixed if the constants γ, μ and ν are given.

Since $u(x,t) \to \eta_0$ exponentially as $|x| \to \infty$, we may approximate the initial value problem (6.90) by an initial boundary value problem imposed in $(-L,L)$ as long as the soliton does not reach the boundary $x = L$. Hence, in the following computations, we turn our attentions to the problem (6.90) in $(-L,L)$ with the boundary conditions:

$$u(\pm L,t) = u'(\pm L,t) = u''(L,t) = 0.$$

As in the third-order KdV case, we can use a fully discrete Crank-Nicolson leap-frog dual-Petrov-Galerkin scheme for this problem. We refer to Yuan et al. (2008) for more details on the implementation and error analysis of this scheme, and simulations of several other physically relevant equations such as the Kawahara equation (cf. Kawahara (1972)).

Example 1. Single soliton solution. We consider the problem (6.90)–(6.92) with

$$\mu = \gamma = 1, \quad \nu = 1.1, \quad \eta_0 = 0, \quad x_0 = -10.$$

In the computation, we take $L = 50, N = 120$ and $\tau = 0.001$. In Fig. 6.6a, we plot the time evaluation of the solution, and in (b), we plot the pointwise maximum errors against various N at $t = 1, 50, 100$. It is clear from the figure that the convergence rate behaves like e^{-cN}.

Example 2. Parallel propagation of two solitons. We start with the following initial condition:

$$u_0(x) = u_1(x,0) + u_2(x,0), \tag{6.93}$$

with

$$u_i(x,t) = \eta_i + A\,\mathrm{sech}^4(\kappa(x - c_i t - x_i)), \quad c_i = \gamma\eta_i + \frac{36\nu^2}{169\mu}, \quad i = 1,2$$

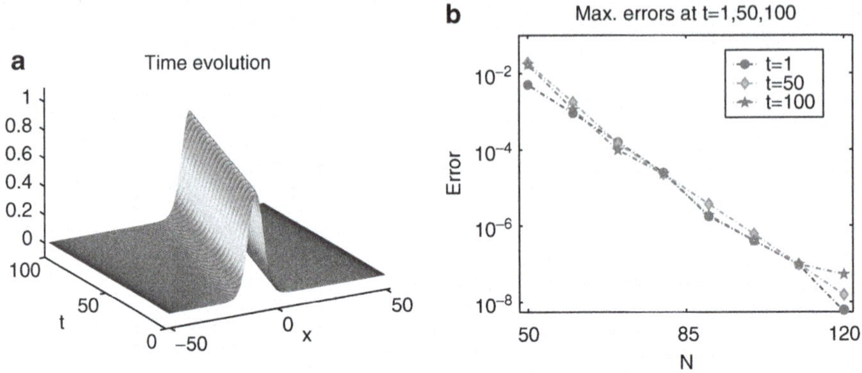

Fig. 6.6 Time evolution of the numerical solution for the fifth-order KdV equation and the maximum errors at $t = 1, 50, 100$ with different N

being the solutions of (6.90). Let

$$\mu = \gamma = 1,\ \ \nu = 1.1,\ \ \eta_1 = 1.5,\ \ \eta_2 = -0.5,\ \ L = 150,\ \ N = 256,\ \ \tau = 0.01.$$

We take $x_1 = -90$ and $x_2 = -60$ so that the peaks of two solitary waves are initially separated. It is expected that the two solitons would propagate in parallel with the same speed:

$$C = \gamma(\eta_1 + \eta_2) + \frac{36\nu^2}{169\mu}. \tag{6.94}$$

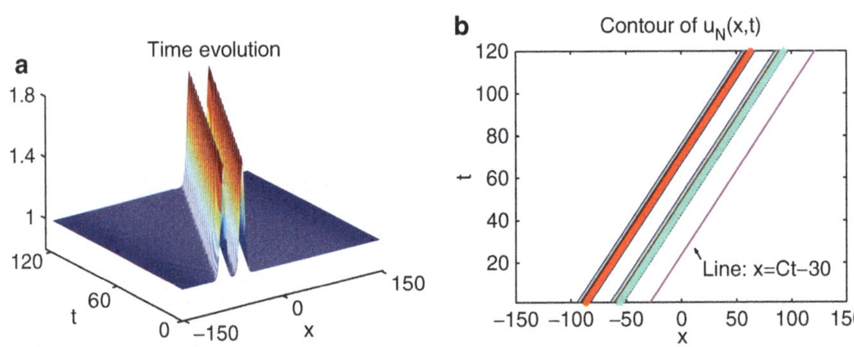

Fig. 6.7 Solitons generated by (6.93)

In Fig. 6.7a, we plot the time evolution of the solution in the (x,t)-plane, and its contour is depicted in (b). To check the propagation speed, a reference line: $x = Ct - 30$ is plotted as well. One observes that two solitary waves propagate in parallel at the same speed C, and preserve their amplitudes as expected.

Example 3. Interaction of two solitons. We define

$$G(x;\lambda,x_0) = 2.65758756\exp(-0.16z^2)\big(1.000794208 - 0.006761592432z^2 - 0.001355732644z^4 + 2.520234609\times 10^{-5}z^6 - 4.782592684\times 10^{-6}z^8\big), \tag{6.95}$$

where $z = (\lambda/\mu)^{1/4}(x-x_0)$. This function describes the wave profile, in which λ represents the velocity of a solitary wave and x_0 is the initial position. The amplitude of the wave is proportional to the velocity.

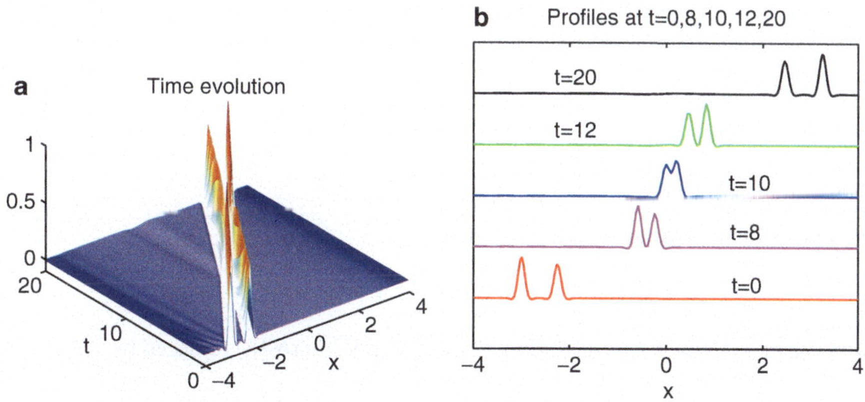

Fig. 6.8 Solitons generated by (6.95)

The initial state is given by

$$u(x,0) = \lambda_1 G(x;\lambda_1,x_1) + \lambda_2 G(x;\lambda_2,x_2), \tag{6.96}$$

and in the computation, we take

$$\mu = 7\times 10^{-7},\ \nu = 0,\ \gamma = 1,\ \lambda_1 = 0.3,\ \lambda_2 = 0.25,$$
$$x_1 = -3,\ x_2 = -2.25,\ L = 4,\ N = 256,\ \tau = 0.01.$$

In Fig. 6.8a, we plot the time evolution solution $u_N(x,t)$ in the (x,t)-plane, while in Fig. 6.8b, we plot the initial profile (6.96), and profiles at $t = 8, 10, 12, 20$. We observe that the solitary wave with higher amplitude travels at a faster speed, and then emerges into the solitary wave with lower amplitude. After the interaction, the two solitary waves propagate with their own amplitudes preserved. Notice that some small ripples (non-solitary waves) are generated after the interaction (Note: the initial condition (6.96) is not the initial profile of the exact solution). This phenomenon was also observed in Nagashima and Kawahara (1981), Djidjeli et al. (1995).

Problems

6.1. Determine the non-zero entries of B and C in (6.21) and implement the scheme (6.18). Take the exact solution of (6.14) to be $u(x) = \sin^2(2\pi x)$. Plot the l^∞-error versus N in semi-log scale.

6.2. Consider the problem

$$\begin{aligned} &u^{(4)} - \alpha u'' + \beta u = f, \quad x \in I = (-1,1), \\ &u'(\pm 1) = u'''(\pm 1) = 0. \end{aligned} \tag{6.97}$$

Construct an efficient Legendre-Galerkin method for solving the above equation. Take the exact solution of (6.97) to be $u(x) = \cos(4\pi x)$. Plot the l^∞-error versus N in semi-log scale.

6.3. Prove Theorem 6.3.

6.4.

- Let V_N be defined in (6.32). Show that

$$-\big(\partial_x^2 u, \partial_x(\omega^{-1,0} u)\big) \geq c\big(\|\partial_x u\|^2 + |\partial_x u(-1)|^2\big), \quad \forall u \in V_N.$$

- Consider a new Petrov-Galerkin method with V_N^* in (6.18) replaced by

$$W_N := \big\{u \in P_{N-1} \ : \ u(\pm 1) = 0\big\}.$$

  Using the above inequality, show that this new Petrov-Galerkin method admits a unique solution.
- Choose suitable basis functions for V_N and W_N to show that the corresponding linear system has a sparse matrix.
- Derive an error estimate for this method.

6.5. Prove Theorem 6.5.

6.6. Consider the KdV equation (6.77). Replace the dual-Petrov-Galerkin method in (6.80) by the collocation method using the nodes of the quadrature formula (6.58) with $l = 1$ and $r = 2$ as collocation points. Test it on **Example 1: Single soliton solution** and compare it with that obtained by the scheme (6.80).

Chapter 7
Unbounded Domains

We study in this chapter spectral approximations by orthogonal polynomials/ functions on unbounded intervals, such as Laguerre and Hermite polynomials/ functions and rational functions. Considerable progress has been made in the last two decades in using these orthogonal systems for solving PDEs in unbounded domains (cf. Chap. 17 in Boyd (2001) and a more recent review article Shen and Wang (2009)).

Spectral methods for unbounded domains can be essentially classified into four categories:

(a) Domain truncation: truncate unbounded domains to bounded domains and solve PDEs on bounded domains supplemented with artificial or transparent boundary conditions (see, e.g., Engquist and Majda (1977), Grote and Keller (1995));
(b) Approximation by classical orthogonal systems on unbounded domains, e.g., Laguerre or Hermite polynomials/functions (see, e.g., Boyd (1980), Funaro and Kavian (1990), Guo (1999), Shen (2000), Guo et al. (2003));
(c) Approximation by other non-classical orthogonal systems (see, e.g., Christov (1982)), or by mapped orthogonal systems, e.g., image of classical Jacobi polynomials through a suitable mapping (see, e.g., Guo et al. (2000), Guo and Shen (2001));
(d) Mapping: map unbounded domains to bounded domains and use standard spectral methods to solve the mapped PDEs in bounded domains (see, e.g., Grosch and Orszag (1977), Boyd (1987b,a), Cloot and Weideman (1992), Guo (1998b), Boyd (2001)).

In general, the domain truncation approach is only viable for problems with rapidly (exponentially) decaying solutions or when accurate non-reflecting or exact boundary conditions are available at the truncated boundary. On the other hand, with a proper choice of mappings and/or scaling parameters, the other three approaches can all be effectively applied to a variety of problems with decaying (or even growing) solutions. Since there exists a vast literature on domain truncations, particularly

J. Shen et al., *Spectral Methods: Algorithms, Analysis and Applications*, Springer Series in Computational Mathematics 41, DOI 10.1007/978-3-540-71041-7_7,

for the Helmholtz equation and Maxwell equations for scattering problems, and the analysis involved is very different from the other three approaches, we shall restrict our attentions to the last three approaches.

7.1 Laguerre Polynomials/Functions

Following the general framework established in Chap. 3, we present in this section some basic properties of Laguerre polynomials/functions, and introduce the Laguerre-Gauss-type quadrature formulas and the associated interpolation, discrete transforms and spectral differentiation.

7.1.1 Basic Properties

Since the properties of Laguerre polynomials can be derived in a similar fashion as for the Jacobi polynomials, we just collect the relevant formulas without providing their derivations.

7.1.1.1 Generalized Laguerre Polynomials

The generalized Laguerre polynomials (GLPs), denoted by $\mathscr{L}_n^{(\alpha)}(x)$ (with $\alpha > -1$), are orthogonal with respect to the weight function $\omega_\alpha(x) = x^\alpha e^{-x}$ on the half line $\mathbb{R}_+ := (0, +\infty)$, i.e.,

$$\int_0^{+\infty} \mathscr{L}_n^{(\alpha)}(x)\mathscr{L}_m^{(\alpha)}(x)\omega_\alpha(x)dx = \gamma_n^{(\alpha)}\delta_{mn}, \tag{7.1}$$

where

$$\gamma_n^{(\alpha)} = \frac{\Gamma(n+\alpha+1)}{n!}. \tag{7.2}$$

In particular, $\mathscr{L}_n^{(0)}(x)$ is the usual Laguerre polynomial which will be denoted by $\mathscr{L}_n(x)$. It is clear that $\{\mathscr{L}_n\}$ are orthonormal with respect to the weight function $\omega(x) = e^{-x}$, i.e.,

$$\int_0^{+\infty} \mathscr{L}_n(x)\mathscr{L}_m(x)e^{-x}dx = \delta_{mn}. \tag{7.3}$$

The three-term recurrence formula that generates the GLPs reads

$$(n+1)\mathscr{L}_{n+1}^{(\alpha)}(x) = (2n+\alpha+1-x)\mathscr{L}_n^{(\alpha)}(x) - (n+\alpha)\mathscr{L}_{n-1}^{(\alpha)}(x), \tag{7.4}$$

and the first few members are

$$\begin{aligned}
\mathscr{L}_0^{(\alpha)}(x) &= 1,\\
\mathscr{L}_1^{(\alpha)}(x) &= -x+\alpha+1,\\
\mathscr{L}_2^{(\alpha)}(x) &= \frac{1}{2}\left(x^2-2(\alpha+2)x+(\alpha+1)(\alpha+2)\right),\\
\mathscr{L}_3^{(\alpha)}(x) &= \frac{1}{6}\big(-x^3+3(\alpha+3)x^2-3(\alpha+2)(\alpha+3)x\\
&\quad +(\alpha+1)(\alpha+2)(\alpha+3)\big).
\end{aligned}$$

The leading coefficient of $\mathscr{L}_n^{(\alpha)}(x)$ is

$$k_n^{(\alpha)} = \frac{(-1)^n}{n!}, \tag{7.5}$$

and we have the formula

$$\mathscr{L}_n^{(\alpha)}(0) = \frac{\Gamma(n+\alpha+1)}{n!\Gamma(\alpha+1)} = \frac{\gamma_n^{(\alpha)}}{\Gamma(\alpha+1)}. \tag{7.6}$$

By the Stirling's formula (A.7), $\mathscr{L}_n^{(\alpha)}(0) \sim n^\alpha$ for $n \gg 1$. Notice that $\mathscr{L}_n(0)=1$.

The generalized Laguerre polynomial satisfies the Sturm-Liouville equation

$$x^{-\alpha}e^x\partial_x\left(x^{\alpha+1}e^{-x}\partial_x\mathscr{L}_n^{(\alpha)}(x)\right) + \lambda_n\mathscr{L}_n^{(\alpha)}(x) = 0, \tag{7.7}$$

or equivalently,

$$x\partial_x^2\mathscr{L}_n^{(\alpha)}(x) + (\alpha+1-x)\partial_x\mathscr{L}_n^{(\alpha)}(x) + \lambda_n\mathscr{L}_n^{(\alpha)}(x) = 0, \tag{7.8}$$

with the corresponding eigenvalue $\lambda_n = n$. We emphasize that λ_n grows linearly as opposed to quadratical growth in the Jacobi case. This has two important implications: (a) the convergence rate of the expansions by GLPs will only be half of the expansions by Jacobi polynomials with "similar" regularities (but in different weighted spaces); on the other hand, (b) the minimum distance between adjacent Laguerre-Gauss type points is of order $O(n^{-1})$ instead of $O(n^{-2})$ in the Jacobi case.

By (7.1) and (7.7), we have the orthogonality of $\{\partial_x\mathscr{L}_n^{(\alpha)}\}$, namely,

$$\int_0^{+\infty}\partial_x\mathscr{L}_n^{(\alpha)}(x)\partial_x\mathscr{L}_m^{(\alpha)}(x)\omega_{\alpha+1}(x)dx = \lambda_n\gamma_n^{(\alpha)}\delta_{mn}. \tag{7.9}$$

The Rodrigues' formula for the GLPs takes the form:

$$\mathscr{L}_n^{(\alpha)}(x) = \frac{x^{-\alpha}e^x}{n!}\frac{d^n}{dx^n}\left\{x^{n+\alpha}e^{-x}\right\}. \tag{7.10}$$

Furthermore, we have the explicit expression

$$\mathscr{L}_n^{(\alpha)}(x) = \sum_{k=0}^{n} \frac{(-1)^k}{k!} \binom{n+\alpha}{n-k} x^k. \tag{7.11}$$

The GLPs satisfy the following recurrence relations:

$$\partial_x \mathscr{L}_n^{(\alpha)}(x) = -\mathscr{L}_{n-1}^{(\alpha+1)}(x) = -\sum_{k=0}^{n-1} \mathscr{L}_k^{(\alpha)}(x), \tag{7.12a}$$

$$\mathscr{L}_n^{(\alpha)}(x) = \partial_x \mathscr{L}_n^{(\alpha)}(x) - \partial_x \mathscr{L}_{n+1}^{(\alpha)}(x), \tag{7.12b}$$

$$x\partial_x \mathscr{L}_n^{(\alpha)}(x) = n\mathscr{L}_n^{(\alpha)}(x) - (n+\alpha)\mathscr{L}_{n-1}^{(\alpha)}(x). \tag{7.12c}$$

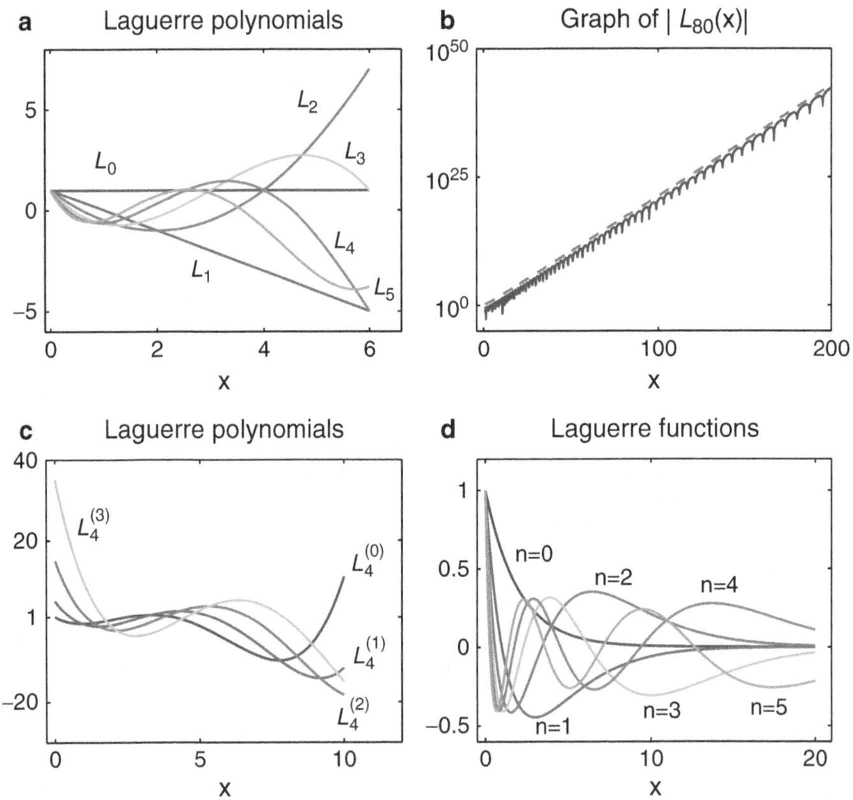

Fig. 7.1 (**a**) Graphs of the first six Laguerre polynomials $\mathscr{L}_n(x)$ with $n = 0, 1, \ldots, 5$ and $x \in [0, 6]$; (**b**) Growth of $|\mathscr{L}_{80}(x)|$ against the upper bound $\pi^{-1/2}(80x)^{-1/4}e^{x/2}$ (*dashed line*); (**c**) Graphs of the generalized Laguerre polynomials $\mathscr{L}_4^{(\alpha)}(x)$ with $\alpha = 0, 1, 2, 3$ and $x \in [0, 10]$; (**d**) Graphs of the first six Laguerre functions $\widehat{\mathscr{L}}_n(x)$ with $n = 0, 1, \ldots, 5$ and $x \in [0, 20]$

Next, we study the asymptotic properties of the GLPs. It is clear that for large x, the value of $\mathscr{L}_n^{(\alpha)}(x)$ is dominated by the leading term and grows like $\frac{(-1)^n}{n!}x^n$. Moreover, according to Theorem 7.6.2 of Szegö (1975), the successive relative maxima of $e^{-x/2}x^{(\alpha+1)/2}|\mathscr{L}_n^{(\alpha)}(x)|$ form an increasing sequence, provided that $x > x_0$, where

$$x_0 = 0, \quad \text{if} \quad \alpha \le 1; \quad x_0 = \frac{\alpha^2 - 1}{2n+\alpha+1}, \quad \text{if} \quad \alpha > 1.$$

The GLPs have a rapid growth as x approaches to infinity and as n increases. Indeed, by Theorem 8.22.5 of Szegö (1975),

$$\begin{aligned}\mathscr{L}_n^{(\alpha)}(x) = \pi^{-1/2}e^{x/2}x^{-\alpha/2-1/4}n^{\alpha/2-1/4}\times \\ \Big[\cos\big(2(nx)^{1/2} - \alpha\pi/2 - \pi/4\big) + (nx)^{-1/2}O(1)\Big],\end{aligned} \tag{7.13}$$

which holds uniformly for all $x \in [cn^{-1}, b]$ with c and b being fixed positive numbers. We observe from Fig. 7.1 the growth of the Laguerre polynomials with respect to the degree n and/or x. In particular, we plot in Fig. 7.1b, the graph of $|\mathscr{L}_{80}(x)|$ against the asymptotic estimate $\pi^{-1/2}(80x)^{-1/4}e^{x/2}$, which shows

$$|\mathscr{L}_n(x)| \approx \pi^{-1/2}(nx)^{-1/4}e^{x/2}, \tag{7.14}$$

for $n \gg 1$ and all x in a finite interval. On the other hand, we have the following uniform upper bounds for GLPs (see the Appendix of Davis (1975)):

$$|\mathscr{L}_n^{(\alpha)}(x)| \le \begin{cases} \mathscr{L}_n^{(\alpha)}(0)e^{x/2}, & \text{if } \alpha \ge 0, \\ \big(2 - \mathscr{L}_n^{(\alpha)}(0)\big)e^{x/2}, & \text{if } \alpha > 0, \end{cases} \tag{7.15}$$

for all $x \in [0, +\infty)$.

7.1.1.2 Generalized Laguerre Functions

In many applications, the underlying solutions decay algebraically or exponentially at infinity, it is certainly not a good idea to approximate these functions by GLPs which grow rapidly at infinity. It is advisable to approximate them by spectral expansions of generalized Laguerre functions (GLFs).

The generalized Laguerre functions (GLFs) are defined by

$$\widehat{\mathscr{L}}_n^{(\alpha)}(x) := e^{-x/2}\mathscr{L}_n^{(\alpha)}(x), \quad x \in \mathbb{R}_+, \ \alpha > -1. \tag{7.16}$$

By (7.1)–(7.2), the GLFs are orthogonal with respect to the weight function $\hat{\omega}_\alpha = x^\alpha$, i.e.,

$$\int_0^{+\infty} \widehat{\mathscr{L}}_n^{(\alpha)}(x)\widehat{\mathscr{L}}_m^{(\alpha)}(x)\hat{\omega}_\alpha(x)dx = \gamma_n^{(\alpha)}\delta_{mn}. \tag{7.17}$$

In particular, the usual Laguerre functions defined by

$$\widehat{\mathscr{L}_n}(x) = e^{-x/2}\mathscr{L}_n(x), \quad n \geq 0, \tag{7.18}$$

are orthonormal with respect to the uniform weight function $\hat{\omega}_0 \equiv 1$.

The following properties of the GLFs can be derived directly from those of the GLPs. To simplify the notation, we introduce the derivative operator:

$$\hat{\partial}_x = \partial_x + \frac{1}{2}. \tag{7.19}$$

It is clear that

$$\partial_x \mathscr{L}_n^{(\alpha)}(x) = e^{x/2}\hat{\partial}_x \widehat{\mathscr{L}_n^{(\alpha)}}(x). \tag{7.20}$$

The basic properties of the GLFs are summarized below.

- Three-term recurrence relation:

$$\begin{aligned}(n+1)\widehat{\mathscr{L}_{n+1}^{(\alpha)}} &= (2n+\alpha+1-x)\widehat{\mathscr{L}_n^{(\alpha)}} - (n+\alpha)\widehat{\mathscr{L}_{n-1}^{(\alpha)}},\\ \widehat{\mathscr{L}_0^{(\alpha)}} = e^{-x/2}, &\quad \widehat{\mathscr{L}_1^{(\alpha)}} = (\alpha+1-x)e^{-x/2}.\end{aligned} \tag{7.21}$$

  This formula allows for a stable evaluation of the GLFs. Indeed, in contrast to the GLPs (cf. (7.13)), the GLFs are well-behaved with the decay property (see Fig. 7.1d):

$$\left|\widehat{\mathscr{L}_n^{(\alpha)}}(x)\right| \to 0, \quad \text{as } x \to +\infty. \tag{7.22}$$

  By (7.15) and the definition (7.16), the GLFs are uniformly bounded. In particular, we have

$$\left|\widehat{\mathscr{L}_n}(x)\right| \leq 1, \quad \forall x \in [0,+\infty). \tag{7.23}$$

- Sturm-Liouville equation:

$$x^{-\alpha}e^{x/2}\partial_x\left(x^{\alpha+1}e^{-x/2}\hat{\partial}_x\widehat{\mathscr{L}_n^{(\alpha)}}(x)\right) + n\widehat{\mathscr{L}_n^{(\alpha)}}(x) = 0. \tag{7.24}$$

- Orthogonality:

$$\int_0^{+\infty} \hat{\partial}_x\widehat{\mathscr{L}_n^{(\alpha)}}(x)\hat{\partial}_x\widehat{\mathscr{L}_m^{(\alpha)}}(x)\hat{\omega}_{\alpha+1}(x)dx = \lambda_n\gamma_n^{(\alpha)}\delta_{mn}. \tag{7.25}$$

- Recurrence formulas:

$$\hat{\partial}_x\widehat{\mathscr{L}_n^{(\alpha)}}(x) = -\widehat{\mathscr{L}_{n-1}^{(\alpha+1)}}(x) = -\sum_{k=0}^{n-1}\widehat{\mathscr{L}_k^{(\alpha)}}(x), \tag{7.26a}$$

$$\widehat{\mathscr{L}_n^{(\alpha)}}(x) = \hat{\partial}_x\widehat{\mathscr{L}_n^{(\alpha)}}(x) - \hat{\partial}_x\widehat{\mathscr{L}_{n+1}^{(\alpha)}}(x), \tag{7.26b}$$

$$x\hat{\partial}_x\widehat{\mathscr{L}_n^{(\alpha)}}(x) = n\widehat{\mathscr{L}_n^{(\alpha)}}(x) - (n+\alpha)\widehat{\mathscr{L}_{n-1}^{(\alpha)}}(x). \tag{7.26c}$$

Note that by using the derivative operator (7.19), many formulas relative to GJFs can be expressed in the same forms as those for GLPs. This notation will also greatly simplify the analysis later.

7.1.2 Laguerre-Gauss-Type Quadratures

The Laguerre-Gauss-type quadrature formulas, including Laguerre-Gauss and Laguerre-Gauss-Radau rules, can be derived from the general framework in Sect. 3.1.

Theorem 7.1. *Let* $\{x_j^{(\alpha)}, \omega_j^{(\alpha)}\}_{j=0}^N$ *be the set of Laguerre-Gauss or Laguerre-Gauss-Radau quadrature nodes and weights.*

- *For the Laguerre-Gauss quadrature,*

$$
\begin{aligned}
&\{x_j^{(\alpha)}\}_{j=0}^N \text{ are the zeros of } \mathscr{L}_{N+1}^{(\alpha)}(x);\\
&\omega_j^{(\alpha)} = -\frac{\Gamma(N+\alpha+1)}{(N+1)!}\frac{1}{\mathscr{L}_N^{(\alpha)}(x_j^{(\alpha)})\partial_x\mathscr{L}_{N+1}^{(\alpha)}(x_j^{(\alpha)})}\\
&\quad = \frac{\Gamma(N+\alpha+1)}{(N+\alpha+1)(N+1)!}\frac{x_j^{(\alpha)}}{[\mathscr{L}_N^{(\alpha)}(x_j^{(\alpha)})]^2}, \quad 0 \le j \le N.
\end{aligned}
\tag{7.27}
$$

- *For the Laguerre-Gauss-Radau quadrature,*

$$
\begin{aligned}
&x_0^{(\alpha)} = 0 \text{ and } \{x_j^{(\alpha)}\}_{j=1}^N \text{ are the zeros of } \partial_x\mathscr{L}_{N+1}^{(\alpha)}(x);\\
&\omega_0^{(\alpha)} = \frac{(\alpha+1)\Gamma^2(\alpha+1)N!}{\Gamma(N+\alpha+2)},\\
&\omega_j^{(\alpha)} = \frac{\Gamma(N+\alpha+1)}{N!(N+\alpha+1)}\frac{1}{[\partial_x\mathscr{L}_N^{(\alpha)}(x_j^{(\alpha)})]^2}\\
&\quad = \frac{\Gamma(N+\alpha+1)}{N!(N+\alpha+1)}\frac{1}{[\mathscr{L}_N^{(\alpha)}(x_j^{(\alpha)})]^2}, \quad 1 \le j \le N.
\end{aligned}
\tag{7.28}
$$

With the above nodes and weights, we have

$$
\int_0^{+\infty} p(x)x^\alpha e^{-x}dx = \sum_{j=0}^N p(x_j^{(\alpha)})\omega_j^{(\alpha)}, \quad \forall p \in P_{2N+\delta}, \tag{7.29}
$$

where $\delta = 1, 0$ *for the Laguerre-Gauss quadrature and the Laguerre-Gauss-Radau quadrature, respectively.*

Proof. We first consider the Laguerre-Gauss case. In view of Theorem 3.5, it suffices to derive the second formula for the quadrature weights in (7.27). Note that $\mathscr{L}_{N+1}^{(\alpha)}(x_j^{(\alpha)}) = 0$ for $0 \le j \le N$, so we have from (7.12c) that

$$\partial_x \mathscr{L}_{N+1}^{(\alpha)}(x_j^{(\alpha)}) = -\frac{N+\alpha+1}{x_j^{(\alpha)}} \mathscr{L}_N^{(\alpha)}(x_j^{(\alpha)}), \quad 0 \le j \le N.$$

Plugging it into the first formula in (7.27) leads to the second one.

For the Laguerre-Gauss-Radau case, we deduce from (7.6) and (7.12c) that the quadrature polynomial (3.48) (with $a=0$) turns out to be

$$\begin{aligned} q_N(x) &= \frac{1}{x}\Big[\mathscr{L}_{N+1}^{(\alpha)}(x) - \frac{\mathscr{L}_{N+1}^{(\alpha)}(0)}{\mathscr{L}_N^{(\alpha)}(0)} \mathscr{L}_N^{(\alpha)}(x)\Big] \\ &= \frac{1}{(N+1)x}\Big[(N+1)\mathscr{L}_{N+1}^{(\alpha)}(x) - (N+\alpha+1)\mathscr{L}_N^{(\alpha)}(x)\Big] \\ &= \frac{\partial_x \mathscr{L}_{N+1}^{(\alpha)}(x)}{N+1}. \end{aligned}$$

Hence, the interior Laguerre-Gauss-Radau points $\{x_j^{(\alpha)}\}_{j=1}^N$ are zeros of $\partial_x \mathscr{L}_{N+1}^{(\alpha)}(x)$.

We now derive the weight expressions in (7.28). It is clear that the Lagrange basis polynomial corresponding to $x_0^{(\alpha)} = 0$ is $\partial_x \mathscr{L}_{N+1}^{(\alpha)}(x)/\partial_x \mathscr{L}_{N+1}^{(\alpha)}(0)$, so by definition,

$$\begin{aligned} \omega_0^{(\alpha)} &= \int_0^{+\infty} \frac{\partial_x \mathscr{L}_{N+1}^{(\alpha)}(x)}{\partial_x \mathscr{L}_{N+1}^{(\alpha)}(0)} \omega_\alpha(x)dx \\ &\overset{(7.12a)}{=} \frac{1}{\partial_x \mathscr{L}_{N+1}^{(\alpha)}(0)} \int_0^{+\infty} \Big(-\sum_{k=0}^{N} \mathscr{L}_k^{(\alpha)}(x)\Big) \omega_\alpha(x)dx \\ &\overset{(7.1)}{=} -\frac{\gamma_0^{(\alpha)}}{\partial_x \mathscr{L}_{N+1}^{(\alpha)}(0)} \overset{(7.12a)}{=} \frac{\gamma_0^{(\alpha)}}{\mathscr{L}_N^{(\alpha+1)}(0)} \\ &\overset{(7.2)}{\underset{(7.6)}{=}} \frac{(\alpha+1)\Gamma^2(\alpha+1)N!}{\Gamma(N+\alpha+2)}. \end{aligned}$$

Next, we turn to the expressions for the interior weights. Like the Jacobi-Gauss-type quadratures, the Laguerre-Gauss and Laguerre-Gauss-Radau nodes and weights have a close relation:

$$x_j^{(\alpha)} = \xi_{j-1}^{(\alpha+1)}, \quad \omega_j^{(\alpha)} = (x_j^{(\alpha)})^{-1} \rho_{j-1}^{(\alpha+1)}, \quad 1 \le j \le N, \tag{7.30}$$

where we denoted by $\big\{\xi_j^{(\alpha+1)}, \rho_j^{(\alpha+1)}\big\}_{j=0}^{N-1}$ the N nodes (zeros of $\mathscr{L}_N^{(\alpha+1)}$) and weights of the Laguerre-Gauss quadrature associated with the weight function

$\omega_{\alpha+1}$, as defined in (7.27). To justify (7.30), we find from the formula $\partial_x \mathscr{L}_{N+1}^{(\alpha)}(x) = -\mathscr{L}_N^{(\alpha+1)}(x)$, the relation between the nodes, and by the definition (3.36),

$$\begin{aligned}\omega_j^{(\alpha)} &= \int_0^{+\infty} \frac{x\partial_x \mathscr{L}_{N+1}^{(\alpha)}(x)}{\partial_x(x\partial_x \mathscr{L}_{N+1}^{(\alpha)}(x))\Big|_{x=x_j^{(\alpha)}}(x - x_j^{(\alpha)})}\omega_\alpha(x)dx \\ &\overset{(7.12a)}{=} \frac{1}{x_j^{(\alpha)}}\int_0^{+\infty} \frac{\mathscr{L}_N^{(\alpha+1)}(x)}{\partial_x \mathscr{L}_N^{(\alpha+1)}(\xi_{j-1}^{(\alpha+1)})(x-\xi_{j-1}^{(\alpha+1)})}\omega_{\alpha+1}(x)dx \\ &= \frac{\rho_{j-1}^{(\alpha+1)}}{x_j^{(\alpha)}}, \quad 1 \le j \le N.\end{aligned}$$

Therefore, by the first formula of (7.27) with $\{N-1, \alpha+1\}$ in place of $\{N, \alpha\}$, we obtain

$$\omega_j^{(\alpha)} = -\frac{\Gamma(N+\alpha+1)}{N!x_j^{(\alpha)}} \frac{1}{\mathscr{L}_{N-1}^{(\alpha+1)}(x_j^{(\alpha)})\partial_x \mathscr{L}_N^{(\alpha+1)}(x_j^{(\alpha)})}, \quad 1 \le j \le N. \tag{7.31}$$

Using the fact

$$\partial_x \mathscr{L}_{N+1}^{(\alpha)}(x_j^{(\alpha)}) = -\mathscr{L}_N^{(\alpha+1)}(x_j^{(\alpha)}) = 0, \quad 1 \le j \le N,$$

we derive

$$\begin{aligned}x_j^{(\alpha)}\partial_x \mathscr{L}_N^{(\alpha+1)}(x_j^{(\alpha)}) &\overset{(7.12c)}{=} -(N+\alpha+1)\mathscr{L}_{N-1}^{(\alpha+1)}(x_j^{(\alpha)}) \\ &\overset{(7.12a)}{=} (N+\alpha+1)\partial_x \mathscr{L}_N^{(\alpha)}(x_j^{(\alpha)}) \\ &\overset{(7.12b)}{=} (N+\alpha+1)\mathscr{L}_N^{(\alpha)}(x_j^{(\alpha)}),\end{aligned}$$

which implies

$$-\mathscr{L}_{N-1}^{(\alpha+1)}(x_j^{(\alpha)}) = \partial_x \mathscr{L}_N^{(\alpha)}(x_j^{(\alpha)}) = \mathscr{L}_N^{(\alpha)}(x_j^{(\alpha)}).$$

A combination of the above identities leads to the weight formulas in (7.28). □

Remark 7.1. *If $\alpha = 0$, we simply denote the usual Laguerre-Gauss-type nodes and weights by $\{x_j, \omega_j\}_{j=0}^N$.*

- *In the Gauss case, $\{x_j\}_{j=0}^N$ are the zeros of $\mathscr{L}_{N+1}(x)$, and the weights have the representations:*

$$\omega_j = \frac{x_j}{(N+1)^2 \mathscr{L}_N^2(x_j)}, \quad 0 \le j \le N. \tag{7.32}$$

- *In the Gauss-Radau case,* $x_0 = 0$ *and* $\{x_j\}_{j=1}^N$ *are the zeros of* $\partial_x \mathscr{L}_{N+1}(x)$, *and the weights are expressed by*

$$\omega_j = \frac{1}{(N+1)\mathscr{L}_N^2(x_j)}, \quad 0 \le j \le N. \tag{7.33}$$

We find from (7.13) *and* (7.14) *that the weights are exponentially small for large* x_j (see Table 7.1).

With a slight modification of the quadrature weights in Theorem 7.1, we can derive the quadrature formulas associated with the generalized Laguerre functions.

Theorem 7.2. *Let* $\{x_j^{(\alpha)}, \omega_j^{(\alpha)}\}_{j=0}^N$ *be the set of Laguerre-Gauss or Laguerre-Gauss-Radau quadrature nodes and weights given in Theorem* 7.1. *Define*

$$\hat{\omega}_j^{(\alpha)} = e^{x_j^{(\alpha)}} \omega_j^{(\alpha)}, \quad 0 \le j \le N, \tag{7.34}$$

and

$$\widehat{P}_N := \{\phi : \phi = e^{-x/2}\psi, \ \forall \psi \in P_N\}. \tag{7.35}$$

Then we have the modified quadrature formula

$$\int_0^{+\infty} p(x)q(x)x^\alpha dx = \sum_{j=0}^N p(x_j^{(\alpha)})q(x_j^{(\alpha)})\hat{\omega}_j^{(\alpha)}, \quad \forall p \cdot q \in \widehat{P}_{2N+\delta}, \tag{7.36}$$

where $\delta = 1, 0$ *for the modified Laguerre-Gauss rule and the modified Laguerre-Gauss-Radau rule, respectively.*

Thanks to (7.16), (7.20) and (7.34), the formulas for the weights $\hat{\omega}_j^{(\alpha)}$ are obtained by replacing the derivative operator ∂_x by the new operator $\hat{\partial}_x$, and the GLP $\mathscr{L}_k^{(\alpha)}(x)$ by the GLF $\widehat{\mathscr{L}}_k^{(\alpha)}(x)$, respectively.

In particular, for $\alpha = 0$, we derive from (7.33) the modified Laguerre-Gauss-Radau quadrature weights:

$$\hat{\omega}_j = \frac{1}{(N+1)\left[\widehat{\mathscr{L}}_N(x_j)\right]^2}, \quad 0 \le j \le N. \tag{7.37}$$

Thanks to (7.14), we have $\hat{\omega}_j = O\big((x_j/N)^{1/2}\big)$, as opposed to the exponential decay of $\{\omega_j\}$ as N increases (see Table 7.1).

7.1.3 Computation of Nodes and Weights

Thanks to the relation (7.30), it suffices to compute the Laguerre-Gauss quadrature nodes and weights. We find from Theorem 3.4 and (7.4) that the zeros $\{x_j^{(\alpha)}\}_{j=0}^N$ of $\mathscr{L}_{N+1}^{(\alpha)}(x)$ are the eigenvalues of the symmetric tridiagonal matrix

$$A_{N+1} = \begin{bmatrix} a_0 & -\sqrt{b_1} & & & \\ -\sqrt{b_1} & a_1 & -\sqrt{b_2} & & \\ & \ddots & \ddots & \ddots & \\ & & -\sqrt{b_{N-1}} & a_{N-1} & -\sqrt{b_N} \\ & & & -\sqrt{b_N} & a_N \end{bmatrix}, \tag{7.38}$$

whose entries are derived from (7.4):

$$a_j = 2j+\alpha+1,\ 0 \le j \le N; \quad b_j = j(j+\alpha),\ 1 \le j \le N. \tag{7.39}$$

As shown in Theorem 3.6, the quadrature weights $\{\omega_j^{(\alpha)}\}_{j=0}^N$ can be computed from the first component of the orthonormal eigenvectors of A_{N+1}. Alternatively, they can be evaluated by using the weight formulas given in Theorem 7.1. However, this process usually suffers from numerical instability for large N, due to the exponential growth of GLPs. However, the modified weights $\{\hat{\omega}_j^{(\alpha)}\}_{j=0}^N$ can be evaluated in a stable manner. Consequently, it is desirable to compute $\{\omega_j^{(\alpha)}\}_{j=0}^N$ by

$$\omega_j^{(\alpha)} = e^{-x_j^{(\alpha)}} \hat{\omega}_j^{(\alpha)}, \quad 0 \le j \le N. \tag{7.40}$$

Another approach to locate the zeros is the iterative method as described in Sect. 3.1.3. Once again, to avoid ill-conditioned operations involving the GLPs, we work with GLFs. Funaro (1992) suggested an initial guess of the zeros as follows:

(a) find the roots of the equation

$$y_j^{(\alpha)} - \sin y_j^{(\alpha)} = 2\pi \frac{N-j+3/4}{2N+\alpha+3}, \quad 0 \le j \le N;$$

(b) set

$$\hat{y}_j^{(\alpha)} = \Big(\cos \frac{1}{2} y_j^{(\alpha)}\Big)^2, \quad 0 \le j \le N,$$

and

$$\begin{aligned} z_j^{(\alpha)} &= 2(2N+\alpha+3)\hat{y}_j^{(\alpha)} \\ &\quad - \frac{1}{6(2N+\alpha+3)}\Big(\frac{5}{4(1-\hat{y}_j^{(\alpha)})^2} - \frac{1}{1-\hat{y}_j^{(\alpha)}} - 1 + 3\alpha^2\Big), \end{aligned} \tag{7.41}$$

which provides a good approximation of $x_j^{(\alpha)}$.

To understand better the behavior of the zeros of GJPs, we provide some asymptotic estimates in Szegö (1975). Assume that $\{x_j^{(\alpha)}\}_{j=0}^N$ are arranged in ascending order. By Theorem 8.9.2 of Szegö (1975),

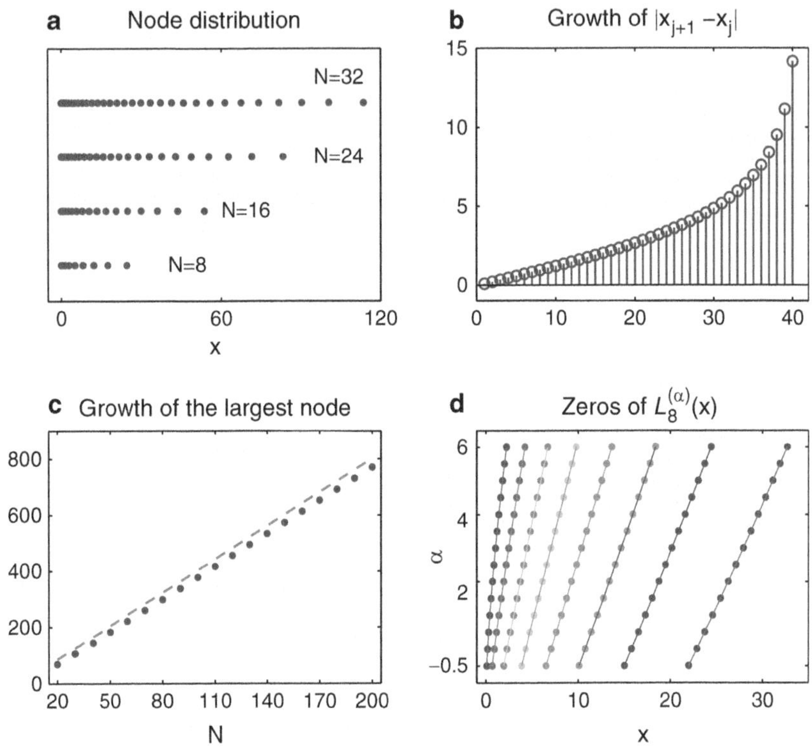

Fig. 7.2 (**a**) Distribution of Laguerre-Gauss-Radau nodes $\{x_j\}_{j=0}^N$ with $N = 8, 16, 24, 32$; (**b**) Growth of $\{|x_{j+1} - x_j|\}_{j=0}^{40}$; (**c**) Growth of the largest node x_N against the asymptotic estimate: $4N + 6 - 4N^{1/3}$ (cf. (7.43)) with various N; (**d**) Distribution of zeros of $\mathscr{L}_8^{(\alpha)}(x)$ with various α

$$2\sqrt{x_j^{(\alpha)}} = \frac{(j+1)\pi + O(1)}{\sqrt{N+1}}, \quad \text{for } x_j^{(\alpha)} \in (0, \eta], \ \alpha > -1, \tag{7.42}$$

where $\eta > 0$ is a fixed constant. Moreover, by Theorem 6.31.3 of Szegö (1975),

$$x_j^{(\alpha)} < \big(j + (\alpha+3)/2\big) \frac{2j + \alpha + 3 + \sqrt{(2j+\alpha+3)^2 + 1/4 - \alpha^2}}{N + (\alpha+3)/2}, \tag{7.43}$$

for all $0 \le j \le N$ and $\alpha > -1$. In particular,

$$x_N^{(\alpha)} = 4N + 2\alpha + 6 + O(N^{1/3}). \tag{7.44}$$

By (6.31.13) of Szegö (1975), we have that for $\alpha = 0$,

$$x_j = C_{j,N} \frac{(j+2)^2}{N+2}, \quad 0 \le j \le N, \tag{7.45}$$

where $1/4 < C_{j,N} < 4$.

We see that the largest zero grows like $4N$, while the smallest one behaves like $O(N^{-1})$. These properties are illustrated in Fig. 7.2a–c. We see from Fig. 7.2a, b that the nodes are clustered near the endpoint $x = 0$, with $\min_j |x_{j+1} - x_j| \sim N^{-1}$ (Recall that in the Jacobi case, this minimum distance is $O(N^{-2})$), whereas the distance between consecutive nodes $|x_{j+1} - x_j|$ increases dramatically as the index j increases. Moreover, Fig. 7.2c shows that the largest node x_N grows like $4N$ as N increases. We observe from Fig. 7.2d that, for a fixed j, the zero $x_j^{(\alpha)}$ moves away from the endpoint $x = 0$ as the index α increases, which is similar to that of the Jacobi polynomials (cf. Fig. 3.2).

In Table 7.1, we present some samples of the Laguerre-Gauss-Radau quadrature nodes and weights (with $\alpha = 0$ and $N = 16$).

Table 7.1 Laguerre-Gauss-Radau quadrature nodes and weights with $\alpha = 0$ and $N = 16$

Nodes x_j	Weights ω_j	Weights $\hat{\omega}_j$
0.000000000000000e+00	5.882352941176471e-02	5.882352941176471e-02
2.161403052394536e-01	2.927604493268249e-01	3.633966730173862e-01
7.263882432518047e-01	3.181362909815314e-01	6.577784169175935e-01
1.533593160373541e+00	2.066607692008763e-01	9.578314405922921e-01
2.644970998611911e+00	8.994208934619415e-02	1.266657740603195e+00
4.070978160880192e+00	2.708753007297033e-02	1.587715767327549e+00
5.825855515105604e+00	5.679781839868921e-03	1.925167659218618e+00
7.928504185306668e+00	8.230703112809221e-04	2.284244322226874e+00
1.040380828995104e+01	8.100028502120838e-05	2.671794664046113e+00
1.328466107070703e+01	5.266367668946434e-06	3.097178050984928e+00
1.661517321686662e+01	2.173968333033556e-07	3.573808337030715e+00
2.045600602002722e+01	5.384928271901132e-09	4.122023458991009e+00
2.489384702535191e+01	7.374041106876865e-11	4.774918361810553e+00
3.005986292020259e+01	4.928461815736925e-13	5.591701623302517e+00
3.617069454367918e+01	1.309028045707531e-15	6.693942819894268e+00
4.364036518417683e+01	9.358643168465394e-19	8.394265017253366e+00
5.352915116026845e+01	6.770058713668848e-23	1.196760936707925e+01

7.1.4 Interpolation and Discrete Laguerre Transforms

We first consider the interpolation and discrete transforms using generalized Laguerre *functions*.

Let $\{x_j^{(\alpha)}, \hat{\omega}_j^{(\alpha)}\}_{j=0}^N$ be a set of modified Laguerre-Gauss or Laguerre-Gauss-Radau quadrature nodes and weights given in Theorem 7.2. We define the associated discrete inner product and discrete norm as

$$\langle u, v\rangle_{N,\hat{\omega}_\alpha} = \sum_{j=0}^N u(x_j^{(\alpha)}) v(x_j^{(\alpha)}) \hat{\omega}_j^{(\alpha)}, \quad \|u\|_{N,\hat{\omega}_\alpha} = \sqrt{\langle u, u\rangle_{N,\hat{\omega}_\alpha}}.$$

The exactness of the quadrature formula (7.36) implies

$$\langle p, q\rangle_{N,\hat{\omega}_\alpha} = (p, q)_{\hat{\omega}_\alpha}, \quad \forall p \cdot q \in \widehat{P}_{2N+\delta}, \tag{7.46}$$

where $\delta = 1, 0$ for the modified Laguerre-Gauss and Laguerre-Gauss-Radau quadratures, respectively.

Let $\widehat{P}_N$ be the finite dimensional space defined in (7.35). Define the corresponding interpolation operator $\hat{I}_N^{(\alpha)} : C[0, +\infty) \to \widehat{P}_N$ such that

$$(\hat{I}_N^{(\alpha)} u)(x_j^{(\alpha)}) = u(x_j^{(\alpha)}), \quad 0 \le j \le N,$$

which can be expressed by

$$(\hat{I}_N^{(\alpha)} u)(x) = \sum_{n=0}^N \tilde{u}_n^{(\alpha)} \widehat{\mathscr{L}}_n^{(\alpha)}(x) \in \widehat{P}_N.$$

Given the physical values $\{u(x_j^{(\alpha)})\}_{j=0}^N$, the coefficients $\{\tilde{u}_n^{(\alpha)}\}_{n=0}^N$ can be determined by the *forward discrete transform*

$$\tilde{u}_n^{(\alpha)} = \frac{1}{\gamma_n^{(\alpha)}} \sum_{j=0}^N u(x_j^{(\alpha)}) \widehat{\mathscr{L}}_n^{(\alpha)}(x_j^{(\alpha)}) \hat{\omega}_j^{(\alpha)}, \quad 0 \le n \le N,$$

where $\gamma_n^{(\alpha)}$ is given by (7.2). On the other hand, given the expansion coefficients $\{\tilde{u}_n^{(\alpha)}\}_{n=0}^N$, the physical values $\{u(x_j^{(\alpha)})\}_{j=0}^N$ can be computed by the *backward discrete transform*

$$u(x_j^{(\alpha)}) = \sum_{n=0}^N \tilde{u}_n^{(\alpha)} \widehat{\mathscr{L}}_n^{(\alpha)}(x_j^{(\alpha)}), \quad 0 \le j \le N,$$

The above definitions and transforms can be extended to the set of Laguerre-Gauss-type nodes and weights $\{x_j^{(\alpha)}, \omega_j^{(\alpha)}\}_{j=0}^N$ given by Theorem 7.1 by removing "$\widehat{\ }$" from the corresponding ones.

7.1.5 Differentiation in the Physical Space

Let $\{h_j\}_{j=0}^N$ be the Lagrange basis polynomials associated with the Laguerre-Gauss-Radau points $\{x_j^{(\alpha)}\}_{j=0}^N$. Given $u \in P_N$ and its physical values at $\{x_j^{(\alpha)}\}_{j=0}^N$, the derivative values can be evaluated *exactly* by the general formula given in Sect. 3.1.6:

$$\mathbf{u}^{(m)} = D^m \mathbf{u}, \quad m \ge 1, \tag{7.47}$$

where

$$D = (d_{kj})_{k,j=0,1,\ldots,N} = (h_j'(x_k))_{k,j=0,1,\ldots,N};$$
$$\mathbf{u}^{(m)} = \big(u^{(m)}(x_0), u^{(m)}(x_1), \ldots, u^{(m)}(x_N)\big)^T, \quad \mathbf{u} := \mathbf{u}^{(0)}.$$

Hence, it suffices to evaluate the first-order differentiation matrix D.

Using Theorem 3.11 and the properties of GLPs, we can determine the entries of D.

(i) Laguerre-Gauss:

$$d_{kj} = \begin{cases} \dfrac{1}{2} - \dfrac{\alpha+1}{2x_j^{(\alpha)}}, & k = j, \\[2ex] \dfrac{x_j^{(\alpha)} \mathscr{L}_N^{(\alpha)}(x_k^{(\alpha)})}{x_k^{(\alpha)} \mathscr{L}_N^{(\alpha)}(x_j^{(\alpha)})} \dfrac{1}{x_k^{(\alpha)} - x_j^{(\alpha)}}, & k \ne j. \end{cases} \tag{7.48}$$

(ii) Laguerre-Gauss-Radau:

$$d_{kj} = \begin{cases} -\dfrac{N}{\alpha+2}, & k = j = 0, \\[2ex] -\dfrac{\Gamma(N+\alpha+1)}{\Gamma(\alpha+2)N!\mathscr{L}_N^{(\alpha)}(x_j^{(\alpha)})} \dfrac{1}{x_j^{(\alpha)}}, & k = 0,\ 1 \le j \le N, \\[2ex] \dfrac{\Gamma(\alpha+2)N!\mathscr{L}_N^{(\alpha)}(x_k^{(\alpha)})}{\Gamma(N+\alpha+1)} \dfrac{1}{x_k^{(\alpha)}}, & 1 \le k \le N,\ j = 0, \\[2ex] \dfrac{1}{2} - \dfrac{\alpha}{2x_j^{(\alpha)}}, & 1 \le k = j \le N, \\[2ex] \dfrac{\mathscr{L}_N^{(\alpha)}(x_k^{(\alpha)})}{\mathscr{L}_N^{(\alpha)}(x_j^{(\alpha)})} \dfrac{1}{x_k^{(\alpha)} - x_j^{(\alpha)}}, & 1 \le k \ne j \le N. \end{cases} \tag{7.49}$$

Now, we turn to the differentiation process related to the generalized Laguerre function approach. Define

$$\hat{h}_j(x) = \frac{e^{-x/2}}{e^{-x_j^{(\alpha)}/2}} h_j(x). \tag{7.50}$$

One verifies readily that

$$\hat{h}_j(x_k^{(\alpha)}) = \delta_{kj},\ 0 \le k,j \le N; \quad \widehat{P}_N = \mathrm{span}\{\hat{h}_j\ :\ 0 \le j \le N\}. \tag{7.51}$$

Moreover, we find that $\hat{h}'_j \in \widehat{P}_N$ and

$$\hat{d}_{kj} := \hat{h}'_j(x_k^{(\alpha)}) = \frac{e^{-x_k^{(\alpha)}/2}}{e^{-x_j^{(\alpha)}/2}} d_{kj} - \frac{1}{2}\delta_{kj}, \quad 0 \le j,k \le N. \tag{7.52}$$

Therefore, the entries of $\widehat{D} = (\hat{d}_{kj})_{0 \le k,j \le N}$ can be computed by

(i) modified Laguerre-Gauss:

$$\hat{d}_{kj} = \begin{cases} -\dfrac{\alpha+1}{2x_j^{(\alpha)}}, & k = j, \\ \dfrac{x_j^{(\alpha)} \widehat{\mathscr{L}}_N^{(\alpha)}(x_k^{(\alpha)})}{x_k^{(\alpha)} \widehat{\mathscr{L}}_N^{(\alpha)}(x_j^{(\alpha)})} \dfrac{1}{x_k^{(\alpha)} - x_j^{(\alpha)}}, & k \ne j. \end{cases} \tag{7.53}$$

(ii) modified Laguerre-Gauss-Radau:

$$\hat{d}_{kj} = \begin{cases} -\dfrac{N}{\alpha+2} - \dfrac{1}{2}, & k = j = 0, \\ -\dfrac{\Gamma(N+\alpha+1)}{\Gamma(\alpha+2)N!\widehat{\mathscr{L}}_N^{(\alpha)}(x_j^{(\alpha)})} \dfrac{1}{x_j^{(\alpha)}}, & k = 0,\ 1 \le j \le N, \\ \dfrac{\Gamma(\alpha+2)N!\widehat{\mathscr{L}}_N^{(\alpha)}(x_k^{(\alpha)})}{\Gamma(N+\alpha+1)} \dfrac{1}{x_k^{(\alpha)}}, & 1 \le k \le N,\ j = 0, \\ -\dfrac{\alpha}{2x_j^{(\alpha)}}, & 1 \le k = j \le N, \\ \dfrac{\widehat{\mathscr{L}}_N^{(\alpha)}(x_k^{(\alpha)})}{\widehat{\mathscr{L}}_N^{(\alpha)}(x_j^{(\alpha)})} \dfrac{1}{x_k^{(\alpha)} - x_j^{(\alpha)}}, & 1 \le k \ne j \le N. \end{cases} \tag{7.54}$$

Note that higher-order differentiation can be performed by consecutively differentiating (7.50).

7.1.6 Differentiation in the Frequency Space

Given $u \in P_N$, we write

$$u(x) = \sum_{n=0}^{N} \hat{u}_n \mathscr{L}_n^{(\alpha)}(x) \quad \text{with} \quad \hat{u}_n = \frac{1}{\gamma_n^{(\alpha)}} \int_0^{+\infty} u(x) \mathscr{L}_n^{(\alpha)}(x) \omega_\alpha(x) dx,$$

and

$$u'(x) = \sum_{n=1}^{N} \hat{u}_n \partial_x \mathscr{L}_n^{(\alpha)}(x) = \sum_{n=0}^{N} \hat{u}_n^{(1)} \mathscr{L}_n^{(\alpha)}(x) \in P_{N-1} \text{ with } \hat{u}_N^{(1)} = 0.$$

Thanks to (7.12b), we can compute $\{\hat{u}_n^{(1)}\}_{n=0}^{N-1}$ from $\{\hat{u}_n\}_{n=0}^{N}$ by the following backward relation (cf. Theorem 3.12):

$$\begin{cases} \hat{u}_{n-1}^{(1)} = \hat{u}_n^{(1)} - \hat{u}_n, \quad n = N, N-1, \ldots, 1, \\ \hat{u}_N^{(1)} = 0. \end{cases} \tag{7.55}$$

Higher-order derivatives can be computed by repeatedly using the above formula.

Now, we turn to the GLF approach. For any $v \in \widehat{P}_N$, we write

$$v(x) = \sum_{n=0}^{N} \hat{v}_n \widehat{\mathscr{L}}_n^{(\alpha)}(x).$$

It is clear that $v' \in \widehat{P}_N$ if $v \in \widehat{P}_N$ (compared with $v' \in P_{N-1}$ if $v \in P_N$). Hence, we can write

$$v'(x) = \sum_{n=0}^{N} \hat{v}_n \partial_x \widehat{\mathscr{L}}_n^{(\alpha)}(x) = \sum_{n=0}^{N} \hat{v}_n^{(1)} \widehat{\mathscr{L}}_n^{(\alpha)}(x).$$

Using (7.19) and (7.26a) leads to

$$\begin{aligned} v'(x) &= \sum_{n=0}^{N} \hat{v}_n \partial_x \widehat{\mathscr{L}}_n^{(\alpha)}(x) = \sum_{n=0}^{N} \hat{v}_n \Big(\frac{1}{2} \widehat{\mathscr{L}}_n^{(\alpha)}(x) - \sum_{k=0}^{n} \widehat{\mathscr{L}}_k^{(\alpha)}(x) \Big) \\ &= \frac{1}{2} \sum_{n=0}^{N} \hat{v}_n \widehat{\mathscr{L}}_n^{(\alpha)}(x) - \sum_{k=0}^{N} \Big(\sum_{n=k}^{N} \hat{v}_n \Big) \widehat{\mathscr{L}}_k^{(\alpha)}(x) \\ &= \sum_{n=0}^{N} \Big(\frac{1}{2} \hat{v}_n - \sum_{k=n}^{N} \hat{v}_k \Big) \widehat{\mathscr{L}}_n^{(\alpha)}(x). \end{aligned}$$

Hence, we compute $\{\hat{v}_n^{(1)}\}_{n=0}^{N}$ from $\{\hat{v}_n\}_{n=0}^{N}$ by the formula:

$$v_n^{(1)} = \frac{1}{2} \hat{v}_n - \sum_{k=n}^{N} \hat{v}_k, \quad n = 0, 1, \ldots, N,$$

or equivalently, by the backward relation:

$$\begin{cases} \hat{v}_n^{(1)} = \hat{v}_{n+1}^{(1)} - \frac{1}{2} \big(\hat{v}_n + \hat{v}_{n+1} \big), \quad n = N-1, \ldots, 0, \\ \hat{v}_N^{(1)} = -\frac{1}{2} \hat{v}_N. \end{cases} \tag{7.56}$$

7.2 Hermite Polynomials/Functions

We present in this section basic properties of the Hermite polynomials/functions, and derive Hermite-Gauss quadrature and the associated interpolation, discrete transforms and spectral differentiation techniques.

7.2.1 Basic Properties

7.2.1.1 Hermite Polynomials

The Hermite polynomials, defined on the whole line $\mathbb{R} := (-\infty, +\infty)$, are orthogonal with respect to the weight function $\omega(x) = e^{-x^2}$, namely,

$$\int_{-\infty}^{+\infty} H_m(x)H_n(x)\omega(x)dx = \gamma_n \delta_{mn}, \quad \gamma_n = \sqrt{\pi}2^n n!. \tag{7.57}$$

The Hermite polynomials satisfy the three-term recurrence relation:

$$H_{n+1}(x) = 2xH_n(x) - 2nH_{n-1}(x), \;\; n \geq 1, \tag{7.58}$$

and the first few members are

$$\begin{aligned} H_0(x) &= 1, \\ H_1(x) &= 2x, \\ H_2(x) &= 4x^2 - 2, \\ H_3(x) &= 8x^3 - 12x, \\ H_4(x) &= 16x^4 - 48x^2 + 12. \end{aligned}$$

One verifies by induction that the leading coefficient of $H_n(x)$ is 2^n.

The Hermite polynomials have a close connection with the generalized Laguerre polynomials:

$$\begin{aligned} H_{2n}(x) &= (-1)^n 2^{2n} n! \mathscr{L}_n^{(-1/2)}(x^2), \\ H_{2n+1}(x) &= (-1)^n 2^{2n+1} n! x \mathscr{L}_n^{(1/2)}(x^2). \end{aligned} \tag{7.59}$$

Hence, $H_n(x)$ is odd (resp. even) for n odd (resp. even), that is,

$$H_n(-x) = (-1)^n H_n(x). \tag{7.60}$$

Moreover, by (7.6),

$$H_{2n}(0) = (-1)^n \frac{(2n)!}{n!}, \quad H_{2n+1}(0) = 0. \tag{7.61}$$

The Hermite polynomials satisfy the Sturm-Liouville equation:

$$e^{x^2}\left(e^{-x^2}H_n'(x)\right)' + \lambda_n H_n(x) = 0, \quad \lambda_n = 2n, \tag{7.62}$$

or equivalently,

$$H_n''(x) - 2xH_n'(x) + \lambda_n H_n(x) = 0. \tag{7.63}$$

As with the Laguerre polynomials, the eigenvalue λ_n grows linearly with respect to n.

The Rodrigues' formula of the Hermite polynomial takes the form

$$H_n(x) = (-1)^n e^{x^2}\frac{d^n}{dx^n}\{e^{-x^2}\}, \tag{7.64}$$

and we have the explicit expression

$$H_n(x) = \sum_{k=0}^{[n/2]} \frac{(-1)^k n!}{k!(n-2k)!}(2x)^{n-2k}. \tag{7.65}$$

The following upper bound can be found in Abramowitz and Stegun (1964):

$$|H_n(x)| < c2^{n/2}\sqrt{n!}e^{x^2/2}, \quad c \approx 1.086435. \tag{7.66}$$

The Hermite polynomials satisfy

$$H_n'(x) = \lambda_n H_{n-1}(x), \quad n \geq 1. \tag{7.67}$$

Consequently, for any $k \in \mathbb{N}$, $\{H_n^{(k)}\}_{n=k}^{\infty}$ are orthogonal with respect to the same weight function e^{-x^2}. In particular, we have

$$\int_{-\infty}^{+\infty} H_n'(x)H_m'(x)e^{-x^2}dx = \lambda_n \gamma_n \delta_{mn}. \tag{7.68}$$

Another recurrence relation is

$$H_n'(x) = 2xH_n(x) - H_{n+1}(x), \quad n \geq 0. \tag{7.69}$$

In view of (7.59), one obtains from (7.13) the asymptotic behavior of the Hermite polynomials on a finite interval for large n. Moreover, By Formula (8.22.8) of Szegö (1975), we have

$$\begin{aligned}\frac{\Gamma(n/2+1)}{n!}e^{-x^2/2}H_n(x) &= \cos\left(\sqrt{2n+1}x - n\pi/2\right)\\ &+ \frac{x^3}{6\sqrt{2n+1}}\sin\left(\sqrt{2n+1}x - n\pi/2\right) + O(n^{-1}),\end{aligned} \tag{7.70}$$

which holds uniformly on a finite interval for sufficiently large n. Consequently, the rapid growth of $|H_n(x)|$ with respect to n and $|x|$ (cf. Fig. 7.3a) may cause severe numerical instability in the evaluation of the Hermite polynomials.

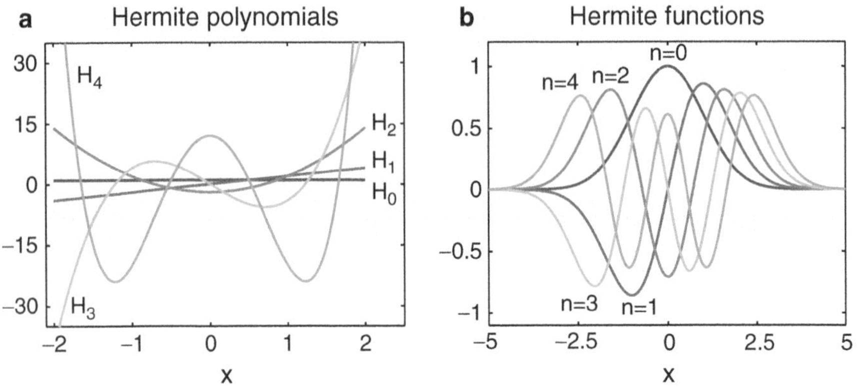

Fig. 7.3 (**a**) The first five Hermite polynomials $H_n(x)$ with $n = 0,\ldots,4$; (**b**) The first five Hermite functions $\widehat{H}_n(x)$ with $n = 0,\ldots,4$

7.2.1.2 Hermite Functions

As in the Laguerre case, the Hermite polynomials are not very useful in practice due to its wild behavior at infinity. Therefore, we consider the Hermite functions defined by

$$\widehat{H}_n(x) = \frac{1}{\pi^{1/4}\sqrt{2^n n!}} e^{-x^2/2} H_n(x), \quad n \ge 0, \ x \in \mathbb{R}, \tag{7.71}$$

which are normalized so that

$$\int_{-\infty}^{+\infty} \widehat{H}_n(x)\widehat{H}_m(x)dx = \delta_{mn}. \tag{7.72}$$

By the three-term recurrence relation (7.58), we obtain

$$\begin{aligned} &\widehat{H}_{n+1}(x) = x\sqrt{\frac{2}{n+1}}\widehat{H}_n(x) - \sqrt{\frac{n}{n+1}}\widehat{H}_{n-1}(x), \ n \ge 1, \\ &\widehat{H}_0(x) = \pi^{-1/4}e^{-x^2/2}, \quad \widehat{H}_1(x) = \sqrt{2}\pi^{-1/4}xe^{-x^2/2}, \end{aligned} \tag{7.73}$$

which allows for a stable evaluation of the Hermite functions.

We find from (7.62) that the Hermite functions satisfy the second-order equation:

$$\widehat{H}_n''(x) + (2n+1-x^2)\widehat{H}_n(x) = 0. \tag{7.74}$$

Moreover, by (7.67) and (7.73),

$$\begin{aligned}\widehat{H}_n'(x) &= \sqrt{2n}\widehat{H}_{n-1}(x) - x\widehat{H}_n(x)\\ &= \sqrt{\frac{n}{2}}\widehat{H}_{n-1}(x) - \sqrt{\frac{n+1}{2}}\widehat{H}_{n+1}(x),\end{aligned} \tag{7.75}$$

which, together with the orthogonality (7.72), yields

$$\int_{-\infty}^{+\infty} \widehat{H}_n'(x)\widehat{H}_m'(x)dx = \begin{cases} -\dfrac{\sqrt{n(n-1)}}{2}, & m = n-2,\\ n+\dfrac{1}{2}, & m = n,\\ -\dfrac{\sqrt{(n+1)(n+2)}}{2}, & m = n+2,\\ 0, & \text{otherwise.}\end{cases} \tag{7.76}$$

In addition, by (7.65) and (7.71),

$$\widehat{H}_n(x) = \pi^{-1/4}\sqrt{2^n n!}\sum_{k=0}^{[n/2]} \frac{(-1)^k}{2^{2k}k!(n-2k)!}(x^{n-2k}e^{-x^2/2}), \tag{7.77}$$

which indicates that for any n, the Hermite functions decay rapidly (cf. Fig. 7.3b). On the other hand, by (7.70) and the Stirling's formula (A.7), the Hermite functions on a finite interval with sufficiently large n behave like

$$\widehat{H}_n(x) = \pi^{-1/2}\Big(\frac{n}{2}+1\Big)^{-\frac{1}{4}} \cos\Big(\sqrt{2n+1}x - \frac{n\pi}{2}\Big) + O(n^{-1/2}). \tag{7.78}$$

7.2.2 Hermite-Gauss Quadrature

The Hermite-Gauss quadrature follows directly from the general formula in Sect. 3.1.

Theorem 7.3. *Let $\{x_j\}_{j=0}^N$ be the zeros of $H_{N+1}(x)$, and let $\{\omega_j\}_{j=0}^N$ be given by*

$$\omega_j = \frac{\sqrt{\pi}2^N N!}{(N+1)H_N^2(x_j)}, \quad 0 \le j \le N. \tag{7.79}$$

Then, we have

$$\int_{-\infty}^{+\infty} p(x)e^{-x^2}dx = \sum_{j=0}^{N} p(x_j)\omega_j, \quad \forall p \in P_{2N+1}. \tag{7.80}$$

Proof. We derive from Theorem 3.5 that

$$\omega_j = \frac{\sqrt{\pi}2^{N+1}N!}{H_N(x_j)H'_{N+1}(x_j)}, \quad 0 \le j \le N,$$

so the representation (7.79) follows from (7.67) directly. □

To approximate functions by Hermite functions, it is suitable to use the modified Hermite-Gauss quadrature formula.

Theorem 7.4. *Let $\{x_j, \omega_j\}_{j=0}^N$ be the Hermite-Gauss quadrature nodes and weights given in Theorem 7.3. Define the modified weights*

$$\hat{\omega}_j = e^{x_j^2}\omega_j = \frac{1}{(N+1)\widehat{H}_N^2(x_j)}, \quad 0 \le j \le N. \tag{7.81}$$

Then, we have

$$\int_{-\infty}^{+\infty} p(x)q(x)dx = \sum_{j=0}^{N} p(x_j)q(x_j)\hat{\omega}_j, \quad \forall p \cdot q \in \widehat{P}_{2N+1}, \tag{7.82}$$

where

$$\widehat{P}_M := \{\phi \,:\, \phi = e^{-x^2/2}\psi,\ \forall \psi \in P_M\}. \tag{7.83}$$

This modified Hermite-Gauss quadrature can be derived from Theorem 7.3 directly.

7.2.3 Computation of Nodes and Weights

According to Theorem 3.4, the zeros $\{x_j\}_{j=0}^N$ of $H_{N+1}(x)$ are the eigenvalues of the symmetric tridiagonal matrix

$$A_{N+1} = \begin{bmatrix} a_0 & \sqrt{b_1} & & & \\ \sqrt{b_1} & a_1 & \sqrt{b_2} & & \\ & \ddots & \ddots & \ddots & \\ & & \sqrt{b_{N-1}} & a_{N-1} & \sqrt{b_N} \\ & & & \sqrt{b_N} & a_N \end{bmatrix}, \tag{7.84}$$

whose entries are derived from (7.58):

$$a_j = 0,\ 0 \le j \le N; \quad b_j = j/2,\ 1 \le j \le N.$$

The quadrature weights $\{\omega_j\}_{j=0}^N$ might be computed from the first component of the orthonormal eigenvectors of A_{N+1}, or from the weight formula (7.79). However, due to the rapid growth of the Hermite polynomials, the latter approach is unstable for large N or x_j, whereas the evaluation of $\{\hat{\omega}_j\}_{j=0}^N$ by (7.81) is always stable.

Alternatively, to apply the iterative approach described in Sect. 3.1.3 to locate the zeros, it is advisable to compute the zeros of $\widehat{H}_{N+1}(x)$. Note from (7.61) and (7.79) that the nodes and weights are symmetric, namely,

$$x_j = -x_{N-j}, \quad \omega_j = \omega_{N-j}, \quad 0 \le j \le N, \tag{7.85}$$

and $x_{N/2} = 0$, if N even. Hence, the computational cost can be halved. In view of (7.59), we might take the initial approximation to be $\{(z_k^{(\alpha)})^{1/2}\}_{k=0}^{[N/2]-1}$ in (7.41) with $\alpha = 1/2$ and $-1/2$ for odd N and even N, respectively.

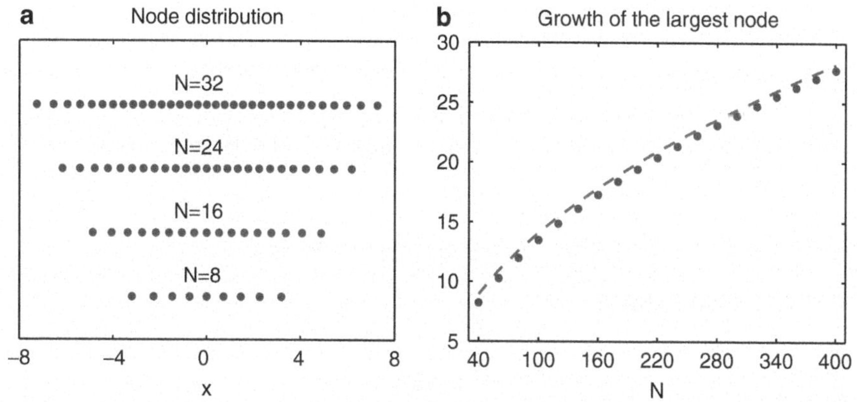

Fig. 7.4 (**a**) Distribution of the Hermite-Gauss nodes $\{x_j\}_{j=0}^N$ with $N = 8, 16, 24, 32$; (**b**) Growth of the largest node against the asymptotic estimate: $\sqrt{2(N+1)} - (2(N+1))^{1/3}$ (*dashed line*) with various N

As a direct consequence of (7.43) and (7.59), we find that

$$\max_j |x_j| \sim \sqrt{2N}, \quad \min_j |x_j - x_{j-1}| \sim N^{-1/2}. \tag{7.86}$$

We see from Fig. 7.4 (right) that the largest node grows at a rate as predicted. Moreover, Levin and Lubinsky (1992) pointed out that

$$\omega_j \sim \frac{1}{N} e^{-x_j^2} \Big(1 - \frac{|x_j|}{\sqrt{2(N+1)}}\Big), \quad 0 \le j \le N. \tag{7.87}$$

In Table 7.2, we tabulate half (due to the symmetry (7.85)) of the Hermite-Gauss nodes, weights and the modified weights: $\hat{\omega}_j = e^{x_j^2} \omega_j$ with $N = 8, 16$.

Table 7.2 Hermite-Gauss nodes and weights

Nodes x_j	Weights ω_j	Weights $\hat{\omega}_j$
0.000000000000000e+00	7.202352156060509e-01	7.202352156060509e-01
7.235510187528376e-01	4.326515590025558e-01	7.303024527450919e-01
1.468553289216668e+00	8.847452739437670e-02	7.646081250945510e-01
2.266580584531843e+00	4.943624275536940e-03	8.417527014786689e-01
3.190993201781524e+00	3.960697726326615e-05	1.047003580976707e+00
0.000000000000000e+00	5.309179376248636e-01	5.309179376248636e-01
5.316330013426551e-01	4.018264694704117e-01	5.330706545735971e-01
1.067648725743451e+00	1.726482976700968e-01	5.397631139084976e-01
1.612924314221230e+00	4.092003414975658e-02	5.517773530781673e-01
2.173502826666621e+00	5.067349957627506e-03	5.707392941244530e-01
2.757762915703891e+00	2.986432866977458e-04	5.998927326677652e-01
3.378932091141491e+00	7.112289140021636e-06	6.462917002128916e-01
4.061946675875474e+00	4.977078981630849e-08	7.287483705871041e-01
4.871345193674399e+00	4.580578930798994e-11	9.262541399895513e-01

7.2.4 Interpolation and Discrete Hermite Transforms

Let I_N^h be the interpolation operator associated with the Hermite-Gauss points $\{x_j\}_{j=0}^N$ such that for any $v \in C(\mathbb{R})$,

$$I_N^h v \in P_N; \quad (I_N^h v)(x_j) = v(x_j), \quad 0 \le j \le N,$$

which can be expanded as

$$(I_N^h v)(x) = \sum_{n=0}^{N} \tilde{v}_n H_n(x).$$

Given $\{v(x_j)\}_{j=0}^N$, the expansion coefficients $\{\tilde{v}_n\}_{n=0}^N$ can be computed by the *forward discrete transform*

$$\tilde{v}_n = \frac{1}{\gamma_n} \sum_{j=0}^{N} v(x_j) H_n(x_j) \omega_j, \quad 0 \le n \le N. \tag{7.88}$$

On the other hand, given $\{\tilde{v}_n\}_{n=0}^N$, the physical values $\{v(x_j)\}_{j=0}^N$ can be evaluated by the *backward discrete transform*

$$v(x_j) = \sum_{n=0}^{N} \tilde{v}_n H_n(x_j), \quad 0 \le j \le N. \tag{7.89}$$

Accordingly, for the Hermite function approach, we define the interpolant $\hat{I}_N^h u \in \widehat{P}_N$, which interpolates u at $\{x_j\}_{j=0}^N$ with the expansion

$$(\hat{I}_N^h u)(x) = \sum_{n=0}^{N} \tilde{u}_n \widehat{H}_n(x). \tag{7.90}$$

The forward and backward discrete transforms are performed by

$$\tilde{u}_n = \sum_{j=0}^{N} u(x_j)\widehat{H}_n(x_j)\hat{\omega}_j, \ 0 \le n \le N; \quad u(x_j) = \sum_{n=0}^{N} \tilde{u}_n \widehat{H}_n(x_j), \ 0 \le j \le N.$$

7.2.5 Differentiation in the Physical Space

Let $\{h_j\}_{j=0}^N$ be the Lagrange basis polynomials associated with the Hermite-Gauss points $\{x_j\}_{j=0}^N$. Given $u \in P_N$, we obtain from the general formula in Sect. 3.1.6 that

$$\mathbf{u}^{(m)} = D^m \mathbf{u}, \quad m \ge 1, \tag{7.91}$$

where the notation is the same as before. By Theorem 3.11 and the properties of the Hermite polynomials, the entries of the first-order differentiation matrix D can be computed by

$$d_{kj} = \begin{cases} \dfrac{H_N(x_k)}{H_N(x_j)} \dfrac{1}{x_k - x_j}, & \text{if } k \ne j, \\ x_k, & \text{if } k = j. \end{cases} \tag{7.92}$$

We now consider differentiation associated with the Hermite function approach. Observe that for $v \in \widehat{P}_N$, we have $v' \in \widehat{P}_{N+1}$. For any $v \in \widehat{P}_N$, we write $v = e^{-x^2/2}u$ with $u \in P_N$. We can compute its derivative values by

$$\begin{aligned} v'(x_k) &= -x_k v(x_k) + e^{-x_k^2/2} u'(x_k) = -x_k v(x_k) + e^{-x_k^2/2} \sum_{j=0}^{N} u(x_j) d_{kj} \\ &= -x_k v(x_k) + e^{-x_k^2/2} \sum_{j=0}^{N} e^{x_j^2/2} v(x_j) d_{kj} := \sum_{j=0}^{N} v(x_j) \hat{d}_{kj}, \end{aligned}$$

where by (7.92),

$$\hat{d}_{kj} = -x_k \delta_{kj} + e^{-x_k^2/2} d_{kj} e^{x_j^2/2} = \begin{cases} \dfrac{\widehat{H}_N(x_k)}{\widehat{H}_N(x_j)} \dfrac{1}{x_k - x_j}, & \text{if } k \ne j, \\ 0, & \text{if } k = j. \end{cases} \tag{7.93}$$

Higher order derivatives can be evaluated in a similar recursive manner.

7.2.6 Differentiation in the Frequency Space

Given $u \in P_N$ with the expansion:

$$u(x) = \sum_{n=0}^{N} \hat{u}_n H_n(x) \quad \text{where} \quad \hat{u}_n = \frac{1}{\gamma_n}(u, H_n)_\omega,$$

we have

$$u'(x) = \sum_{n=1}^{N} \hat{u}_n H_n'(x) = \sum_{n=0}^{N} \hat{u}_n^{(1)} H_n(x) \in P_{N-1} \quad \text{with} \quad \hat{u}_N^{(1)} = 0.$$

As a direct consequence of (7.67), the coefficients $\{\hat{u}_n^{(1)}\}$ can be evaluated by the backward recursive formula:

$$\hat{u}_N^{(1)} = 0; \quad \hat{u}_n^{(1)} = 2(n+1)\hat{u}_{n+1}, \; n = N-1, N-2, \ldots, 0. \tag{7.94}$$

We now consider the differentiation associated with the Hermite functions. Given $v \in \widehat{P}_N$, we can write $v = e^{-x^2/2}u$ with $u \in P_N$. Therefore,

$$v(x) = \sum_{n=0}^{N} \hat{v}_n \widehat{H}_n(x) \quad \overset{(7.71)}{\Longrightarrow} \quad u(x) = \sum_{n=0}^{N} \hat{v}_n \frac{H_n(x)}{\sqrt{\gamma_n}}. \tag{7.95}$$

Then we have from (7.67), (7.71) and (7.73) that

$$\begin{aligned}
v' &= e^{-x^2/2}u' - xv = e^{-x^2/2} \sum_{n=0}^{N} \hat{v}_n \frac{H_n'}{\sqrt{\gamma_n}} - \sum_{n=0}^{N} \hat{v}_n (x\widehat{H}_n) \\
&= \sum_{n=0}^{N} \hat{v}_n \sqrt{2n}\widehat{H}_{n-1} - \sum_{n=0}^{N} \hat{v}_n \left(\sqrt{\frac{n}{2}}\widehat{H}_{n-1} + \sqrt{\frac{n+1}{2}}\widehat{H}_{n+1}\right) \\
&= \sum_{n=0}^{N} \hat{v}_n \left(\sqrt{\frac{n}{2}}\widehat{H}_{n-1} - \sqrt{\frac{n+1}{2}}\widehat{H}_{n+1}\right) \\
&= \frac{\hat{v}_1}{\sqrt{2}}\widehat{H}_0 + \sum_{n=1}^{N-1} \left(\sqrt{\frac{n+1}{2}}\hat{v}_{n+1} - \sqrt{\frac{n}{2}}\hat{v}_{n-1}\right)\widehat{H}_n - \sum_{n=N}^{N+1} \sqrt{\frac{n}{2}}\hat{v}_{n-1}\widehat{H}_n.
\end{aligned}$$

Since $v' \in \widehat{P}_{N+1}$, we can expand it as

$$v'(x) = \sum_{n=0}^{N+1} \hat{v}_n^{(1)} \widehat{H}_n(x),$$

and the coefficients $\{\hat{v}_n^{(1)}\}_{n=0}^{N+1}$ can be computed by

$$\hat{v}_n^{(1)} = \sqrt{\frac{n+1}{2}}\hat{v}_{n+1} - \sqrt{\frac{n}{2}}\hat{v}_{n-1}, \quad n = N+1, N, \ldots, 0, \tag{7.96}$$

with $\hat{v}_{-1} = \hat{v}_{N+1} = \hat{v}_{N+2} = 0$.

7.3 Approximation by Laguerre and Hermite Polynomials/Functions

This section is devoted to the analysis of approximations by (generalized) Laguerre and Hermite polynomials/functions. These results will be useful for error analysis of spectral methods for unbounded domains.

7.3.1 Inverse Inequalities

We first present several inverse inequalities associated with the Laguerre polynomials/functions. Recall the weight functions $\omega_\alpha = x^\alpha e^{-x}$ and $\hat{\omega}_\alpha = x^\alpha$.

Theorem 7.5. *For $\alpha > -1$ and any $\phi \in P_N$,*

$$\|\partial_x^m \phi\|_{\omega_{\alpha+m}} \lesssim N^{m/2} \|\phi\|_{\omega_\alpha}, \quad m \geq 0. \tag{7.97}$$

Proof. In view of the orthogonality (7.1) and (7.9), this inequality can be proved by following the same procedure as in the proof of (3.236). □

Let $\hat{\partial}_x = \partial_x + \frac{1}{2}$ be the same differential operator as before. For any $\psi \in \widehat{P}_N$ (defined in (7.35)), we have $\psi = e^{-x/2}\phi$ and

$$\partial_x^m \phi = \partial_x^m (e^{x/2}\psi) = e^{x/2}\Big(\partial_x + \frac{1}{2}\Big)^m \psi = e^{x/2}\hat{\partial}_x^m \psi. \tag{7.98}$$

Therefore, a direct consequence of (7.97) is as follows.

Corollary 7.1. *For $\alpha > -1$ and any $\psi \in \widehat{P}_N$,*

$$\|\hat{\partial}_x^m \psi\|_{\hat{\omega}_{\alpha+m}} \lesssim N^{m/2} \|\psi\|_{\hat{\omega}_\alpha}, \quad m \geq 0. \tag{7.99}$$

Next, we derive an inverse inequality involving the same weight function for derivatives of different order.

Theorem 7.6. *For $\alpha \geq 0$ and any $\phi \in P_N$,*

$$\|\partial_x^m \phi\|_{\omega_\alpha} \leq N^m \|\phi\|_{\omega_\alpha}, \quad m \geq 0. \tag{7.100}$$

Proof. For any $\phi \in P_N$, we have

$$\phi = \sum_{n=0}^{N} \hat{\phi}_n^{(\alpha)} \mathscr{L}_n^{(\alpha)} \quad \Rightarrow \quad \|\phi\|_{\omega_\alpha}^2 = \sum_{n=0}^{N} \gamma_n^{(\alpha)} |\hat{\phi}_n^{(\alpha)}|^2.$$

By (7.12a),

$$\phi' = \sum_{n=1}^{N} \hat{\phi}_n^{(\alpha)} \Big(-\sum_{k=0}^{n-1} \mathscr{L}_k^{(\alpha)}\Big) = \sum_{k=0}^{N-1}\Big(-\sum_{n=k+1}^{N} \hat{\phi}_n^{(\alpha)}\Big) \mathscr{L}_k^{(\alpha)}.$$

Thus,

$$\|\phi'\|_{\omega_\alpha}^2 = \sum_{k=0}^{N-1} \left(\sum_{n=k+1}^{N} \hat{\phi}_n^{(\alpha)} \right)^2 \gamma_k^{(\alpha)}.$$

By the Cauchy–Schwarz inequality,

$$\left(\sum_{n=k+1}^{N} \hat{\phi}_n^{(\alpha)} \right)^2 \leq \left(\sum_{n=k+1}^{N} \gamma_n^{(\alpha)} |\hat{\phi}_n^{(\alpha)}|^2 \right) \left(\sum_{n=k+1}^{N} (\gamma_n^{(\alpha)})^{-1} \right).$$

By (7.2),

$$\frac{\gamma_{j+1}^{(\alpha)}}{\gamma_j^{(\alpha)}} = \frac{j+\alpha+1}{j+1},$$

which implies that if $\alpha \geq 0$, $\{\gamma_j^{(\alpha)}\}$ is an increasing sequence. A combination of the above facts leads to that for $\alpha \geq 0$,

$$\begin{aligned} \|\phi'\|_{\omega_\alpha}^2 &\leq \|\phi\|_{\omega_\alpha}^2 \sum_{k=0}^{N-1} \gamma_k^{(\alpha)} \left(\sum_{n=k+1}^{N} (\gamma_n^{(\alpha)})^{-1} \right) \\ &\leq N \|\phi\|_{\omega_\alpha}^2 \sum_{k=0}^{N-1} \frac{\gamma_k^{(\alpha)}}{\gamma_{k+1}^{(\alpha)}} \leq N^2 \|\phi\|_{\omega_\alpha}^2. \end{aligned}$$

Therefore, applying the above inequality repeatedly leads to (7.100). □

Corollary 7.2. *For* $\alpha \geq 0$ *and any* $\psi \in \widehat{P}_N$,

$$\|\hat{\partial}_x^m \psi\|_{\hat{\omega}_\alpha} \lesssim N^m \|\psi\|_{\hat{\omega}_\alpha}, \quad m \geq 0. \tag{7.101}$$

Unlike the Jacobi and generalized Laguerre polynomials, there is only one weight function, $\omega(x) = e^{-x^2}$, associated with the Hermite polynomials. Consequently, there is no analogue of Theorem 7.5 in the Hermite case. However, we can prove the following:

Theorem 7.7. *For any* $\phi \in P_N$,

$$\|\partial_x \phi\|_\omega \lesssim N \|\phi\|_\omega \quad \textit{where} \quad \omega = e^{-x^2}.$$

Moreover, let $\hat{\partial}_x = \partial_x + x$. *Then for any* $\psi \in \widehat{P}_N$ *(defined in* (7.83)*),*

$$\|\hat{\partial}_x \psi\| \lesssim N \|\psi\|.$$

We leave the proof as an excise (see Problem 7.1).

7.3.2 Orthogonal Projections

We first consider the approximations by (generalized) Laguerre polynomials/ functions. Consider the $L^2_{\omega_\alpha}$-orthogonal projection $\Pi_{N,\alpha} : L^2_{\omega_\alpha}(\mathbb{R}_+) \to P_N$, defined by

$$(\Pi_{N,\alpha}u - u, v_N)_{\omega_\alpha} = 0, \quad \forall v_N \in P_N, \tag{7.102}$$

so we have

$$\Pi_{N,\alpha}u(x) = \sum_{n=0}^{N} \hat{u}_n^{(\alpha)} \mathscr{L}_n^{(\alpha)}(x) \quad \text{with} \quad \hat{u}_n^{(\alpha)} = \frac{1}{\gamma_n^{(\alpha)}} \int_{\mathbb{R}_+} u(x)\mathscr{L}_n^{(\alpha)}(x)\omega_\alpha(x)dx.$$

Similar to the Jacobi approximations, we introduce the space

$$B^m_\alpha(\mathbb{R}_+) := \{u : \partial_x^k u \in L^2_{\omega_{\alpha+k}}(\mathbb{R}_+), \ 0 \le k \le m\}, \tag{7.103}$$

equipped with the norm and semi-norm

$$\|u\|_{B^m_\alpha} = \Big(\sum_{k=0}^{m} \|\partial_x^k u\|^2_{\omega_{\alpha+k}}\Big)^{1/2}, \quad |u|_{B^m_\alpha} = \|\partial_x^m u\|_{\omega_{\alpha+m}}.$$

As usual, we will drop the subscript α if $\alpha = 0$. Notice that the weight function corresponding to the derivative of different order is different in $B^m_\alpha(\mathbb{R}_+)$, as opposed to the Sobolev space $H^m_{\omega_\alpha}(\mathbb{R}_+)$.

Observe from (7.12a) that

$$\partial_x^k \mathscr{L}_n^{(\alpha)}(x) = (-1)^k \mathscr{L}_{n-k}^{(\alpha+k)}(x), \quad n \ge k, \tag{7.104}$$

which shows that $\{\partial_x^k \mathscr{L}_n^{(\alpha)}\}$ are orthogonal with respect to the weight $\omega_{\alpha+k}$, i.e.,

$$\int_0^{+\infty} \partial_x^k \mathscr{L}_l^{(\alpha)} \partial_x^k \mathscr{L}_n^{(\alpha)} \omega_{\alpha+k}\, dx = \gamma_{n-k}^{(\alpha+k)} \delta_{ln}. \tag{7.105}$$

Thanks to (7.105), the following fundamental results can be proved by using an argument similar to that for Theorem 3.35, and the proof is left as an exercise (see Problem 7.2).

Theorem 7.8. *Let $\alpha > -1$. If $u \in B^m_\alpha(\mathbb{R}_+)$ and $0 \le m \le N+1$, we have*

$$\|\partial_x^l(\Pi_{N,\alpha}u - u)\|_{\omega_{\alpha+l}} \le \sqrt{\frac{(N-m+1)!}{(N-l+1)!}} \|\partial_x^m u\|_{\omega_{\alpha+m}}, \quad 0 \le l \le m. \tag{7.106}$$

We observe that the above result is valid for $u \subset B^m_\alpha(\mathbb{R}_+)$ which includes functions that do not decay at infinity, however, the error estimate is given in a weighted space with an exponentially decay rate. In particular, a fast convergence rate in the norm $\|\cdot\|_{\omega_{\alpha+l}}$ does not mean that the error would decay rapidly for large x.

Consider, for example, the expansion of an entire function $\sin x$ for which the above result is valid for any $m > 0$. However, its expansion in terms of Laguerre polynomials is:

$$\sin x = \sum_{n=0}^{\infty} \frac{1}{2^{(n+1)/2}} \cos\left[\frac{\pi}{4}(n+1)\right] \mathscr{L}_n(x). \tag{7.107}$$

In view of (7.14), the decay of the N-term truncation is roughly

$$\frac{1}{\sqrt{2\pi}} \frac{e^{x/2}}{2^{N/2}(Nx)^{1/4}}, \tag{7.108}$$

which implies that the error is small only if $N \ln 2 > x$ or $N > 1.44x$.

Next, we consider the projection error of the Laguerre function expansions. Recall that $\hat{\omega}_\alpha = x^\alpha$. For any $u \in L^2_{\hat{\omega}_\alpha}(\mathbb{R}_+)$, we have $ue^{x/2} \in L^2_{\omega_\alpha}(\mathbb{R}_+)$. Define the operator

$$\hat{\Pi}_{N,\alpha} u = e^{-x/2} \Pi_{N,\alpha}(ue^{x/2}) \in \widehat{P}_N. \tag{7.109}$$

Clearly, we obtain from (7.102) that for any $v_N \in \widehat{P}_N$,

$$(\hat{\Pi}_{N,\alpha} u - u, v_N)_{\hat{\omega}_\alpha} = \big(\Pi_{N,\alpha}(ue^{x/2}) - (ue^{x/2}), (v_N e^{x/2})\big)_{\omega_\alpha} = 0,$$

which shows that $\hat{\Pi}_{N,\alpha} u$ is the $L^2_{\hat{\omega}_\alpha}$-orthogonal projection of u.

Let us define

$$\hat{B}^m_\alpha(\mathbb{R}_+) := \{u : \hat{\partial}^k_x u \in L^2_{\hat{\omega}_{\alpha+k}}(\mathbb{R}_+),\ 0 \le k \le m\}, \tag{7.110}$$

equipped with the norm and semi-norm

$$\|u\|_{\hat{B}^m_\alpha} = \Big(\sum_{k=0}^m \|\hat{\partial}^k_x u\|^2_{\hat{\omega}_{\alpha+k}}\Big)^{1/2}, \quad |u|_{\hat{B}^m_\alpha} = \|\hat{\partial}^m_x u\|_{\hat{\omega}_{\alpha+m}}.$$

It is straightforward to extend Theorem 7.8 to the Laguerre function case.

Theorem 7.9. *Let $\hat{\partial}_x = \partial_x + \frac{1}{2}$ and $\alpha > -1$. Then for any $u \in \hat{B}^m_\alpha(\mathbb{R}_+)$ and $0 \le m \le N+1$,*

$$\|\hat{\partial}^l_x(\hat{\Pi}_{N,\alpha} u - u)\|_{\hat{\omega}_{\alpha+l}} \le \sqrt{\frac{(N-m+1)!}{(N-l+1)!}} \|\hat{\partial}^m_x u\|_{\hat{\omega}_{\alpha+m}}, \quad 0 \le l \le m. \tag{7.111}$$

Proof. Let $v = ue^{x/2}$. It is clear that

$$\partial^l_x(\Pi_{N,\alpha} v - v) = \partial^l_x(e^{x/2}(\hat{\Pi}_{N,\alpha} u - u)) = e^{x/2}\hat{\partial}^l_x(\hat{\Pi}_{N,\alpha} u - u),$$

and likewise, $\partial^m_x v = e^{x/2}\hat{\partial}^m_x u$. Hence, the desired result is a direct consequence of (7.106). □

Remark 7.2. *Like Remark* 3.7, *we find that*

$$(N-l+1)^{(l-m)/2} \leq \sqrt{\frac{(N-m+1)!}{(N-l+1)!}} \leq (N-m+2)^{(l-m)/2}, \tag{7.112}$$

for $0 \leq l \leq m \leq N+1$ *and fixed* l. *In particular, if* m *is fixed, then*

$$\sqrt{\frac{(N-m+1)!}{(N-l+1)!}} \cong N^{(l-m)/2}. \tag{7.113}$$

Remark 7.3. *When comparing the error estimate in the above theorem with the corresponding result for classical Jacobi approximation* (*see Theorem* 3.35), *we notice that the convergence rate of the Laguerre approximation is only half of the classical Jacobi approximation. This is a direct consequence of the linear growth of the eigenvalues in the Laguerre Sturm-Liouville problem, as opposed to the quadratic growth in the Jacobi Sturm-Liouville problem.*

Note that $u \in \hat{B}^m_\alpha(\mathbb{R}_+)$ requires that u decays at infinity so the above results do not apply to functions like $\sin x$, despite the fact that it is an entire function. Consider, on the other hand, $u(x) = (1+x)^{-h}$ and $u(x) = \frac{\sin kx}{(1+x)^h}$ with $h > 0$. It can be easily checked that for both functions $\|\hat{\partial}^m_x u\|_{\hat{\omega}_{\alpha+m}} < \infty$ if $m < 2h - \alpha - 1$, so Theorem 7.9 implies that

$$\|u - \hat{\Pi}_{N,\alpha} u\|_{\hat{\omega}_\alpha} \lesssim N^{-(2h-\alpha-1)/2}. \tag{7.114}$$

Numerical evidences exhibiting the above convergence rate are provided in Figs. 7.5 and 7.6.

We now consider the H^1-type orthogonal projections. For simplicity, we restrict the analysis to the approximations by usual Laguerre polynomials/functions. Let $\omega(x) = e^{-x}$ be the usual Laguerre weight function, and denote

$$\begin{aligned} &H^1_{0,\omega}(\mathbb{R}_+) = \{u \in H^1_\omega(\mathbb{R}_+) \;:\; u(0) = 0\},\\ &P^0_N = \{\phi \in P_N \;:\; \phi(0) = 0\}. \end{aligned} \tag{7.115}$$

The orthogonal projection $\Pi^{1,0}_N : H^1_{0,\omega}(\mathbb{R}_+) \to P^0_N$ is defined by

$$\big((u - \Pi^{1,0}_N u)', v_N'\big)_\omega = 0, \quad \forall v_N \in P^0_N. \tag{7.116}$$

Theorem 7.10. *If* $u \in H^1_{0,\omega}(\mathbb{R}_+)$ *and* $\partial_x u \in B^{m-1}_0(\mathbb{R}_+)$, *then for* $1 \leq m \leq N+1$,

$$\|\Pi^{1,0}_N u - u\|_{1,\omega} \leq c\sqrt{\frac{(N-m+1)!}{N!}}\|\partial^m_x u\|_{\omega_{m-1}}, \tag{7.117}$$

where c *is a positive constant independent of* m, N *and* u.

Proof. Let

$$\phi(x) = \int_0^x \Pi_{N-1,0}u'(y)dy.$$

Then $u - \phi \in H^1_{0,\omega}(\mathbb{R}_+)$. Using (B.35b) and Theorem 7.8 with $\alpha = 0$ yields

$$\begin{aligned}\|\Pi_N^{1,0}u - u\|_{1,\omega} &\le \|\phi - u\|_{1,\omega} \le c\|\partial_x(\phi - u)\|_\omega \\ &\le c\sqrt{\frac{(N-m+1)!}{N!}}\|\partial_x^m u\|_{\omega_{m-1}}.\end{aligned}$$

This ends the proof. □

Since for any $u \in H_0^1(\mathbb{R}_+)$, we have $ue^{x/2} \in H^1_{0,\omega}(\mathbb{R}_+)$. Define the operator

$$\hat{\Pi}_N^{1,0}u = e^{-x/2}\Pi_N^{1,0}(ue^{x/2}) \in \widehat{P}_N^0,$$

whose approximation property is characterized by the following theorem.

Theorem 7.11. *For any $u \in H_0^1(\mathbb{R}_+)$, we have*

$$((u - \hat{\Pi}_N^{1,0}u)', v_N') + \frac{1}{4}(u - \hat{\Pi}_N^{1,0}u, v_N) = 0, \quad \forall v_N \in \widehat{P}_N^0. \tag{7.118}$$

Let $\hat{\partial}_x = \partial_x + \frac{1}{2}$. If $u \in H_0^1(\mathbb{R}_+)$ and $\hat{\partial}_x u \in \widehat{B}_0^{m-1}(\mathbb{R}_+)$, then for $1 \le m \le N+1$,

$$\|\hat{\Pi}_N^{1,0}u - u\|_1 \le c\sqrt{\frac{(N-m+1)!}{N!}}\|\hat{\partial}_x^m u\|_{\hat{\omega}_{m-1}}, \tag{7.119}$$

where c is a positive constant independent of m, N and u.

Proof. Using the definition of $\Pi_N^{1,0}$ and integration by parts, we find that for any $v_N = w_N e^{-x/2}$ with $w_N \in P_N^0$,

$$\begin{aligned}&((u - \hat{\Pi}_N^{1,0}u)', v_N') \\ &= \Big([(ue^{x/2}) - \Pi_N^{1,0}(ue^{x/2})]' - \frac{1}{2}[(ue^{x/2}) - \Pi_N^{1,0}(ue^{x/2})], w_N' - \frac{1}{2}w_N\Big)_\omega \\ &= -\frac{1}{2}\int_0^{+\infty}[(ue^{x/2}) - \Pi_N^{1,0}(ue^{x/2})w_N]'e^{-x}dx + \frac{1}{4}((ue^{x/2}) - \Pi_N^{1,0}(ue^{x/2}), w_N)_\omega \\ &= -\frac{1}{4}((ue^{x/2}) - \Pi_N^{1,0}(ue^{x/2}), v_N)_\omega = -\frac{1}{4}(u - \hat{\Pi}_N^{1,0}u, v_N),\end{aligned}$$

which yields the identity (7.118).

Let $v = ue^{x/2}$. Clearly,

$$\partial_x(\hat{\Pi}_N^{1,0}u - u) = -\frac{1}{2}e^{-x/2}(\Pi_N^{1,0}v - v) + e^{-x/2}\partial_x(\Pi_N^{1,0}v - v).$$

Hence, using Theorem 7.10 and the fact $\partial_x^m v = e^{x/2}\hat{\partial}_x^m u$ leads to

$$\begin{aligned}\|\partial_x(\hat{\Pi}_N^{1,0} u - u)\| &\le c\big(\|\Pi_N^{1,0} v - v\|_\omega + \|\partial_x(\Pi_N^{1,0} v - v)\|_\omega\big)\\ &\le c\sqrt{\frac{(N-m+1)!}{N!}}\|\partial_x^m v\|_{\omega_{m-1}} \le c\sqrt{\frac{(N-m+1)!}{N!}}\|\hat{\partial}_x^m u\|_{\hat{\omega}_{m-1}}.\end{aligned}$$

Similarly, by Theorem 7.10,

$$\|\hat{\Pi}_N^{1,0} u - u\| = \|\Pi_N^{1,0}(ue^{x/2}) - (ue^{x/2})\|_\omega \le c\sqrt{\frac{(N-m+1)!}{N!}}\|\hat{\partial}_x^m u\|_{\hat{\omega}_{m-1}}.$$

This completes the proof. □

In the analysis of Laguerre spectral methods for fourth-order problems, it is necessary to consider H^2-type orthogonal projections. For this purpose, we denote

$$\begin{aligned} &H^2_{0,\omega}(\mathbb{R}_+) = \{v \in H^2_\omega(\mathbb{R}_+) \,:\, v(0) = v'(0) = 0\}, \quad X_N = H^2_{0,\omega}(\mathbb{R}_+) \cap P_N,\\ &H^2_0(\mathbb{R}_+) = \{v \in H^2(\mathbb{R}_+) \,:\, v(0) = v'(0) = 0\}, \quad \widehat{X}_N = H^2_0(\mathbb{R}_+) \cap \widehat{P}_N,\end{aligned}$$

Consider the orthogonal projection: $\Pi_N^{2,0} : H^2_{0,\omega}(\mathbb{R}_+) \to X_N$, defined by

$$\big((v - \Pi_N^{2,0} v)'', v_N''\big)_\omega = 0, \quad \forall v_N \in X_N, \tag{7.120}$$

and define the mapping $\hat{\Pi}_N^{2,0} : H^2_0(\mathbb{R}_+) \to \widehat{X}_N$ by

$$\hat{\Pi}_N^{2,0} u = e^{-x/2}\Pi_N^{2,0}(ue^{x/2}). \tag{7.121}$$

We have the following approximation result, and leave its proof as an excise (see Problem 7.3).

Theorem 7.12. *If $v \in H^2_{0,\omega}(\mathbb{R}_+)$ and $\partial_x^2 v \in B_0^{m-2}(\mathbb{R}_+)$ with $2 \le m \le N+1$, then we have*

$$\|\Pi_N^{2,0} v - v\|_{2,\omega} \le c\sqrt{\frac{(N-m+1)!}{(N-1)!}}\|\partial_x^m v\|_{\omega_{m-2}}. \tag{7.122}$$

Let $\hat{\Pi}_N^{2,0}$ be the operator defined by (7.121)*. For any $u \in H^2_0(\mathbb{R}_+)$, we have that for all $u_N \in \widehat{X}_N$,*

$$\big((u - \hat{\Pi}_N^{2,0} u)'', u_N''\big) + \frac{1}{2}\big((u - \hat{\Pi}_N^{2,0} u)', u_N'\big) + \frac{1}{16}\big(u - \hat{\Pi}_N^{2,0} u, u_N\big) = 0. \tag{7.123}$$

Moreover, if $u \in H^2_0(\mathbb{R}_+)$ and $\hat{\partial}_x^2 u \in \hat{B}_0^{m-2}(\mathbb{R}_+)$ with $2 \le m \le N+1$, then we have

$$\|\hat{\Pi}_N^{2,0} u - u\|_{2,\hat{\omega}} \le c\sqrt{\frac{(N-m+1)!}{(N-1)!}}\|\hat{\partial}_x^m u\|_{\hat{\omega}_{m-2}}. \tag{7.124}$$

We now consider approximations by Hermite polynomials/functions. To this end, let $\omega = e^{-x^2}$ be the Hermite weight function as before. Consider the L^2_ω-orthogonal projection $\Pi_N : L^2_\omega(\mathbb{R}) \to P_N$, defined by

$$(u - \Pi_N u, v_N)_\omega = 0, \quad \forall v_N \in P_N. \tag{7.125}$$

It is clear that

$$\Pi_N u(x) = \sum_{n=0}^{N} \hat{u}_n H_n(x) \text{ with } \hat{u}_n = \frac{1}{\gamma_n} \int_{\mathbb{R}} u(x) H_n(x) \omega(x) dx,$$

where γ_n is given by (7.57).

Using (7.67) repeatedly leads to

$$\partial_x^k H_n(x) = \frac{2^k n!}{(n-k)!} H_{n-k}(x), \quad n \geq k, \tag{7.126}$$

which implies that $\{H_n\}$ are orthogonal with respect to the inner product of the Sobolev space $H^m_\omega(\mathbb{R})$. Thanks to (7.126), we can derive the approximation results by following the same argument as for Theorem 3.35.

Theorem 7.13. *For any $u \in H^m_\omega(\mathbb{R})$ with $0 \leq m \leq N+1$,*

$$\|\partial_x^l(\Pi_N u - u)\|_\omega \leq 2^{(l-m)/2} \sqrt{\frac{(N-m+1)!}{(N-l+1)!}} \|\partial_x^m u\|_\omega, \quad 0 \leq l \leq m. \tag{7.127}$$

This result shows that the convergence order of the L^2_ω-orthogonal projection is simultaneously optimal in the H^l_ω-norm with $l \geq 1$.

We next consider the extension of Theorem 7.13 to the Hermite function approximations. Notice that for any $u \in L^2(\mathbb{R})$, we have $ue^{x^2/2} \in L^2_\omega(\mathbb{R})$. Define

$$\hat{\Pi}_N u := e^{-x^2/2} \Pi_N (ue^{x^2/2}) \in \widehat{P}_N, \tag{7.128}$$

which satisfies

$$(u - \hat{\Pi}_N u, v_N) = \left(ue^{x^2/2} - \Pi_N(ue^{x^2/2}), v_N e^{x^2/2}\right)_\omega = 0, \quad \forall v_N \in \widehat{P}_N. \tag{7.129}$$

The following result is a direct consequence of Theorem 7.13.

Corollary 7.3. *Let $\hat{\partial}_x = \partial_x + x$. For any $\hat{\partial}_x^m u \in L^2(\mathbb{R})$ with $0 \leq m \leq N+1$,*

$$\|\hat{\partial}_x^l(\hat{\Pi}_N u - u)\| \leq c 2^{(l-m)/2} \sqrt{\frac{(N-m+1)!}{(N-l+1)!}} \|\hat{\partial}_x^m u\|, \quad 0 \leq l \leq m, \tag{7.130}$$

where c is a positive constant independent of m, N and u.

The above result does not imply the estimate $\|\partial_x^l(\hat{\Pi}_N u - u)\|$, which is concluded in the following theorem.

Theorem 7.14. *Let $\hat{\partial}_x = \partial_x + x$. If $\hat{\partial}_x^m u \in L^2(\mathbb{R})$ and $2 \le m \le N+1$, then*

$$\|\partial_x^l(\hat{\Pi}_N u - u)\| \le c\sqrt{\frac{(N-m+1)!}{2^m(N-l+1)!}}\|\hat{\partial}_x^m u\|, \quad l = 0,1,2, \tag{7.131}$$

where c is a positive constant independent of m, N and u.

Proof. The estimate (7.131) with $l = 0$ follows from Corollary 7.3 with $l = 0$.

For $l = 1$, we find

$$\begin{aligned}\partial_x(\hat{\Pi}_N u - u) &= e^{-x^2/2}\partial_x\Big(\Pi_N(e^{x^2/2}u) - (e^{x^2/2}u)\Big)\\ &\quad - xe^{-x^2/2}\Big(\Pi_N(e^{x^2/2}u) - (e^{x^2/2}u)\Big).\end{aligned}$$

Hence, by (B.36b) and Theorem 7.13,

$$\begin{aligned}\|\partial_x(\hat{\Pi}_N u - u)\| &\le |\Pi_N(e^{x^2/2}u) - (e^{x^2/2}u)|_{1,\omega} + \|x(\Pi_N(e^{x^2/2}u) - (e^{x^2/2}u))\|_\omega\\ &\le c\|\Pi_N(e^{x^2/2}u) - (e^{x^2/2}u)\|_{1,\omega} \le c\sqrt{\frac{(N-m+1)!}{2^m N!}}\|\partial_x^m(e^{x^2/2}u)\|_\omega.\end{aligned}$$

The case with $l = 2$ can be proved in the same fashion. □

Remark 7.4. *Like Remark 7.2, we have that for fixed m, the order of convergence is $O(N^{(l-m)/2})$.*

Remark 7.5. *As in the Laguerre case, the eigenvalues of the Sturm-Liouville problem associated with the Hermite polynomials also grows linearly, so the convergence rate of the Hermite approximation is similar to that of the Laguerre approximation. Consider, for example, $u(x) = (1+x^2)^{-h}$ and $u(x) = \frac{\sin kx}{(1+x^2)^h}$ with $h > 0$. It can be checked that for both functions $\|\hat{\partial}_x^m u\| < \infty$ if $m < 2h - 1/2$, which implies*

$$\|\hat{\Pi}_N u - u\| \lesssim N^{-(h-1/4)}. \tag{7.132}$$

7.3.3 Interpolations

Let $\{x_j^{(\alpha)}\}_{j=0}^N$ be the Laguerre-Gauss points given in Theorem 7.1, and denote by $I_N^{(\alpha)}$ the corresponding polynomial interpolation operator. Its approximation property is stated below, and the proof can be found in Guo et al. (2006b).

Theorem 7.15. *Let $\alpha > -1$. If $u \in C(\mathbb{R}_+) \cap B^m_\alpha(\mathbb{R}_+)$ and $\partial_x u \in B^{m-1}_\alpha(\mathbb{R}_+)$ with $1 \le m \le N+1$, then*

$$\|I_N^{(\alpha)} u - u\|_{\omega_\alpha} \le c\sqrt{\frac{(N-m+1)!}{N!}}\left(\|\partial_x^m u\|_{\omega_{\alpha+m-1}} + (\ln N)^{1/2}\|\partial_x^m u\|_{\omega_{\alpha+m}}\right), \quad (7.133)$$

where c is a positive constant independent of m, N and u.

We also refer to Guo et al. (2006b) for a similar estimate for the Laguerre-Gauss-Radau interpolation.

Remark 7.6. *Compared with Theorem 7.8 (with $l = 0$), the estimate for the interpolation has an extra $(\ln N)^{1/2}$ term. This results improve previous estimates in* Maday et al. (1985), Mastroianni and Occorsio (2001a) and Xu and Guo (2002). *Note however that in* Maday et al. (1985) (*also see* Bernardi and Maday (1997)), *the following estimate was derived for the case $\alpha = 0$,*

$$\|I_N^{(0)} u - u\|_{\omega_0} \lesssim N^{(1-m)/2}\|u\|_{H^m_{\omega_\tau}}, \quad (7.134)$$

where the weight function $\omega_\tau(x) = e^{-(1-\tau)x}$ with $0 < \tau < 1$. Mastroianni and Occorsio (2001a) *studied the generalized Laguerre-Gauss interpolation* (see Formula (3.8) of Mastroianni and Occorsio (2001a)) *and showed that*

$$\|x^\gamma e^{-x/2}(I_N^{(\alpha)} u - u)\|_\infty \lesssim N^{-m/2}\ln N\|x^{m/2+\gamma}e^{-x/2}\partial_x^m u\|_\infty, \quad (7.135)$$

for fixed $m \ge 1$, $\alpha > -1$ and some $\gamma \ge 0$ satisfying

$$2\gamma - \frac{5}{2} \le \alpha \le 2\gamma - \frac{1}{2}.$$

In Xu and Guo (2002), *the usual Laguerre interpolation was analyzed in the weighted Sobolev space, and the main result is*

$$\|I_N^{(0)} u - u\|_{\omega_0} \lesssim N^{(1-m)/2+\varepsilon}\|u\|_{H^m_{\omega_m}}, \quad m \ge 1, \quad 0 < \varepsilon \le 1/2. \quad (7.136)$$

This result was improved in Guo et al. (2006b) *with $\ln N$ in place of N^ε.*

It is straightforward to extend the interpolation error estimate (7.133) to the modified Laguerre-Gauss-type interpolation $\hat{I}_N^{(\alpha)}$, associated with the quadrature in Theorem 7.2. Observe that

$$(\hat{I}_N^{(\alpha)} u)(x) = e^{-x/2} I_N^{(\alpha)}(u e^{x/2}) \in \widehat{P}_N,$$

so we derive immediately from Theorem 7.15 the following result.

Theorem 7.16. *Let $\alpha > -1$. If $u \in C(\mathbb{R}_+) \cap \widehat{B}_\alpha^m(\mathbb{R}_+)$ and $\hat{\partial}_x u \in \widehat{B}_\alpha^{m-1}(\mathbb{R}_+)$ with $1 \le m \le N+1$, then*

$$\|\hat{I}_N^{(\alpha)} u - u\|_{\hat{\omega}_\alpha} \le c\sqrt{\frac{(N-m+1)!}{N!}}\left(\|\hat{\partial}_x^m u\|_{\hat{\omega}_{\alpha+m-1}} + (\ln N)^{1/2}\|\hat{\partial}_x^m u\|_{\hat{\omega}_{\alpha+m}}\right), \quad (7.137)$$

where c is a positive constant independent of m, N and u.

Now, we turn to the interpolation error estimates associated with the Hermite-Gauss quadrature in Theorem 7.3. Let $I_N^h : C(\mathbb{R}) \to P_N$ be the interpolation operator associated with the Hermite-Gauss points. By combing Theorem 7.13 and the results in Guo and Xu (2000) and Aguirre and Rivas (2005), we can prove the following result, which is just a more concise form of Theorem 2.1 in Guo and Xu (2000) (see also Aguirre and Rivas (2005)):

Theorem 7.17. *For $u \in C(\mathbb{R}) \cap H_\omega^m(\mathbb{R})$ with $m \ge 1$, we have*

$$\|\partial_x^l (I_N^h u - u)\|_\omega \lesssim N^{\frac{1}{6}+\frac{l-m}{2}} \|\partial_x^m u\|_\omega, \quad 0 \le l \le m. \quad (7.138)$$

In the above, m is assumed to be a fixed integer. This result improved that in Guo and Xu (2000) (with a convergence order $N^{\frac{1}{3}+\frac{l-m}{2}}$).

Finally, we consider the interpolation associated with the Hermite functions in (7.90). Note that $(\hat{I}_N^h u) = e^{-x^2/2} I_N^h (u e^{x^2/2})$. The following estimate follows from Theorem 7.17.

Theorem 7.18. *Let $\hat{\partial}_x = \partial_x + x$. For $u \in C(\mathbb{R})$ and $\hat{\partial}_x^m u \in L^2(\mathbb{R})$ with fixed $m \ge 1$, we have*

$$\|\hat{\partial}_x^l (\hat{I}_N^h u - u)\| \lesssim N^{\frac{1}{6}+\frac{l-m}{2}} \|\hat{\partial}_x^m u\|, \quad 0 \le l \le m. \quad (7.139)$$

7.4 Spectral Methods Using Laguerre and Hermite Functions

In this section, we consider spectral-Galerkin methods using Laguerre and Hermite functions. An advantage of using Laguerre functions is that they are mutually orthogonal in the usual (non-weighted) L^2-space, so we can work with the usual (i.e. non-weighted) variational formulation.

7.4.1 Laguerre-Galerkin Method

Consider the model equation:

$$-u_{xx} + \gamma u = f, \quad x \in \mathbb{R}_+, \ \gamma > 0; \quad u(0) = 0, \ \lim_{x \to +\infty} u(x) = 0. \quad (7.140)$$

Let $H_0^1(\mathbb{R}_+)$ and $\widehat{P}_N^0$ be the spaces as defined before. Then, a weak formulation for (7.140) is

$$\begin{cases} \text{Find } u \in H_0^1(\mathbb{R}_+) \text{ such that} \\ a(u,v) := (u',v') + \gamma(u,v) = (f,v), \quad \forall v \in H_0^1(\mathbb{R}_+), \end{cases} \tag{7.141}$$

for $f \in \big(H_0^1(\mathbb{R}_+)\big)'$. Note that $u \in H_0^1(\mathbb{R}_+)$ implies $\lim_{x\to\infty} u(x) = 0$. It is clear that for $\gamma > 0$, the problem admits a unique solution, since

$$a(u,u) = |u|_1^2 + \gamma\|u\|^2 \geq \min(1,\gamma)\|u\|_1^2, \quad \forall u \in H_0^1(\mathbb{R}_+). \tag{7.142}$$

The Laguerre spectral-Galerkin approximation to (7.140) is

$$\begin{cases} \text{Find } u_N \in \widehat{P}_N^0 \text{ such that} \\ a(u_N,v_N) = (\hat{I}_N f, v_N), \quad \forall v_N \in \widehat{P}_N^0, \end{cases} \tag{7.143}$$

where $\hat{I}_N$ is the Laguerre-Gauss-Radau interpolation operator. Thanks to (7.142), the unique approximate solution u_N satisfies $\|u_N\|_1 \lesssim \|\hat{I}_N f\|$.

Defining

$$\hat{\phi}_k(x) = \big(\mathscr{L}_k(x) - \mathscr{L}_{k+1}(x)\big)e^{-x/2} = \widehat{\mathscr{L}_k}(x) - \widehat{\mathscr{L}_{k+1}}(x), \tag{7.144}$$

one verifies that

$$\widehat{P}_N^0 = \text{span}\big\{\hat{\phi}_0, \hat{\phi}_1, \ldots, \hat{\phi}_{N-1}\big\}. \tag{7.145}$$

Hence, by setting

$$\begin{aligned} &u_N = \sum_{k=0}^{N-1} \hat{u}_k \hat{\phi}_k, \quad \mathbf{u} = \big(\hat{u}_0, \hat{u}_1, \ldots, \hat{u}_{N-1}\big)^T; \\ &f_j = (\hat{I}_N f, \hat{\phi}_j), \quad \mathbf{f} = \big(f_0, f_1, \ldots, f_{N-1}\big)^T; \\ &s_{jk} = (\hat{\phi}_k', \hat{\phi}_j'), \ S = \big(s_{jk}\big)_{0\le j,k\le N-1}, \ c_{jk} = (\hat{\phi}_k, \hat{\phi}_j), \ C = \big(c_{jk}\big)_{0\le j,k\le N-1}, \end{aligned}$$

we find that C is a symmetric tridiagonal matrix and $S = I - \frac{1}{4}C$, so the system (7.143) reduces to the linear system

$$\Big(I + (\gamma - \frac{1}{4})C\Big)\mathbf{u} = \mathbf{f}. \tag{7.146}$$

Note that the entries of the coefficient matrix can be evaluated exactly, and this system is easy to invert.

Using the approximation results established in the previous section, we can derive the following convergence result.

Theorem 7.19. *Let $\gamma > 0$. If $u \in H_0^1(\mathbb{R}_+)$, $\hat{\partial}_x u \in \widehat{B}_0^{m-1}(\mathbb{R}_+)$, $f \in C(\bar{\mathbb{R}}_+) \cap \widehat{B}_0^k(\mathbb{R}_+)$ and $\hat{\partial}_x f \in \widehat{B}_0^{k-1}(\mathbb{R}_+)$ with $1 \le k, m \le N+1$, then we have*

$$\begin{aligned} \|u - u_N\|_1 \le\, & c\sqrt{\frac{(N-m+1)!}{N!}}\|\hat{\partial}_x^m u\|_{\hat{\omega}_{m-1}} \\ & + c\sqrt{\frac{(N-k+1)!}{N!}}\big(\|\hat{\partial}_x^k f\|_{\hat{\omega}_{k-1}} + (\ln N)^{1/2}\|\hat{\partial}_x^k f\|_{\hat{\omega}_k}\big), \end{aligned} \tag{7.147}$$

where $\hat{\omega}_{m-1} = x^{m-1}$, and c is a positive constant independent of m, k, N, u and f.

Proof. Let $\hat{\Pi}_N^{1,0}$ be the orthogonal projection operator defined in Theorem 7.11. Let $e_N = u_N - \hat{\Pi}_N^{1,0} u$ and $\tilde{e}_N = u - \hat{\Pi}_N^{1,0} u$. Then by (7.141)-(7.143),

$$a(u_N - u, v_N) = (\hat{I}_N f - f, v_N), \quad \forall v_N \in \widehat{P}_N^0,$$

which implies

$$a(e_N, v_N) = a(\tilde{e}_N, v_N) + (\hat{I}_N f - f, v_N), \quad \forall v_N \in \widehat{P}_N^0.$$

Taking $v_N = e_N$ in the above, we find

$$\|e_N\|_1 \le c\big(\|\tilde{e}_N\|_1 + \|\hat{I}_N f - f\|\big).$$

Then, the desired estimate follows from Theorems 7.11 and 7.16 (with $\alpha = 0$) and the triangle inequality. □

7.4.2 Hermite-Galerkin Method

As an example, we consider the following model problem:

$$-u_{xx} + \gamma u = f, \;\; x \in \mathbb{R}, \;\; \gamma > 0; \quad \lim_{|x| \to \infty} u(x) = 0. \tag{7.148}$$

A weak formulation for (7.148) is

$$\begin{cases} \text{Find } u \in H^1(\mathbb{R}) \text{ such that} \\ (\partial_x u, \partial_x v) + \gamma(u, v) = (f, v), \quad \forall v \in H^1(\mathbb{R}), \end{cases} \tag{7.149}$$

for given $f \in \big(H^1(\mathbb{R})\big)'$. Notice that the decay of u at infinity is incorporated into the space $H^1(\mathbb{R})$.

The Hermite-Galerkin method for (7.149) is

$$\begin{cases} \text{Find } u_N \in \widehat{P}_N \text{ such that} \\ (\partial_x u_N, \partial_x v_N) + \gamma(u_N, v_N) = (\hat{I}_N^h f, v_N), \quad \forall v_N \in \widehat{P}_N, \end{cases} \tag{7.150}$$

where $\hat{I}_N^h$ is the (modified) Hermite-Gauss interpolation operator.

In view of (7.76), the system (7.150) has a similar structure as (7.146).

By using a standard argument as for the scheme (7.143), the following error estimate is a straightforward consequence of Theorems 7.14 and 7.18.

Theorem 7.20. *Let* $\gamma > 0$ *and* $\hat{\partial}_x = \partial_x + x$. *If* $u \in H^1(\mathbb{R})$ *with* $\hat{\partial}_x^m u \in L^2(\mathbb{R})$, *and* $f \in C(\mathbb{R})$ *with* $\hat{\partial}_x^k f \in L^2(\mathbb{R})$ *and fixed* $k, m \geq 1$, *then we have*

$$\|u_N - u\|_1 \lesssim N^{\frac{1-m}{2}} \|\hat{\partial}_x^m u\| + N^{\frac{1}{6}-\frac{k}{2}} \|\hat{\partial}_x^k f\|. \tag{7.151}$$

7.4.3 Numerical Results and Discussions

Now, we present some numerical results to illustrate the convergence behavior of the proposed schemes. We consider (7.140) and (7.148) with three sets of exact solutions having different decay properties.

Set 1. Exponential decay with oscillation at infinity

$$u(x) = e^{-x} \sin kx \text{ for } x \in (0, \infty) \text{ or } u(x) = e^{-x^2} \sin kx \text{ for } x \in (-\infty, \infty). \tag{7.152}$$

Set 2. Algebraic decay without oscillation at infinity

$$u(x) = (1+x)^{-h} \text{ for } x \in (0, \infty) \text{ or } u(x) = (1+x^2)^{-h} \text{ for } x \in (-\infty, \infty). \tag{7.153}$$

Set 3. Algebraic decay with oscillation at infinity

$$u(x) = \frac{\sin kx}{(1+x)^h} \text{ for } x \in (0, \infty) \text{ or } u(x) = \frac{\sin kx}{(1+x^2)^h} \text{ for } x \in (-\infty, \infty). \tag{7.154}$$

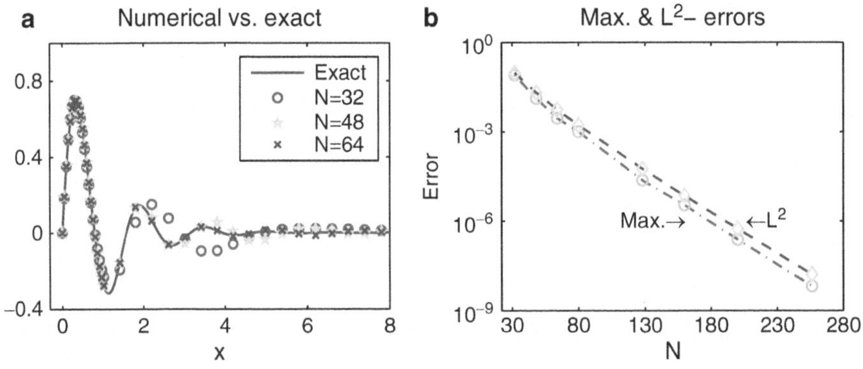

Fig. 7.5 Example 1: (**a**) Exact solution against the numerical solutions obtained by (7.146) with $N = 32, 48, 64$; (**b**) Maximum and L^2-errors vs. various N

In Fig. 7.5a, we plot the exact solution (7.152) with $k = 4$ against the numerical solution obtained by the scheme (7.143) with $\gamma = 1$ and different numbers of modes N, and in Fig. 7.5b, we depict the rate of convergence. A geometric convergence rate (i.e. $\exp(-cN)$ with $c > 0$) is observed, which is consistent with the estimate in (7.147).

In Shen (2000), numerical results are reported for the scheme (7.143) using the functions in (7.153)-(7.154) as exact solutions. Sub-geometric convergence of order $\exp(-c\sqrt{N})$ for (7.153) are observed (cf. Fig. 3.2 in Shen (2000)), while a convergence rate consistent with the estimate in (7.147) and (7.114) is observed for (7.154). The sub-geometric convergence for (7.153) was puzzling since the error estimate in (7.114) only predicts a rate of order about N^{-h}. In order to explain this surprising disagreement, we performed additional tests with different h and with N much larger than what was used in Shen (2000). The numerical results are reported in Fig. 7.6. On the left, we plot the results with $h = 3$ and 4.5 for N up to 128, and we observe again the sub-geometric convergence rate as reported in Shen (2000). However, when we increased N further, the convergence rates eventually became algebraic. This indicates that the sub-geometric convergence reported in Shen (2000) was still in the pre-asymptotic range. To illustrate this behavior, we plot the results with $h = 1.5$ and 2 (so the asymptotic range can be reached faster) for N up to 256 on the right of Fig. 7.6. It is clear that after a pre-asymptotic range, the convergence rates settle down to the algebraic rates consistent with (7.147) and (7.114).

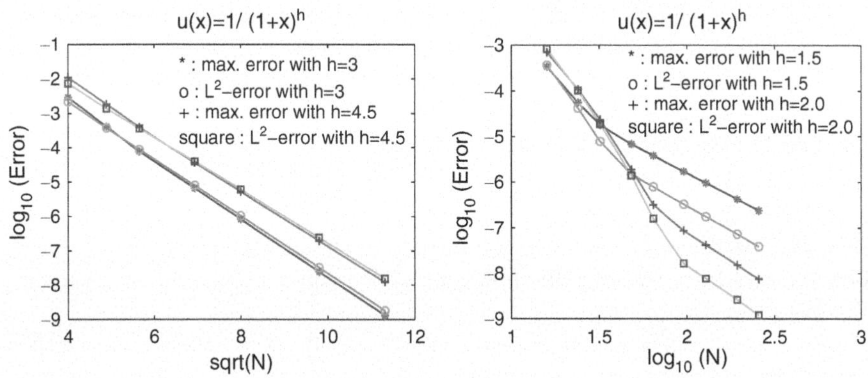

Fig. 7.6 Convergence rates of the scheme (7.143)

We now present numerical results using the scheme (7.150) with the exact solutions (7.152)-(7.154). On the left of Fig. 7.7, we observe a geometric convergence for (7.152). For (7.153), we observe essentially the same behavior as in the Laguerre case (cf. the right of Fig. 7.6), i.e., there is a pre-asymptotic range where one observes a sub-geometric convergence, but after the pre-asymptotic range, the convergence rates become algebraic as predicted in (7.132) and (7.151) (cf. the right of Fig. 7.7).

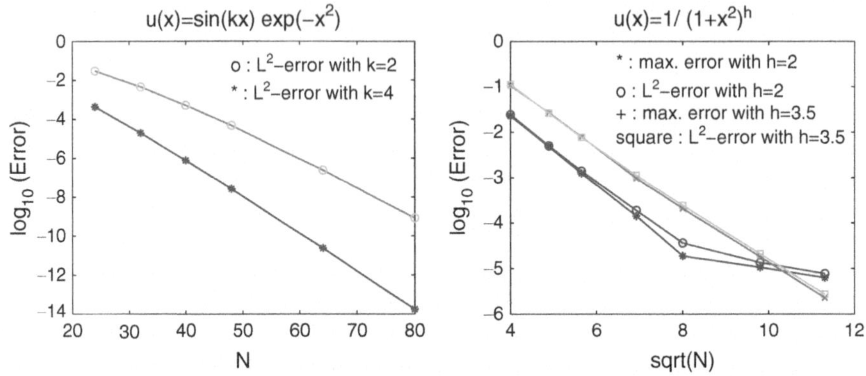

Fig. 7.7 Convergence rates of the scheme (7.150)

7.4.4 Scaling Factor

For a problem with solution decaying at infinity, there is an effective interval outside of which the solution is negligible, and collocation points which fall outside of this interval are essentially wasted. On the other hand, if the solution is still far from negligible at the collocation point(s) with largest magnitude, one can not expect a very good approximation. Hence, the performance of spectral methods in unbounded domains can be significantly enhanced by choosing a proper scaling parameter such that the extreme collocation points are at or close to the endpoints of the effective interval. For Laguerre and Hermite spectral methods, one usually needs to determine a suitable scaling parameter β and then make a coordinate transform $y = \beta x$ (cf. Tang (1993), Shen (2000), Guo et al. (2006b)).

To illustrate the idea, let us consider (7.140). Given an accuracy threshold ε, we estimate an M such that $|u(x)| \le \varepsilon$ for $x > M$. Then, we set the scaling factor $\beta_N = x_N^{(N)}/M$ where $x_N^{(N)}$ is the largest Laguerre-Gauss-Lobatto point. Now instead of solving (7.140), we solve the following scaled equation with the new variable $y = \beta_N x$:

$$-\beta_N^2 v_{yy} + \gamma v = g(y); \quad v(0) = 0, \ \lim_{y \to +\infty} v(y) = 0, \tag{7.155}$$

where $v(y) = u(\beta_N x)$ and $g(y) = f(\beta_N x)$. Thus, the effective collocation points $x_j = y_j/\beta_N$ (with $\{y_j\}_{j=0}^N$ being the Laguerre Gauss-Radau points) are all located in $[0, M]$.

An an illustrative example, we consider (7.140) with the exact solution: $u(x) = \sin(10x)/(1+x)^5$. In Fig. 7.8a, we plot the exact solution and the approximations without scaling using 128 points and with a scaling factor $= 15$ using 32 points. Notice from Fig. 7.8a that if no scaling is used, the approximation with $N = 128$ still exhibits an observable error, while the approximation with a scaling factor of 15 using only 32 modes is virtually indistinguishable from the exact solution. This simple example demonstrates that a proper scaling will greatly enhance the resolution capabilities of the Laguerre functions (cf. Fig. 7.8a). Similar ideas can be applied to the Hermite spectral approximations. In Ma et al. (2005), a Hermite spectral method with time-dependent scaling is proposed for parabolic problems.

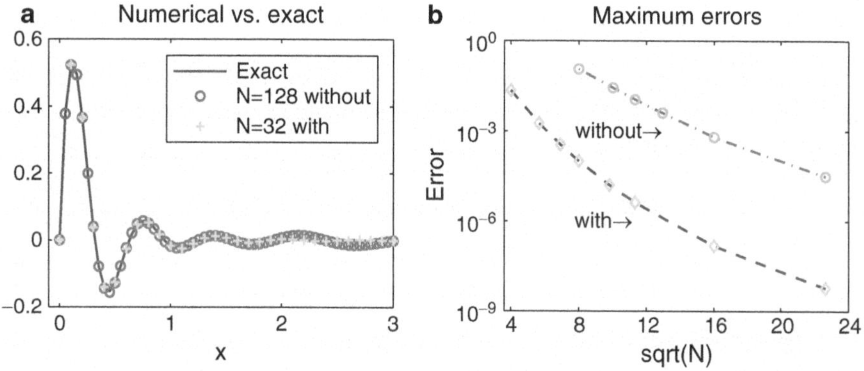

Fig. 7.8 Example 2: (**a**) Exact solution against numerical solutions obtained by the scheme (7.146) with $N = 128$ (without scaling), and by solving (7.155) with $N = 32$ and the scaling factor $\beta_N = 15$; (**b**) Maximum errors of two approaches vs. $\sqrt{N}$ (note: $N \in [16, 512]$)

7.5 Mapped Spectral Methods and Rational Approximations

A common and effective strategy in dealing with unbounded domains is to use a suitable mapping that transforms an infinite domain to a finite domain. Then, images of classical orthogonal polynomials under the inverse mapping will form a set of orthogonal basis functions which can be used to approximate solutions of PDEs in the infinite domains. Early practitioners of this approach include Grosch and Orszag (1977) and Boyd (1982). The book by Boyd (2001) contains an extensive review on many practical aspects of the mapped spectral methods. In the last couple of years, a series of papers has been devoted to the convergence analysis of the mapped spectral methods (see Shen and Wang (2009) and the references therein).

The purpose of this section is to present a general framework for the analysis and implementation of the mapped spectral methods.

7.5.1 Mappings

Consider a family of mappings of the form:

$$x = g(y; s), \quad s > 0, \quad y \in I := (-1, 1), \quad x \in \Lambda := (0, +\infty) \text{ or } (-\infty, +\infty), \qquad (7.156)$$

such that

$$\begin{aligned}&\frac{dx}{dy}=g'(y;s)>0, \quad s>0,\ y\in I,\\&g(-1;s)=0,\ g(1;s)=+\infty, \quad \text{if } \Lambda=(0,+\infty),\\&g(\pm 1;s)=\pm\infty, \quad \text{if } \Lambda=(-\infty,+\infty).\end{aligned} \tag{7.157}$$

In this one-to-one transform, the parameter s is a positive scaling factor. Without loss of generality, we further assume that the mapping is explicitly invertible, and denote its inverse mapping by

$$y=g^{-1}(x;s):=h(x;s), \quad x\in\Lambda,\ y\in I,\ s>0. \tag{7.158}$$

Several typical mappings that have been proposed and used in practice are of the above type (see, e.g., Boyd (2001) and the references therein):

(i) Mappings between $x\in\Lambda=(-\infty,+\infty)$ and $y\in I=(-1,1)$ with $s>0$:

- Algebraic mapping:

$$x=\frac{sy}{\sqrt{1-y^2}}, \quad y=\frac{x}{\sqrt{x^2+s^2}}. \tag{7.159}$$

- Logarithmic mapping:

$$x=s\,\mathrm{arctanh}(y)=\frac{s}{2}\ln\frac{1+y}{1-y}, \quad y=\tanh(s^{-1}x). \tag{7.160}$$

- Exponential mapping:

$$x=\sinh(sy),\ y=\frac{1}{s}\ln\left(x+\sqrt{x^2+1}\right),\ y\in(-1,1),\ x\in(-L_s,L_s), \tag{7.161}$$

where $L_s=\sinh(s)$.

(ii) Mappings between $x\in\Lambda=(0,+\infty)$ and $y\in I=(-1,1)$ with $s>0$:

- Algebraic mapping:

$$x=\frac{s(1+y)}{1-y}, \quad y=\frac{x-s}{x+s}. \tag{7.162}$$

- Logarithmic mapping:

$$x=s\,\mathrm{arctanh}\left(\frac{y+1}{2}\right)=\frac{s}{2}\ln\frac{3+y}{1-y}, \quad y=1-2\tanh(s^{-1}x). \tag{7.163}$$

- Exponential mapping:

$$x=\sinh\left(\frac{s}{2}(1+y)\right), \quad y=\frac{2}{s}\ln\left(x+\sqrt{x^2+1}\right)-1, \tag{7.164}$$

where $y\in(-1,1)$ and $x\in(0,L_s)$ with $L_s=\sinh(s)$.

The special feature which distinguishes these mappings is that, as $|y| \to \pm 1$, x varies algebraically, logarithmically or exponentially for algebraic, logarithmic or exponential mappings, respectively. The parameter s is a scaling/stretching factor which can be used to tune the spacing of collocation points. We also notice that the image of the exponential mappings (7.161) and (7.164) is a finite interval, so they combine both mapping and domain truncation.

7.5.2 Approximation by Mapped Jacobi Polynomials

Given a mapping $x = g(y;s)$ satisfying (7.156)–(7.158) and a family of orthogonal polynomials $\{p_k(y)\}$ with $y \in I = (-1,1)$, $\{p_k(h(x;s))\}$ forms a new family of orthogonal functions in $\Lambda = (0,\infty)$ or $(-\infty,\infty)$. In particular, the algebraic mappings (7.159) or (7.162) with the Chebyshev or Legendre polynomials lead to orthogonal rational basis functions which have been studied in Boyd (1982), Christov (1982), Boyd (1987a), Liu et al. (1994) and Guo et al. (2000, 2002).

For the sake of generality, we consider the mapped Jacobi approximations. Let $J_k^{\alpha,\beta}(y)$ $(\alpha,\beta > -1)$ be the classical Jacobi polynomial of degree k as defined in Chap. 3. Define the mapped Jacobi polynomials as

$$j_{n,s}^{\alpha,\beta}(x) := J_n^{\alpha,\beta}(y) = J_n^{\alpha,\beta}(h(x;s)), \quad x \in \Lambda,\ y \in I. \tag{7.165}$$

We infer from (3.88) that (7.165) defines a new family of orthogonal functions $\{j_{n,s}^{\alpha,\beta}\}$ satisfying

$$\int_\Lambda j_{n,s}^{\alpha,\beta}(x)\, j_{m,s}^{\alpha,\beta}(x)\, \omega_s^{\alpha,\beta}(x)dx = \gamma_n^{\alpha,\beta}\delta_{mn}, \tag{7.166}$$

where the constant $\gamma_n^{\alpha,\beta}$ is given in (3.109), and the weight function

$$\omega_s^{\alpha,\beta}(x) = \omega^{\alpha,\beta}(y)\frac{dy}{dx} = \omega^{\alpha,\beta}(y)(g'(y;s))^{-1} > 0, \tag{7.167}$$

with $y = h(x;s)$ and $\omega^{\alpha,\beta}(y) = (1-y)^\alpha(1+y)^\beta$ being the Jacobi weight function.

In Fig. 7.9, we plot some samples of $j_{n,s}^{0,0}(x)$ for different s and n under the mappings (7.159) and (7.162).

We now present some approximation properties of these mapped Jacobi polynomials. Let us define the finite dimensional approximation space

$$V_{N,s}^{\alpha,\beta} = \text{span}\{j_{n,s}^{\alpha,\beta}(x):\ n = 0,1,\ldots,N\}, \quad s > 0, \tag{7.168}$$

and consider the orthogonal projection $\pi_{N,s}^{\alpha,\beta} : L^2_{\omega_s^{\alpha,\beta}}(\Lambda) \to V_{N,s}^{\alpha,\beta}$ such that

$$\left(\pi_{N,s}^{\alpha,\beta}u - u, v_N\right)_{\omega_s^{\alpha,\beta}} = 0, \quad \forall v_N \in V_{N,s}^{\alpha,\beta}. \tag{7.169}$$

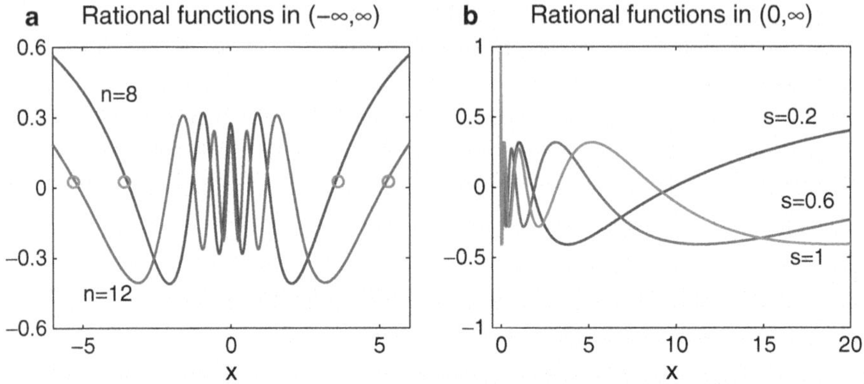

Fig. 7.9 (**a**) Graphs of $j_{n,1}^{0,0}(x)$ with $n = 8, 12$ under the mapping (7.159); (**b**) Graphs of $j_{8,s}^{0,0}(x)$ with $s = 0.2, 0.6, 1$ under the mapping (7.162)

Thanks to the orthogonality, we can write

$$(\pi_{N,s}^{\alpha,\beta} u)(x) = \sum_{n=0}^{N} \hat{u}_{n,s}^{\alpha,\beta} j_{n,s}^{\alpha,\beta}(x), \tag{7.170}$$

where

$$\hat{u}_{n,s}^{\alpha,\beta} = \frac{1}{\gamma_n^{\alpha,\beta}} \int_\Lambda u(x) j_{n,s}^{\alpha,\beta}(x) \omega_s^{\alpha,\beta}(x) dx.$$

We now introduce a weighted space which is particularly suitable to describe the L^2-projection errors. Given a mapping satisfying (7.156)-(7.158), we set

$$a_s(x) := \frac{dx}{dy} (> 0), \quad U_s(y) := u(x) = u(g(y;s)). \tag{7.171}$$

The key to express the error estimates in a concise form is to introduce a differential operator $D_x u := a_s \frac{du}{dx}$. One verifies readily that

$$\frac{dU_s}{dy} = a_s \frac{du}{dx} = D_x u, \quad \frac{d^2 U_s}{dy^2} = a_s \frac{d}{dx}\Big(a_s \frac{du}{dx}\Big) = D_x^2 u,$$

and an induction leads to

$$\frac{d^k U_s}{dy^k} = \underbrace{a_s \frac{d}{dx}\Big(a_s \frac{d}{dx}\Big(\cdots\cdots \Big(a_s \frac{du}{dx}\Big) \cdots\cdots \Big)}_{k-1 \text{ parentheses}} := D_x^k u. \tag{7.172}$$

Let us define

$$\widetilde{B}_{\alpha,\beta}^m(\Lambda) = \big\{ u : u \text{ is measurable in } \Lambda \text{ and } \|u\|_{\widetilde{B}_{\alpha,\beta}^m} < \infty \big\}, \tag{7.173}$$

equipped with the norm and semi-norm

$$\|u\|_{\widetilde{B}^m_{\alpha,\beta}} = \Big(\sum_{k=0}^{m} \|D^k_x u\|^2_{\omega_s^{\alpha+k,\beta+k}}\Big)^{1/2}, \quad |u|_{\widetilde{B}^m_{\alpha,\beta}} = \|D^m_x u\|_{\omega_s^{\alpha+m,\beta+m}},$$

where the weight function $\omega_s^{\alpha+k,\beta+k}$ is defined in (7.167). It turns out to be the mapped version of the anisotropic Jacobi-weighted Sobolev space in (3.251)–(3.252).

We have the following fundamental results for the mapped Jacobi approximations.

Theorem 7.21. *Let $\alpha,\beta > -1$. If $u \in \widetilde{B}^m_{\alpha,\beta}(\Lambda)$, we have that for $0 \le m \le N+1$,*

$$\|\pi^{\alpha,\beta}_{N,s} u - u\|_{\omega_s^{\alpha,\beta}} \le c\sqrt{\frac{(N-m+1)!}{(N+1)!}}(N+m)^{-m/2}\|D^m_x u\|_{\omega_s^{\alpha+m,\beta+m}}, \tag{7.174}$$

and for $1 \le m \le N+1$,

$$\begin{aligned}&\|\partial_x(\pi^{\alpha,\beta}_{N,s} u - u)\|_{\tilde{\omega}_s^{\alpha,\beta}}\\ &\quad \le c\sqrt{\frac{(N-m+1)!}{N!}}(N+m)^{(1-m)/2}\|D^m_x u\|_{\omega_s^{\alpha+m,\beta+m}},\end{aligned} \tag{7.175}$$

where c is a positive constant independent of m, N and u, and

$$\tilde{\omega}_s^{\alpha,\beta}(x) = \omega^{\alpha+1,\beta+1}(y) g'(y;s), \quad y = h(x;s).$$

Proof. Let $U_s(y) = u(x) = u(h(y;s))$, whose Jacobi expansion is

$$U_s(y) = \sum_{n=0}^{\infty} \widehat{U}^{\alpha,\beta}_{n,s} J^{\alpha,\beta}_n(y).$$

Then, by the definition (7.165), we have the relation between the coefficients of the Jacobi and mapped Jacobi expansions:

$$\hat{u}^{\alpha,\beta}_{n,s} = \frac{1}{\gamma^{\alpha,\beta}_n}(u, j^{\alpha,\beta}_{n,s})_{\omega_s^{\alpha,\beta}} = \frac{1}{\gamma^{\alpha,\beta}_n}(U_s, J^{\alpha,\beta}_n)_{\omega^{\alpha,\beta}} = \widehat{U}^{\alpha,\beta}_{n,s}. \tag{7.176}$$

Let $\hat{\pi}^{\alpha,\beta}_N$ be the $L^2_{\omega^{\alpha,\beta}}$-orthogonal projection operator associated with the Jacobi polynomials (see (3.249), and here we put '^' on the original notation for clarity). By (3.88), (7.166) and Theorem 3.35,

$$\begin{aligned}\|\pi^{\alpha,\beta}_{N,s} u - u\|^2_{\omega_s^{\alpha,\beta}} &= \sum_{n=N+1}^{\infty} (\hat{u}^{\alpha,\beta}_{n,s})^2 \gamma^{\alpha,\beta}_n = \sum_{n=N+1}^{\infty} (\widehat{U}^{\alpha,\beta}_{n,s})^2 \gamma^{\alpha,\beta}_n\\ &= \|\hat{\pi}^{\alpha,\beta}_N U_s - U_s\|^2_{\omega^{\alpha,\beta}} \le c\frac{(N-m+1)!}{(N+1)!}(N+m)^{-m}\|\partial^m_y U_s\|^2_{\omega^{\alpha+m,\beta+m}}\\ &\le c\frac{(N-m+1)!}{(N+1)!}(N+m)^{-m}\|D^m_x u\|^2_{\omega_s^{\alpha+m,\beta+m}}.\end{aligned} \tag{7.177}$$

Next, we deduce from (7.165) and the orthogonality of $\{\partial_y J_n^{\alpha,\beta}\}$ that $\{\partial_x j_{n,s}^{\alpha,\beta}\}$ is orthogonal with respect to $\tilde{\omega}_s^{\alpha,\beta}$, and

$$\|\partial_x j_{n,s}^{\alpha,\beta}\|^2_{\tilde{\omega}_s^{\alpha,\beta}} = \|\partial_y J_n^{\alpha,\beta}\|^2_{\omega^{\alpha+1,\beta+1}} = \lambda_n^{\alpha,\beta}\gamma_n^{\alpha,\beta},$$

where $\lambda_n^{\alpha,\beta}$ is the eigenvalue of the Jacobi Sturm-Liouville problem (cf. (3.91)). Therefore, by (7.176) and Theorem 3.35,

$$\begin{aligned}\|\partial_x(\pi_{N,s}^{\alpha,\beta}u-u)\|^2_{\tilde{\omega}_s^{\alpha,\beta}} &= \sum_{n=N+1}^{\infty}\lambda_n^{\alpha,\beta}\gamma_n^{\alpha,\beta}(\hat{u}_{n,s}^{\alpha,\beta})^2 = \sum_{n=N+1}^{\infty}\lambda_n^{\alpha,\beta}\gamma_n^{\alpha,\beta}(\widehat{U}_{n,s}^{\alpha,\beta})^2\\ &= \|\partial_y(\hat{\pi}_N^{\alpha,\beta}U_s-U_s)\|^2_{\omega^{\alpha+1,\beta+1}}\\ &\le c\frac{(N-m+1)!}{N!}(N+m)^{1-m}\|\partial_y^m U_s\|^2_{\omega^{\alpha+m,\beta+m}}\\ &\le c\frac{(N-m+1)!}{N!}(N+m)^{1-m}\|D_x^m u\|^2_{\omega_s^{\alpha+m,\beta+m}}.\end{aligned}$$

This ends the proof. □

Remark 7.7. *It should be pointed out that under the above general settings, the approximation results on the higher-order projections, such as the* $H^1_{\omega_s^{\alpha,\beta}}$*-orthogonal projection* $\pi_{N,s}^{1,\alpha,\beta}: H^1_{\omega_s^{\alpha,\beta}}(\Lambda)\to V_{N,s}^{\alpha,\beta}$ *can be established by using the existing Jacobi approximations results in Chap.* 3 *and a similar argument as above.*

In particular, applying the above results with $\alpha=\beta=0,-1/2$ *to the algebraic mappings* (7.159) *and* (7.162) *leads to more concise and in some cases improved Chebyshev and Legendre rational approximation results which were developed separately in* Guo et al. (2000, 2002) and Wang and Guo (2002).

The error estimates in the above theorem look very similar to the usual spectral error estimates in a finite interval (cf. Theorem 3.35). First of all, it is clear from the above theorem that the projection error converges faster than any algebraic rate if a function decays exponentially fast at infinity. For a function with singularities inside the domain, the above theorem and Theorem 3.35 lead to the same order of convergence, assuming that the function decays sufficiently fast at infinity. However, for a given smooth function, they may lead to very different convergence rates due to the difference in the norms used to measure the regularity.

We now determine the convergence rates for three sets of functions (7.152)–(7.154) with typical decay properties. We first consider the mapping (7.162). Then,

$$D_x=\Big(\frac{dy}{dx}\Big)^{-1}\frac{d}{dx}=\frac{(x+s)^2}{2s}\frac{d}{dx},\quad \omega_s^{k,l}(x)=\Big(\frac{2s}{x+s}\Big)^k\Big(\frac{2x}{x+s}\Big)^l\frac{2s}{(x+s)^2}.$$

Hence, for $u(x)=(1+x)^{-h}$, it can be easily checked that $\|D_x^m u\|_{\omega_s^{\alpha+m,\beta+m}}<\infty$ if $m<2h+\alpha+1$, which implies that

$$\|u - \pi_{N,s}^{\alpha,\beta} u\|_{\omega_s^{\alpha,\beta}} \lesssim N^{-(2h+\alpha+1)} \qquad (u(x) = (1+x)^{-h}). \tag{7.178}$$

On the other hand, for $u(x) = \frac{\sin kx}{(1+x)^h}$, it can also be easily checked that $\|D_x^m u\|_{\omega_s^{\alpha+m,\beta+m}} < \infty$ if $m < \frac{2h+\alpha+1}{3}$, which implies that

$$\|u - \pi_{N,s}^{\alpha,\beta} u\|_{\omega_s^{\alpha,\beta}} \lesssim N^{-(2h+\alpha+1)/3} \qquad \left(u(x) = \frac{\sin kx}{(1+x)^h}\right). \tag{7.179}$$

Next, we consider the mapping (7.159) which leads to

$$D_x = \left(\frac{dy}{dx}\right)^{-1} \frac{d}{dx} = \frac{(x^2+s^2)^{3/2}}{s^2} \frac{d}{dx},$$
$$\omega_s^{k,l}(x) = \left(\frac{\sqrt{x^2+s^2}-x}{\sqrt{x^2+s^2}}\right)^k \left(\frac{\sqrt{x^2+s^2}+x}{\sqrt{x^2+s^2}}\right)^l \frac{s^2}{(x^2+s^2)^{3/2}}.$$

Hence, for $u(x) = (1+x^2)^{-h}$, we have $\|D_x^m u\|_{\omega_s^{\alpha+m,\beta+m}} < \infty$ if $m < 2h+\alpha+1$, which implies that

$$\|u - \pi_{N,s}^{\alpha,\beta} u\|_{\omega_s^{\alpha,\beta}} \lesssim N^{-(2h+\alpha+1)} \qquad (u(x) = (1+x^2)^{-h}). \tag{7.180}$$

On the other hand, for $u(x) = \frac{\sin kx}{(1+x^2)^h}$, we have $\|D_x^m u\|_{\omega_s^{\alpha+m,\beta+m}} < \infty$ if $m < \frac{2h+\alpha+1}{2}$, which implies that

$$\|u - \pi_{N,s}^{\alpha,\beta} u\|_{\omega_s^{\alpha,\beta}} \lesssim N^{-(2h+\alpha+1)/2} \qquad \left(u(x) = \frac{\sin kx}{(1+x^2)^h}\right). \tag{7.181}$$

Remark 7.8. (i) *If h is a positive integer, then $u(x) = (1+x)^{-h}$ and $u(x) = (1+x^2)^{-h}$ are rational functions and they can be expressed* exactly *by a finite sum of mapped rational functions. For other cases, the algebraic convergence rate is related to the decay rate of the solution. In both cases, the convergence rates are* **faster** *than the approximations by the Laguerre functions* (cf. (7.114)) *and Hermite functions* (cf. (7.132)), *respectively.*

(ii) *The convergence rate for solutions with oscillation at infinity is much slower than that for solutions without oscillation at infinities. For example, for $u(x) = \frac{\sin kx}{(1+x)^h}$ and $u(x) = \frac{\sin kx}{(1+x^2)^h}$, the convergence rates are* **slower** *than the approximations by the Laguerre functions* (cf. (7.114)) *and Hermite functions* (cf. (7.132)), *respectively.*

(iii) *For solutions with exponential decay at infinity, the convergence rate will be faster than any algebraic rate, and numerical results in* Guo et al. (2000, 2002) and Wang and Guo (2002) (also see Boyd (2001)) *indicate that the convergence rate is sub-geometrical as $e^{-c\sqrt{N}}$.*

(iv) *Numerical results performed in* Guo et al. (2000, 2002) and Wang and Guo (2002) *are consistent with the estimates in* (7.178)–(7.181).

Next, we consider the Gauss and Gauss-Radau quadrature formulas on unbounded domains based on the mapped Jacobi polynomials. To fix the idea, we only consider the Gauss quadrature, since the Gauss-Radau quadrature (which is useful in the semi-infinite interval) can be obtained by exactly the same means.

Let $\{\xi_{N,j}^{\alpha,\beta}, \omega_{N,j}^{\alpha,\beta}\}_{j=0}^N$ be the Jacobi-Gauss quadrature nodes and weights such that (cf. Theorem 3.25):

$$\int_{-1}^{1} \phi(y)\omega^{\alpha,\beta}(y)dy = \sum_{j=0}^{N} \phi(\xi_{N,j}^{\alpha,\beta})\omega_{N,j}^{\alpha,\beta}, \quad \forall \phi \in P_{2N+1}. \tag{7.182}$$

Applying the mapping (7.156) to the above leads to the mapped Jacobi-Gauss quadrature:

$$\int_{\Lambda} u(x)\omega_s^{\alpha,\beta}(x)dx = \sum_{j=0}^{N} u(\zeta_{N,j,s}^{\alpha,\beta})\rho_{N,j,s}^{\alpha,\beta}, \quad \forall u \in V_{2N+1,s}^{\alpha,\beta}, \tag{7.183}$$

where

$$\zeta_{N,j,s}^{\alpha,\beta} := g(\xi_{N,j}^{\alpha,\beta}; s), \quad \rho_{N,j,s}^{\alpha,\beta} := \omega_{N,j}^{\alpha,\beta}, \quad 0 \le j \le N \tag{7.184}$$

are the mapped Jacobi-Gauss nodes and weights.

Accordingly, we can define the discrete inner product and discrete norm:

$$(u,v)_{N,\omega_s^{\alpha,\beta}} = \sum_{j=0}^{N} u(\zeta_{N,j,s}^{\alpha,\beta})v(\zeta_{N,j,s}^{\alpha,\beta})\rho_{N,j,s}^{\alpha,\beta}, \quad \|u\|_{N,\omega_s^{\alpha,\beta}} = (u,u)_{N,\omega_s^{\alpha,\beta}}^{1/2}, \quad \forall u,v \in C(\Lambda).$$

The mapped Jacobi-Gauss interpolation operator $I_{N,s}^{\alpha,\beta} : C(\Lambda) \to V_{N,s}^{\alpha,\beta}$, is defined by

$$I_{N,s}^{\alpha,\beta} u \in V_{N,s}^{\alpha,\beta} \quad \text{such that} \quad (I_{N,s}^{\alpha,\beta} u)(\zeta_{N,j,s}^{\alpha,\beta}) = u(\zeta_{N,j,s}^{\alpha,\beta}), \ j = 0,1,\ldots,N. \tag{7.185}$$

Let $I_N^{\alpha,\beta}$ be the Jacobi-Gauss interpolation operator as in Chap. 3. By definition, we have

$$I_{N,s}^{\alpha,\beta} u(x) = (I_N^{\alpha,\beta} U_s)(y) = (I_N^{\alpha,\beta} U_s)(h(x;s)). \tag{7.186}$$

Then, we can easily derive the following results by combining Theorems 3.41 and 7.21.

Theorem 7.22. *Let $\alpha, \beta > -1$. If $u \in \widetilde{B}^m_{\alpha,\beta}(\Lambda)$ with $1 \le m \le N+1$, then*

$$\begin{aligned} \|\partial_x(I_{N,s}^{\alpha,\beta} u - u)\|_{\tilde{\omega}_s^{\alpha,\beta}} &+ N\|I_{N,s}^{\alpha,\beta} u - u\|_{\omega_s^{\alpha,\beta}} \\ &\le c\sqrt{\frac{(N-m+1)!}{N!}}(N+m)^{(1-m)/2}\|D_x^m u\|_{\omega_s^{\alpha+m,\beta+m}}, \end{aligned} \tag{7.187}$$

where c is a positive constant independent of m, N and u.

We now examine how the mapping parameter s affects the distribution of the nodes. Assume that the nodes $\{\zeta_{N,j,s}^{\alpha,\beta}\}_{j=0}^{N}$ are arranged in ascending order. We first observe that by the mean value theorem,

$$\zeta_{N,j+1,s}^{\alpha,\beta} - \zeta_{N,j,s}^{\alpha,\beta} = g'(\xi;s)(\xi_{N,j+1,s}^{\alpha,\beta} - \xi_{N,j,s}^{\alpha,\beta}), \tag{7.188}$$

for certain $\xi \in (\xi_{N,j,s}^{\alpha,\beta}, \xi_{N,j+1,s}^{\alpha,\beta})$. Hence, the intensity of stretching essentially depends on the derivative values of the mapping. For the mappings (7.160), (7.159), (7.163) and (7.162), we have

$$\frac{dx}{dy} = g'(y;s) = \frac{s}{1-y^2}, \quad \frac{s}{(1-y^2)^{3/2}}, \quad \frac{2s}{(3+y)(1-y)}, \quad \frac{2s}{(1-y)^2}, \tag{7.189}$$

respectively. Therefore, the grid is stretched more and more as s increases.

In Fig. 7.10, we plot sample grid distributions for different scaling factors with various numbers of nodes for the mapped Legendre-Gauss (or Gauss-Radau) points (see the caption for details).

A comparison with Hermite-Gauss points is also presented in Fig. 7.10a. We notice that the mapped Legendre-Gauss points are mostly clustered near the origin and spread further, while the Hermite-Gauss points are more evenly distributed. It should be observed that the distribution of mapped Legendre-Gauss points is more favorable since a much larger effective interval is covered. However, it can be shown that in both cases, the smallest distance between neighboring points is $O(N^{-1})$, as opposed to $O(N^{-2})$ for Jacobi-Gauss type nodes in a finite interval.

A comparison of mapped Legendre- and Laguerre-Gauss-Radau nodes is shown in Fig. 7.10c. The mapped Legendre-Gauss-Radau points are much more clustered near the origin, and one can check that the smallest distance between neighboring points is $O(N^{-2})$, as opposed to $O(N^{-1})$ for the Laguerre-Gauss-Radau nodes. Hence, the distribution of mapped Legendre-Gauss-Radau points is more favorable as far as resolution/accuracy is concerned but it will lead to a more restrictive CFL condition if explicit schemes are used for time-dependent problems.

7.5.3 Spectral Methods Using Mapped Jacobi Polynomials

7.5.3.1 A Generic Example

Consider the model equation

$$\gamma u - \partial_x(a(x)\partial_x u) = f, \quad x \in \Lambda = (-\infty, +\infty), \quad \gamma \geq 0, \tag{7.190}$$

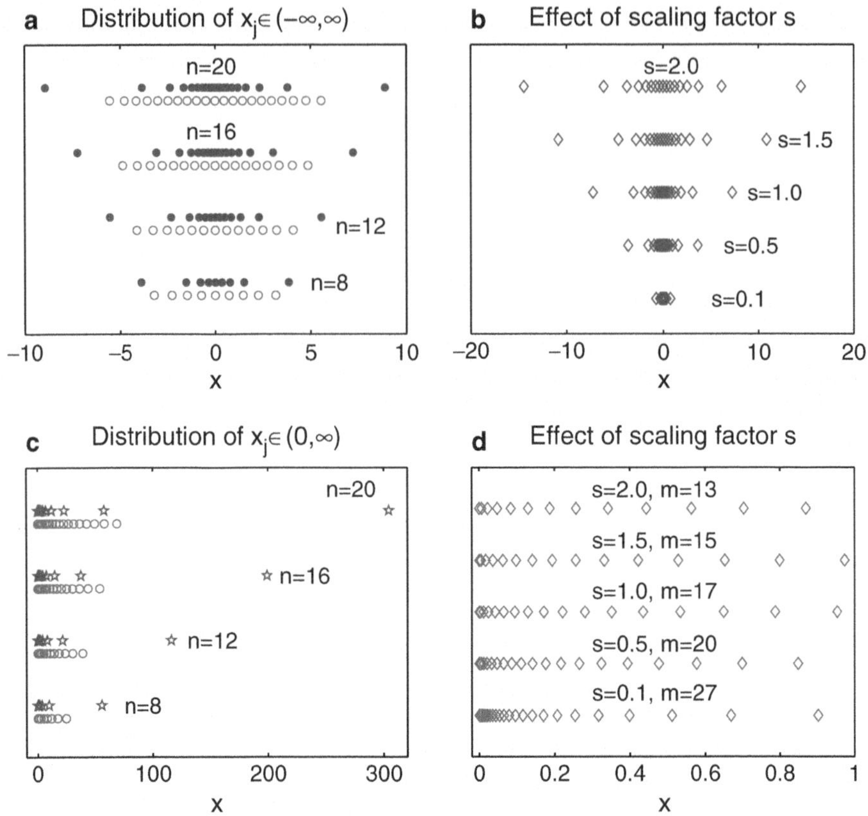

Fig. 7.10 (**a**) Hermite-Gauss points ("∘") vs. mapped Legendre-Gauss points using the algebraic map (7.159) with $s = 1$ ("•") for various n; (**b**) Mapped Legendre-Gauss points with $n = 16$ and various scaling factor s; (**c**) Laguerre-Gauss-Radau points ("∘") vs. mapped Legendre-Gauss-Radau points using the algebraic map (7.162) with $s = 1$ ("⋆") for various n; (**d**) Mapped Legendre-Gauss-Radau points with $n = 32$ and various scaling factor s (m is the number of points in the subinterval $[0, 1)$)

with suitable decay conditions at $\pm\infty$, which will depend on the weight function in the following weighted weak formulation: For a given mapping $x = g(y; s)$ with $x \in \Lambda$ and $y \in (-1, 1)$,

$$\begin{cases} \text{Find } u \in \widetilde{B}^1_{\alpha,\beta}(\Lambda) \text{ such that} \\ \gamma(u,v)_{\omega_s^{\alpha,\beta}} + \big(a(x)\partial_x u, \partial_x(v\omega_s^{\alpha,\beta})\big) \\ \qquad\qquad = (f,v)_{\omega_s^{\alpha,\beta}}, \quad \forall v \in \widetilde{B}^1_{\alpha,\beta}(\Lambda), \end{cases} \tag{7.191}$$

where $\omega_s^{\alpha,\beta}$ and $\widetilde{B}^1_{\alpha,\beta}(\Lambda)$ are defined in (7.167) and (7.173), respectively.

Then, the mapped Jacobi-Galerkin method for (7.190) is

$$\begin{cases} \text{Find } u_N \in V_{N,s}^{\alpha,\beta} \text{ such that} \\ \gamma(u_N, v_N)_{\omega_s^{\alpha,\beta}} + \big(a(x)\partial_x u_N, \partial_x(v_N \omega_s^{\alpha,\beta})\big) \\ \qquad\qquad = (I_{N,s}^{\alpha,\beta} f, v_N)_{\omega_s^{\alpha,\beta}}, \quad \forall v_N \in V_{N,s}^{\alpha,\beta}. \end{cases} \tag{7.192}$$

A second approach is to first transform (7.190) into the finite interval $(-1,1)$, and then apply a Jacobi approximation for the transformed problem. More precisely, (7.190) is first mapped into

$$\gamma U_s - \frac{1}{g'(y;s)}\partial_y\Big(\frac{a(g(y;s))}{g'(y;s)}\partial_y U_s\Big) = F_s, \tag{7.193}$$

where $U_s(y) = u(g(y;s))$ and $F_s(y) = f(g(y;s))$.

Let $\hat{\omega}_s^{\alpha,\beta}(y) = \omega^{\alpha,\beta}(y)g'(y;s)$. Then the Jacobi-Galerkin method for (7.193) is

$$\begin{cases} \text{Find } \tilde{u}_N \in P_N \text{ such that} \\ \gamma(\tilde{u}_N, \tilde{v}_N)_{\omega^{\alpha,\beta}} + \Big(\frac{a(g(y;s))}{g'(y;s)}\partial_y \tilde{u}_N, \partial_y(\tilde{v}_N \hat{\omega}_s^{\alpha,\beta})\Big) \\ \qquad\qquad = (I_N^{\alpha,\beta} F_s, \tilde{v}_N)_{\omega^{\alpha,\beta}}, \quad \forall \tilde{v}_N \in P_N. \end{cases} \tag{7.194}$$

One can verify easily that $\tilde{u}_N(y) = u_N(g(y;s))$. Hence, the above two approaches are mathematically equivalent.

We remark that the formulation (7.194) is in general more difficult to analyze due to the singular nature of $g'(y;s)$, while the analysis for the formulation (7.192) becomes standard once we established the basic approximation properties of the mapped Jacobi polynomials.

On the other hand, (7.193) can be easily implemented using the standard Jacobi-collocation (or more specifically Chebyshev-collocation) method. Indeed, let $\{h_{j,N}(y)\}_{1\le j\le N}$ be the Lagrange basis polynomials associated with the Jacobi-Gauss points $\{y_j\}_{1\le j\le N}$, the Jacobi-collocation approximation to (7.193) is

$$\begin{cases} \text{Find } U_{N,s}(y) = \sum_{j=1}^{N} u_j h_{j,N}(y) \text{ such that} \\ \gamma U_{N,s}(y_j) - \Big(\frac{1}{g'(y;s)}\partial_y\Big(\frac{a(g(y;s))}{g'(y;s)}\partial_y U_{N,s}\Big)\Big)(y_j) \\ \qquad\qquad = F_s(y_j), \quad 1 \le j \le N. \end{cases} \tag{7.195}$$

Let us denote

$$\mathbf{u} = (u_1,\ldots,u_N)^T, \quad \mathbf{f} = (F_s(y_1),\ldots,F_s(y_N))^T, \quad D_{ij} = h_j'(y_i), \quad D = (D_{ij}),$$

$$\Lambda_i = \frac{a(g(y_i;s))}{g'(y_i;s)}, \quad \Lambda = \mathrm{diag}(\Lambda_i), \quad \Sigma_i = \frac{1}{g'(y_i;s)}, \quad \Sigma = \mathrm{diag}(\Sigma_i).$$

Then, (7.195) reduces to the linear system

$$(\gamma I - \Sigma D \Lambda D)\mathbf{u} = \mathbf{f},$$

which can be easily solved by using a standard procedure. Note that in the above procedure, we only need to compute the Jacobi-Gauss points $\{y_j\}_{1\le j\le N}$ and the associated differentiation matrix D whose entries can be found from Chap. 3.

7.5.3.2 Error Estimates for a Model Problem

We consider the Jacobi rational approximation to the following model problem

$$\gamma u(x) - \partial_x^2 u(x) = f(x), \quad x \in \Lambda = (0,\infty), \ \gamma > 0; \quad u(0) = 0, \tag{7.196}$$

with a suitable decay condition at infinity which is to be determined by the weak formulation of (7.196).

For a given mapping, let $\omega = \omega_s^{\alpha,\beta}$ be the weight function associated with the mapped Jacobi polynomials, and denote $H_{0,\omega}^1(\Lambda) = \{u \in H_\omega^1(\Lambda) : u(0) = 0\}$. We define a bilinear form

$$a_\omega(u,v) = \gamma(u,v)_\omega + (\partial_x u, \partial_x(v\omega)), \quad \forall u,v \in H_{0,\omega}^1(\Lambda). \tag{7.197}$$

Then, a weak formulation for (7.196) is

$$\begin{cases} \text{Find } u \in H_{0,\omega}^1(\Lambda) \text{ such that} \\ a_\omega(u,v) = (f,v)_\omega, \quad \forall v \in H_{0,\omega}^1(\Lambda), \end{cases} \tag{7.198}$$

for $f \in \big(H_{0,\omega}^1(\Lambda)\big)'$. Note that $u \in H_{0,\omega}^1(\Lambda)$ implies a decay condition for u at infinity.

Denote $X_N = \{u \in V_{N,s}^{\alpha,\beta} : u(0) = 0\}$. The Jacobi-Galerkin approximation of (7.198) by the mapped Jacobi polynomials is

$$\begin{cases} \text{Find } u_N \in X_N \text{ such that} \\ a_\omega(u_N,v_N) = (I_{N,s}^{\alpha,\beta} f, v_N)_\omega, \quad \forall v_N \in X_N, \end{cases} \tag{7.199}$$

for $f \in L_\omega^2(\Lambda) \cap C(\bar{\Lambda})$.

Unlike the standard spectral method in a finite domain, the well-posedness of (7.198) and of (7.199) is not guaranteed for all cases with $\gamma \ge 0$. A general result for the well-posedness of an abstract equation of the form (7.198) is established in Shen and Wang (2004). For the readers' convenience, we recall this result below (cf. Lemma 2.3 in Shen and Wang (2004)):

Lemma 7.1. *Assume that*

$$d_1 = \max_{x\in\bar{\Lambda}} |\omega^{-1}(x)\partial_x\omega(x)|, \quad d_2 = \max_{x\in\bar{\Lambda}} |\omega^{-1}(x)\partial_x^2\omega(x)|$$

are finite. Then, for any $u, v \in H^1_\omega(\Lambda)$,

$$a_\omega(u,v) \le (d_1+1)|u|_{1,\omega}\|v\|_{1,\omega} + \gamma\|u\|_\omega\|v\|_\omega. \tag{7.200}$$

If, in addition, $v^2(x)\omega'(x)|_{x=0} = 0$ *and* $\lim_{x\to\infty} v^2(x)\omega'(x) \ge 0$, *then for any* $v \in H^1_\omega(\Lambda)$,

$$a_\omega(v,v) \ge |v|^2_{1,\omega} + (\gamma - d_2/2)\|v\|^2_\omega. \tag{7.201}$$

Proof. By (7.197) and the Cauchy–Schwarz inequality,

$$\begin{aligned} a_\omega(u,v) &\le |(\partial_x u, \partial_x v)_\omega + (\partial_x u, v\partial_x\omega)| + \gamma|(u,v)_\omega| \\ &\le |u|_{1,\omega}|v|_{1,\omega} + \max_{x\in\bar{\Lambda}} |\omega^{-1}(x)\partial_x\omega(x)||u|_{1,\omega}\|v\|_\omega + \gamma\|u\|_\omega\|v\|_\omega \\ &\le (d_1+1)|u|_{1,\omega}\|v\|_{1,\omega} + \gamma\|u\|_\omega\|v\|_\omega. \end{aligned} \tag{7.202}$$

On the other hand,

$$\begin{aligned} a_\omega(v,v) &= |v|^2_{1,\omega} + \gamma\|v\|^2_\omega + \frac{1}{2}\int_\Lambda \partial_x(v^2(x))\partial_x\omega(x)dx \\ &= |v|^2_{1,\omega} + \gamma\|v\|^2_\omega - \frac{1}{2}\int_\Lambda v^2(x)\partial_x^2\omega(x)dx \\ &\ge |v|^2_{1,\omega} + (\gamma - d_2/2)\|v\|^2_\omega. \end{aligned}$$

This ends the proof. □

Thanks to the above lemma, it is then straightforward to prove the following general result.

Theorem 7.23. *Assume that the conditions of Lemma 7.1 are satisfied and* $\gamma - d_2/2 > 0$. *Then the problem* (7.198) (*resp.* (7.199)) *admits a unique solution. Furthermore, we have the error estimate:*

$$\|u-u_N\|_{1,\omega} \lesssim \inf_{v_N\in X_N} \|u-v_N\|_{1,\omega} + \|f - I^{\alpha,\beta}_{N,s} f\|_\omega. \tag{7.203}$$

Remark 7.9. *The inequality* (7.201) *is derived under a general framework. For a specific mapping, the constraint* $\gamma - d_2/2 > 0$ *can often be relaxed. On the other hand, with a change of variable* x *to* x/c $(c>0)$ *for* (7.190)*, this restriction can be replaced by* $\gamma > 0$.

Given a mapping and a pair of Jacobi parameters (α,β), we just need to compute upper bounds for d_1 and d_2, verify that the conditions of Theorem 7.23 are satisfied, and apply the approximation results in Theorems 7.21 and 7.22 to (7.203) to get the desired error estimates.

Consider for example the mapped Legendre method for (7.198) with the mapping (7.162). It can be shown that for this mapping, we have $d_1 \le 2$ and $d_2 \le 6$. Applying Theorems 7.21 and 7.22 to (7.203) with $(\alpha,\beta) = (0,0)$ leads to the following results.

Corollary 7.4. *Let u and u_N be respectively the solutions of* (7.198) *and* (7.199) *with* $(\alpha,\beta)=(0,0)$ *and the mapping* (7.162) *with* $s=1$. *Assuming that* $u\in \widetilde{B}^m_{0,0}(\Lambda)$ *and* $f\in \widetilde{B}^k_{0,0}(\Lambda)$ *and* $\gamma>3$, *we have*

$$\|u-u_N\|_{1,\omega_1^{0,0}} \lesssim N^{1-m}\|D_x^m u\|_{\omega_1^{m,m}} + N^{-k}\|D_x^k f\|_{\omega_1^{k,k}}, \tag{7.204}$$

where $k,m\geq 1$ *are fixed integers.*

We note that a slightly improved condition on γ was derived in Guo et al. (2000) using a refined estimate for (7.201).

A similar procedure can be applied to the mapped Chebyshev method for (7.198) with the mapping (7.162). Note however that in this case we have $d_1,d_2=\infty$. Nevertheless, one can still show that $a_\omega(\cdot,\cdot)$ is continuous and coercive (cf. Guo et al. (2002)). Applying Theorems 7.21 and 7.22 to (7.203) with $(\alpha,\beta)=(-1/2,-1/2)$ leads to the following results (cf. Guo et al. (2002)).

Corollary 7.5. *Let u and u_N be the solutions of* (7.198) *and* (7.199) *with* $(\alpha,\beta)=(-1/2,-1/2)$ *and the mapping* (7.162) *with* $s=1$. *Assuming that* $u\in \widetilde{B}^m_{-1/2,-1/2}(\Lambda)$ *and* $f\in \widetilde{B}^k_{-1/2,-1/2}(\Lambda)$ *and that* $\gamma>\frac{14}{27}$, *we have*

$$\|u-u_N\|_{1,\omega_1^{-1/2,-1/2}} \lesssim N^{1-m}\|D_x^m u\|_{\omega_1^{m-1/2,m-1/2}} + N^{-k}\|D_x^k f\|_{\omega_1^{k-1/2,k-1/2}}, \tag{7.205}$$

where $k,m\geq 1$ *are fixed integers.*

Remark 7.10. *Error estimates which are essentially equivalent to* (7.204) *and* (7.205) *but in different forms were derived in* Guo et al. (2000, 2002).

The same procedure can be used to derive error estimates on mapped Jacobi methods for problems on the whole line (cf. Wang and Guo (2002)).

7.5.3.3 Implementations and A Comparison Study

We briefly discuss the implementation of the mapped Jacobi spectral method, and compare its convergence behavior with the approaches by Hermite functions and Laguerre functions. Let $\{\phi_j\}_{j=0}^{N-1}$ be a set of basis functions of X_N. We set

$$\begin{aligned}
&u_N=\sum_{k=0}^{N-1}\hat u_k\phi_k(x), \quad \mathbf{u}=\big(\hat u_0,\hat u_1,\ldots,\hat u_{N-1}\big)^T;\\
&f_j=(I_N^{\alpha,\beta}f,\phi_j)_\omega, \quad \mathbf{f}=\big(f_0,f_1,\ldots,f_{N-1}\big)^T;\\
&s_{jk}=(\phi_k',\,(\phi_j\omega)'), \quad S=\big(s_{jk}\big)_{0\leq j,k\leq N-1},\\
&m_{jk}=(\phi_k,\,\phi_j)_\omega, \quad M=\big(m_{jk}\big)_{0\leq j,k\leq N-1},
\end{aligned}$$

Thus, the system (7.199) reduces to

$$(\gamma M + S)\mathbf{u} = \mathbf{f}. \tag{7.206}$$

For example, we consider the mapping (7.162) with $s = 1$. This is a special case of the general setting analyzed in Sect. 7.5.3. As shown in Chap. 4, it is advantageous to construct basis functions using compact combinations of orthogonal functions. In this case, we define $\phi_k(x) = j_{s,k}^{0,0}(x) + j_{s,k+1}^{0,0}(x)$ with $s = 1$, which satisfies $\phi_k(0) = 0$. Then, we have $\omega(x) = \frac{2}{(x+1)^2}$, and

$$m_{jk} = \int_0^\infty \phi_k(x)\phi_j(x)\omega(x)dx = \int_{-1}^1 (L_k(y) + L_{k+1}(y))(L_j(y) + L_{j+1}(y))dy,$$

and

$$\begin{aligned} s_{jk} &= \int_0^\infty \phi_k'(x)(\phi_j(x)\omega(x))'dx = -\int_0^\infty \phi_k''(x)\phi_j(x)\omega(x)dx \\ &= -\frac{1}{4s}\int_{-1}^1 (1-y)^2\partial_y((1-y)^2\partial_y(L_k(y) + L_{k+1}(y)))(L_j(y) + L_{j+1}(y))dy, \end{aligned}$$

where L_k is the Legendre polynomial of degree k. By using the properties of Legendre polynomials, it is then easy to see that M is a symmetric tridiagonal matrix and S is a non-symmetric seven diagonal matrix. Hence, the system (7.206) can be solved efficiently. However, we note that a disadvantage of the mapped Legendre method is that it leads to a non-symmetric system even though the original problem (7.196) is symmetric.

The convergence behaviors of the mapped Jacobi, Laguerre and Hermite spectral methods have been discussed in detail using the three sets of functions (7.152)-(7.154) as examples. In order to provide a quantitative assessment, we now present some direct comparisons of the mapped Legendre method (using mapping (7.162) or (7.159) with $s = 1$) against the Laguerre or Hermite method for the same model equation.

In the following computations, we fix $\gamma = 2$ in (7.196) and (7.148). The parameters in the three sets of exact solutions are set as follows: $k = 2$ in (7.152), $h = 2.5$ in (7.153) and $k = 2$, $h = 3.5$ in (7.154). The numerical results are plotted in Figs. 7.11–7.13 in which "Max-ML", "Max-Lag" and "Max-Hmt" denote respectively the errors in maximum norm for mapped Legendre, Laguerre and Hermite methods (likewise for the L^2-notation).

Several remarks are in order (see also Remark 7.8 for the general mapped Jacobi case):

(a) For exact solutions in (7.152), Laguerre and Hermite methods converge faster.
(b) For exact solutions in (7.153), the mapped Legendre method performs much better.
(c) For exact solutions in (7.154), the Laguerre method is slightly better than the mapped Legendre method, while the Hermite method is still worse than the

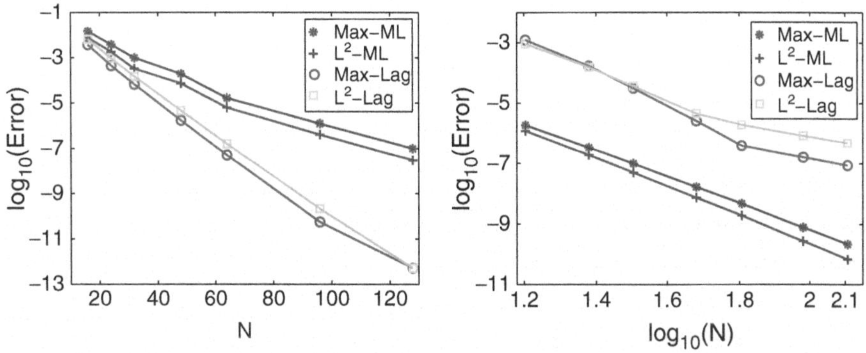

Fig. 7.11 Convergence rates with exact solution: $u(x) = \sin(2x)\,e^{-x}$ (left) and $u(x) = 1/(1+x)^{5/2}$ (right)

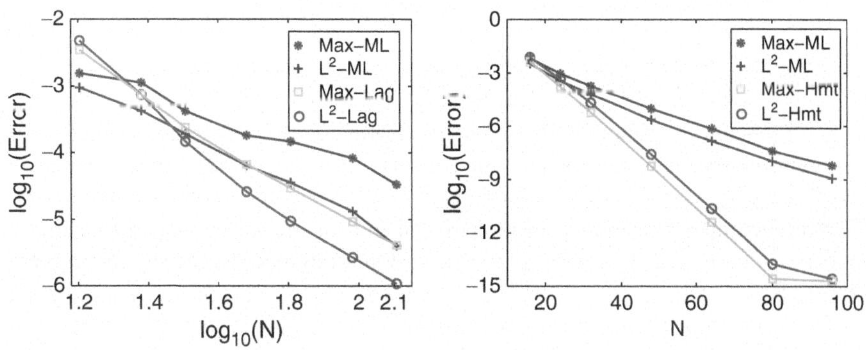

Fig. 7.12 Convergence rates with exact solution: $u(x) = \sin 2x/(1+x)^{7/2}$ (left) and $u(x) = \sin(2x)\,e^{-x^2}$ (right)

mapped Legendre method. We note however that the performance of Laguerre and Hermite methods can be significantly improved using a proper scaling (cf. Tang (1993), Shen (2000) and the discussions in the previous section).

7.5.4 Modified Legendre-Rational Approximations

Notice that the mapped Jacobi polynomials, including the mapped Legendre polynomials, are mutually orthogonal in a weighted Sobolev space. Thus, their applications involve weighted formulations which are, on the one hand, difficult to analyze and implement, and on the other hand, not suitable for certain problems which are only well-posed in non-weighted Sobolev spaces. Therefore, it is sometimes useful to construct (non-weighted) orthogonal systems from mapped Jacobi

polynomials. Next, let us consider one of such examples. We define the modified Legendre-rational functions of degree l by

$$R_l(x) = \frac{\sqrt{2}}{x+1} L_l\left(\frac{x-1}{x+1}\right), \quad l = 0, 1, 2, \ldots .$$

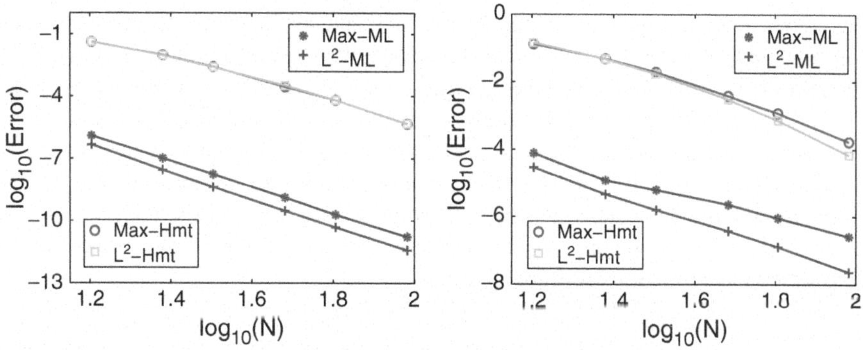

Fig. 7.13 Convergence rates with exact solution: $u(x) = 1/(1+x^2)^{5/2}$ (left) and $u(x) = \sin 2x/(1+x^2)^{7/2}$ (right)

By (3.171), $\{R_l\}$ are the eigenfunctions of the singular Sturm-Liouville problem

$$(x+1)\partial_x(x(\partial_x((x+1)v(x))) + \lambda v(x) = 0, \quad x \in \Lambda,$$

with the corresponding eigenvalues $\lambda_l = l(l+1)$, $l = 0, 1, 2, \ldots$. Thanks to (3.168) and (3.174a), they satisfy the recurrence relations

$$R_{l+1}(x) = \frac{2l+1}{l+1}\frac{x-1}{x+1} R_l(x) - \frac{l}{l+1} R_{l-1}(x), \quad l \geqslant 1,$$

and

$$\begin{aligned} 2(2l+1)R_l(x) = &(x+1)^2\left(\partial_x R_{l+1}(x) - \partial_x R_{l-1}(x)\right) \\ &+ (x+1)\left(R_{l+1}(x) - R_{l-1}(x)\right). \end{aligned}$$

Furthermore,

$$\lim_{x\to\infty}(x+1)R_l(x) = \sqrt{2}, \quad \lim_{x\to\infty} x\partial_x((x+1)R_l(x)) = 0. \tag{7.207}$$

By the orthogonality of the Legendre polynomials,

$$\int_\Lambda R_l(x)R_m(x)dx = \left(l + \frac{1}{2}\right)^{-1} \delta_{l,m}. \tag{7.208}$$

We refer to Guo and Shen (2001) and to Wang and Guo (2004) for the analysis and applications of the modified Legendre-rational spectral approximations on the half line and on the whole line, respectively.

7.5.5 Irrational Mappings

For many applications, e.g., in fluid dynamics and in financial mathematics, the solutions may tend to a constant or even grow with a specified rate at infinity. For such problems, variational formulations in Sobolev spaces with uniform weight or a given non-matching weight are usually not well-posed. Therefore, it becomes necessary to construct orthogonal systems which match the asymptotic behaviors of the underlying problem. The first effort of such kind is carried out in Boyd (2001) where a rational Chebyshev method with polynomial growth basis functions is developed. A more general approach is presented in Guo and Shen (2008) where they considered the following orthogonal system:

$$I_l^{(\gamma,\delta)}(r) := \frac{1}{r^\gamma} J_l^{\alpha,0}\Big(1 - \frac{2}{r^\delta}\Big). \tag{7.209}$$

In the above, $J_l^{\alpha,0}(r)$ is the Jacobi polynomial of degree l with index $(\alpha,0)$. The parameter γ is chosen to match, as closely as possible, the asymptotic behavior of the function to be approximated; the parameter $\delta > 0$ is a mapping parameter which affects the accuracy of the approximation (see Guo and Shen (2008) for the details); α is determined in such a way that $\{I_k^{(\gamma,\delta)}(r)\}$ form an orthogonal system in $L^2_{\omega_\sigma}(\Lambda)$, where σ is another parameter, $\Lambda = (1,\infty)$ and $\omega_\sigma = r^\sigma$. This latter condition requires that $\alpha = \frac{1}{\delta}(2\gamma - \delta - \sigma - 1)$. Hence, α is *not* a free parameter. Therefore, the proposed family of orthogonal systems $\{I_k^{(\gamma,\delta)}(r)\}$ is very general and includes in particular many special cases already studied in the literature. The flexibility afforded by the free parameters γ, δ (and σ) allows us to design suitable approximations for a large class of partial differential equations.

7.5.6 Miscellaneous Issues and Extensions

We discuss below some miscellaneous issues and extensions related to spectral methods in unbounded domains.

7.5.6.1 Other One-Dimensional Applications

While we have only presented analysis and implementation details for second-order model equations, the basic approximation results presented in this chapter can be used for many other applications. We refer to Boyd (2001) for a review on the work before the year 2000, which includes in particular many applications in oceanography, and list some of the more recent work below.

In Fok et al. (2002), a combined Hermite-finite difference method is proposed for a Fokker-Planck equation with one spatial and one phase dimension; in Guo et al. (2003), the authors applied the Hermite spectral method for solving the Dirac equation on the whole line; in Guo and Shen (2001), a modified Legendre rational method is presented for the KdV equation in a semi-infinite interval; the same problem is also studied in Shen and Wang (2006) where a single domain Laguerre and two-domain Legendre-Laguerre method are introduced and analyzed.

7.5.6.2 Multidimensional Problems

Although only one-dimensional problems are discussed in the previous sections, these one-dimensional orthogonal systems can be easily used for multidimensional problems through the usual tensor product approach. While it is possible to use mapped Jacobi methods for multidimensional problems, the analysis and implementation become complicated due to the non-uniform weights involved in the weak formulation. Consequently, most of the work for multidimensional problems in unbounded domains use either Laguerre or Hermite functions combined with Fourier series or Jacobi polynomials.

Consider, for example, problems in an infinite (resp. semi-infinite) channel. It is natural to use Hermite (resp. Laguerre) functions in the infinite direction and Jacobi polynomials in the finite direction. In Xu and Guo (2002), the authors studied a Laguerre-Legendre approximation to the 2-D Navier–Stokes equations in the streamline diffusion-vorticity formulation in a semi-infinite channel, while in Azaiez et al. (2008), the authors studied approximation of the 2-D Stokes equations in primitive variables by a Laguerre-Legendre method, and derived a complete error analysis with an explicit estimate of the inf-sup condition.

Consider, as another example, problems in exterior domains. It is convenient, for a 2-D domain exterior to a circle, to use polar coordinates and a Laguerre-Fourier approximation (cf. Guo et al. (2005)); and for a 3-D domain exterior to a sphere, to use spherical coordinates and a Laguerre-spherical harmonic approximation (cf. Zhang and Guo (2006)). In these cases, the analysis is a bit more complicated due to the coordinate transforms, but still can be carried out using essentially the approximation results presented in this chapter.

Problems

7.1. Prove the inverse inequality in Theorem 7.7.

7.2. Derive the approximation results in Theorem 7.8.

7.3. Prove Theorem 7.12.

7.4. Consider the problem:

$$-u_{xx}+u=f,\quad x\in\mathbb{R}_+;\quad u(0)-u'(0)=0,\quad \lim_{x\to+\infty}u(x)=0. \tag{7.210}$$

Denote

$$\widehat{X}_N=\{\phi\in\widehat{P}_N\,:\,\phi(0)-\phi'(0)=0\}, \tag{7.211}$$

where $\widehat{P}_N$ is defined in (7.83).

(i) Write down the non-weighted Laguerre spectral-Galerkin scheme for (7.210) by seeking the approximation solution in $\widehat{X}_N$.

(ii) Determine a_k so that

$$\phi_k(x)=\widehat{\mathscr{L}}_k(x)-a_k\widehat{\mathscr{L}}_{k+1}(x),\quad 0\le k\le N-1 \tag{7.212}$$

form a basis of $\widehat{X}_N$.

(iii) Implement the proposed scheme and test it on the three sets of functions (7.152)–(7.154) by subtracting a suitable function to meet the homogeneous boundary condition. Plot the numerical errors as in Figs. 7.5 and 7.6.

7.5. Consider the third-order equation on the half line:

$$u_{xxx}+u=f,\ x\in(0,\infty);\quad u(0)=0,\ \lim_{x\to+\infty}u(x)=\lim_{x\to+\infty}u_x(x)=0. \tag{7.213}$$

We introduce the "dual" approximation space

$$X_N:=\{u\in\widehat{P}_N:u(0)=0\},\ X_N^*:=\{u\in\widehat{P}_{N+1}:u(0)=u_x(0)=0\}, \tag{7.214}$$

and consider the Laguerre dual-Petrov-Galerkin approximation to (7.213), that is,

$$\begin{cases}\text{Find } u_N\in X_N \text{ such that}\\ (\partial_x u_N,\partial_x^2 v_N)+(u_N,v_N)=(\hat{I}_N f,v_N),\quad \forall v_N\in X_N^*,\end{cases} \tag{7.215}$$

where $\hat{I}_N$ is the Laguerre-Gauss-Radau interpolation associated with the Laguerre function approach.

(i) Show that

$$\|u_N\|_{\hat{\omega}}^2+3\|\partial_x u_N\|^2\le\|\hat{I}_N f\|_{\hat{\omega}}^2, \tag{7.216}$$

where $\hat{\omega}=x$.

(ii) Choose the basis functions consisting of compact combinations of Laguerre functions (cf. (7.212)) for X_N and X_N^*, and implement the proposed scheme by testing it on the three sets of exact solutions (7.152)–(7.154). Present the numerical errors as in Figs. 7.5 and 7.6.

(iii) Analyze the convergence of this dual-Petrov-Galerkin scheme (cf. Shen and Wang (2006)).

Chapter 8
Separable Multi-Dimensional Domains

The main goals of this chapter are (a) to design efficient spectral algorithms for solving second-order elliptic equations in separable geometries; and (b) to provide a basic framework for error analysis of multi-dimensional spectral methods. More specifically, we shall concentrate on the following topics.

(a) We shall present in detail spectral-Galerkin algorithms for the model equation

$$\alpha u - \Delta u = f \text{ in } \Omega, \tag{8.1}$$

with suitable boundary conditions in rectangular, cylindrical or spherical geometries. These algorithms are based on the tensor product approach, and the resulting linear systems are solved by the so-called matrix decomposition/diagonalization method (cf. Lynch et al. (1964), Haidvogel and Zang (1979)) with partial or full diagonalizations.

(b) We shall extend the one-dimensional Jacobi approximation results presented in Chap. 3 to the multi-dimensional case, and carry out error analysis for some typical multi-dimensional spectral schemes.

(c) We shall also consider sparse spectral approximations for high dimensional problems (with dimension ≥ 4). In particular, we shall discuss Jacobi approximations in hyperbolic cross, and sparse spectral-Galerkin methods based on Chebyshev sparse grids.

We shall limit our attention in this chapter to several typical separable domains. There are, however, other types of separable domains, such as the 2-D elliptical and 3-D ellipsoidal domains. Separation of variables in these domains leads to Mathieu functions and spheroidal wave functions (cf. McLachlan (1951), Abramowitz and Stegun (1964)). The approximation properties of Mathieu functions have been established recently in Shen and Wang (2008), and a Legendre-Mathieu method was developed in Fang et al. (2009) for acoustic scattering in 2-D elongated domains. On the other hand, a special family of spheroidal wave functions, the so-called prolate spheroidal wave functions of degree zero (PSWFs) which are in particular bandlimited (cf. Slepian and Pollak (1961)), has attracted some recent attention (see Boyd (2004), Chen et al. (2005), Wang (2010) and the references therein).

J. Shen et al., *Spectral Methods: Algorithms, Analysis and Applications*,
Springer Series in Computational Mathematics 41, DOI 10.1007/978-3-540-71041-7_8,

8.1 Two- and Three-Dimensional Rectangular Domains

To fix the idea, we consider (8.1) with $\Omega = (-1,1)^d (d = 2,3)$ and the homogeneous Dirichlet boundary conditions: $u|_{\partial\Omega} = 0$. Note that non-homogeneous boundary conditions can be treated by a lifting of boundary data. We refer to Shen (1994) (resp. Auteri and Quartapelle (2000)) for details of the two (resp. three) dimensional lifting.

Let us denote $\boldsymbol{X}_N = (P_N^0)^d$. With a slight abuse of notation, we shall use hereafter $(\cdot,\cdot)_\omega$ to denote one-dimensional or multi-dimensional inner product in $L^2_\omega(\Omega)$. The weighted spectral-Galerkin approximation to (8.1) with $u|_{\partial\Omega} = 0$ is

$$\begin{cases} \text{Find } u_N \in \boldsymbol{X}_N \text{ such that} \\ \alpha(u_N, v_N)_\omega + a_\omega(u_N, v_N) = (I_N f, v_N)_\omega, \quad \forall v_N \in \boldsymbol{X}_N, \end{cases} \tag{8.2}$$

where $I_N : C(\bar{\Omega}) \to P_N^d$ is the interpolation operator associated with the tensor product of Gauss-Lobatto points, and

$$a_\omega(u,v) = \left(\nabla u, \omega^{-1}\nabla(v\omega)\right)_\omega. \tag{8.3}$$

8.1.1 Two-Dimensional Case

We shall construct multi-dimensional basis functions by using the tensor product of one-dimensional basis functions. More precisely, let $\{\phi_k\}_{k=0}^{N-2}$ be a set of basis functions of P_N^0. Then,

$$\boldsymbol{X}_N = (P_N^0)^2 = \text{span}\big\{\phi_i(x)\phi_j(y) : i,j = 0,1,\ldots,N-2\big\}.$$

Let us denote

$$\begin{aligned} &u_N = \sum_{k,j=0}^{N-2} \tilde{u}_{kj}\phi_k(x)\phi_j(y), \quad U = (\tilde{u}_{kj})_{k,j=0,1,\ldots,N-2}; \\ &a_{kj} = \int_I \phi_j'(x)(\phi_k(x)\omega(x))'dx, \quad A = (a_{kj})_{k,j=0,1,\ldots,N-2}; \\ &b_{kj} = \int_I \phi_j(x)\phi_k(x)\omega(x)dx, \quad B = (b_{kj})_{k,j=0,1,\ldots,N-2}; \\ &f_{kj} = (I_N f, \phi_k(x)\phi_j(y))_\omega, \quad F = (f_{kj})_{k,j=0,1,\ldots,N-2}. \end{aligned} \tag{8.4}$$

Taking $v = \phi_l(x)\phi_m(y)$ in (8.2) for $l,m = 0,1\ldots,N-2$, we find that (8.2) is equivalent to the following linear system:

$$\alpha BUB + AUB + BUA^T = F. \tag{8.5}$$

We can also rewrite (8.5) in the following form using the tensor product notation:

$$\left(\alpha B \otimes B + A \otimes B + B \otimes A^T\right)\boldsymbol{u} = \boldsymbol{f}, \tag{8.6}$$

where $\boldsymbol{f}$ and $\boldsymbol{u}$ are vectors of length $(N-1)^2$ formed by the columns of U and F, i.e.,

$$\boldsymbol{f} = (f_{00}, f_{10}, \ldots, f_{q0}; f_{01} \ldots, f_{q1}; \ldots; f_{0q}, \ldots, f_{qq})^T,$$

and $\otimes$ denotes the tensor product operator, i.e., $A \otimes B = (Ab_{ij})_{i,j=0,1,\ldots,q}$ with $q = N-2$.

8.1.1.1 Matrix Diagonalization Method

The linear system (8.5) can be solved in particular by the matrix decomposition method described in Lynch et al. (1964), which is also known in the field of spectral methods as the matrix diagonalization method (cf. Haidvogel and Zang (1979)). To this end, we consider the generalized eigenvalue problem:

$$B\bar{x} = \lambda A\bar{x}. \tag{8.7}$$

In the Legendre case, A and B are symmetric positive definite matrices so all the eigenvalues are real and positive. In the Chebyshev case, A is no longer symmetric but it is still positive definite. Furthermore, it is shown in Gottlieb and Lustman (1983) that all the eigenvalues are real, positive and distinct. Let Λ be the diagonal matrix whose diagonal entries $\{\lambda_p\}$ are the eigenvalues of (8.7), and let E be the matrix whose columns are the corresponding eigenvectors of (8.7), i.e.,

$$BE = AE\Lambda. \tag{8.8}$$

8.1.1.2 Partial Diagonalization

Setting $U = EV$, thanks to (8.8), (8.5) becomes

$$\alpha AE\Lambda VB + AEVB + AE\Lambda VA^T = F.$$

Multiplying the above equation by $E^{-1}A^{-1}$, we arrive at

$$\alpha\Lambda VB + VB + \Lambda VA^T = E^{-1}A^{-1}F := G. \tag{8.9}$$

The transpose of the above equation reads

$$\alpha BV^T\Lambda + BV^T + AV^T\Lambda = G^T. \tag{8.10}$$

Let $\boldsymbol{v}_p = (v_{p0}, v_{p1}, \ldots, v_{pq})^T$ and $\boldsymbol{g}_p = (g_{p0}, g_{p1}, \ldots, g_{pq})^T$ (with $q = N-2$) for $p = 0, 1, \ldots, N-2$. Then the p-th column of (8.10) can be written as

$$\big((\alpha\lambda_p + 1)B + \lambda_p A\big)\boldsymbol{v}_p = \boldsymbol{g}_p, \quad p = 0, 1, \ldots, N-2, \tag{8.11}$$

which is equivalent to $N-1$ systems of the form (4.17)[1] from the spectral-Galerkin approximation of the one-dimensional equation considered in Sect. 4.1.

In summary, the solution of (8.5) consists of the following steps:

1. Pre-processing: compute the eigenvalues and eigenvectors of the generalized eigenvalue problem (8.7) and compute E^{-1} (if A is not symmetric);
2. Compute $G = E^{-1}A^{-1}F$;
3. Obtain V by solving (8.11);
4. Set $U = EV$.

The above procedure corresponds to the diagonalization in the x direction, and one may of course choose to diagonalize in the y direction. In fact, if different numbers of modes are used in each direction, one should choose to diagonalize in the direction with fewer modes so as to minimize the operational counts of the two matrix–matrix multiplications in the above procedure.

- **Legendre case:** Let $\phi_k(x) = \frac{1}{\sqrt{4k+6}}(L_k(x) - L_{k+2}(x))$. Then, we have $A = I$ and B can be split into two symmetric tridiagonal sub-matrices, so the eigenvalues and eigenvectors of B can be easily computed in $O(N^2)$ operations by standard procedures. Furthermore, we have $E^{-1} = E^T$. Step 3 consists of solving $N-1$ tridiagonal systems of order $N-1$. Therefore, for each right-hand side, the cost of solving the system (8.5) is dominated by the two matrix–matrix multiplications in Steps 2 and 4, which can be carried out in a small multiple of N^3 operations.
- **Chebyshev case:** Let $\phi_k(x) = T_k(x) - T_{k+2}(x)$. Then, A is a special upper triangular matrix given in (4.30) and B is a symmetric positive definite matrix with three non-zero diagonals. Similar to the Legendre case, A and B can be split into two sub-matrices so that the eigen-problem (8.7) can be split into four subproblems which can be solved directly by using a QR method. Note that an interesting $O(N^2)$ algorithm for solving (8.11) was developed in Shen (1995) (cf. Remark 4.3). Once again, the cost of solving system (8.5) in the Chebyshev case is also dominated by the two matrix–matrix multiplications in Steps 2 and 4 which can be carried out in a small multiple of N^3 operations.

8.1.1.3 Full Diagonalization

One can also diagonalize in both directions. To this end, we set $U = EWE^T$ in (8.5) to get

$$\alpha AE\Lambda WE^TB + AEWE^TB + AE\Lambda WE^TA^T = F.$$

Multiplying the left (resp. right) of the above equation by $(EA)^{-1}$ (resp. $(EA)^{-T}$) and using the fact $E^TB = \Lambda E^TA^T$ (cf. (8.8)), we arrive at

$$\alpha\Lambda W\Lambda + W\Lambda + \Lambda W = (EA)^{-1}F(EA)^{-T} := H, \tag{8.12}$$

[1] The matrices A and B in (8.11) are respectively the matrices S and M in (4.17).

which is equivalent to

$$(\alpha\lambda_i\lambda_j+\lambda_i+\lambda_j)w_{ij}=h_{ij}, \quad 0\le i,j\le N-2, \tag{8.13}$$

where w_{ij} and h_{ij} are entries of W and H.

In summary, the full diagonalization for solving (8.5) consists of the following steps:

1. Pre-processing: compute the eigenvalues and eigenvectors of the generalized eigenvalue problem (8.7), and compute $(EA)^{-1}$;
2. Compute $H=(EA)^{-1}F(EA)^{-T}$;
3. Obtain W from (8.13);
4. Set $U=EWE^T$.

Therefore, for both the Legendre-Galerkin and Chebyshev-Galerkin methods, the cost of the above algorithm is essentially four matrix–matrix multiplications, as opposed to two matrix–matrix multiplications for the partial diagonalization. Notice that the full diagonalization procedure is simpler to implement than the partial one.

Remark 8.1. *The full diagonalization procedure can also be applied to the matrix system obtained from a collocation approach* (cf. Haldenwang et al. (1984)). *In this case, let $\phi_k(x)=h_{k+1}(x)$, where $\{h_j\}_{j=1}^{N-1}$ are Lagrange basis polynomials associated with the interior Gauss-Lobatto points. Then, the mass matrix B in* (8.5) *is diagonal and the stiffness matrix A is full.*

However, in addition to the fact that A is full, the ill-conditioning of A makes the eigenvalue problem (8.7) *prone to round-off errors as N becomes large. In fact, a loss of* 4-5 *digits was observed for $N=256$ in* Shen (1994).

Hereafter, we shall only consider the partial diagonalization procedure based on the modal basis.

Remark 8.2. *The matrix diagonalization approach applies directly to separable elliptic equations with general boundary conditions including in particular the Neumann boundary conditions. However, the case*

$$-\Delta u=f \ \ in\ \Omega; \quad \frac{\partial u}{\partial n}\Big|_{\partial\Omega}=0 \tag{8.14}$$

needs some special care due to the fact that the solution u of (8.14) *is only determined up to an additive constant* (see Problem 8.1).

8.1.1.4 An Equivalent Approach Based on Separation of Variables

It is worthwhile to note that the matrix decomposition algorithm described above can be interpreted as a discrete version of the *separation of variables* for partial differential equations. Indeed, consider the eigenvalue problem:

$$\begin{cases} \text{Find } u_N\in P_N^0 \text{ and } \mu \text{ such that} \\ a_\omega(u_N,v_N)=\mu(u_N,v_N)_\omega, \quad \forall v_N\in P_N^0. \end{cases} \tag{8.15}$$

It is clear that (8.15) is the weighted Galerkin approximation of the eigenvalue problem

$$-u_{xx} = \mu u, \quad u(\pm 1) = 0. \tag{8.16}$$

The formulation (8.15) reduces to the generalized eigenvalue problem

$$\lambda A\bar{x} = B\bar{x}, \tag{8.17}$$

with $\lambda = 1/\mu$. Let Λ be the diagonal matrix whose diagonal entries $\{\lambda_p\}$ are the eigenvalues of (8.17), and let $E = (e_{jk})$ be the matrix whose columns are the eigenvectors of (8.17), i.e.,

$$BE = AE\Lambda. \tag{8.18}$$

Then the functions

$$\psi_k(x) := \sum_{j=0}^{N-2} e_{jk}\phi_j(x), \quad k = 0, 1, \ldots, N-2, \tag{8.19}$$

where $\mu_k = 1/\lambda_k$ are solutions of (8.15). The fact that E is non-singular implies that $\{\psi_k\}_{k=0}^{N-2}$ also form a basis of P_N^0. Furthermore, we have

$$\begin{aligned} m_{ln} := (\psi_n, \psi_l)_\omega &= \sum_{k,j=0}^{N-2} e_{kn}e_{jl}(\phi_k, \phi_j)_\omega \\ &= \sum_{k,j=0}^{N-2} e_{kn}b_{jk}e_{jl} = (E^TBE)_{ln} = (E^TAE\Lambda)_{ln}, \end{aligned} \tag{8.20}$$

and

$$\begin{aligned} s_{ln} := a_\omega(\psi_n, \psi_l) &= \sum_{k,j=0}^{N-2} e_{kn}e_{jl}a_\omega(\phi_k, \phi_j) \\ &= \sum_{k,j=0}^{N-2} e_{kn}a_{jk}e_{jl} = (E^TAE)_{ln}. \end{aligned} \tag{8.21}$$

Remark 8.3. *In the Legendre case, $A = I$ and $BE = E\Lambda$. Therefore, $\{\psi_k\}_{k=0}^{N-2}$ are mutually orthogonal with respect to the inner products (u,v) and (u',v'). More precisely,*

$$m_{ln} = (\psi_n, \psi_l) = \lambda_l\delta_{ln}, \;\; s_{ln} = (\psi_n', \psi_l') = \delta_{ln}.$$

Therefore, this set of basis functions are very similar to the Fourier basis functions for periodic problems and can be very attractive in many situations (cf. Shen and Wang (2007b)).

In the Chebyshev case, $\{\psi_k\}_{k=0}^{N-2}$ are no longer mutually orthogonal, but the associated mass matrix M with entries $m_{ln} = (\psi_n, \psi_l)_\omega$ and stiffness matrix S with entries $s_{ln} = a_\omega(\psi_n, \psi_l)$ are related by $M = S\Lambda$.

Hence, setting

$$u_N = \sum_{n,l=0}^{N-2} \tilde{v}_{nl} \psi_n(x)\phi_l(y), \quad V = (\tilde{v}_{nl})_{n,l=0,1,\ldots,N-2},$$

$$h_{kj} = (I_N f, \psi_k(x)\phi_j(y)), \quad H = (h_{kj})_{k,j=0,1,\ldots,N-2},$$

we find that (8.2) is equivalent to

$$\alpha MVB + SVB + MVA^T = H. \tag{8.22}$$

We derive from $M = S\Lambda$ that

$$\alpha \Lambda VB + VB + \Lambda VA^T = S^{-1}H, \tag{8.23}$$

which is exactly the same as (8.9) by noting that $U = EV$ and $H = E^T F$.

Remark 8.4. *In the Legendre case, it is particularly interesting to use* $\{\psi_k(x)\psi_j(y)\}$ *as basis functions. The resulting algorithm corresponds to the diagonalization in both x and y directions, and leads to diagonal mass and stiffness matrices. However, this algorithm is not as efficient as the partial diagonalization algorithm since two more matrix–matrix multiplications are needed.*

Remark 8.5. *One can also consider a spectral-Galerkin method for solving fourth-order equations of the form*

$$\alpha u - \beta \Delta u + \Delta^2 u = f, \quad \text{in } \Omega = (-1,1)^2; \quad u|_{\partial\Omega} = \frac{\partial u}{\partial n}\Big|_{\partial\Omega} = 0 \tag{8.24}$$

by using a tensor-product approach as in the second-order case. Unfortunately, due to the fact that the above equation is not separable, one can not directly apply the partial or full diagonalization technique. However, by using $\{J_k^{-2,-2}(x)J_j^{-2,-2}(y)\}$ *as basis functions, the resulting linear system can still be solved, with essentially the same computational complexity as for a second-order equation, by using the Sherman-Morrison-Woodbury formula* (cf. Golub and Van Loan (1996)). *We refer to* Shen (1994) *for the Legendre case and* Bjørstad and Tjostheim (1997) *for the Chebyshev case.*

8.1.2 Three-Dimensional Case

The three-dimensional case can be treated most straightforwardly by using the full diagonalization procedure (see Problem 8.2). However, a more efficient approach is to diagonalize two of the three directions successively and solve the third direction directly. To fix the idea, we consider the expansion

$$\boldsymbol{X}_N = \text{span}\{\phi_n(x)\phi_m(y)\psi_k(z) : n,m,k = 0,1,\ldots,N-2\},$$

where $\{\phi_j\}_{j=0}^{N-2}$ is a set of basis functions of P_N^0 and $\{\psi_j\}_{j=0}^{N-2}$ are defined in (8.19). This expansion corresponds to the diagonalization in the z direction. Denote

$$u_N = \sum_{n,m,k=0}^{N-2} \tilde{u}_{nm}^{(k)} \phi_n(x)\phi_m(y)\psi_k(z), \quad U^{(k)} = (\tilde{u}_{nm}^{(k)})_{0\le n,m\le N-2};$$
$$f_{nm}^{(k)} = (I_N f, \phi_n(x)\phi_m(y)\psi_k(z))_\omega, \quad F^{(k)} = (f_{nm}^{(k)})_{0\le n,m\le N-2}. \tag{8.25}$$

We also recall that $A=(a_{ij})$ with $a_{ij} = -(\phi_j'', \phi_i)_\omega$, and $B=(b_{ij})$ with $b_{ij} = (\phi_j, \phi_i)_\omega$. Then, by (8.20)–(8.21) and Remark 8.3, we have

$$\begin{aligned} &-(\partial_{xx}u_N, \phi_i(x)\phi_j(y)\psi_l(z))_\omega \\ &\quad = -\sum_{n,m,k=0}^{N-2} \tilde{u}_{nm}^{(k)} (\phi_n''(x), \phi_i(x))_\omega (\phi_m(y), \phi_j(y))_\omega (\psi_k(z), \psi_l(z))_\omega \\ &\quad = \sum_{n,m,k=0}^{N-2} \tilde{u}_{nm}^{(k)} a_{in} b_{jm} m_{lk} = \sum_{n,m,k=0}^{N-2} \tilde{u}_{nm}^{(k)} a_{in} b_{jm} s_{lk} \lambda_k. \end{aligned}$$

Similarly,

$$\begin{aligned} -(\partial_{yy}u_N, \phi_i(x)\phi_j(y)\psi_l(z))_\omega &= \sum_{n,m,k=0}^{N-2} \tilde{u}_{nm}^{(k)} b_{in} a_{jm} s_{lk} \lambda_k, \\ -(\partial_{zz}u_N, \phi_i(x)\phi_j(y)\psi_l(z))_\omega &= \sum_{n,m,k=0}^{N-2} \tilde{u}_{nm}^{(k)} b_{in} b_{jm} s_{lk}, \\ (u_N, \phi_i(x)\phi_j(y)\psi_l(z))_\omega &= \sum_{n,m,k=0}^{N-2} \tilde{u}_{nm}^{(k)} b_{in} b_{jm} s_{lk} \lambda_k. \end{aligned}$$

Therefore, (8.2) in the three dimensional case is equivalent to

$$\begin{aligned} &\sum_{n,m,k=0}^{N-2} \tilde{u}_{nm}^{(k)} \{ \alpha b_{in} b_{jm} \lambda_k + a_{in} b_{jm} \lambda_k + b_{in} a_{jm} \lambda_k + b_{in} b_{jm} \} s_{lk} \\ &\qquad = f_{ij}^{(l)}, \quad i,j,l = 0,1,\ldots,N-2, \end{aligned}$$

which can be rewritten in the following compact form

$$\lambda_l (AU^{(l)}B + BU^{(l)}A^T) + (\alpha\lambda_l + 1) BU^{(l)}B = F^{(l)} S^{-T} := G^{(l)}, \tag{8.26}$$

for all $l = 0,1,\ldots,N-2$. Note that for each l, (8.26) is of the form (8.5) which corresponds to a two-dimensional problem and can be solved in $O(N^3)$ operations. In summary, the solution of (8.2) in the three dimensional case consists of the following steps:

1. Pre-processing: compute the eigenvalues and eigenvectors (Λ, E) of the generalized eigenvalue problem (8.7), and compute $g_{ij} = \phi_j(x_i)$, $h_{ij} = \psi_j(x_i)$ and S^{-1}

2. Compute $f_{nm}^{(k)} = (I_N f, \phi_n(x)\phi_m(y)\psi_k(z))_\omega$
3. Compute $G^{(l)}$ and obtain $U^{(l)}$ by solving (8.26) for $l = 0, 1, \ldots, N-2$
4. Compute

$$u_N(x_i, y_j, z_l) = \sum_{n,m,k=0}^{N-2} \tilde{u}_{nm}^{(k)} \phi_n(x_i)\phi_m(y_j)\psi_k(z_l) = \sum_{n,m,k=0}^{N-2} \tilde{u}_{nm}^{(k)} g_{in} g_{jm} h_{lk},$$

for $i, j, l = 0, 1, \ldots, N-2$.

The cost for each of Steps 2 to 4 is a small multiple of N^4 operations, which consist of mainly matrix–matrix multiplications.

Remark 8.6. *Step 2 can be most efficiently computed by using the Legendre-Gauss-Lobatto quadrature. However, in the Chebyshev case, it is more efficient to carry out Steps 2 and 4 using fast discrete Chebyshev transforms.*

8.2 Circular and Cylindrical Domains

We shall develop in this section spectral-Galerkin algorithms for two-dimensional circular and three-dimensional cylindrical domains. Most of the material below is taken from Shen (1997, 2000) from which one can find a more detailed presentation. For a comprehensive discussion of the spectral methods for three dimensional axisymmetric domains, we refer to the book by Bernardi et al. (1999).

8.2.1 Dimension Reduction and Pole Conditions

We first consider the following model equation on a unit disk:

$$\begin{aligned} &\alpha U - \Delta U = F \quad \text{in } \Omega = \{(x,y) : x^2 + y^2 < 1\}, \\ &U = 0 \ \text{on } \partial\Omega. \end{aligned} \tag{8.27}$$

A weak formulation of (8.27) is

$$\begin{cases} \text{Find } U \in H_0^1(\Omega) \text{ such that} \\ A(U,V) := \alpha \displaystyle\int_\Omega UV\,dxdy + \int_\Omega \nabla U \cdot \nabla V\,dxdy \\ \qquad = \displaystyle\int_\Omega FV\,dxdy, \quad \forall V \in H_0^1(\Omega). \end{cases} \tag{8.28}$$

Applying the polar transformation $x = r\cos\theta$, $y = r\sin\theta$ to (8.27), and setting

$$u(r,\theta) = U(r\cos\theta, r\sin\theta), \quad f(r,\theta) = F(r\cos\theta, r\sin\theta),$$

we obtain

$$\begin{aligned} &\alpha u - \frac{1}{r}(ru_r)_r - \frac{1}{r^2}u_{\theta\theta} = f, \quad (r,\theta) \in Q := (0,1) \times [0,2\pi), \\ &u(1,\theta) = 0, \ \ \theta \in [0,2\pi); \quad u \text{ is periodic in } \theta. \end{aligned} \tag{8.29}$$

Correspondingly, the weak formulation (8.28) becomes

$$\begin{cases} \text{Find } u \in X \text{ such that} \\ a(u,v) := \displaystyle\int_Q u_r v_r r dr d\theta + \int_Q \frac{1}{r} u_\theta v_\theta dr d\theta \\ \quad + \alpha \displaystyle\int_Q u v r dr d\theta = \int_Q f v r dr d\theta, \quad \forall v \in X, \end{cases} \tag{8.30}$$

where

$$\begin{aligned} X = \Big\{ u : & u(1,\theta) = 0 \text{ for } \theta \in [0,2\pi), u \text{ is } 2\pi\text{-periodic, and} \\ & \int_Q \big(u^2 r + \frac{1}{r} u_\theta^2\big) dr d\theta < \infty \Big\}. \end{aligned} \tag{8.31}$$

Since the polar transformation is singular at the pole $r = 0$, additional pole conditions should be imposed for the solution of (8.29) so as to have desired regularity in the Cartesian coordinates. In fact, if the function

$$u(r,\theta) = \sum_{m=0}^{\infty} \big(u_{1m}(r)\cos(m\theta) + u_{2m}(r)\sin(m\theta) \big) \tag{8.32}$$

(hereafter, we assume that $u_{20}(r) \equiv 0$) were to be infinitely differentiable in the Cartesian coordinates, the following pole conditions would have to be imposed (cf. Orszag and Patera (1983)):

$$u_{1m}(r) = O(r^m), \ u_{2m}(r) = O(r^m) \quad \text{as } r \to 0 \ \text{ for } \ m = 1,2,\ldots. \tag{8.33}$$

Obviously, it is not computationally efficient to impose all the pole conditions in (8.33). Since our approximations will be based on the weak formulation (8.30) which is well defined if $u_\theta(0,\theta) = 0$ for $\theta \in [0,2\pi)$, or equivalently,

$$u_{1m}(0) = u_{2m}(0) = 0 \quad \text{for } \ m \neq 0, \tag{8.34}$$

(8.34) will be referred to as the *essential* pole conditions for (8.29), while all other conditions in (8.33) will be called *natural* or *nonessential* pole conditions. Although it is possible to impose any given number of pole conditions in (8.33) in a numerical scheme, it is generally inefficient, and may lead to ill-posed linear systems if *more than necessary* pole conditions are imposed so that the total number of boundary conditions in the radial direction *exceeds* the order of the underlying differential equation. On the other hand, ignoring the *essential* pole condition(s) will lead to inaccurate results.

We emphasize that the accuracy of a spectral approximation to (8.30) is only affected by the smoothness of the solution u in the polar coordinates. In particular, the singularity of the solution u at the pole in Cartesian coordinates will not degrade the accuracy of the spectral Galerkin schemes presented below.

We now describe our spectral approximations to (8.29). We shall first reduce the two-dimensional problem (8.29) to a sequence of one-dimensional problems by using the Fourier expansion in θ direction. More precisely, given a cut-off number $M>0$, let $(f^{1m}(r), f^{2m}(r))$ be the discrete Fourier coefficients of f in the expansion:

$$f(r,\theta_j) = \sum_{m=0}^{M} \left(f^{1m}(r)\cos(m\theta_j) + f^{2m}(r)\sin(m\theta_j)\right), \tag{8.35}$$

where $\left\{\theta_j = j\pi/M\right\}_{j=0}^{2M-1}$ are the Fourier-collocation points. We define a Fourier-spectral approximation to the solution u of (8.29) by

$$u_M(r,\theta) = \sum_{m=0}^{M} \left(u^{1m}(r)\cos(m\theta) + u^{2m}(r)\sin(m\theta)\right), \tag{8.36}$$

where $(u^{1m}(r), u^{2m}(r))$ $(m = 0,1,\ldots,M)$ satisfy

$$\begin{aligned} &-\frac{1}{r}(ru_r^{1m})_r + \left(\frac{m^2}{r^2}+\alpha\right)u^{1m} = f^{1m}(r), \quad 0<r<1,\\ &-\frac{1}{r}(ru_r^{2m})_r + \left(\frac{m^2}{r^2}+\alpha\right)u^{2m} = f^{2m}(r), \quad 0<r<1,\\ &u^{1m}(0) = u^{2m}(0) = 0 \text{ if } m\neq 0, \quad u^{1m}(1) = u^{2m}(1) = 0. \end{aligned} \tag{8.37}$$

Remark 8.7. *The extra pole conditions $u_r^{10}(0) = u_r^{20}(0) = 0$ used by many authors (see, for instance,* Gottlieb and Orszag (1977), Canuto et al. (1987), Eisen et al. (1991), Huang and Sloan (1993), Fornberg (1995)) *are derived from the parity argument on the expansion* (8.36)*. It is, however, not part of the essential pole condition for* (8.29)*. Although in most cases, there is no harm to impose extra pole conditions, we choose not to do so since its implementation is more complicated and it may fail to give accurate results in some extreme* (*but still legitimate*) *cases, e.g., when the exact solution is a function of $r-1$.*

8.2.2 Spectral-Galerkin Method for a Bessel-Type Equation

It is now clear that after the Fourier transform in the θ direction, we only have to consider the approximation of the following one-dimensional Bessel-type equation:

$$\begin{aligned} &-\frac{1}{r}(ru_r)_r + \left(\frac{m^2}{r^2}+\alpha\right)u = f, \quad 0<r<1;\\ &u(0) = 0 \text{ if } m\neq 0, \quad u(1) = 0, \end{aligned} \tag{8.38}$$

where u and f now represent generic functions. It is important to note that a direct treatment of (8.38) is not quite appropriate, since the measure $rdrd\theta$ related to the polar coordinate transformation is not taken into account.

Now let us derive a weighted weak formulation, which is suitable for both the Legendre and Chebyshev methods. We first make a coordinate transformation $r=(1+t)/2$ in (8.38). Setting $v(t)=u\big((1+t)/2\big)$, we get

$$\begin{aligned}&-\frac{1}{1+t}((1+t)v_t)_t+\Big(\frac{m^2}{(1+t)^2}+\frac{\alpha}{4}\Big)v=\frac{1}{4}f\Big(\frac{1+t}{2}\Big),\quad t\in I=(-1,1),\\&v(-1)=0\ \text{ if } m\neq 0,\quad v(1)=0.\end{aligned}\tag{8.39}$$

Thus, a weighted weak formulation for (8.39) is

$$\begin{cases}\text{Find } v\in X(m) \text{ such that}\\ \big((1+t)v_t,(w\omega)_t\big)+\Big(\dfrac{m^2}{1+t}v,w\Big)_\omega+\beta\big((1+t)v,w\big)_\omega\\ \qquad=(I_Ng,w)_\omega,\quad \forall w\in X(m),\end{cases}\tag{8.40}$$

where for all $m\neq 0$,

$$X(m)=\Big\{v: v(\pm 1)=0 \text{ and } \int_I(1+t)v_t(v\omega)_t dt+\int_I\frac{1}{1+t}v^2\omega dt<\infty\Big\},$$

and

$$X(0)=\Big\{v: v(1)=0 \text{ and } \int_I(1+t)v_t(v\omega)_t dt+\int_I(1+t)v^2\omega dt<\infty\Big\},$$

$\beta=\alpha/4$, $g(t)=\frac{1}{4}(1+t)f\big((1+t)/2\big)$, and I_N is the interpolation operator relative to the Gauss-Lobatto points.

Given a set of basis functions $\{\phi_j\}_{j=0}^{q}$ of $X_N(m):=X(m)\cap P_N$, where $q=N-2$ (resp. $q=N-1$) if $m\neq 0$ (resp. $m=0$), we denote

$$\begin{aligned}a_{ij}&=\int_I(1+t)\,\phi_j'\,(\phi_i\omega)'\,dt,\quad A=(a_{ij})_{i,j=0,1,\ldots,q};\\ b_{ij}&=\int_I\frac{1}{1+t}\,\phi_j\,\phi_i\,\omega\,dt,\quad B=(B_{ij})_{i,j=0,1,\ldots,q};\\ c_{ij}&=\int_I(1+t)\,\phi_j\,\phi_i\,\omega\,dt,\quad C=(C_{ij})_{i,j=0,1,\ldots,q};\\ f_i&=\int_I I_Ng\,\phi_i\,\omega\,dt,\quad \mathbf{f}=(f_0,f_i,\ldots,f_q)^T;\\ v_N&=\sum_{i=0}^{q}x_i\phi_i(t),\quad \mathbf{x}=(x_0,x_i,\ldots,x_q)^T,\end{aligned}\tag{8.41}$$

then (8.40) reduces to the linear system

$$(A+m^2B+\beta C)\mathbf{x}=\mathbf{f}. \tag{8.42}$$

Next, we determine exactly the entries of matrices A, B and C for the Legendre ($\omega=1$) and Chebyshev ($\omega=(1-x^2)^{-1/2}$) cases separately.

8.2.2.1 Legendre-Galerkin Approximation

Case $m\neq 0$: In this case, we take $\phi_i(t)=L_i(t)-L_{i+2}(t)$ so that

$$X_N(m)=\text{span}\{\phi_i(t):\ i=0,1,\ldots,N-2\}.$$

It turns out that the variable coefficients of the form $(1+t)^{\pm 1}$ do not lead to dense matrices. In fact, we have

Lemma 8.1. *The matrices A and B are symmetric tridiagonal with*

$$a_{ij}=\begin{cases}2i+4, & j=i+1,\\ 4i+6, & j=i,\end{cases}\qquad b_{ij}=\begin{cases}-\dfrac{2}{i+2}, & j=i+1,\\ \dfrac{2(2i+3)}{(i+1)(i+2)}, & j=i.\end{cases} \tag{8.43}$$

The matrix C is symmetric seven-diagonal with

$$c_{ij}=\begin{cases}-\dfrac{2(i+3)}{(2i+5)(2i+7)}, & j=i+3,\\ -\dfrac{2}{2i+5}, & j=i+2,\\ \dfrac{2}{(2i+1)(2i+5)}+\dfrac{2(i+3)}{(2i+5)(2i+7)}, & j=i+1,\\ \dfrac{2}{2i+1}+\dfrac{2}{2i+5}, & j=i.\end{cases} \tag{8.44}$$

Proof. It is obvious from the definition that the matrices A, B and C are symmetric positive definite. The formula for a_{ij} can be easily obtained by using the following properties of the Legendre polynomials derived from (3.168), (3.176a) and (3.176d):

$$\phi_i'(t)=-(2i+3)L_{i+1}(t), \tag{8.45}$$

$$(i+1)L_{i+1}(t)=(2i+1)tL_i(t)-iL_{i-1}(t), \tag{8.46}$$

and

$$\phi_i(t)=\frac{2i+3}{(i+1)(i+2)}(1-t^2)L_{i+1}'(t). \tag{8.47}$$

Therefore, setting temporarily $\alpha_j = \frac{2j+3}{(j+1)(j+2)}$, and using (8.47) successively, integration by parts and (8.45), we have

$$\begin{aligned} b_{ij} &= \int_I \frac{1}{1+t}\phi_j\phi_i\,dt = \alpha_j\int_I (1-t)L'_{j+1}\phi_i\,dt \\ &= -\alpha_j\int_I L_{j+1}\left((1-t)\phi_i\right)'dt \\ &= \alpha_j\int_I L_{j+1}\left((2i+3)(1-t)L_{i+1}+\phi_i\right)dt. \end{aligned}$$

The formula for b_{ij} can be easily established by using (8.46), and the formula for c_{ij} can be derived similarly. □

Case $m=0$: In this case, we take $\phi_i(t) = L_i(t) - L_{i+1}(t)$ so that

$$X_N(0) = \text{span}\{\phi_i(t) : i = 0, 1, \ldots, N-1\}.$$

Lemma 8.2. *The matrix A is diagonal with $a_{ii} = 2i+2$. The matrix C is symmetric penta-diagonal with*

$$c_{ij} = \begin{cases} -\dfrac{2(i+2)}{(2i+3)(2i+5)}, & j = i+2, \\ \dfrac{4}{(2i+1)(2i+3)(2i+5)}, & j = i+1, \\ \dfrac{4(i+1)}{(2i+1)(2i+3)}, & j = i. \end{cases}$$

Proof. It is easy to see that $a_{ij} = 0$ for $i \neq j$. On the other hand, using interpolation by parts gives

$$a_{ii} = \left((1+t)\phi'_i, \phi'_i\right) = -(\phi'_i, \phi_i) - \left((1+t)\phi''_i, \phi_i\right).$$

Direct computations using (8.46) lead to $a_{ii} = 2i+2$ and the formula for c_{ij}. □

8.2.2.2 Chebyshev-Galerkin Approximation

Case $m \neq 0$: We take $\phi_i(t) = T_i(t) - T_{i+2}(t)$ so that

$$X_N(m) = \text{span}\{\phi_i(t) : i = 0, 1, \ldots, N-2\}.$$

The direct computation of the elements of A, B and C is very involved, but it can be substantially simplified by using the following results:

$$\tilde{a}_{ij} = -\int_I \phi''_j\phi_i\,\omega\,dt = \begin{cases} 2\pi(i+1)(i+2), & j = i, \\ 4\pi(i+1), & j = i+2, i+4, i+6, \ldots, \\ 0, & \text{otherwise}; \end{cases} \tag{8.48}$$

$$\tilde{b}_{ij} = \int_I \phi_j \phi_i \omega\, dt = \begin{cases} \frac{d_i+1}{2}\pi, & j = i, \\ -\frac{\pi}{2}, & j = i-2 \text{ or } i+2, \\ 0, & \text{otherwise}; \end{cases} \tag{8.49}$$

$$\tilde{c}_{ij} = \int_I \phi_j' \phi_i \omega\, dt = \begin{cases} \pi(i+1), & j = i+1, \\ -\pi(i+1), & j = i-1, \\ 0, & \text{otherwise}, \end{cases} \tag{8.50}$$

where $d_0 = 2$ and $d_i = 1$ for $i \geq 1$. In the above, (8.48) was derived in Lemma 4.4, while (8.49) and (8.50) can be easily established using (3.214) and (3.216a).

Lemma 8.3.

- *A is an upper Hessenberg matrix with*

$$a_{ij} = \begin{cases} (i+1)^2\pi, & j = i-1, \\ 2(i+1)(i+2)\pi, & j = i, \\ (i+1)(i+5)\pi, & j = i+1, \\ 4(i+1)\pi, & j \geq i+2. \end{cases} \tag{8.51}$$

- *B is a symmetric tridiagonal matrix with*

$$b_{ij} = \begin{cases} 2\pi, & j = i, \\ -\pi, & j = i+1, \\ 0, & \text{otherwise}. \end{cases} \tag{8.52}$$

- *C is a symmetric seven-diagonal matrix with*

$$c_{ij} = \begin{cases} \frac{d_i+1}{2}\pi, & j = i, \\ \frac{d_{i-1}}{4}\pi, & j = i-1, \\ -\frac{\pi}{2}, & j = i-2, \\ -\frac{\pi}{4}, & j = i-3. \end{cases} \tag{8.53}$$

Proof. We have

$$\begin{aligned} a_{ij} &= -\int_I ((1+t)\phi_j')' \phi_i \omega\, dt = -\int_I \phi_j'' \phi_i \omega\, dt - \int_I (t\phi_j')' \phi_i \omega\, dt \\ &= \tilde{a}_{ij} - \tilde{c}_{ij} - \int_I \phi_j'' t \phi_i \omega\, dt. \end{aligned} \tag{8.54}$$

It is clear from the definition that $a_{ij} = 0$ if $j < i-1$. On the other hand, by (3.207),

$$t\phi_i = t(T_i - T_{i+2}) = \frac{1}{2}(T_{i-1} - T_{i+3}) = \frac{1}{2}(\phi_{i-1} + \phi_{i+1}), \quad i \geq 1, \tag{8.55}$$

which, together with (8.54), implies

$$a_{ij} = \tilde{a}_{ij} - \tilde{c}_{ij} + \frac{1}{2}(\tilde{a}_{i-1,j} + \tilde{a}_{i+1,j}).$$

The formula (8.51) is then a direct consequence of the above relation, (8.48) and (8.50).

It is easy to see from the definition that $b_{ij} = 0$ if $|i-j| > 1$. By (3.212c) and (3.216a),

$$\phi_j(t) = (1-t^2)\frac{2}{j+1}T'_{j+1}(t) = (1-t^2)\sum_{\substack{k=0\\ k+j \text{ even}}}^{j} \frac{4}{d_k}T_k(t), \tag{8.56}$$

where $d_0 = 2$ and $d_k = 1$ for $k \geq 1$. Therefore,

$$\begin{aligned} b_{ij} &= \int_I \frac{1}{1+t}\phi_j\,\phi_i\,\omega\,dt = \sum_{\substack{k=0\\ k+j \text{ even}}}^{j} \frac{4}{d_k}\int_I (1-t)\,\phi_i\,T_k\,\omega\,dt \\ &= \sum_{\substack{k=0\\ k+j \text{ even}}}^{j} \frac{4}{d_k}\int_I \Big(\phi_i - \frac{1}{2}(\phi_{i-1} + \phi_{i+1})\Big)\,T_k\,\omega\,dt. \end{aligned} \tag{8.57}$$

The elements b_{ii} and $b_{i,i+1}$ can then be easily computed from the above relation.

Finally, by using (8.55), we find

$$c_{ij} = \tilde{b}_{ij} + \frac{1}{2}(\tilde{b}_{i-1,j} + \tilde{b}_{i+1,j}).$$

Hence, (8.53) is a direct consequence of the above relation and (8.49). □

Remark 8.8. *Although the matrix A is not sparse,* (8.42) *can still be solved in* $O(N)$ *operations by taking advantage of the special structure of A, namely,* $a_{ij} = 4(i+1)\pi$ *for* $j \geq i+2$.

An alternative is to use a new set of basis functions $\phi_i(t) = (1-t^2)T_i(t)$ (cf. Heinrichs (1989)). *It is easy to verify that in this case B and C are symmetric sparse matrices with* $b_{ij} = 0$ *for* $|i-j| > 3$ *and* $c_{ij} = 0$ *for* $|i-j| > 5$. *One can also show by using integration by parts that A is a non-symmetric sparse matrix with* $a_{ij} = 0$ *for* $|i-j| > 3$. *Thus,* (8.42) *can also be solved in* $O(N)$ *operations* (see Problem 8.3).

Case $m = 0$: We take $\phi_i(t) = T_i(t) - T_{i+1}(t)$ so that

$$X_N(0) = \text{span}\{\phi_i(t) : i = 0, 1, \ldots, N-1\}.$$

Lemma 8.4.

- *A is an upper-triangular matrix with*

$$a_{ij} = \begin{cases} (i+1)\pi^2, & j = i, \\ (i-j)\pi, & j = i+1, i+3, i+5\ldots, \\ (i+j+1)\pi, & j = i+2, i+4, i+6\ldots. \end{cases} \tag{8.58}$$

- *C is a symmetric penta-diagonal matrix with non-zero elements*

$$\begin{aligned} c_{ii} &= \frac{\pi}{2}, \quad i = 0,1,\ldots,N-1, \\ c_{i,i+2} = c_{i+2,i} &= -\frac{\pi}{4}, \quad i = 0,1,\ldots,N-3, \\ c_{01} = c_{10} &= \frac{\pi}{4}. \end{aligned} \tag{8.59}$$

Proof. The computation of c_{ij} is straightforward by using the orthogonality of the Chebyshev polynomials and the relation derived from (3.207):

$$t\phi_i(t) = t\big(T_i(t) - T_{i+1}(t)\big) = \frac{1}{2}\big(\phi_{i-1}(t) + \phi_{i+1}(t)\big),\ i \geq 1.$$

However, the computation of a_{ij} is quite involved. The idea is to use the relations (3.216a) and (3.216b) to expand $\big((1+t)\phi_j'(t)\big)'$ in Chebyshev series. The details are left to the interested readers. □

Remark 8.9. *Once again, the matrix A is not sparse. But* (8.42) (*with* $m = 0$) *can still be solved in* $O(N)$ *operations by exploring the special structure of A. We refer to* P. 80-81 *in* Shen (1995) *for more details on this procedure for a similar problem.*

8.2.3 Another Fourier-Chebyshev Galerkin Approximation

The algorithms in the previous section were developed without taking into account the inherent parity of the Fourier coefficients $u_m(r) := (u_{1m}(r), u_{2m}(r))$ in (8.32). In fact, the expansion coefficients in (8.32) can not be arbitrary since it is well-known (see, e.g., Orszag and Patera (1983), Canuto et al. (1987), Fornberg (1995)) that $u_m(r)$ has the same parity as m and can be expanded *smoothly* to the interval $[-1,0]$, i.e., if u_m is in $H^k(0,1)$, then the expanded function is in $H^k(-1,1)$. In particular, for $u \in C(\Omega)$ we have

$$u_m \in Y^{(m)} := \big\{v \in C(-1,1) : v(-r) = (-1)^m v(r),\ r \in (0,1)\big\},$$

and consequently,

$$u \in Y := \Big\{v = \sum_{|m|=0}^{\infty} v_m(r) e^{im\theta} :\ v_{-m}(r) = \bar{v}_m(r) \ \text{and}\ v_m \in Y^{(m)}\Big\}.$$

A consequence of not taking into account the parity of $u_m(r)$ is that we can not *smoothly* extend it to the interval $[-1,0]$. So we are forced to use the Gauss-Lobatto points based on the interval $[0,1]$, resulting a severe clustering of points near the origin (see the second row of Fig. 8.1 and Remark 8.10 below).

We construct below approximations of $Y^{(m)}$ and Y which preserve the odd-even parity and lead to better distribution of collocation points (see the first row of Fig. 8.1). We shall consider only the Chebyshev case and leave the Legendre case to the interested readers.

8.2.3.1 A Fourier-Chebyshev Interpolation Operator on the Unit Disk

Define

$$\psi_j^{(m)}(r) = \begin{cases} T_{2j}(r), & \text{if } m \text{ is even}, \\ T_{2j+1}(r), & \text{if } m \text{ is odd}. \end{cases} \tag{8.60}$$

Given a pair of even integers (N,M), we introduce the spaces:

$$Y_N^{(m)} := \Big\{ v = \sum_{j=0}^{N/2-\mathrm{mod}(m,2)} v_j \psi_j^{(m)}(r) : v_j \text{ are complex numbers} \Big\}, \tag{8.61}$$

and

$$Y_{NM} := \Big\{ v = \sum_{|m|=0}^{M} v_m(r) e^{\mathrm{i}m\theta} : v_m \in Y_N^{(m)},\ v \text{ is real} \Big\}. \tag{8.62}$$

We further define a set of collocation points on $\bar{Q}$ relative to Y_{NM} by

$$\Sigma_{NM} := \left\{ (r_k, \theta_j) : \begin{array}{r} k = 0,1,\ldots,N/2-1,\ j = 0,1,\ldots,2M-1 \\ k = N/2,\ j = 0,1,\ldots,M-1 \end{array} \right\}, \tag{8.63}$$

where $r_k = \cos(k\pi/N)$ and $\theta_j = j\pi/M$. Note that for $v \in Y$ or Y_{NM}, $v(0,\theta) = v(0,\pi+\theta)$ for all θ. Hence, the points (r_k,θ_j) with $k = N/2$ and $j = M, M+1,\ldots,2M-1$ are excluded from Σ_{NM}.

One can now readily check that there exists a unique interpolation operator $I_{NM} : Y \cap C(\bar{Q}) \to Y_{NM}$, defined by

$$I_{NM} g(r,\theta) = \sum_{|m|=0}^{M} \sum_{n=0}^{N/2-\mathrm{mod}(m,2)} g_{nm} \psi_n^{(m)}(r) e^{\mathrm{i}m\theta} \in Y_{NM}, \tag{8.64}$$

such that

$$(I_{NM} g)(r_k,\theta_j) = g(r_k,\theta_j), \quad \forall (r_k,\theta_j) \in \Sigma_{NM}. \tag{8.65}$$

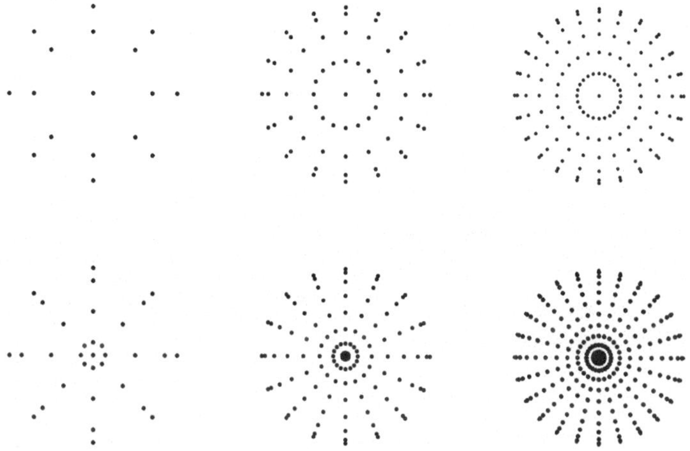

Fig. 8.1 Distribution of collocation points: first row - -Σ_{NM}; second row – $\tilde{\Sigma}_{NM}$

Remark 8.10. *In the previous section, the change of variable $r = (1+t)/2$ is applied so that the functions in the transformed spaces no longer satisfy the odd-even parity condition and that the corresponding set of collocation points in $\bar{Q}$ is*

$$\tilde{\Sigma}_{NM} := \{((t_k+1)/2, \theta_j) : k = 0,1,\ldots,N,\ j = 0,1,\ldots,2M-1\}$$

with $t_k = \cos(k\pi/N)$ and $\theta_j = j\pi/M$. Not only $\tilde{\Sigma}_{NM}$ has twice as many points as Σ_{NM}, but also the points are unnecessarily clustered in the radial direction near the pole ($r = 0$), see the second row of Fig. 8.1. Indeed, the smallest distance in the Cartesian coordinates between two adjacent points in $\tilde{\Sigma}_{NM}$ (resp. Σ_{NM}) near the pole is of order $O(N^{-2}M^{-1})$ (resp. $O(N^{-1}M^{-1})$). Besides being wasteful, this unnecessary clustering may also lead to severe time step constraints when an explicit scheme is used for time discretization.

8.2.3.2 Description of the Algorithm

Denoting $u_m = (u^{1m}, u^{2m})$ and $f_m = (f^{1m}, f^{2m})$, we rewrite (8.37) as

$$-\frac{1}{r}\frac{d}{dr}\Big(r\frac{d}{dr}u_m\Big) + \Big(\frac{m^2}{r^2} + \alpha\Big)u_m = f_m, \quad r \in (0,1), \quad u_m(1) = 0. \tag{8.66}$$

Note that we have dropped off the pole condition $u_m(0) = 0$ for $m \neq 0$ which was *essential* for the formulation (8.38). However, this pole condition will become *natural* for the formulation (8.68) below.

We now seek approximation of u_m in the space

$$X_N^{(m)} := \{v \in Y_N^{(m)} : v(1) = 0\}. \tag{8.67}$$

We consider the following weighted (with the weight function $r^2\omega(r)$) spectral-Galerkin approximation to (8.66):

$$\begin{cases} \text{Find } u_N^{(m)} \in X_N^{(m)} \text{ such that} \\ -\int_0^1 \frac{d}{dr}\Big(r\frac{d}{dr}u_N^{(m)}\Big) rv\omega\, dr + m^2 \int_0^1 u_N^{(m)} v\omega\, dr \\ \qquad + \alpha \int_0^1 r^2 u_N^{(m)} v\omega\, dr = \int_0^1 r^2 f_N^{(m)} v\omega\, dr, \quad \forall v \in X_N^{(m)}, \end{cases} \tag{8.68}$$

for $m = 0, 1, \ldots, M$, where $\omega(r) = (1-r^2)^{-1/2}$ is the Chebyshev weight function, and $f_N^{(m)}$ is the m-th component of

$$I_{NM} f = \sum_{|m|=0}^{M} \sum_{n=0}^{N/2-\text{mod}(m,2)} f_{nm} \psi_n^{(m)}(r) e^{\text{i}m\theta},$$

namely,

$$f_N^{(m)} = \sum_{n=0}^{N/2-\text{mod}(m,2)} f_{nm} \psi_n^{(m)}(r).$$

Then, the approximation to u is given by

$$u_{NM}(r,\theta) = \sum_{|m|=0}^{M} u_N^{(m)}(r) e^{\text{i}m\theta}, \quad u_N^{(-m)} = \bar{u}_N^{(m)}.$$

Note that u_{NM} is neither necessarily single valued nor differentiable at the pole in the Cartesian coordinates. However, u_{NM} still converges to u exponentially provided that f is smooth in the Cartesian coordinates.

It is clear that $X_N^{(m)}$ is a $N/2$ (resp. $N/2-1$) dimensional space if m is even (resp. odd) and that

$$\phi_j^{(m)}(r) := (1-r^2)\psi_j^{(m)}(r) \in X_N^{(m)}. \tag{8.69}$$

Therefore,

$$X_N^{(m)} = \text{span}\{\phi_j^{(m)} : j = 0, 1, \ldots, q := N/2 - 1 - \text{mod}(m,2)\}. \tag{8.70}$$

Thus, setting

$$\begin{aligned} u_N^{(m)} &= \sum_{k=0}^{q} x_k^{(m)} \phi_k^{(m)}, \quad \mathbf{x}^{(m)} = (x_0^{(m)}, \ldots, x_q^{(m)})^T; \\ a_{kj}^{(m)} &:= -\int_0^1 \frac{d}{dr}\Big(r\frac{d}{dr}\phi_j^{(m)}\Big) r\phi_k^{(m)} \omega\, dr, \quad A^{(m)} = (a_{kj}^{(m)})_{0\le k,j\le q}; \\ b_{kj}^{(m)} &:= \int_0^1 \phi_j^{(m)} \phi_k^{(m)} \omega dr, \quad B^{(m)} = (b_{kj}^{(m)})_{0\le k,j\le q}; \\ c_{kj}^{(m)} &:= \int_0^1 r^2 \phi_j^{(m)} \phi_k^{(m)} \omega dr, \quad C^{(m)} = (c_{kj}^{(m)})_{0\le k,j\le q}; \\ f_k^{(m)} &:= \int_0^1 r^2 f_N^{(m)} \phi_k^{(m)} \omega\, dr, \quad \mathbf{f}^{(m)} = (f_0^{(m)}, \ldots, f_q^{(m)})^T, \end{aligned} \tag{8.71}$$

the formulation (8.68) reduces to the linear system

$$\big(A^{(m)} + m^2 B^{(m)} + \alpha C^{(m)}\big)\mathbf{x}^{(m)} = \mathbf{f}^{(m)}. \tag{8.72}$$

Note that although an index m is used in $A^{(m)}$, $B^{(m)}$, $C^{(m)}$ and $X_N^{(m)}$, these matrices only depend on the parity of m, rather than the actual value of m.

Lemma 8.5. *For m even or odd, $A^{(m)}$ and $B^{(m)}$ are penta-diagonal matrices, and $C^{(m)}$ is a seven-diagonal matrix.*

Proof. Notice that all the integrands in $a_{kj}^{(m)}$, $b_{kj}^{(m)}$ and $c_{kj}^{(m)}$ are even functions. Therefore, we can replace the integral $\int_0^1$ by $\frac{1}{2}\int_{-1}^{1}$. Then, thanks to the orthogonality relation of the Chebyshev polynomials and the special form of the basis functions (8.69), one derives immediately that $B^{(m)}$ and $C^{(m)}$ are respectively penta- and seven-diagonal symmetric matrices. By the same argument, we have

$$a_{kj}^{(m)} = -\frac{1}{2}\int_{-1}^{1} \frac{d}{dr}\Big(r\frac{d}{dr}\phi_j^{(m)}\Big) r\phi_k^{(m)}\omega\, dr = 0 \text{ for } j < k-2.$$

On the other hand, integrating by parts twice, using (8.69) and the identity $\omega'(r) = \frac{r}{1-r^2}\omega(r)$, we have

$$\begin{aligned}
a_{kj}^{(m)} &= \frac{1}{2}\int_{-1}^{1}\Big(r\frac{d}{dr}\phi_j^{(m)}\Big)\frac{d}{dr}\big(r\phi_k^{(m)}\omega\big)dr \\
&= \frac{1}{2}\int_{-1}^{1}\frac{d}{dr}\phi_j^{(m)}\Big(r^2\frac{d}{dr}\phi_k^{(m)} + \frac{r}{1-r^2}\phi_k^{(m)}\Big)\omega\, dr \\
&= -\frac{1}{2}\int_{-1}^{1}\phi_j^{(m)}\Big\{\frac{d}{dr}\Big(r^2\frac{d}{dr}\phi_k^{(m)} + \frac{r}{1-r^2}\phi_k^{(m)}\Big) \\
&\qquad + \Big(r^2\frac{d}{dr}\phi_k^{(m)} + \frac{r}{1-r^2}\phi_k^{(m)}\Big)\frac{r}{1-r^2}\Big\}\omega\, dr \\
&= -\frac{1}{2}\int_{-1}^{1}\psi_j^{(m)}\Big\{(1-r^2)\frac{d}{dr}\Big(r^2\frac{d}{dr}\phi_k^{(m)} + r\psi_k^{(m)}\Big) \\
&\qquad + \Big(r^2\frac{d}{dr}\phi_k^{(m)} + r\psi_k^{(m)}\Big)r\Big\}\omega\, dr.
\end{aligned}$$

Then, thanks to the special form of the basis functions in (8.69), we find that the function between the pair of brackets is a polynomial of degree $2k+4$ (resp. $2k+5$) for m even (resp. odd). Therefore, we have $a_{kj}^{(m)} = 0$ if $k < j-2$. □

The entries of $A^{(m)}$, $B^{(m)}$ and $C^{(m)}$ can be evaluated exactly, but this process can be quite tedious. Alternatively, one can compute these entries automatically by using the Chebyshev-Gauss-Lobatto quadrature with $N+2$ nodes.

8.2.4 Numerical Results and Discussions

We now present some numerical results using the two Fourier-Chebyshev algorithms, which we shall refer to as CFG1 and CFG2, presented in Sects. 8.2.2.2 and 8.2.3, respectively.

We consider the Poisson equation on a unit disk with the exact solution

$$U(x,y) = (x^2+y^2-1)\big(\cos(\beta(x+y))+\sin(\beta(x+y))\big). \tag{8.73}$$

The maximum errors of the two algorithms for the exact solution (8.73) with $\beta = 16$ are listed in Table 8.1. This exact solution is smooth in both the Cartesian and polar coordinates, so both algorithms converge exponentially fast. Note that to achieve the same accuracy as CFG1 with the pair (N,N), CFG2 should be used, roughly speaking, with the pair $(N+10, N+10)$ for this particular example. We recall that *for a fixed pair of (N,M), the number of unknowns and the CPU time of* CFG2 *are about half of* CFG1 (cf. Shen (1997)). Thus, CFG2 could be significantly more efficient, in terms of CPU and memory, than CFG1. Another advantage of CFG2 is that the collocation points are not unnecessarily clustered in the radial direction near the pole.

Table 8.1 Maximum errors: exact solution being (8.73) with $\beta = 16$

$N=M$	22	26	30	34	38	42
CFG1	8.68E-3	3.55E-4	8.52E-6	1.17E-7	9.99E-10	5.52E-12
$N=M$	32	36	40	44	48	52
CFG2	1.98E-2	3.98E-4	4.63E-6	3.34E-8	1.59E-10	4.58E-13

The computational complexity for solving (8.29) using each of the methods presented above is $O(NM)+2T(NM)$, where N and M are respectively the cut-off number of the spectral expansion in radial and axial directions, and $T(NM)$ is the cost of one forward or inverse discrete transform of the form

$$\begin{aligned} g(t_i,\theta_j) &= \sum_{n=0}^{N}\left(\sum_{m=0}^{M}\left(g_n^{1m}\cos(m\theta_j)+g_n^{2m}\sin(m\theta_j)\right)\right)p_n(t_j), \\ &i=0,1,\ldots,N,\ \ j=0,1,\ldots,2M-1, \end{aligned} \tag{8.74}$$

where $p_n(t)$ is the Chebyshev or Legendre polynomial of degree n. Therefore,

$$T(NM) = N^2M + O(NM\log_2 M)$$

for the Fourier-Legendre Galerkin method, while

$$T(NM) = O(NM\log_2 M) + O(NM\log_2 N)$$

for the Fourier-Chebyshev Galerkin method. Thus, the computational complexity of the Fourier-Chebyshev Galerkin method is quasi-optimal.

8.2.5 Three-Dimensional Cylindrical Domains

Consider the model equation in a cylinder:

$$\begin{aligned} & -\Delta U + \alpha U = F \quad \text{in } \Omega = \{(x,y,z) : x^2 + y^2 < 1,\ z \in I := (-1,1)\}, \\ & U = 0 \quad \text{on } \partial\Omega. \end{aligned} \tag{8.75}$$

Applying the cylindrical transformation

$$x = r\cos\theta, \quad y = r\sin\theta, \quad z = z,$$

and setting

$$u(r,\theta,z) = U(r\cos\theta, r\sin\theta, z), \quad f(r,\theta,z) = F(r\cos\theta, r\sin\theta, z),$$

we obtain

$$\begin{aligned} & -\frac{1}{r}(ru_r)_r - \frac{1}{r^2}u_{\theta\theta} - u_{zz} + \alpha u = f, \quad (r,\theta,z) \in (0,1) \times [0,2\pi) \times I, \\ & u = 0 \ \text{at } r = 1 \text{ or } z = \pm 1, \quad u \text{ is periodic in } \theta, \end{aligned} \tag{8.76}$$

with the essential pole conditions

$$\frac{\partial}{\partial\theta}u(0,\theta,z) = 0, \quad (\theta,z) \in [0,2\pi) \times I. \tag{8.77}$$

As before, let $(f^{1m}(r,z), f^{2m}(r,z))$ be defined by

$$f(r,z,\theta_j) = \sum_{m=0}^{M} \left(f^{1m}(r,z)\cos(m\theta_j) + f^{2m}(r,z)\sin(m\theta_j)\right), \tag{8.78}$$

where $\{\theta_j = j\pi/M\}_{j=0}^{2M-1}$. Then, a Fourier-spectral approximation to u is given by

$$u_M(r,\theta,z) = \sum_{m=0}^{M} \left(u^{1m}(r,z)\cos(m\theta) + u^{2m}(r,z)\sin(m\theta)\right)$$

with $u^{im}(i=1,2)$ satisfying the following two-dimensional equation

$$\begin{aligned} & -u^{im}_{zz} - \frac{1}{r}\left(ru^{im}_r\right)_r + \left(\frac{m^2}{r^2} + \alpha\right)u^{im} = f^{im}(r,z) \quad \text{in } \Omega = (0,1) \times I, \\ & u^{im} = 0 \ \text{at } r = 0 \text{ if } m \neq 0, \quad u^{im} = 0 \text{ at } r = 1 \ \text{ or } z = \pm 1. \end{aligned} \tag{8.79}$$

We then make a coordinate transformation $r = (1+t)/2$. Denoting

$$v(t,z) = u^{im}(r,z), \quad g(t,z) = \frac{1}{4}(1+t)f^{im}(r,z), \quad \beta = \frac{\alpha}{4},$$

we obtain the prototypical two-dimensional equation

$$\begin{aligned} &-\frac{1+t}{4}v_{zz}-((1+t)v_t)_t+\Big(m^2\frac{1}{1+t}+\beta(1+t)\Big)v=g, \quad (t,z)\in I\times I,\\ &v=0 \ \text{ at } t=-1 \text{ if } m\neq 0, \quad v=0 \ \text{ at } t=1 \text{ or } z=\pm 1. \end{aligned} \tag{8.80}$$

Let us denote $\psi_i(z)=p_i(z)-p_{i+2}(z)$ and $\phi_i(t)=p_i(t)-p_{i+s(m)}(t)$ where $s(m)=2$ if $m\neq 0$ and $s(0)=1$, and p_j is either the Legendre or Chebyshev polynomial of degree j. Let

$$X_N(m)=\text{span}\{\phi_i(t)\psi_j(z): 0\le i\le N-s(m),\, 0\le j\le N-2\}.$$

Then a spectral-Galerkin approximation to (8.80) is

$$\begin{cases} \text{Find } v_N\in X_N(m) \text{ such that}\\ \frac{1}{4}((1+t)\partial_z v_N,\partial_z(w\omega))+((1+t)\partial_t v_N,\partial_t(w\omega))\\ +m^2\Big(\frac{1}{1+t}v_N,w\Big)_\omega+\beta\big((1+t)v_N,w\big)_\omega=(I_N g,w)_\omega, \quad \forall w\in X_N(m), \end{cases} \tag{8.81}$$

where $\omega\equiv 1$ in the Legendre case and $\omega=\big((1-t^2)(1-z^2)\big)^{-1/2}$ in the Chebyshev case, $(\cdot,\cdot)_\omega$ is the weighted L^2-inner product in $I\times I$, and I_N is the interpolation operator relative to the tensor product of the Legendre- or Chebyshev-Gauss type points. Setting $q=N-s(m)$ and

$$\begin{aligned} a_{ij}&=\frac{1}{4}\int_I(1+t)\,\phi_j'\,(\phi_i\omega(t))'\,dt, \quad A=(a_{ij})_{i,j=0,1,\ldots,q};\\ b_{ij}&=\int_I\frac{1}{1+t}\,\phi_j\,\phi_i\,\omega(t)\,dt, \quad B=(B_{ij})_{i,j=0,1,\ldots,q};\\ c_{ij}&=\int_I(1+t)\,\phi_j\,\phi_i\,\omega(t)\,dt, \quad C=(C_{ij})_{i,j=0,1,\ldots,q};\\ d_{ij}&=\int_I\psi_j\,\psi_i\,\omega(z)\,dz, \quad D=(D_{ij})_{i,j=0,1,\ldots,N-2};\\ e_{ij}&=\int_I\psi_j'\,(\psi_i\,\omega(z))'\,dz, \quad E=(E_{ij})_{i,j=0,1,\ldots,N-2}, \end{aligned} \tag{8.82}$$

and

$$\begin{aligned} f_{ij}&=\int_{I\times I}I_N g\,\phi_i\psi_j\omega\,dt\,dz, \quad F=(f_{ij})_{0\le i\le q,0\le j\le N-2};\\ v_N&=\sum_{i=0}^{q}\sum_{j=0}^{N-2}x_{ij}\phi_i\psi_j, \quad X=(x_{ij})_{0\le i\le q,0\le j\le N-2}, \end{aligned} \tag{8.83}$$

then (8.81) reduces to

$$CXE+(A+m^2B+\beta C)XD=F. \tag{8.84}$$

The entries of A, B and C in the Legendre or Chebyshev case are explicitly given in the previous section, while those of D and E can be computed by using the properties of the Legendre or Chebyshev polynomials given in Chap. 3 (also see Shen (1994, 1995)). This matrix equation can be efficiently solved, in particular, by using the matrix decomposition method (cf. Sect. 8.1). More precisely, we consider the following generalized eigenvalue problem $E^T \mathbf{g} = \lambda D \mathbf{g}$, and let Λ be the diagonal matrix formed by the eigenvalues and by G be the matrix formed by the corresponding eigenvectors. Then,

$$E^T G = D G \Lambda \quad \text{or} \quad G^T E = \Lambda G^T D. \tag{8.85}$$

It is well-known that the eigenvalues are all real positive (the Legendre case is trivial while the Chebyshev case can be proved as in Gottlieb and Lustman (1983)). Making a change of variable $X = Y G^T$ in (8.84), we find

$$C Y G^T E + \big(A + m^2 B + \beta C\big) Y G^T D = F.$$

We then derive from (8.85) that

$$C Y \Lambda + \big(A + m^2 B + \beta C\big) Y = F D^{-1} G^{-T}. \tag{8.86}$$

The above matrix equation is nothing but a sequence of $N-1$ one-dimensional equation (8.42). In summary, after the pre-processing for the computation of the eigen-pair (Λ, G) and G^{-1} (in the Legendre case, G is an orthonormal matrix, i.e., $G^{-1} = G^T$), the solution of (8.81) for each m consists of three steps:

1. Compute $F D^{-1} G^{-T}$ with $N^3 + O(N^2)$ flops;
2. Solving Y from (8.86) with $O(N^2)$ flops;
3. Set $X = Y G^T$ with N^3 flops.

8.3 Spherical Domains

We consider in this section the spectral-Galerkin method for solving the model problem (8.1) in spherical domains.

8.3.1 Spectral Methods on the Surface of a Sphere

We start with the following model equation on the surface of a unit sphere

$$\alpha U - \Delta U = F \quad \text{on } S := \big\{(x,y,z) : x^2 + y^2 + z^2 = 1\big\}. \tag{8.87}$$

Applying the spherical transformation

$$x = \cos\phi \sin\theta, \quad y = \sin\phi \sin\theta, \quad z = \cos\theta \tag{8.88}$$

to (8.87), and setting

$$u(\theta,\phi) = U(x,y,z), \quad f(\theta,\phi) = F(x,y,z), \quad D := (0,\pi) \times [0,2\pi),$$

we obtain

$$\alpha u - \Delta_S u := \alpha u - \frac{1}{\sin\theta}\partial_\theta(\sin\theta\,\partial_\theta u) - \frac{1}{\sin^2\theta}\partial_\phi^2 u = f, \quad (\theta,\phi) \in D, \tag{8.89}$$

where Δ_S is the so-called Laplace-Beltrami operator.

If $\alpha > 0$, the above equation has a unique solution, while for $\alpha = 0$, the compatibility condition

$$\int_0^{2\pi} d\phi \int_0^{\pi} f(\theta,\phi)\sin\theta\, d\theta = 0 \tag{8.90}$$

should be satisfied, and the solution u is only determined up to an additive constant.

The most straightforward way to solve (8.89) is to use spherical harmonic functions. We recall that the spherical harmonic functions $\{Y_l^m\}$ are defined by

$$Y_l^m(\theta,\phi) = \sqrt{\frac{(2l+1)(l-m)!}{4\pi(l+m)!}} P_l^m(\cos\theta)e^{im\phi}, \quad l \geq |m| \geq 0, \tag{8.91}$$

where P_l^m is the associated Legendre functions given by

$$P_l^m(x) = \frac{(-1)^m}{2^l l!}(1-x^2)^{m/2}\frac{d^{l+m}}{dx^{l+m}}\{(x^2-1)^l\}, \quad m \geq 0,$$

and

$$P_l^{-m}(x) = (-1)^m \frac{(l-m)!}{(l+m)!} P_l^m(x).$$

The set of spherical harmonic functions forms a complete orthonormal system in $L^2(S)$, i.e.,

$$\int_{\theta=0}^{\pi}\int_{\phi=0}^{2\pi} Y_l^m(\theta,\phi)\overline{Y}_{l'}^{m'}(\theta,\phi)\sin\theta\, d\phi\, d\theta = \delta_{ll'}\delta_{mm'}. \tag{8.92}$$

An important property of the spherical harmonic functions is that they are eigenfunctions of the Laplace-Beltrami operator (cf. Morse and Feshback (1953)). More precisely,

$$-\Delta_S Y_l^m(\theta,\phi) = l(l+1)Y_l^m(\theta,\phi). \tag{8.93}$$

Let us set

$$X_N = \Big\{ v : v(\theta,\phi) = \sum_{l=0}^{N} \sum_{|m|=0}^{l} v_{lm} P_l^m(\cos\theta) e^{im\phi}, \ v(\theta,\phi) \text{ real} \Big\}, \tag{8.94}$$

and let

$$(\pi_N f)(\theta,\phi) = \sum_{l=0}^{N} \sum_{|m|=0}^{l} f_{lm} P_l^m(\cos\theta) e^{im\phi}$$

be the L^2-orthogonal projection of f onto X_N. Then, thanks to (8.93), the spectral-Galerkin approximation of the solution u of (8.89) in X_N is given by

$$u_N(\theta,\phi) = \sum_{l=0}^{N} \sum_{|m|=0}^{l} \frac{f_{lm}}{\alpha + l(l+1)} P_l^m(\cos\theta) e^{im\phi}. \tag{8.95}$$

Note that given the values of f at the set of collocation points

$$\Sigma_N = \Big\{ (\theta_k,\phi_j) : \theta_k = \frac{k\pi}{N}, \ 0 \le k \le N; \ \ \phi_j = \frac{2j\pi}{N}, \ 0 \le j \le N-1 \Big\}, \tag{8.96}$$

one can compute $\{f_{nm}\}$ *approximately* by using a discrete spherical harmonic transform software, e.g., SpherePack (cf. Swarztrauber and Spotz (2000)).

The main disadvantage of using spherical harmonics is that the usual discrete spherical harmonic transform is not of (quasi) optimal computational complexity. We note however that substantial progress has been made in recent years on fast discrete spherical harmonic transforms (cf. Rokhlin and Tygert (2006), Tygert (2010) and the references therein) using ideas stemming from the fast multipole method (cf. Greengard and Rokhlin (1987)). However, these algorithms are usually competitive with FFT for N quite large. An alternative is to expand functions on the sphere by using double Fourier series (cf. Orszag (1974), Boyd (1978), Shen (1999)) which enjoys fast discrete transforms thanks to FFT, but may suffer from the unnecessary clustering of points near the north and south poles.

8.3.2 Spectral Methods in a Spherical Shell

We consider the model equation

$$\begin{aligned} &\alpha U - \Delta U = F \quad \text{in } \Omega = \{(x,y,z) : 0 \le R_1 < x^2+y^2+z^2 < R_2\}, \\ &U|_{\partial\Omega} = 0. \end{aligned} \tag{8.97}$$

Applying the spherical transformation (8.88) to (8.97), and setting

$$u(r,\theta,\phi) = U(x,y,z), \quad f(r,\theta,\phi) = F(x,y,z),$$

we obtain

$$
\begin{aligned}
&\alpha u - \frac{1}{r^2}\partial_r(r^2\partial_r u) - \frac{1}{r^2\sin\theta}\partial_\theta(\sin\theta\,\partial_\theta u) - \frac{1}{r^2\sin^2\theta}\partial_\phi^2 u = f,\\
&(r,\theta,\phi)\in(R_1,R_2)\times(0,\pi)\times[0,2\pi),\\
&u=0 \ \text{ at } \ r=R_1 \ (\text{if } R_1\neq 0) \text{ and } r=R_2.
\end{aligned}
\tag{8.98}
$$

Let

$$
(\pi_N f)(r,\theta,\phi) = \sum_{l=0}^{N}\sum_{|m|=0}^{l} f_{lm}(r)P_l^m(\cos\theta)e^{im\phi}
$$

be the L^2-orthogonal projection of $f(r,\cdot,\cdot)$ onto X_N (defined in (8.94)), and let

$$
u_N(r,\theta,\phi) = \sum_{l=0}^{N}\sum_{|m|=0}^{l} u_{lm}(r)Y_l^m(\theta,\phi)
$$

be the spectral-Galerkin approximation of u in X_N. Thanks to (8.93), we find that the expansion coefficients $\{u_{lm}\}$ satisfy the following sequence of equations:

$$
\begin{aligned}
&\alpha u_{lm} - \frac{1}{r^2}(r^2u_{lm}')' + \frac{l(l+1)}{r^2}u_{lm} = f_{lm}, \quad 0\le|m|\le l\le N,\\
&u_{lm}(R_1)=0, \ \text{ if } R_1\neq 0, \ u_{lm}(R_2)=0.
\end{aligned}
\tag{8.99}
$$

Since the interval $[R_1,R_2]$ can be mapped to $[-1,1]$ by using the transform

$$
r = \frac{R_2-R_1}{2}(t+\beta) \ \text{ with } \ \beta = \frac{R_2+R_1}{R_2-R_1}\ge 1,
\tag{8.100}
$$

we only have to consider the following prototypical one-dimensional problem (after multiplying r^2 on both sides of (8.99)):

$$
\begin{aligned}
&(\alpha(t+\beta)^2+\gamma)u - ((t+\beta)^2u')' = (t+\beta)^2 f,\\
&u(-1)=0 \ \text{ if } \ \beta>1; \ u(1)=0,
\end{aligned}
\tag{8.101}
$$

where $\gamma = l(l+1)$.

Let P_K be the space of polynomials of degree $\le K$, and let

$$
Z_K = Z_K(\beta) := \{v\in P_K : \ v(-1)=0 \text{ if } \beta>1, \ v(1)=0\}.
\tag{8.102}
$$

Then, the weighted spectral-Galerkin approximation to (8.101) is

$$
\begin{cases}
\text{Find } u_K\in Z_K \text{ such that}\\
\alpha((t+\beta)^2u_K, v\omega) + ((t+\beta)^2u_K', (v\omega)') + \gamma(u_K, v\omega)\\
\qquad = ((t+\beta)^2 J_K f, v\omega), \quad \forall v\in Z_K,
\end{cases}
\tag{8.103}
$$

where $\omega\equiv 1$ in the Legendre case and $\omega(t) = (1-t^2)^{-1/2}$ in the Chebyshev case, $(u,v) = \int_{-1}^{1} uv\,dt$, and J_K is the interpolation operator based on the Legendre- or Chebyshev-Gauss-Lobatto points.

Let $\{\phi_k\}_{k=0}^{p}$with $p = K - I_\beta$ be a set of basis functions of Z_K, where $I_\beta = 1$ if $\beta = 1$ while $I_\beta = 2$ if $\beta > 1$. Set

$$\begin{aligned} q_{kj} &= \int_{-1}^{1} (t+\beta)^2 \phi_j \phi_k \omega \, dt, \quad Q = (q_{kj})_{0\le k,j\le p}; \\ r_{kj} &= \int_{-1}^{1} (t+\beta)^2 \phi_j' (\phi_k \omega)' \, dt, \quad R = (r_{kj})_{0\le k,j\le p}; \\ s_{kj} &= \int_{-1}^{1} \phi_j \phi_k \omega \, dt, \quad S = (s_{kj})_{0\le k,j\le p}; \\ f_j &= \int_{-1}^{1} (t+\beta)^2 J_K f \phi_j \omega \, dt, \quad \mathbf{f} = (f_0, \ldots, f_p)^T; \\ u_N &= \sum_{j=0}^{p} x_j \phi_j(t), \quad \mathbf{x} = (x_0, \ldots, x_p)^T. \end{aligned} \tag{8.104}$$

Then, (8.103) becomes

$$(\alpha Q + R + \gamma S)\mathbf{x} = \mathbf{f}. \tag{8.105}$$

The efficiency of the method depends on the choice of the basis functions which in turn determine the structure of the matrices Q, R and S.

To simplify the notation, we shall only consider the case $R_1 > 0$ (i.e., $\beta > 1$). The case $R_1 = 0$ (i.e., $\beta = 1$) can be treated similarly.

We present below the Legendre- and Chebyshev-Galerkin method for (8.103).

- **Legendre-Galerkin**: In this case, we set $\omega = 1$ and $\phi_j(t) = L_j(t) - L_{j+2}(t)$. By using the identities (8.45)–(8.47), one can readily derive that Q, R and S are positive definite symmetric matrices with $q_{ij} = 0$ for $|i-j| > 4$, $r_{ij} = 0$ for $|i-j| > 2$ and $s_{ij} = 0$ for $j \neq i, i \pm 2$ (see Chap. 4).
- **Chebyshev-Galerkin**: We set $\omega = (1-t^2)^{-1/2}$ and $\phi_j(t) = (1-t^2)T_j(t)$. It can be easily shown that Q and S are positive definite symmetric matrices with $q_{ij} = 0$ for $|i-j| > 6$, $s_{ij} = 0$ for $|i-j| > 4$ and $|i-j|$ odd. Although R is non-symmetric, it can be shown that R is banded with $r_{ij} \neq 0$ for $i-4 \le j \le i+4$. Indeed, it is easy to see that

$$r_{ij} = -\int_I ((t+\beta)^2 \phi_j')'(1-t^2)T_i \omega dt = 0 \text{ for } i > j+4.$$

On the other hand, thanks to identity $\omega'(t) = t\omega^3(t)$ and integration by parts,

$$\begin{aligned} r_{ij} &= \int_I (t+\beta)^2 \phi_j' ((1-t^2)T_i \omega)' dt \\ &= \int_I (t+\beta)^2 \phi_j' \big(((1-t^2)T_i)' + tT_i\big) \omega dt \\ &= \int_I \phi_j' P_{i+3} \omega dt = -\int_I T_j (1-t^2)(P_{i+3}\omega)' dt \\ &= -\int_I T_j \big((1-t^2)P_{i+3}' + tP_{i+3}\big) \omega dt = \int_I T_j P_{i+4} \omega dt, \end{aligned}$$

where P_{i+3} (resp. P_{i+4}) is a polynomial of degree less than or equal to $i+3$ (resp. $i+4$). Hence, $r_{ij}=0$ for $j>i+4$.

Although it is very tedious to determine their non-zero entries by hand, one can easily compute them by using appropriate Gaussian quadratures.

Remark 8.11. *In case* $R_1=0$ (*i.e.*, $\beta=1$), *the appropriate basis functions are* $\phi_j(t)=L_j(t)-L_{j+1}(t)$ *in the Legendre case and* $\phi_j(t)=(1-t)T_j(t)$ *in the Chebyshev case.*

Higher-order equations can be solved in a similar fashion. For instance, by using the expansion in spherical harmonics, the biharmonic equation would reduce to a set of one-dimensional fourth-order equations which can be solved efficiently by using a spectral-Galerkin method (*see* Shen (1997) *for a similar case*).

8.4 Multivariate Jacobi Approximations

In this section, we extend the one-dimensional polynomial approximation results in Chap. 3 to d-dimensional tensor product spaces.

8.4.1 Notation and Preliminary Properties

Let us first introduce some notation.

- Let $\mathbb{R}$ (resp. $\mathbb{N}$) be the set of all real numbers (resp. non-negative integers), and let $\mathbb{N}_0=\mathbb{N}\cup\{0\}$.
- For $d\in\mathbb{N}$, we use boldface lowercase letters to denote d-dimensional multi-indexes and vectors, e.g., $\boldsymbol{k}=(k_1,\ldots,k_d)\in\mathbb{N}_0^d$ and $\boldsymbol{\alpha}=(\alpha_1,\ldots,\alpha_d)\in\mathbb{R}^d$. Also, let $\mathbf{1}=(1,1,\ldots,1)\in\mathbb{N}^d$, and let $\boldsymbol{e}_i=(0,\ldots,1,\ldots,0)$ be the ith unit vector in $\mathbb{R}^d$. For a scalar $s\in\mathbb{R}$, we define the following component-wise operations:

$$\boldsymbol{\alpha}+\boldsymbol{k}=(\alpha_1+k_1,\ldots,\alpha_d+k_d),\quad \boldsymbol{\alpha}+s:=\boldsymbol{\alpha}+s\mathbf{1}=(\alpha_1+s,\ldots,\alpha_d+s),\quad (8.106)$$

  and use the following conventions:

$$\boldsymbol{\alpha}\ge\boldsymbol{k}\ \Leftrightarrow\ \forall_{1\le j\le d}\ \ \alpha_j\ge k_j;\ \ \boldsymbol{\alpha}\ge s\ \Leftrightarrow\ \boldsymbol{\alpha}\ge s\mathbf{1}\ \Leftrightarrow\ \forall_{1\le j\le d}\ \ \alpha_j\ge s.\quad (8.107)$$

- Denote

$$|\boldsymbol{k}|_1=\sum_{j=1}^{d}k_j,\quad |\boldsymbol{k}|_\infty=\max_{1\le j\le d}k_j.\qquad (8.108)$$

- Let $I:=(-1,1)$ and $I^d:=(-1,1)^d$. Given a multivariate function $u(\boldsymbol{x})$, we denote the $|\boldsymbol{k}|_1$-th (mixed) partial derivative by

$$\partial_{\boldsymbol{x}}^{\boldsymbol{k}}u=\frac{\partial^{|\boldsymbol{k}|_1}u}{\partial x_1^{k_1}\ldots\partial x_d^{k_d}}=\partial_{x_1}^{k_1}\ldots\partial_{x_d}^{k_d}u.\qquad (8.109)$$

In particular, we denote $\partial_{\boldsymbol{x}}^{s} u := \partial_{\boldsymbol{x}}^{s\mathbf{1}} u = \partial_{\boldsymbol{x}}^{(s,s,\ldots,s)} u$.
Given a generic weight function $\boldsymbol{\omega}(\boldsymbol{x})$ in I^d, we define the weighted Sobolev spaces $H^r_{\boldsymbol{\omega}}(I^d)$ with the norm $\|\cdot\|_{r,\boldsymbol{\omega},I^d}$ as in Adams (1975). In particular, we have $L^2_{\boldsymbol{\omega}}(I^d) = H^0_{\boldsymbol{\omega}}(I^d)$, and denote its inner product and norm by $(\cdot,\cdot)_{\boldsymbol{\omega},I^d}$ and $\|\cdot\|_{\boldsymbol{\omega},I^d}$, respectively. If $\boldsymbol{\omega}(\boldsymbol{x}) \equiv 1$, we drop $\boldsymbol{\omega}$ in the above notations.

- As before, let $P_N(I)$ be the set of all real polynomials of degree $\le N$ in I.
- The notation $A \simeq B$ means that the ratio A/B with $B \ne 0$ approaches to 1 in certain limiting process.

Hereafter, we consider a normalization of the Jacobi polynomials different from that in Chap. 3. More precisely, let $J_n^{\alpha,\beta}(x)$ be the normalized Jacobi polynomials such that

$$\int_{-1}^{1} J_n^{\alpha,\beta}(x) J_{n'}^{\alpha,\beta}(x)\omega^{\alpha,\beta}(x)dx = \delta_{nn'}. \tag{8.110}$$

One derives from (3.98) that the normalized Jacobi polynomials satisfy

$$\partial_x J_n^{\alpha,\beta}(x) = \sqrt{\lambda_n^{\alpha,\beta}}\, J_{n-1}^{\alpha+1,\beta+1}(x), \quad n \ge 1. \tag{8.111}$$

Applying this formula recursively leads to

$$\partial_x^k J_n^{\alpha,\beta}(x) = \sqrt{\chi_{n,k}^{\alpha,\beta}}\, J_{n-k}^{\alpha+k,\beta+k}(x), \quad n \ge k \ge 1, \tag{8.112}$$

where the factor

$$\chi_{n,k}^{\alpha,\beta} = \prod_{j=0}^{k-1} \lambda_{n-j}^{\alpha+j,\beta+j} = \frac{n!\Gamma(n+k+\alpha+\beta+1)}{(n-k)!\Gamma(n+\alpha+1+t)}, \quad n \ge k \ge 1, \tag{8.113}$$

One verifies readily that for all $n \ge j+1 \ge 1$ and $\alpha,\beta > -1$,

$$\lambda_{n-j-1}^{\alpha+j+1,\beta+j+1} - \lambda_{n-j}^{\alpha+j,\beta+j} = -(2j+\alpha+\beta+2) < 0, \tag{8.114}$$

which indicates that $\lambda_{n-j}^{\alpha+j,\beta+j}$ is strictly descending with respect to j. Hence, there holds

$$\left(\lambda_{n-k+1}^{\alpha+k-1,\beta+k-1}\right)^k \le \chi_{n,k}^{\alpha,\beta} \le \left(\lambda_n^{\alpha,\beta}\right)^k, \quad n \ge k \ge 1,\ \alpha,\beta > -1. \tag{8.115}$$

For notational convenience, we extend the definition of $\chi_{n,k}^{\alpha,\beta}$ to all $n,k \ge 0$ by defining

$$\chi_{n,0}^{\alpha,\beta} = 1, \quad \text{if } n \ge 0; \qquad \chi_{n,k}^{\alpha,\beta} = 0, \quad \text{for } k > n \ge 0. \tag{8.116}$$

We deduce from (8.110), (8.112) and (8.116) that $\{\partial_x^k J_n^{\alpha,\beta}\}_{n=k}^{\infty}$ are mutually orthogonal with respect to the weight function $\omega^{\alpha+k,\beta+k}$, and

$$\|\partial_x^k J_n^{\alpha,\beta}\|^2_{\omega^{\alpha+k,\beta+k},I} = \chi_{n,k}^{\alpha,\beta}, \quad n,k \ge 0,\ \alpha,\beta > -1. \tag{8.117}$$

Defining the d-dimensional tensorial Jacobi polynomial and Jacobi weight function as

$$\boldsymbol{J}_{\boldsymbol{n}}^{\boldsymbol{\alpha},\boldsymbol{\beta}}(\boldsymbol{x})=\prod_{j=1}^{d} J_{n_j}^{\alpha_j,\beta_j}(x_j),\ \ \boldsymbol{\omega}^{\boldsymbol{\alpha},\boldsymbol{\beta}}(\boldsymbol{x})=\prod_{j=1}^{d}\omega^{\alpha_j,\beta_j}(x_j),\ \forall\,\boldsymbol{\alpha},\boldsymbol{\beta}>-1,\ \boldsymbol{x}\in I^d, \tag{8.118}$$

we derive from (8.112) and (8.117) that

$$\partial_{\boldsymbol{x}}^{\boldsymbol{k}}\boldsymbol{J}_{\boldsymbol{n}}^{\boldsymbol{\alpha},\boldsymbol{\beta}}(\boldsymbol{x})=\sqrt{\boldsymbol{\chi}_{\boldsymbol{n},\boldsymbol{k}}^{\boldsymbol{\alpha},\boldsymbol{\beta}}}\,\boldsymbol{J}_{\boldsymbol{n}-\boldsymbol{k}}^{\boldsymbol{\alpha}+\boldsymbol{k},\boldsymbol{\beta}+\boldsymbol{k}}(\boldsymbol{x})\quad\text{with}\quad \boldsymbol{\chi}_{\boldsymbol{n},\boldsymbol{k}}^{\boldsymbol{\alpha},\boldsymbol{\beta}}=\prod_{j=1}^{d}\chi_{n_j,k_j}^{\alpha_j,\beta_j}, \tag{8.119}$$

and

$$\int_{I^d}\partial_{\boldsymbol{x}}^{\boldsymbol{k}}\boldsymbol{J}_{\boldsymbol{n}}^{\boldsymbol{\alpha},\boldsymbol{\beta}}(\boldsymbol{x})\,\partial_{\boldsymbol{x}}^{\boldsymbol{k}}\boldsymbol{J}_{\boldsymbol{m}}^{\boldsymbol{\alpha},\boldsymbol{\beta}}(\boldsymbol{x})\,\boldsymbol{\omega}^{\boldsymbol{\alpha}+\boldsymbol{k},\boldsymbol{\beta}+\boldsymbol{k}}(\boldsymbol{x})\,d\boldsymbol{x}=\boldsymbol{\chi}_{\boldsymbol{n},\boldsymbol{k}}^{\boldsymbol{\alpha},\boldsymbol{\beta}}\,\boldsymbol{\delta}_{\boldsymbol{n}\boldsymbol{m}}, \tag{8.120}$$

where $\boldsymbol{n},\boldsymbol{k}\ge 0$, $\boldsymbol{\alpha},\boldsymbol{\beta}>-1$ and $\boldsymbol{\delta}_{\boldsymbol{n}\boldsymbol{m}}=\prod_{j=1}^{d}\delta_{n_j m_j}$.

For any $u\in L^2_{\boldsymbol{\omega}^{\boldsymbol{\alpha},\boldsymbol{\beta}}}(I^d)$, we write

$$u(\boldsymbol{x})=\sum_{\boldsymbol{n}\ge 0}\hat u_{\boldsymbol{n}}^{\boldsymbol{\alpha},\boldsymbol{\beta}}\boldsymbol{J}_{\boldsymbol{n}}^{\boldsymbol{\alpha},\boldsymbol{\beta}}(\boldsymbol{x})\quad\text{with}\quad \hat u_{\boldsymbol{n}}^{\boldsymbol{\alpha},\boldsymbol{\beta}}=\int_{I^d}u(\boldsymbol{x})\boldsymbol{J}_{\boldsymbol{n}}^{\boldsymbol{\alpha},\boldsymbol{\beta}}(\boldsymbol{x})\boldsymbol{\omega}^{\boldsymbol{\alpha},\boldsymbol{\beta}}(\boldsymbol{x})d\boldsymbol{x}. \tag{8.121}$$

Formally, we have $\partial_{\boldsymbol{x}}^{\boldsymbol{k}}u=\sum_{\boldsymbol{n}\ge\boldsymbol{k}}\hat u_{\boldsymbol{n}}^{\boldsymbol{\alpha},\boldsymbol{\beta}}\partial_{\boldsymbol{x}}^{\boldsymbol{k}}\boldsymbol{J}_{\boldsymbol{n}}^{\boldsymbol{\alpha},\boldsymbol{\beta}}$, and by the orthogonality (8.120),

$$\|\partial_{\boldsymbol{x}}^{\boldsymbol{k}}u\|^2_{\boldsymbol{\omega}^{\boldsymbol{\alpha}+\boldsymbol{k},\boldsymbol{\beta}+\boldsymbol{k}},I^d}=\sum_{\boldsymbol{n}\ge\boldsymbol{k}}\boldsymbol{\chi}_{\boldsymbol{n},\boldsymbol{k}}^{\boldsymbol{\alpha},\boldsymbol{\beta}}|\hat u_{\boldsymbol{n}}^{\boldsymbol{\alpha},\boldsymbol{\beta}}|^2\overset{(8.116)}{=}\sum_{\boldsymbol{n}\in\mathbb{N}_0^d}\boldsymbol{\chi}_{\boldsymbol{n},\boldsymbol{k}}^{\boldsymbol{\alpha},\boldsymbol{\beta}}|\hat u_{\boldsymbol{n}}^{\boldsymbol{\alpha},\boldsymbol{\beta}}|^2. \tag{8.122}$$

8.4.2 Orthogonal Projections

Consider the orthogonal projection $\boldsymbol{\pi}_N^{\boldsymbol{\alpha},\boldsymbol{\beta}}:L^2_{\boldsymbol{\omega}^{\boldsymbol{\alpha},\boldsymbol{\beta}}}(I^d)\to P_N^d$, defined by

$$\int_{I^d}(\boldsymbol{\pi}_N^{\boldsymbol{\alpha},\boldsymbol{\beta}}u-u)v_N\boldsymbol{\omega}^{\boldsymbol{\alpha},\boldsymbol{\beta}}d\boldsymbol{x}=0,\quad \forall v_N\in P_N^d, \tag{8.123}$$

or equivalently,

$$(\boldsymbol{\pi}_N^{\boldsymbol{\alpha},\boldsymbol{\beta}}u)(\boldsymbol{x})=\sum_{\boldsymbol{n}\in\Upsilon_N}\hat u_{\boldsymbol{n}}^{\boldsymbol{\alpha},\boldsymbol{\beta}}\boldsymbol{J}_{\boldsymbol{n}}^{\boldsymbol{\alpha},\boldsymbol{\beta}}(\boldsymbol{x}), \tag{8.124}$$

where $\Upsilon_N=\{\boldsymbol{n}\in\mathbb{N}_0^d:|\boldsymbol{n}|_\infty\le N\}$. We define the d-dimensional Jacobi-weighted Sobolev space as an extension of the one-dimensional setting in (3.251):

$$B^m_{\boldsymbol{\alpha},\boldsymbol{\beta}}(I^d):=\Big\{u:\ \partial_{\boldsymbol{x}}^{\boldsymbol{k}}u\in L^2_{\boldsymbol{\omega}^{\boldsymbol{\alpha}+\boldsymbol{k},\boldsymbol{\beta}+\boldsymbol{k}}}(I^d),\ 0\le|\boldsymbol{k}|_1\le m\Big\},\quad \forall m\in\mathbb{N}_0, \tag{8.125}$$

equipped with the norm and semi-norm

$$\begin{aligned} \|u\|_{B^m_{\alpha,\beta}(I^d)} &= \Big(\sum_{0\le |\mathbf{k}|_1 \le m} \|\partial_{\mathbf{x}}^{\mathbf{k}} u\|^2_{\omega^{\alpha+\mathbf{k},\beta+\mathbf{k}},I^d} \Big)^{1/2}, \\ |u|_{B^m_{\alpha,\beta}(I^d)} &= \Big(\sum_{j=1}^{d} \|\partial_{x_j}^m u\|^2_{\omega^{\alpha+m\mathbf{e}_j,\beta+m\mathbf{e}_j},I^d} \Big)^{1/2}, \end{aligned} \tag{8.126}$$

where $\mathbf{e}_j$ is the j-th unit vector of $\mathbb{R}^d$. It is clear that $B^m_{\alpha,\beta}(I^d) \subseteq H^m_{\omega^{\alpha,\beta}}(I^d)$ and $B^0_{\alpha,\beta}(I^d) = L^2_{\omega^{\alpha,\beta}}(I^d)$.

Theorem 8.1. *Let $\alpha, \beta > -1$. For any $u \in B^m_{\alpha,\beta}(I^d)$, we have that for $0 \le l \le m \le N+1$,*

$$|\pi_N^{\alpha,\beta} u - u|_{B^l_{\alpha,\beta}(I^d)} \le c\sqrt{\frac{(N-m)!}{(N-l)!}}(N+m)^{(l-m)/2}|u|_{B^m_{\alpha,\beta}(I^d)}, \tag{8.127}$$

where $c \sim \sqrt{2}$ for $N \gg 1$.

Proof. By (8.120)-(8.122) and (8.124), we have that for any $1 \le j \le d$,

$$\begin{aligned} \|\partial_{x_j}^l (\pi_N^{\alpha,\beta} u - u)\|^2_{\omega^{\alpha+l\mathbf{e}_j,\beta+l\mathbf{e}_j},I^d} &= \sum_{|\mathbf{n}|_\infty > N, n_j \ge l} \chi^{\alpha_j,\beta_j}_{n_j,l} |\hat{u}^{\alpha,\beta}_{\mathbf{n}}|^2 \\ &= \sum_{\mathbf{n}\in\Lambda_N^{1,j}} \chi^{\alpha_j,\beta_j}_{n_j,l} |\hat{u}^{\alpha,\beta}_{\mathbf{n}}|^2 + \sum_{\mathbf{n}\in\Lambda_N^{2,j}} \chi^{\alpha_j,\beta_j}_{n_j,l} |\hat{u}^{\alpha,\beta}_{\mathbf{n}}|^2, \quad 1 \le j \le d, \end{aligned} \tag{8.128}$$

where the index sets are

$$\Lambda_N^{1,j} := \big\{ \mathbf{n} \in \mathbb{N}_0^d : |\mathbf{n}|_\infty > N,\ l \le n_j \le N \big\}, \quad \Lambda_N^{2,j} := \big\{ \mathbf{n} \in \mathbb{N}_0^d : |\mathbf{n}|_\infty > N,\ n_j > N \big\}.$$

Now, we deal with the first summation. Clearly, for any $\mathbf{n} \in \Lambda_N^{1,j}$, there exists at least one index k $(k \ne j)$ such that $n_k > N$, so we obtain from (8.122) that

$$\begin{aligned} \sum_{\mathbf{n}\in\Lambda_N^{1,j}} \chi^{\alpha_j,\beta_j}_{n_j,l} |\hat{u}^{\alpha,\beta}_{\mathbf{n}}|^2 &\le \max_{\mathbf{n}\in\Lambda_N^{1,j}} \left\{ \frac{\chi^{\alpha_j,\beta_j}_{n_j,l}}{\chi^{\alpha_k,\beta_k}_{n_k,m}} \right\} \sum_{\mathbf{n}\in\Lambda_N^{1,j}} \chi^{\alpha_k,\beta_k}_{n_k,m} |\hat{u}^{\alpha,\beta}_{\mathbf{n}}|^2 \\ &\le \max_{\mathbf{n}\in\Lambda_N^{1,j}} \left\{ \frac{\chi^{\alpha_j,\beta_j}_{n_j,l}}{\chi^{\alpha_k,\beta_k}_{n_k,m}} \right\} \|\partial_{x_k}^m u\|^2_{\omega^{\alpha+m\mathbf{e}_k,\beta+m\mathbf{e}_k},I^d} \\ &\le \frac{\chi^{\alpha_j,\beta_j}_{N,l}}{\chi^{\alpha_k,\beta_k}_{N+1,m}} \|\partial_{x_k}^m u\|^2_{\omega^{\alpha+m\mathbf{e}_k,\beta+m\mathbf{e}_k},I^d}. \end{aligned} \tag{8.129}$$

Similarly, we treat the second summation in (8.128) as

$$\sum_{\boldsymbol{n}\in\Lambda_N^{2,j}} \chi_{n_j,l}^{\alpha_j,\beta_j} |\hat{u}_{\boldsymbol{n}}^{\boldsymbol{\alpha},\boldsymbol{\beta}}|^2 \le \max_{\boldsymbol{n}\in\Lambda_N^{2,j}} \left\{ \frac{\chi_{n_j,l}^{\alpha_j,\beta_j}}{\chi_{n_j,m}^{\alpha_j,\beta_j}} \right\} \sum_{\boldsymbol{n}\in\Lambda_N^{2,j}} \chi_{n_j,m}^{\alpha_j,\beta_j} |\hat{u}_{\boldsymbol{n}}^{\boldsymbol{\alpha},\boldsymbol{\beta}}|^2$$
$$\le \frac{\chi_{N+1,l}^{\alpha_j,\beta_j}}{\chi_{N+1,m}^{\alpha_j,\beta_j}} \|\partial_{x_j}^m u\|^2_{\boldsymbol{\omega}^{\boldsymbol{\alpha}+m\boldsymbol{e}_j,\boldsymbol{\beta}+m\boldsymbol{e}_j},I^d}. \tag{8.130}$$

Therefore, we obtain from (8.113) and the Stirling's formula (A.7) that for all $1 \le j \le d$,

$$\|\partial_{x_j}^l (\boldsymbol{\pi}_N^{\boldsymbol{\alpha},\boldsymbol{\beta}} u - u)\|^2_{\boldsymbol{\omega}^{\boldsymbol{\alpha}+l\boldsymbol{e}_j,\boldsymbol{\beta}+l\boldsymbol{e}_j},I^d}$$
$$\le \hat{c}\frac{(N-m)!}{(N-l)!}(N+m)^{l-m}\left(\|\partial_{x_j}^m u\|^2_{\boldsymbol{\omega}^{\boldsymbol{\alpha}+m\boldsymbol{e}_j,\boldsymbol{\beta}+m\boldsymbol{e}_j},I^d} + \|\partial_{x_k}^m u\|^2_{\boldsymbol{\omega}^{\boldsymbol{\alpha}+m\boldsymbol{e}_k,\boldsymbol{\beta}+m\boldsymbol{e}_k},I^d}\right),$$

where $\hat{c} \simeq 1$. By the definition (8.126), summing $1 \le j \le d$ leads to the desired result. □

Remark 8.12. *Error estimates for the multi-dimensional polynomial approximations have been derived previously by several authors in various situations* (see, e.g., Bernardi and Maday (1997), Canuto et al. (2006)). *The above proof appears to be much simpler and leads to more precise results in terms of the norms on the left hand and right hand of* (8.127).

Remark 8.13. *As pointed out in Remark* 3.7, *the order of convergence is* $O(N^{l-m})$ *for fixed* l *and* m.

Remark 8.14. *It can be shown that the result in Theorem* 8.1 *is also valid if some components of* $\boldsymbol{\alpha}$ *and* $\boldsymbol{\beta}$ *are negative integers. Indeed, by replacing the classical Jacobi polynomials* (8.118) *by the corresponding generalized Jacobi polynomials* (6.1)(*which are orthonormal*), *we verify* (8.127) *by using the derivative relation*

$$\partial_x J_n^{\alpha_k,\beta_k}(x_k) = d_n J_{n-1}^{\alpha_k+1,\beta_k+1}(x_k), \quad x_k \in I, \tag{8.131}$$

where α_k *or* β_k *is a negative integer, and the explicit expression of* d_n (*behaves like* $O(n)$) *can be worked out by using* (6.12).

Next, we study the orthogonal projection in $H^1_{\boldsymbol{\omega}^{\boldsymbol{\alpha},\boldsymbol{\beta}}}(I^d)$. Let us denote

$$H^1_{0,\boldsymbol{\omega}^{\boldsymbol{\alpha},\boldsymbol{\beta}}}(I^d) = \{u \in H^1_{\boldsymbol{\omega}^{\boldsymbol{\alpha},\boldsymbol{\beta}}}(I^d) \ : \ u|_{\partial I^d} = 0\}.$$

We have the following weighted Poincaré inequality.

Lemma 8.6. *If there exists a pair of* (α_k, β_k) *such that* $|\alpha_k| < 1$ *and* $|\beta_k| < 1$, *then we can find a positive constant* c *such that*

$$\|u\|_{\boldsymbol{\omega}^{\boldsymbol{\alpha},\boldsymbol{\beta}}} \le c\|\nabla u\|_{\boldsymbol{\omega}^{\boldsymbol{\alpha},\boldsymbol{\beta}}}, \quad \forall u \in H^1_{0,\boldsymbol{\omega}^{\boldsymbol{\alpha},\boldsymbol{\beta}}}(I^d), \tag{8.132}$$

that is, the semi-norm $|\cdot|_{1,\boldsymbol{\omega}^{\boldsymbol{\alpha},\boldsymbol{\beta}}}$ *is equivalent to the norm* $\|\cdot\|_{1,\boldsymbol{\omega}^{\boldsymbol{\alpha},\boldsymbol{\beta}}}$ *in* $\in H^1_{0,\boldsymbol{\omega}^{\boldsymbol{\alpha},\boldsymbol{\beta}}}(I^d)$.

Proof. As a direct consequence of Lemma B.7, we have that for $|\alpha_k|, |\beta_k| < 1$,

$$\int_{I^d} u^2(\boldsymbol{x})\boldsymbol{\omega}^{\boldsymbol{\alpha},\boldsymbol{\beta}}(\boldsymbol{x})d\boldsymbol{x} \le c\int_{I^d} (\partial_{x_k} u(\boldsymbol{x}))^2 \omega^{\boldsymbol{\alpha},\boldsymbol{\beta}}(\boldsymbol{x})d\boldsymbol{x} \le c\|\nabla u\|^2_{\boldsymbol{\omega}^{\boldsymbol{\alpha},\boldsymbol{\beta}}}.$$

This ends the proof. □

Consider now the orthogonal projection $\boldsymbol{\pi}^{1,0}_{N,\boldsymbol{\alpha},\boldsymbol{\beta}} : H^1_{0,\boldsymbol{\omega}^{\boldsymbol{\alpha},\boldsymbol{\beta}}}(I^d) \to (P_N^0)^d$, defined by

$$a_{\boldsymbol{\alpha},\boldsymbol{\beta}}(\boldsymbol{\pi}^{1,0}_{N,\boldsymbol{\alpha},\boldsymbol{\beta}} u - u, v) = 0, \quad \forall v \in (P_N^0)^d, \tag{8.133}$$

where the bilinear form $a_{\boldsymbol{\alpha},\boldsymbol{\beta}}(u,v) := (\nabla u, \nabla(v\boldsymbol{\omega}^{\boldsymbol{\alpha},\boldsymbol{\beta}}))$.

As in the one-dimensional case, we need to show the coercivity of $a_{\boldsymbol{\alpha},\boldsymbol{\beta}}(\cdot,\cdot)$.

Lemma 8.7. *If* $-1 < \boldsymbol{\alpha}, \boldsymbol{\beta} < 1$, *then the bilinear form* $a_{\boldsymbol{\alpha},\boldsymbol{\beta}}(u,v)$ *is continuous and coercive in* $H^1_{0,\boldsymbol{\omega}^{\boldsymbol{\alpha},\boldsymbol{\beta}}}(I^d) \times H^1_{0,\boldsymbol{\omega}^{\boldsymbol{\alpha},\boldsymbol{\beta}}}(I^d)$, *namely,*

$$|a_{\boldsymbol{\alpha},\boldsymbol{\beta}}(u,v)| \le c_1\|\nabla u\|_{\omega^{\boldsymbol{\alpha},\boldsymbol{\beta}}}\|\nabla v\|_{\omega^{\boldsymbol{\alpha},\boldsymbol{\beta}}}, \quad \forall u,v \in H^1_{0,\boldsymbol{\omega}^{\boldsymbol{\alpha},\boldsymbol{\beta}}}(I^d), \tag{8.134a}$$

$$a_{\boldsymbol{\alpha},\boldsymbol{\beta}}(u,u) \ge \|\nabla u\|^2_{\omega^{\boldsymbol{\alpha},\boldsymbol{\beta}}}, \quad \forall u \in H^1_{0,\boldsymbol{\omega}^{\boldsymbol{\alpha},\boldsymbol{\beta}}}(I^d), \tag{8.134b}$$

where c_1 *is a positive constant.*

Proof. Using the Cauchy–Schwarz inequality gives

$$\begin{aligned}|a_{\boldsymbol{\alpha},\boldsymbol{\beta}}(u,v)| &\le |(\nabla u, \nabla v)_{\omega^{\boldsymbol{\alpha},\boldsymbol{\beta}}}| + |(\nabla u, v\nabla\boldsymbol{\omega}^{\boldsymbol{\alpha},\boldsymbol{\beta}})| \\ &\le \|\nabla u\|_{\omega^{\boldsymbol{\alpha},\boldsymbol{\beta}}}\|\nabla v\|_{\omega^{\boldsymbol{\alpha},\boldsymbol{\beta}}} + \|\nabla u\|_{\omega^{\boldsymbol{\alpha},\boldsymbol{\beta}}}\|v\nabla\boldsymbol{\omega}^{\boldsymbol{\alpha},\boldsymbol{\beta}}\|_{\omega^{-\boldsymbol{\alpha},-\boldsymbol{\beta}}}.\end{aligned}$$

We deduce from Lemma B.7 that

$$\|v\nabla\boldsymbol{\omega}^{\boldsymbol{\alpha},\boldsymbol{\beta}}\|_{\omega^{-\boldsymbol{\alpha},-\boldsymbol{\beta}}} \le c_1\|\nabla v\|_{\omega^{\boldsymbol{\alpha},\boldsymbol{\beta}}}.$$

Then (8.134a) follows.

On the other hand, integration by parts yields

$$\begin{aligned}a_{\boldsymbol{\alpha},\boldsymbol{\beta}}(u,u) &= \|\nabla u\|^2_{\boldsymbol{\omega}^{\boldsymbol{\alpha},\boldsymbol{\beta}}} + \sum_{j=1}^d \int_{I^d} u\partial_{x_j} u\partial_{x_j}\boldsymbol{\omega}^{\boldsymbol{\alpha},\boldsymbol{\beta}} d\boldsymbol{x} \\ &= \|\nabla u\|^2_{\boldsymbol{\omega}^{\boldsymbol{\alpha},\boldsymbol{\beta}}} + \frac{1}{2}\sum_{j=1}^d \int_{I^d} u^2 G_{\alpha_j,\beta_j}(x_j)\boldsymbol{\omega}^{\boldsymbol{\alpha},\boldsymbol{\beta}} d\boldsymbol{x},\end{aligned}$$

where

$$G_{\alpha_j,\beta_j}(x_j) = -\partial^2_{x_j}(\omega^{\alpha_j,\beta_j}(x_j))\omega^{-\alpha_j,-\beta_j}(x_j) = (1-x_j^2)^{-2}W_{\alpha_j,\beta_j}(x_j),$$

with $W_{\alpha_j,\beta_j}(x_j)$ being defined and shown to be non-negative, when $-1 < \alpha_j, \beta_j < 1$, in the proof of Lemma 3.5. Therefore, (8.134b) holds. □

We are now in a position to derive error estimates for the orthogonal projection operator defined in (8.133). To describe the error more precisely, we introduce the space $\widehat{B}^r_{\boldsymbol{\alpha},\boldsymbol{\beta}}(I^d)$ for $r \geq d$ with the semi-norm and norm

$$|u|_{\widehat{B}^r_{\boldsymbol{\alpha},\boldsymbol{\beta}}(I^d)} = \Big(\sum_{j=1}^{d}\sum_{\boldsymbol{r}\in\Upsilon_j} \|\boldsymbol{\partial}_{\boldsymbol{x}}^{\boldsymbol{r}} u\|^2_{\boldsymbol{\omega}^{\boldsymbol{\alpha}+(r_j-1)\boldsymbol{e}_j,\boldsymbol{\beta}+(r_j-1)\boldsymbol{e}_j}}\Big)^{1/2},$$
$$\|u\|_{\widehat{B}^r_{\boldsymbol{\alpha},\boldsymbol{\beta}}(I^d)} = \Big(\|u\|^2_{\boldsymbol{\omega}^{\boldsymbol{\alpha},\boldsymbol{\beta}}} + |u|^2_{\widehat{B}^r_{\boldsymbol{\alpha},\boldsymbol{\beta}}(I^d)}\Big)^{1/2}, \tag{8.135}$$

where for $1 \leq j \leq d$, the index sets

$$\Upsilon_j = \Big\{\boldsymbol{r} \in \mathbb{N}_0^d \,:\, d \leq r_j \leq r;\ r_i \in \{0,1\},\ i \neq j;\ \sum_{k=1}^{d} r_k = r\Big\}.$$

Theorem 8.2. *If* $-1 < \boldsymbol{\alpha},\boldsymbol{\beta} < 1$, *and* $u \in H^1_{0,\boldsymbol{\omega}^{\boldsymbol{\alpha},\boldsymbol{\beta}}}(I^d) \cap \widehat{B}^r_{\boldsymbol{\alpha},\boldsymbol{\beta}}(I^d)$ *with integer* $d \leq r \leq N+1$, *we have*

$$\big\|\nabla(\boldsymbol{\pi}^{1,0}_{N,\boldsymbol{\alpha},\boldsymbol{\beta}} u - u)\big\|_{\boldsymbol{\omega}^{\boldsymbol{\alpha},\boldsymbol{\beta}}} \leq c\sqrt{\frac{(N-r+1)!}{N!}}(N+r)^{(1-r)/2}|u|_{\widehat{B}^r_{\boldsymbol{\alpha},\boldsymbol{\beta}}}. \tag{8.136}$$

Proof. By Lemma 8.7,

$$\|\nabla(\boldsymbol{\pi}^{1,0}_{N,\boldsymbol{\alpha},\boldsymbol{\beta}} u - u)\|_{\boldsymbol{\omega}^{\boldsymbol{\alpha},\boldsymbol{\beta}}} \leq c_1 \inf_{\phi\in(P_N^0)^d} \|\nabla(\phi - u)\|_{\boldsymbol{\omega}^{\boldsymbol{\alpha},\boldsymbol{\beta}}}. \tag{8.137}$$

For $1 \leq j \leq d$, let $\pi^{1,0}_{N,\alpha_j,\beta_j}$ be the one-dimensional orthogonal projection operator as defined in (3.290). We take ϕ in (8.137) as the tensor product

$$\phi = \pi^{1,0}_{N,\alpha_1,\beta_1} \circ \pi^{1,0}_{N,\alpha_2,\beta_2} \circ \ldots \circ \pi^{1,0}_{N,\alpha_d,\beta_d} u,$$

and derive from Theorem 3.39 with $m = 1$ that

$$\|\partial_{x_j}\pi^{1,0}_{N,\alpha_j,\beta_j} u\|_{\boldsymbol{\omega}^{\boldsymbol{\alpha},\boldsymbol{\beta}}} \leq c\|\partial_{x_j} u\|_{\boldsymbol{\omega}^{\boldsymbol{\alpha},\boldsymbol{\beta}}}. \tag{8.138}$$

Hence, using (8.138) and Theorems 3.39 and 3.35 leads to

$$\begin{aligned}
\|\nabla(\boldsymbol{\pi}^{1,0}_{N,\boldsymbol{\alpha},\boldsymbol{\beta}} u - u)\|_{\boldsymbol{\omega}^{\boldsymbol{\alpha},\boldsymbol{\beta}}} &\leq c\sum_{j=1}^{d}\Big(\big\|\partial_{x_j}(\pi^{1,0}_{N,\alpha_j,\beta_j} u - u)\big\|_{\boldsymbol{\omega}^{\boldsymbol{\alpha},\boldsymbol{\beta}}} \\
&+ \big\|\partial_{x_j}\pi^{1,0}_{N,\alpha_j,\beta_j} \circ \big(\pi^{1,0}_{N,\alpha_1,\beta_1} \circ \ldots \circ \pi^{1,0}_{N,\alpha_{j-1},\beta_{j-1}} \circ \pi^{1,0}_{N,\alpha_{j+1},\beta_{j+1}} \circ \ldots \circ \pi^{1,0}_{N,\alpha_d,\beta_d} u - u\big)\big\|_{\boldsymbol{\omega}^{\boldsymbol{\alpha},\boldsymbol{\beta}}}\Big) \\
&\leq c\sqrt{\frac{(N-r+1)!}{N!}}(N+r)^{(1-r)/2}\sum_{j=1}^{d} \|\partial^r_{x_j} u\|_{\boldsymbol{\omega}^{\boldsymbol{\alpha}+(r-1)\boldsymbol{e}_j,\boldsymbol{\beta}+(r-1)\boldsymbol{e}_j}} \\
&+ c\sum_{j=1}^{d} \big\|\pi^{1,0}_{N,\alpha_1,\beta_1} \circ \ldots \circ \pi^{1,0}_{N,\alpha_{j-1},\beta_{j-1}} \circ \pi^{1,0}_{N,\alpha_{j+1},\beta_{j+1}} \circ \ldots \circ \pi^{1,0}_{N,\alpha_d,\beta_d}(\partial_{x_j} u) - (\partial_{x_j} u)\big\|_{\boldsymbol{\omega}^{\boldsymbol{\alpha},\boldsymbol{\beta}}}.
\end{aligned} \tag{8.139}$$

Hence, it remains to estimate the terms in the last summation. Consider for instance $j = d$, we obtain from Theorem 3.39 with $\mu = 0$ that for $r \geq 2$,

$$\begin{aligned}
&\big\|\pi^{1,0}_{N,\alpha_1,\beta_1} \circ \ldots \circ \pi^{1,0}_{N,\alpha_{d-1},\beta_{d-1}}(\partial_{x_d}u) - (\partial_{x_d}u)\big\|_{\omega^{\alpha,\beta}} \\
&\leq \big\|\pi^{1,0}_{N,\alpha_{d-1},\beta_{d-1}}(\partial_{x_d}u) - (\partial_{x_d}u)\big\|_{\omega^{\alpha,\beta}} \\
&\quad + \big\|\pi^{1,0}_{N,\alpha_1,\beta_1} \circ \ldots \circ \pi^{1,0}_{N,\alpha_{d-2},\beta_{d-2}}(\partial_{x_d}u) - (\partial_{x_d}u)\big\|_{\omega^{\alpha,\beta}} \\
&\quad + \big\|(\mathbf{I} - \pi^{1,0}_{N,\alpha_{d-1},\beta_{d-1}}) \circ \big(\pi^{1,0}_{N,\alpha_1,\beta_1} \circ \ldots \circ \pi^{1,0}_{N,\alpha_{d-2},\beta_{d-2}}(\partial_{x_d}u) - (\partial_{x_d}u)\big)\big\|_{\omega^{\alpha,\beta}} \\
&\leq c\sqrt{\frac{(N-r+1)!}{N!}}(N+r)^{(1-r)/2}\|\partial^{r-1}_{x_{d-1}}\partial_{x_d}u\|_{\omega^{\alpha+(r-2)e_{d-1},\beta+(r-2)e_{d-1}}} \\
&\quad + \big\|\pi^{1,0}_{N,\alpha_1,\beta_1} \circ \ldots \circ \pi^{1,0}_{N,\alpha_{d-2},\beta_{d-2}}(\partial_{x_d}u) - (\partial_{x_d}u)\big\|_{\omega^{\alpha,\beta}} \\
&\quad + cN^{-1}\big\|\pi^{1,0}_{N,\alpha_1,\beta_1} \circ \ldots \circ \pi^{1,0}_{N,\alpha_{d-2},\beta_{d-2}}(\partial_{x_{d-1}}\partial_{x_d}u) - (\partial_{x_{d-1}}\partial_{x_d}u)\big\|_{\omega^{\alpha,\beta}},
\end{aligned}$$

where $\mathbf{I}$ is the identity operator. Applying this argument repeatedly leads to that for $d \leq r \leq N+1$,

$$\begin{aligned}
&\big\|\pi^{1,0}_{N,\alpha_1,\beta_1} \circ \ldots \circ \pi^{1,0}_{N,\alpha_{d-1},\beta_{d-1}}(\partial_{x_d}u) - (\partial_{x_d}u)\big\|_{\omega^{\alpha,\beta}} \\
&\qquad\qquad \leq c\sqrt{\frac{(N-r+1)!}{N!}}(N+r)^{(1-r)/2}|u|_{\widehat{B}^r_{\alpha,\beta}}.
\end{aligned}$$

We can estimate the other terms in the last summation of (8.139) in a similar fashion. Then, (8.136) follows. □

Remark 8.15. *Taking* $\phi = 0$ *in* (8.137) *leads to the stability result*

$$\|\nabla(\boldsymbol{\pi}^{1,0}_{N,\boldsymbol{\alpha},\boldsymbol{\beta}}u - u)\|_{\omega^{\alpha,\beta}} \leq c_1\|\nabla u\|_{\omega^{\alpha,\beta}}. \tag{8.140}$$

Observe from (8.135) *that* $\widehat{B}^r_{\alpha,\beta}(I^d) \subseteq H^r_{\omega^{\alpha,\beta}}(I^d)$ *for* $r \geq d$. *Hence, we obtain from* (8.136) *that for* $r = d$,

$$\|\nabla(\boldsymbol{\pi}^{1,0}_{N,\boldsymbol{\alpha},\boldsymbol{\beta}}u - u)\|_{\omega^{\alpha,\beta}} \leq c\sqrt{\frac{(N-d+2)!}{N!}}(N+d)^{-d/2}\|u\|_{H^d_{\omega^{\alpha,\beta}}(I^d)}. \tag{8.141}$$

By using an interpolation argument (see, e.g., Theorems 1.5 and 7.2 in Bernardi and Maday (1997)), *we also have that for* $r \geq 1$,

$$\|\nabla(\boldsymbol{\pi}^{1,0}_{N,\boldsymbol{\alpha},\boldsymbol{\beta}}u - u)\|_{\omega^{\alpha,\beta}} \leq c\sqrt{\frac{(N-r+2)!}{N!}}(N+r)^{-r/2}\|u\|_{H^r_{\omega^{\alpha,\beta}}(I^d)}. \tag{8.142}$$

As usual, in order to obtain the optimal estimate in $L^2_{\omega^{\alpha,\beta}}$-norm, we need to use a duality argument which requires the following result (cf. Bernardi and Maday (1997), Canuto et al. (2006) for the special case $\boldsymbol{\alpha} = \boldsymbol{\beta} = 0$ and $\boldsymbol{\alpha} = \boldsymbol{\beta} = -1/2$):

Lemma 8.8. *Let* $-1/2 \le \boldsymbol{\alpha}, \boldsymbol{\beta} \le 0$ *or* $0 \le \boldsymbol{\alpha}, \boldsymbol{\beta} \le 1/2$. *Then for* $u \in H^1_{0,\boldsymbol{\omega}^{\alpha,\beta}}(I^d) \cap H^2_{\boldsymbol{\omega}^{\alpha,\beta}}(I^d)$, *we have*

$$|v|_{2,\boldsymbol{\alpha},\boldsymbol{\beta}} \le c\|\Delta v\|_{\boldsymbol{\omega}^{\alpha,\beta}}, \tag{8.143}$$

where c is a positive constant independent of v.

Proof. It is clear that

$$\|\Delta v\|^2_{\boldsymbol{\omega}^{\alpha,\beta}} = \sum_{j=1}^{d} \|\partial^2_{x_j} v\|^2_{\boldsymbol{\omega}^{\alpha,\beta}} + \sum_{i,j=1;i\ne j}^{d} \int_{I^d} \partial^2_{x_i} v(\boldsymbol{x}) \partial^2_{x_j} v(\boldsymbol{x}) \boldsymbol{\omega}^{\boldsymbol{\alpha},\boldsymbol{\beta}}(\boldsymbol{x}) d\boldsymbol{x}, \tag{8.144}$$

so we just need to bound the last summation.

Note that the derivatives $\partial^k_{x_i} v$ vanish on the boundaries $x_j = \pm 1 (i \ne j)$. For notational convenience, we denote

$$\hat{\boldsymbol{\omega}}_{i,j}(\boldsymbol{x}) = \boldsymbol{\omega}^{\boldsymbol{\alpha},\boldsymbol{\beta}}(\boldsymbol{x}) \omega^{-\alpha_i,-\beta_i}(x_i) \omega^{-\alpha_j,-\beta_j}(x_j).$$

For $i \ne j$, integration by parts yields

$$\begin{aligned}
\int_{I^d} \partial^2_{x_i} v \partial^2_{x_j} v\, \boldsymbol{\omega}^{\boldsymbol{\alpha},\boldsymbol{\beta}} d\boldsymbol{x} &= \int_{I^d} \partial_{x_i}\partial_{x_j}(v\omega^{\alpha_i,\beta_i}(x_i)) \partial_{x_i}\partial_{x_j}(v\omega^{\alpha_j,\beta_j}(x_j)) \hat{\boldsymbol{\omega}}_{i,j} d\boldsymbol{x} \\
&= \int_{I^d} (\partial_{x_i}\partial_{x_j} v)^2 \boldsymbol{\omega}^{\boldsymbol{\alpha},\boldsymbol{\beta}} d\boldsymbol{x} - \frac{1}{2}\int_{I^d} (\partial^2_{x_j} v)^2 (\omega^{-\alpha_i,-\beta_i}(x_i) \partial^2_{x_i} \omega^{\alpha_i,\beta_i}(x_i)) \boldsymbol{\omega}^{\boldsymbol{\alpha},\boldsymbol{\beta}} d\boldsymbol{x} \\
&\quad - \frac{1}{2}\int_{I^d} (\partial^2_{x_i} v)^2 (\omega^{-\alpha_j,-\beta_j}(x_j) \partial^2_{x_j} \omega^{\alpha_j,\beta_j}(x_j)) \boldsymbol{\omega}^{\boldsymbol{\alpha},\boldsymbol{\beta}} d\boldsymbol{x} \\
&\quad + \int_{I^d} \partial_{x_i} v \partial_{x_j} v \partial_{x_i} \omega^{\alpha_i,\beta_i}(x_i) \partial_{x_j} \omega^{\alpha_j,\beta_j}(x_j) \hat{\boldsymbol{\omega}}_{i,j} d\boldsymbol{x}.
\end{aligned} \tag{8.145}$$

We treat the last term by using the Cauchy–Schwarz inequality as

$$\begin{aligned}
&\left| \int_{I^d} \partial_{x_i} v \partial_{x_j} v \partial_{x_i} \omega^{\alpha_i,\beta_i}(x_i) \partial_{x_j} \omega^{\alpha_j,\beta_j}(x_j) \hat{\boldsymbol{\omega}}_{i,j} d\boldsymbol{x} \right| \\
&\qquad \le \frac{1}{2} \int_{I^d} (\partial_{x_i} v)^2 (\partial_{x_j} \omega^{\alpha_j,\beta_j}(x_j))^2 \omega^{-\alpha_j,-\beta_j}(x_j) \omega^{\alpha_i,\beta_i}(x_i) \hat{\boldsymbol{\omega}}_{i,j} d\boldsymbol{x} \\
&\qquad\quad + \frac{1}{2} \int_{I^d} (\partial_{x_j} v)^2 (\partial_{x_i} \omega^{\alpha_i,\beta_i}(x_i))^2 \omega^{-\alpha_i,-\beta_i}(x_i) \omega^{\alpha_j,\beta_j}(x_j) \hat{\boldsymbol{\omega}}_{i,j} d\boldsymbol{x}.
\end{aligned}$$

In view of this, we obtain from (8.145) that

$$\begin{aligned}
\int_{I^d} \partial^2_{x_i} v \partial^2_{x_j} v\, \boldsymbol{\omega}^{\boldsymbol{\alpha},\boldsymbol{\beta}} d\boldsymbol{x} \ge & \int_{I^d} (\partial_{x_i}\partial_{x_j} v)^2 \boldsymbol{\omega}^{\boldsymbol{\alpha},\boldsymbol{\beta}} d\boldsymbol{x} + \frac{1}{2}\int_{I^d} (\partial_{x_i} v)^2 S(x_j;\alpha_j,\beta_j) \boldsymbol{\omega}^{\boldsymbol{\alpha},\boldsymbol{\beta}} d\boldsymbol{x} \\
& + \frac{1}{2}\int_{I^d} (\partial_{x_j} v)^2 S(x_i;\alpha_i,\beta_i) \boldsymbol{\omega}^{\boldsymbol{\alpha},\boldsymbol{\beta}} d\boldsymbol{x},
\end{aligned}$$

where

$$S(t;a,b) = -(\omega^{a,b}(t))'' \omega^{-a,-b}(t) - ((\omega^{a,b}(t))')^2 \omega^{-2a,-2b}(t), \quad |a|,|b| < 1.$$

We now determine the range of a,b such that $S(t;a,b) \geq 0$ for all $t \in (-1,1)$. A direct computation gives

$$\begin{aligned} W(t) :=& (1-t^2)S(t;a,b) = (a+b)(1-2a-2b)t^2 \\ &+ 2(a-b)(1-2a-2b)t + a + b - 2(a-b)^2. \end{aligned}$$

Similar to the proof of Lemma 8.7, we can show that if $0 \leq a,b \leq \frac{1}{2}$, then $W(t) \geq 0$ for all $t \in [-1,1]$. Consequently, we derive from (8.145) and the above analysis that for $0 \leq \alpha_k, \beta_k \leq \frac{1}{2}$ and $-1 < \alpha_l, \beta_l < 1$ with $k = i, j, i \neq j$ and $l \neq i, j$,

$$\begin{aligned} \int_{I^d} \partial_{x_i}^2 v \partial_{x_j}^2 v \, \boldsymbol{\omega}^{\boldsymbol{\alpha},\boldsymbol{\beta}} d\boldsymbol{x} &= \int_{I^d} \partial_{x_i}\partial_{x_j}(v\omega^{\alpha_i,\beta_i}(x_i))\partial_{x_i}\partial_{x_j}(v\omega^{\alpha_j,\beta_j}(x_j))\hat{\boldsymbol{\omega}}_{i,j} d\boldsymbol{x} \\ &\geq \int_{I^d} (\partial_{x_i}\partial_{x_j} v)^2 \boldsymbol{\omega}^{\boldsymbol{\alpha},\boldsymbol{\beta}} d\boldsymbol{x}. \end{aligned} \tag{8.146}$$

Therefore, (8.143) is valid for $0 \leq \boldsymbol{\alpha}, \boldsymbol{\beta} \leq 1/2$, and it remains to show that it holds for $-1/2 \leq \boldsymbol{\alpha}, \boldsymbol{\beta} \leq 0$.

Indeed, if $-\frac{1}{2} \leq \alpha_k, \beta_k \leq 0\, (k = i, j)$, we set $u(\boldsymbol{x}) = v(\boldsymbol{x})\omega^{\alpha_i,\beta_i}(x_i)\omega^{\alpha_j,\beta_j}(x_j)$, and find from (8.146) that

$$\begin{aligned} \int_{I^d} \partial_{x_i}^2 v \partial_{x_j}^2 v \, \boldsymbol{\omega}^{\boldsymbol{\alpha},\boldsymbol{\beta}} d\boldsymbol{x} &= \int_{I^d} \partial_{x_i}\partial_{x_j}(u\omega^{-\alpha_j,-\beta_j}(x_j))\partial_{x_i}\partial_{x_j}(u\omega^{-\alpha_i,-\beta_i}(x_i))\hat{\boldsymbol{\omega}}_{i,j} d\boldsymbol{x} \\ &\geq \int_{I^d} (\partial_{x_i}\partial_{x_j} u)^2 \tilde{\boldsymbol{\omega}}_{i,j} d\boldsymbol{x}, \end{aligned} \tag{8.147}$$

where we denoted

$$\tilde{\boldsymbol{\omega}}_{i,j}(\boldsymbol{x}) = \omega^{-\alpha_i,-\beta_i}(x_i)\omega^{-\alpha_j,-\beta_j}(x_j)\hat{\boldsymbol{\omega}}_{i,j}(\boldsymbol{x}).$$

Notice that

$$\begin{aligned} \int_{I^d} (\partial_{x_i}\partial_{x_j} v)^2 \boldsymbol{\omega}^{\boldsymbol{\alpha},\boldsymbol{\beta}} d\boldsymbol{x} &= \int_{I^d} \left(\partial_{x_i}\partial_{x_j}(u\omega^{-\alpha_i,-\beta_i}\omega^{-\alpha_j,-\beta_j})\right)^2 \boldsymbol{\omega}^{\boldsymbol{\alpha},\boldsymbol{\beta}} d\boldsymbol{x} \\ \leq c\Big(\int_{I^d} |\partial_{x_i}\partial_{x_j} u|^2 \tilde{\boldsymbol{\omega}}_{i,j} d\boldsymbol{x} &+ \int_{I^d} |\partial_{x_j} u|^2 (1-x_i^2)^{-2} \tilde{\boldsymbol{\omega}}_{i,j} d\boldsymbol{x} \\ + \int_{I^d} |\partial_{x_i} u|^2 (1-x_j^2)^{-2} \tilde{\boldsymbol{\omega}}_{i,j} d\boldsymbol{x} &+ \int_{I^d} |u|^2 (1-x_i^2)^{-2}(1-x_j^2)^{-2} \tilde{\boldsymbol{\omega}}_{i,j} d\boldsymbol{x} \Big). \end{aligned} \tag{8.148}$$

Applying (B.40) to the last three terms in the above summation leads to

$$\int_{I^d} (\partial_{x_i}\partial_{x_j} v)^2 \boldsymbol{\omega}^{\boldsymbol{\alpha},\boldsymbol{\beta}} d\boldsymbol{x} \leq c \int_{I^d} |\partial_{x_i}\partial_{x_j} u|^2 \tilde{\boldsymbol{\omega}}_{i,j} d\boldsymbol{x}. \tag{8.149}$$

Hence, by (8.147) and (8.149), we have that for $-\frac{1}{2} \leq \alpha_k, \beta_k \leq 0$ with $k = i, j\ (i \neq j)$:

$$\int_{I^d} \partial_{x_i}^2 v \partial_{x_j}^2 v \, \boldsymbol{\omega}^{\boldsymbol{\alpha},\boldsymbol{\beta}} d\boldsymbol{x} \geq c \int_{I^d} (\partial_{x_i}\partial_{x_j} v)^2 \boldsymbol{\omega}^{\boldsymbol{\alpha},\boldsymbol{\beta}} d\boldsymbol{x}. \tag{8.150}$$

In view of (8.144), we obtain (8.143) with $-1/2 \leq \boldsymbol{\alpha}, \boldsymbol{\beta} \leq 0$. This completes the proof. □

Theorem 8.3. *If* $-1/2 \le \boldsymbol{\alpha}, \boldsymbol{\beta} \le 0$ *or* $0 \le \boldsymbol{\alpha}, \boldsymbol{\beta} \le 1/2$, *then for* $u \in H^1_{0,\boldsymbol{\omega}^{\alpha,\beta}}(I^d) \cap H^r_{\boldsymbol{\omega}^{\alpha,\beta}}(I^d)$ *with integer* $d \le r \le N+1$, *we have*

$$\|\boldsymbol{\pi}^{1,0}_{N,\boldsymbol{\alpha},\boldsymbol{\beta}} u - u\|_{\boldsymbol{\omega}^{\alpha,\beta}} \le c\sqrt{\frac{(N-r+1)!}{N!}}(N+r)^{-(1+r)/2}\|u\|_{H^r_{\boldsymbol{\omega}^{\alpha,\beta}}(I^d)}, \tag{8.151}$$

where c *is a positive constant independent of* r, N *and* u.

Proof. Given $g \in L^2_{\boldsymbol{\omega}^{\alpha,\beta}}(I^d)$, we consider the auxiliary problem:

$$\begin{cases} \text{Find } v \in H^1_{0,\boldsymbol{\omega}^{\alpha,\beta}}(I^d) \text{ such that} \\ a_{\boldsymbol{\alpha},\boldsymbol{\beta}}(v,z) = (g,z)_{\boldsymbol{\omega}^{\alpha,\beta}}, \quad \forall z \in H^1_{0,\boldsymbol{\omega}^{\alpha,\beta}}(I^d). \end{cases} \tag{8.152}$$

It follows from Lemma 8.7 and the Lax-Milgram lemma that (8.152) has a unique solution in $H^1_{0,\boldsymbol{\omega}^{\alpha,\beta}}(I^d)$ and

$$\|\nabla v\|_{\boldsymbol{\omega}^{\alpha,\beta}} \le c\|g\|_{\boldsymbol{\omega}^{\alpha,\beta}}. \tag{8.153}$$

Next, we derive from (8.152) that $\Delta v(\boldsymbol{x}) = g(\boldsymbol{x})$ in the sense of distribution. Therefore, we obtain from (8.132), (8.153) and (8.143) that for $-1/2 \le \boldsymbol{\alpha}, \boldsymbol{\beta} \le 0$ or $0 \le \boldsymbol{\alpha}, \boldsymbol{\beta} \le 1/2$,

$$\|v\|_{H^2_{\boldsymbol{\omega}^{\alpha,\beta}}(I^d)} \le c\big(\|\nabla v\|_{\boldsymbol{\omega}^{\alpha,\beta}} + \|\Delta v\|_{\boldsymbol{\omega}^{\alpha,\beta}}\big) \le c\|g\|_{\boldsymbol{\omega}^{\alpha,\beta}}. \tag{8.154}$$

Taking $z = \boldsymbol{\pi}^{1,0}_{N,\boldsymbol{\alpha},\boldsymbol{\beta}} u - u$ in (8.152), we derive from (8.142) and (8.154) that

$$\begin{aligned} |(\boldsymbol{\pi}^{1,0}_{N,\boldsymbol{\alpha},\boldsymbol{\beta}} u - u, g)_{\boldsymbol{\omega}^{\alpha,\beta}}| &= |a_{\boldsymbol{\alpha},\boldsymbol{\beta}}(\boldsymbol{\pi}^{1,0}_{N,\boldsymbol{\alpha},\boldsymbol{\beta}} u - u, \boldsymbol{\pi}^{1,0}_{N,\boldsymbol{\alpha},\boldsymbol{\beta}} v - v)| \\ &\le c\sqrt{\frac{(N-r+1)!}{N!}}(N+r)^{-(1+r)/2}\|g\|_{\boldsymbol{\omega}^{\alpha,\beta}}\|u\|_{H^r_{\boldsymbol{\omega}^{\alpha,\beta}}(I^d)}. \end{aligned}$$

Consequently,

$$\begin{aligned} \|\boldsymbol{\pi}^{1,0}_{N,\boldsymbol{\alpha},\boldsymbol{\beta}} u - u\|_{\boldsymbol{\omega}^{\alpha,\beta}} &= \sup_{0 \ne g \in L^2_{\boldsymbol{\omega}^{\alpha,\beta}}(I^d)} \frac{|(\boldsymbol{\pi}^{1,0}_{N,\boldsymbol{\alpha},\boldsymbol{\beta}} u - u, g)_{\boldsymbol{\omega}^{\alpha,\beta}}|}{\|g\|_{\boldsymbol{\omega}^{\alpha,\beta}}} \\ &\le c\sqrt{\frac{(N-r+1)!}{N!}}(N+r)^{-(1+r)/2}\|u\|_{H^r_{\boldsymbol{\omega}^{\alpha,\beta}}(I^d)}. \end{aligned}$$

This ends the proof. □

Remark 8.16. *As pointed out in Remark 8.15, we can use a space interpolation argument to show that the estimate* (8.151) *is also valid for* $r \ge 1$ (see Theorem 7.2 in Bernardi and Maday (1997)).

The above estimate is useful for analysis of multi-dimensional problems with Dirichlet boundary conditions. We now consider an orthogonal projection related to the Neumann boundary conditions.

We define $\bar{\Pi}_N^1 : H^1(I^d) \to P_N^d$ by

$$(\nabla(u - \bar{\Pi}_N^1 u), \nabla u_N) = 0, \quad \forall u_N \in P_N^d; \quad (u - \bar{\Pi}_N^1 u, 1) = 0. \tag{8.155}$$

The estimate of this projection error essentially relies on the H^1-projection operator $\Pi_N^1 : H^1(I^d) \to P_N^d$ which is defined by

$$(\nabla(u - \Pi_N^1 u), \nabla u_N) + (u - \Pi_N^1 u, u_N) = 0, \quad \forall u_N \in P_N^d. \tag{8.156}$$

By using a similar procedure as in the proof of Theorem 8.3 (see also Bernardi and Maday (1997)), we can prove the following:

Theorem 8.4. *For any $u \in H^r(I^d)$ with fixed $r \geq 1$,*

$$\|u - \Pi_N^1 u\|_{H^l(I^d)} \lesssim N^{l-r} \|u\|_{H^r(I^d)}, \quad r \geq l \geq 0, \; r \geq 1. \tag{8.157}$$

By using the above result, it is an easy matter to prove the following estimate:

Theorem 8.5. *Let $\bar{\Pi}_N^1$ be the projection operator defined in* (8.155). *Then for any $u \in H^r(I^d)$ with fixed $r \geq 1$,*

$$\|u - \bar{\Pi}_N^1 u\|_{H^l(I^d)} \lesssim N^{l-r} \|u\|_{H^r(I^d)}, \quad r \geq l \geq 0. \tag{8.158}$$

We leave the proof as an exercise (see Problem 8.4).

8.4.3 Interpolations

We consider here the error analysis of polynomial interpolations on the hypercube $I^d = (-1,1)^d$ using the Jacobi-Gauss-type points introduced in Chap. 3. To simplify the presentation, we shall only discuss the Jacobi-Gauss interpolation. Results for the Jacobi-Gauss-Radau, Jacobi-Gauss-Lobatto interpolations or mixed type interpolations can be established in a similar fashion (see Problem 8.5).

For $1 \leq k \leq d$, let $\{\xi_{j_k}^{\alpha_k,\beta_k}\}_{j_k=0}^N$ be the one-dimensional Jacobi-Gauss interpolation points (cf. Theorem 3.25), and let $I_N^{\alpha_k,\beta_k}$ be the associated interpolation operator. The full d-dimensional tensorial grid

$$\Sigma_N^{\boldsymbol{\alpha},\boldsymbol{\beta}} = \left\{ \boldsymbol{\xi}_{\boldsymbol{j}}^{\boldsymbol{\alpha},\boldsymbol{\beta}} = (\xi_{j_1}^{\alpha_1,\beta_1}, \ldots, \xi_{j_d}^{\alpha_d,\beta_d}) : 0 \leq j_1, \ldots, j_d \leq N \right\}. \tag{8.159}$$

The corresponding polynomial interpolation $\boldsymbol{I}_N^{\boldsymbol{\alpha},\boldsymbol{\beta}} : C(I^d) \to P_N^d$ satisfies

$$(\boldsymbol{I}_N^{\boldsymbol{\alpha},\boldsymbol{\beta}} u)(\boldsymbol{x}) = u(\boldsymbol{x}), \quad \forall \boldsymbol{x} \in \Sigma_N^{\boldsymbol{\alpha},\boldsymbol{\beta}}. \tag{8.160}$$

For simplicity, we assumed that the number of points in each direction is the same (i.e., $N+1$ points). One verifies that

$$\boldsymbol{I}_N^{\boldsymbol{\alpha},\boldsymbol{\beta}} = I_N^{\alpha_1,\beta_1} \circ \ldots \circ I_N^{\alpha_d,\beta_d}. \tag{8.161}$$

Define $\widetilde{B}^r_{\boldsymbol{\alpha},\boldsymbol{\beta}}(I^d)$ with $r \geq d$ the space $\widehat{B}^r_{\boldsymbol{\alpha},\boldsymbol{\beta}}(I^d)$ but with $r_j\boldsymbol{e}_j$ in place of $(r_j-1)\boldsymbol{e}_j$ in (8.135), and denote the corresponding norm and semi-norm by $\|\cdot\|_{\widetilde{B}^r_{\boldsymbol{\alpha},\boldsymbol{\beta}}(I^d)}$ and $|\cdot|_{\widetilde{B}^r_{\boldsymbol{\alpha},\boldsymbol{\beta}}(I^d)}$, respectively.

Theorem 8.6. *For $\boldsymbol{\alpha},\boldsymbol{\beta} > -1$, and $u \in \widetilde{B}^r_{\boldsymbol{\alpha},\boldsymbol{\beta}}(I^d)$ with $d \leq r \leq N+1$,*

$$\|\boldsymbol{I}_N^{\boldsymbol{\alpha},\boldsymbol{\beta}} u - u\|_{\boldsymbol{\omega}^{\boldsymbol{\alpha},\boldsymbol{\beta}}} \leq c\sqrt{\frac{(N-r+1)!}{N!}}(N+r)^{-(r+1)/2}|u|_{\widetilde{B}^r_{\boldsymbol{\alpha},\boldsymbol{\beta}}(I^d)}, \tag{8.162}$$

where c is a positive constant independent of r, N and u.

Proof. Thanks to (8.161), we have

$$\begin{aligned}\|\boldsymbol{I}_N^{\boldsymbol{\alpha},\boldsymbol{\beta}} u - u\|_{\boldsymbol{\omega}^{\boldsymbol{\alpha},\boldsymbol{\beta}}} &= \|I_N^{\alpha_1,\beta_1} u - u\|_{\boldsymbol{\omega}^{\boldsymbol{\alpha},\boldsymbol{\beta}}} + \|I_N^{\alpha_2,\beta_2} \circ \ldots \circ I_N^{\alpha_d,\beta_d} u - u\|_{\boldsymbol{\omega}^{\boldsymbol{\alpha},\boldsymbol{\beta}}} \\ &\quad + \|(I_N^{\alpha_1,\beta_1} - \mathbf{I})(I_N^{\alpha_2,\beta_2} \circ \ldots \circ I_N^{\alpha_d,\beta_d} u - u)\|_{\boldsymbol{\omega}^{\boldsymbol{\alpha},\boldsymbol{\beta}}},\end{aligned}$$

where $\mathbf{I}$ is the identity operator. Therefore, by Theorem 3.41,

$$\begin{aligned}\|\boldsymbol{I}_N^{\boldsymbol{\alpha},\boldsymbol{\beta}} u - u\|_{\boldsymbol{\omega}^{\boldsymbol{\alpha},\boldsymbol{\beta}}} &\leq c\sqrt{\frac{(N-r+1)!}{N!}}(N+r)^{-(r+1)/2}\|\partial_{x_1}^r u\|_{\boldsymbol{\omega}^{\boldsymbol{\alpha}+r\boldsymbol{e}_1,\boldsymbol{\beta}+r\boldsymbol{e}_1}} \\ &\quad + \|I_N^{\alpha_2,\beta_2} \circ \ldots \circ I_N^{\alpha_d,\beta_d} u - u\|_{\boldsymbol{\omega}^{\boldsymbol{\alpha},\boldsymbol{\beta}}} \\ &\quad + cN^{-1}\|I_N^{\alpha_2,\beta_2} \circ \ldots \circ I_N^{\alpha_d,\beta_d}(\partial_{x_1} u) - (\partial_{x_1} u)\|_{\boldsymbol{\omega}^{\boldsymbol{\alpha},\boldsymbol{\beta}}}\end{aligned} \tag{8.163}$$

Iterating this argument (see the proof of Theorem 8.2) gives the desired result. □

8.4.4 Applications of Multivariate Jacobi Approximations

We now apply the multivariate Jacobi approximation results to analyze the convergence of several spectral-Galerkin schemes proposed in this chapter.

8.4.4.1 Rectangular Domains

We begin with the analysis of the scheme (8.2) with $d=2$. The corresponding weak formulation is

$$\begin{cases}\text{Find } u \in H^1_{0,\omega}(\Omega) \text{ such that} \\ \alpha(u,v)_\omega + a_\omega(u,v) = (f,v)_\omega, \quad \forall v \in H^1_{0,\omega}(\Omega),\end{cases} \tag{8.164}$$

where $a_\omega(\cdot,\cdot)$ is defined in (8.3). We find from Lemma 8.7 that the bilinear form is continuous and coercive in $H^1_{0,\omega}(\Omega) \times H^1_{0,\omega}(\Omega)$, so by the Lax-Milgram lemma (8.164) admits a unique solution in $H^1_{0,\omega}(\Omega)$, if $f \in L^2_\omega(\Omega)$. This also applies to the approximate solution u_N of (8.2) with $d = 2$.

Applying Theorem 1.2 with $X = H^1_{0,\omega}(\Omega)$, we find immediately

$$\|u - u_N\|_{1,\omega} \lesssim \inf_{v_N \in (P_N^0)^2} \|u - v_N\|_{1,\omega}.$$

Therefore, taking $v_N = \boldsymbol{\pi}^{1,0}_{N,\boldsymbol{\alpha},\boldsymbol{\beta}} u$ with $\boldsymbol{\alpha} = \boldsymbol{\beta} = 0$ or $-1/2$ in the above estimate, we obtain from Theorem 8.3 the following convergence result.

Theorem 8.7. *Let u and u_N be the solutions of* (8.164) *and* (8.2) *with $d = 2$, respectively. If $\alpha \ge 0$ and $u \in H^1_{0,\omega}(\Omega) \cap H^r_\omega(\Omega)$ with $2 \le r \le N + 1$, then we have*

$$\|u - u_N\|_{1,\omega} \le c\sqrt{\frac{(N-r+1)!}{N!}}(N+r)^{(1-r)/2}\|u\|_{H^r_\omega(\Omega)}, \tag{8.165}$$

where c is a positive constant independent of r, N and u.

8.4.4.2 Circular and Spherical Domains

We analyze here the mixed Fourier/spherical harmonic–Legendre Galerkin methods for the problem (8.1) in a unit disk or ball described in Sects. 8.2.1 and 8.3.2.

For clarity of presentation, we first briefly recall the procedure and present the schemes in a uniform format. Consider the model problem:

$$\begin{aligned} &\alpha U - \Delta U = F \;\text{ in } \Omega = \{\boldsymbol{x} \in \mathbb{R}^d : |\boldsymbol{x}| < 1\}, \\ &U|_{\partial\Omega} = 0, \quad \alpha \ge 0, \; d = 2, 3. \end{aligned} \tag{8.166}$$

We rewrite (8.166) in the polar coordinate (r,ϕ)/spherical coordinate (r,θ,ϕ) as

$$\begin{aligned} &-\frac{1}{r^{d-1}}\partial_r(r^{d-1}\partial_r U) - \frac{1}{r^2}\Delta_S U + \alpha U = F \quad \text{in } \Omega = I \times S, \\ &\partial_\theta U|_{r=0} = U|_{r=1} = 0, \end{aligned} \tag{8.167}$$

where S is the unit circle or sphere, and

$$\Delta_S U = \begin{cases} \partial_\phi^2 U, & \text{if } d = 2, \\ \frac{1}{\sin\theta}\partial_\theta(\sin\theta\,\partial_\theta U) + \frac{1}{\sin^2\theta}\partial_\phi^2 U, & \text{if } d = 3. \end{cases} \tag{8.168}$$

Note that with a little abuse of notation, we still used U and F to denote the transformed functions. By expanding the functions in terms of Fourier/spherical harmonic series:

$$
\begin{aligned}
\{U,F\} &= \sum_{|l|=0}^{\infty} \{\hat{u}_l(r), \hat{f}_l(r)\} e^{\mathrm{i} l\theta}; \\
\{U,F\} &= \sum_{l=0}^{\infty} \sum_{|m|=0}^{l} \{\hat{u}_{lm}(r), \hat{f}_{lm}(r)\} Y_l^m(\theta,\phi),
\end{aligned} \tag{8.169}
$$

(where Y_l^m is defined in (8.91)) we can transform the problem (8.166) to a sequence (for each l in 2-D and (l,m) in 3-D) of 1-D equations (for brevity, we use u to denote $\hat{u}_l$ or $\hat{u}_{lm}$, and likewise f for $\hat{f}_l$ or $\hat{f}_{lm}$):

$$
\begin{aligned}
&-\frac{1}{r^{d-1}} \partial_r \big(r^{d-1} \partial_r u\big) + d_l \frac{u}{r^2} + \alpha u = f, \quad r \in (0,1), \; d = 2,3; \\
&u(0) = 0, \; \text{if } d = 2 \text{ and } l \neq 0; \quad u(1) = 0,
\end{aligned} \tag{8.170}
$$

where $d_l = l^2, l(l+1)$ for $d = 2,3$, respectively. Notice that in the 2-D case, it is sufficient to consider the modes $l \geq 0$, since we have $\hat{u}_{-l} = \bar{\hat{u}}_l$ for real U and F (cf. Sect. 2.1).

For convenience of analysis, we make a coordinate transform (cf. (8.39)): $r = (1+t)/2$, and consider the problem:

$$
\begin{aligned}
&-\frac{1}{(1+t)^{d-1}} \partial_t \big((1+t)^{d-1} \partial_t v\big) + d_l \frac{v}{(1+t)^2} + \hat{\alpha} v = h, \; t \in I := (-1,1); \\
&v(-1) = 0, \; \text{if } d = 2 \text{ and } l \neq 0; \quad v(1) = 0,
\end{aligned} \tag{8.171}
$$

where we set

$$
v(t) = u((1+t)/2), \quad h(t) = \frac{1}{4} f((1+t)/2), \quad \hat{\alpha} = \frac{\alpha}{4}.
$$

Define the space

$$
X(l,d) = \begin{cases} H_0^1(I), & \text{if } d = 2 \text{ and } l \neq 0, \\ \{v \in H^1(I) \,:\, v(1) = 0\}, & \text{otherwise}, \end{cases} \tag{8.172}
$$

and define the approximation space: $X_N(l,d) = X(l,d) \cap P_N$. A weak formulation of (8.171) is

$$
\begin{cases} \text{Find } v \in X(l,d) \text{ such that} \\ a_{l,d}(v,w) + \hat{\alpha}(v,w)_{\omega^{0,d-1}} = (h,w)_{\omega^{0,d-1}}, \quad \forall w \in X(l,d), \end{cases} \tag{8.173}
$$

where the bilinear form

$$
a_{l,d}(v,w) := (v',w')_{\omega^{0,d-1}} + d_l (v,w)_{\omega^{0,d-3}},
$$

and the Jacobi weight function $\omega^{0,b} = (1+t)^b$.

The Legendre spectral-Galerkin approximation of (8.173) is

$$\begin{cases} \text{Find } v_N \in X_N(l,d) \text{ such that} \\ a_{l,d}(v_N,w_N) + \hat{\alpha}(v_N,w_N)_{\omega^{0,d-1}} = (h,w_N)_{\omega^{0,d-1}}, \quad \forall w_N \in X_N(l,d), \end{cases} \tag{8.174}$$

To analyze this scheme, it is necessary to study the orthogonal projection $\pi_{d,N}^{1,l}$: $X(l,d) \to X_N(l,d)$, defined by

$$a_{l,d}(\pi_{d,N}^{1,l}v - v, w_N) = 0, \quad \forall w_N \in X_N(l,d). \tag{8.175}$$

It is essential to analyze its approximation property as summarized below.

Theorem 8.8. *For any $v \in X(l,d) \cap B^s_{-1,-1}(I)$ with $1 \le s \le N+1$,*

$$\begin{aligned} &\|\partial_t(\pi_{d,N}^{1,l}v - v)\|^2_{\omega^{0,d-1}} + d_l\|\pi_{d,N}^{1,l}v - v\|^2_{\omega^{0,d-3}} \\ &\qquad \le c(1+d_lN^{-2})\frac{(N-s+1)!}{N!}(N+s)^{1-s}\|\partial_t^s v\|^2_{\omega^{s-1,s-1}}, \end{aligned} \tag{8.176}$$

where $d_l = l^2, l(l+1)$ for $d = 2,3$, respectively, and c is a positive constant independent of l,d,N and v.

Proof. In the first place, we show that there exists an operator $\Pi_N : H^1(I) \to P_N$ such that $(\Pi_N v)(\pm 1) = v(\pm 1)$, and there holds the estimate, that is, for $1 \le \mu \le s \le N+1$,

$$\|\partial_t^\mu(\Pi_N v - v)\|_{\omega^{\mu-1,\mu-1}} \le c\sqrt{\frac{(N-s+1)!}{(N-\mu+1)!}}(N+s)^{(\mu-s)/2}\|\partial_t^s v\|_{\omega^{s-1,s-1}}. \tag{8.177}$$

For this purpose, let $\pi_N^{-1,-1}$ be the orthogonal projection operator associated with the generalized Jacobi polynomials, defined in (6.65). Set

$$v_*(t) = \frac{1-t}{2}v(-1) + \frac{1+t}{2}v(1) \in P_1, \quad \forall v \in H^1(I).$$

It is clear that $(v - v_*)(\pm 1) = 0$. Define

$$\Pi_N v = \pi_N^{-1,-1}(v - v_*) + v_* \in P_N, \quad \forall v \in H^1(I), \tag{8.178}$$

which satisfies $(\Pi_N v)(\pm 1) = v(\pm 1)$. Moreover, we derive from Theorem 6.1 that for $0 \le \mu \le s \le N+1$,

$$\begin{aligned} \|\partial_t^\mu(\Pi_N v - v)\|_{\omega^{\mu-1,\mu-1}} &= \|\partial_t^\mu(\pi_N^{-1,-1}(v - v_*) - (v - v_*))\|_{\omega^{\mu-1,\mu-1}} \\ &\le c\sqrt{\frac{(N-s+1)!}{(N-\mu+1)!}}(N+s)^{(\mu-s)/2}\|\partial_t^s(v - v_*)\|_{\omega^{s-1,s-1}}. \end{aligned} \tag{8.179}$$

For $s \geq 2$, we have $\partial_t^s v_* \equiv 0$, while for $s = 1$, we have

$$|\partial_t v_*| = \frac{|u(1) - u(-1)|}{2} \leq \frac{1}{2}\int_{-1}^{1} |\partial_t u| dt \leq \frac{\sqrt{2}}{2} \|\partial_t u\|.$$

This implies (8.177).

Next, by the definition (8.175) and (8.177),

$$\begin{aligned}
&\|\partial_t(\pi_{d,N}^{1,l} v - v)\|_{\omega^{0,d-1}}^2 + d_l \|\pi_{d,N}^{1,l} v - v\|_{\omega^{0,d-3}}^2 \\
&\quad \leq \|\partial_t(\Pi_N v - v)\|_{\omega^{0,d-1}}^2 + d_l \|\Pi_N v - v\|_{\omega^{0,d-3}}^2 \\
&\quad \leq \|\partial_t(\Pi_N v - v)\|^2 + d_l \|\Pi_N v - v\|_{\omega^{-1,-1}}^2 \\
&\quad \leq c(1 + d_l N^{-2}) \frac{(N-s+1)!}{N!} (N+s)^{1-s} \|\partial_t^s v\|_{\omega^{s-1,s-1}}^2.
\end{aligned} \tag{8.180}$$

This ends the proof. □

We are now ready to analyze the convergence of the full mixed spectral approximation to (8.166). Let S be the unit circle or sphere, and $\Omega = I \times S$. Given a cut-off mode $M > 0$, we seek the approximate solution of (8.167) in the form:

$$\begin{aligned}
&U_{MN}(r,\phi) = \sum_{|l|=0}^{M} \hat{u}_l^N(r) e^{il\theta}, \quad \text{if } d = 2; \\
&U_{MN}(r,\theta,\phi) = \sum_{l=0}^{M} \sum_{|m|=0}^{l} \hat{u}_{lm}^N(r) Y_l^m(\theta,\phi), \quad \text{if } d = 3,
\end{aligned} \tag{8.181}$$

where $\{\hat{u}_l^N((1+t)/2) := v_N(t)\}$ and $\{\hat{u}_{lm}^N((1+t)/2) := v_N(t)\}$ are the solutions of (8.174) with $d = 2, 3$, respectively.

To describe the errors, we introduce the space $H_{p,d}^{s,s'}(\Omega)$ with $s, s' \geq 1$, which contains functions of partial derivatives up to $(s'-1)$ order being 2π-periodic and is equipped with the norm

$$\begin{aligned}
\|U\|_{H_{p,d}^{s,s'}(\Omega)} &= \sum_{|l|\geq 0} \|(r(1-r))^{(s-1)/2} \partial_r^s \hat{u}_l\|^2 \\
&+ \sum_{|l|\geq 0} d_l^{s'-1} \left(\|r^{(d-1)/2} \partial_r \hat{u}_l\|^2 + d_l \|r^{(d-3)/2} \hat{u}_l\|^2 + \|r^{(d-1)/2} \hat{u}_l\|^2 \right), \quad d = 2,
\end{aligned}$$

and for $d = 3$, we replace $\sum_{|l|\geq 0}$ and $\hat{u}_l$ in the above definition by $\sum_{l\geq 0} \sum_{|m|=0}^{l}$ and $\hat{u}_{lm}$, respectively.

Theorem 8.9. *Let U_{MN} be the approximate solution given by* (8.181)*. If $\alpha \geq 0$ and $U \in H_0^1(\Omega) \cap H_{p,d}^{s,s'}(\Omega)$ with $s' \geq 1$ and $1 \leq s \leq N+1$, we have*

$$\begin{aligned}
&\|\nabla(U - U_{MN})\| + \alpha \|U - U_{MN}\| \\
&\quad \leq c\left((1 + MN^{-1}) \sqrt{\frac{(N-s+1)!}{N!}} (N+s)^{(1-s)/2} + M^{1-s'} \right) \|U\|_{H_{p,d}^{s,s'}(\Omega)},
\end{aligned} \tag{8.182}$$

where c is a positive constant independent of M, N, s, s' and U.

Proof. Since the proofs of $d = 2,3$ are similar, we shall only prove the two-dimensional case. For notational convenience, let $E_{MN} = U - U_{MN}$ and $\hat{e}_l^N = \hat{u}_l - \hat{u}_l^N$. Thanks to the orthogonality of the Fourier basis, we have

$$\begin{aligned}
&\|\nabla E_{MN}\|^2 + \alpha\|E_{MN}\|^2 \\
&\quad \le c\sum_{|l|=0}^{M}\left(\|r^{1/2}\partial_r\hat{e}_l^N\|^2 + d_l\|r^{-1/2}\hat{e}_l^N\|^2 + \alpha\|r^{1/2}\hat{e}_l^N\|^2\right) \\
&\quad + c\sum_{|l|>M}\left(\|r^{1/2}\partial_r\hat{u}_l\|^2 + d_l\|r^{-1/2}\hat{u}_l\|^2 + \alpha\|r^{1/2}\hat{u}_l\|^2\right) \\
&\quad := G_1 + G_2.
\end{aligned} \tag{8.183}$$

It is clear that

$$\begin{aligned}
G_2 &\le cM^{2-2s'}\sum_{|l|>M} l^{2s'-2}\left(\|r^{1/2}\hat{u}_l\|^2 + \|r^{1/2}\partial_r\hat{u}_l\|^2 + l^2\|r^{-1/2}\hat{u}_l\|^2\right) \\
&\le cM^{2-2s'}\|U\|^2_{H^{s,s'}_{p,d}(\Omega)}.
\end{aligned} \tag{8.184}$$

It remains to estimate G_1. Notice that $\{v(t) = \hat{u}_l((1+t)/2)\}$ are the solutions to (8.173) with $d = 2$, and find from (8.173) and (8.174) that

$$a_{l,d}(\pi_{d,N}^{1,l}v - v_N, w_N) + \hat{\alpha}(\pi_{d,N}^{1,l}v - v_N, w_N) = \hat{\alpha}(\pi_{d,N}^{1,l}v - v, l_N), \ \ \forall w_N \in X_N(l,d).$$

Taking $w_N = \pi_{d,N}^{1,l}v - v_N$, we derive from Theorem 8.8 that

$$\begin{aligned}
&\|\partial_t(\pi_{d,N}^{1,l}v - v_N)\|^2_{\omega^{0,1}} + d_l\|\pi_{d,N}^{1,l}v - v_N\|^2_{\omega^{0,-1}} + \hat{\alpha}\|\pi_{d,N}^{1,l}v - v_N\|^2_{\omega^{0,1}} \\
&\quad \le c(1 + d_l N^{-2})\frac{(N-s+1)!}{N!}(N+s)^{1-s}\|\partial_t^s v\|^2_{\omega^{s-1,s-1}}.
\end{aligned}$$

Using the triangle inequality and Theorem 8.8 again yields

$$\begin{aligned}
&\|r^{1/2}\partial_r\hat{e}_l^N\|^2 + d_l\|r^{-1/2}\hat{e}_l^N\|^2 + \alpha\|r^{1/2}\hat{e}_l^N\|^2 \\
&\quad \le c(1 + d_l N^{-2})\frac{(N-s+1)!}{N!}(N+s)^{1-s}\|(r(1-r))^{(s-1)/2}\partial_r^s\hat{u}_l\|^2.
\end{aligned}$$

Consequently,

$$G_1 \le c(1 + M^2N^{-2})\frac{(N-s+1)!}{N!}(N+s)^{1-s}\|U\|^2_{H^{s,s'}_{p,d}(\Omega)}. \tag{8.185}$$

A combination of (8.184) and (8.185) leads to the desired result. □

8.5 Sparse Spectral-Galerkin Methods for High-Dimensional Problems

Note that the result in Theorem 8.1, i.e., the error estimate of the Jacobi polynomial approximations on the full tensorial spaces, suffers from the so-called "curse of dimensionality" (cf. Bellman (1961)), as the error decay rate with respect to the cardinality of P_N^d (i.e., $M = (N+1)^d$), deteriorates rapidly as d increases. More precisely,

$$\|\boldsymbol{\pi}_N^{\boldsymbol{\alpha},\boldsymbol{\beta}} u - u\|_{\boldsymbol{\omega}^{\boldsymbol{\alpha},\boldsymbol{\beta}}, I^d} \lesssim M^{-m/d} |u|_{B^m_{\boldsymbol{\alpha},\boldsymbol{\beta}}(I^d)}, \quad 0 \le m \le N+1.$$

An effective way to circumvent such a curse is to use the so-called hyperbolic cross approximations (cf. Korobov (1992) and the references therein). We present in this section some results established in Shen and Wang (2010) on hyperbolic cross approximations by Jacobi polynomials.

8.5.1 Hyperbolic Cross Jacobi Approximations

Define the hyperbolic cross index set:

$$\Upsilon_N := \Upsilon_N^H = \Big\{ \boldsymbol{n} \in \mathbb{N}_0^d : 1 \le |\boldsymbol{n}|_{\text{mix}} := \prod_{j=1}^d \max\{1, n_j\} \le N \Big\}, \tag{8.186}$$

and the finite dimensional space

$$X_N^{\boldsymbol{\alpha},\boldsymbol{\beta}} := \text{span}\Big\{ \boldsymbol{J}_{\boldsymbol{n}}^{\boldsymbol{\alpha},\boldsymbol{\beta}} \ : \ \boldsymbol{n} \in \Upsilon_N \Big\}. \tag{8.187}$$

For convenience, we denote the $\boldsymbol{k}$-complement of Υ_N in (8.186) by

$$\Upsilon_{N,\boldsymbol{k}}^c = \big\{ \boldsymbol{n} \in \mathbb{N}_0^d \ : \ |\boldsymbol{n}|_{\text{mix}} > N \ \text{and} \ \boldsymbol{n} \ge \boldsymbol{k} \big\}, \quad \forall \boldsymbol{k} \in \mathbb{N}_0^d. \tag{8.188}$$

To illustrate the distribution and sparsity of the grids in Υ_N^H, we plot in Fig. 8.2 the hyperbolic cross Υ_{32}^H with $d = 2$ (left) and $d = 3$ (right).

The following estimate on the cardinality of Υ_N^H can be found in, e.g., Dobrovol'skiĭ and Roshchenya (1998) and Griebel and Hamaekers (2007).

Lemma 8.9.

$$\text{card}\big(\Upsilon_N^H\big) = C_d N \big(\ln N\big)^{d-1}, \tag{8.189}$$

where the constant C_d depends on the dimension d.

To demonstrate the dependence of C_d on d, we plot in Fig. 8.3 $C_d = \frac{\text{card}(\Upsilon_N^H)}{N(\ln N)^{d-1}}$ for various $N \in [24, 128]$ and $d \in [2, 16]$, which indicates that C_d is uniformly bounded, and becomes smaller as d increases.

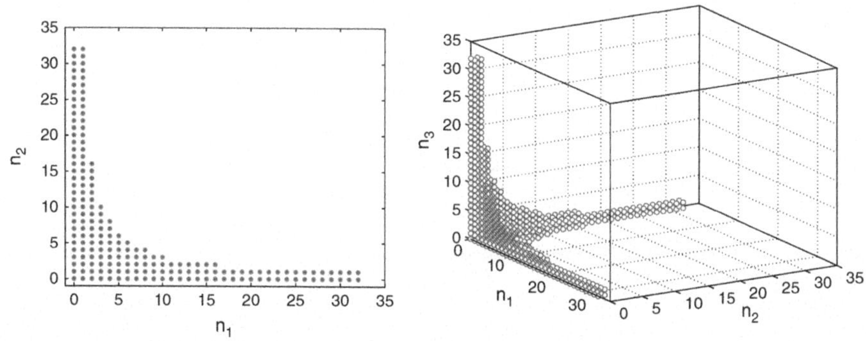

Fig. 8.2 The hyperbolic cross Υ_{32}^{H} with $d = 2$ (*left*) and $d = 3$ (*right*)

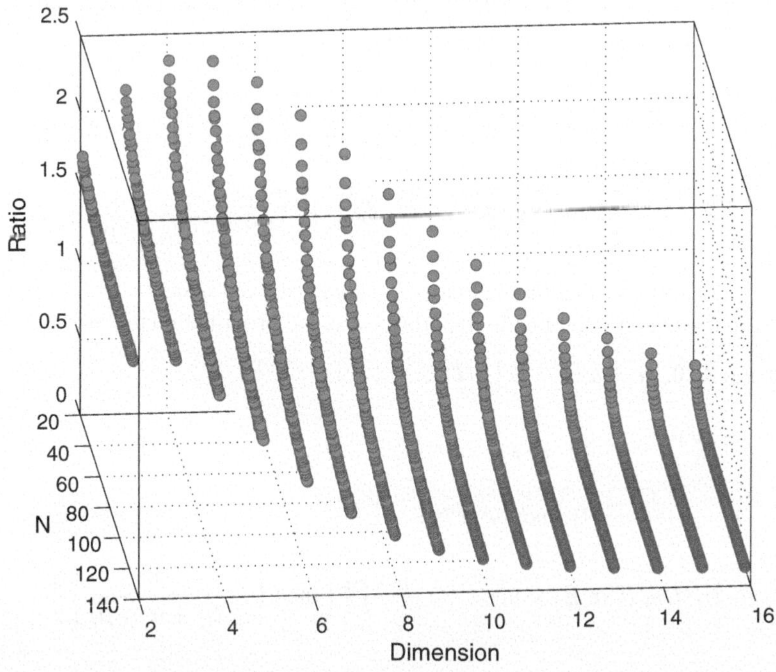

Fig. 8.3 The ratio $\mathrm{card}(\Upsilon_N^H)/\big(N(\ln N)^{d-1}\big)$ against various $N \in [24, 128]$ and $d \in [2, 16]$

We now turn to the estimation of the truncation error $u - \pi_N^{\boldsymbol{\alpha},\boldsymbol{\beta}} u$ of the hyperbolic cross approximation. In contrast to the Sobolev-type space (8.125) for the full grid, a suitable functional space to characterize the hyperbolic cross approximation is the Jacobi-weighted Korobov-type space:

$$\mathbb{K}^m_{\boldsymbol{\alpha},\boldsymbol{\beta}}(I^d) := \Big\{ u \,:\, \partial_{\boldsymbol{x}}^{\boldsymbol{k}} u \in L^2_{\boldsymbol{\omega}^{\boldsymbol{\alpha}+\boldsymbol{k},\boldsymbol{\beta}+\boldsymbol{k}}}(I^d),\ 0 \le |\boldsymbol{k}|_\infty < m \Big\}, \quad \forall m \in \mathbb{N}_0, \qquad (8.190)$$

with the norm and semi-norm

$$\|u\|_{\mathbb{K}^m_{\boldsymbol{\alpha},\boldsymbol{\beta}}(I^d)} = \Big(\sum_{0\le|\boldsymbol{k}|_\infty\le m} \|\partial_{\boldsymbol{x}}^{\boldsymbol{k}} u\|^2_{\boldsymbol{\omega}^{\boldsymbol{\alpha}+\boldsymbol{k},\boldsymbol{\beta}+\boldsymbol{k}},I^d} \Big)^{1/2},$$
$$|u|_{\mathbb{K}^m_{\boldsymbol{\alpha},\boldsymbol{\beta}}(I^d)} = \Big(\sum_{|\boldsymbol{k}|_\infty=m} \|\partial_{\boldsymbol{x}}^{\boldsymbol{k}} u\|^2_{\boldsymbol{\omega}^{\boldsymbol{\alpha}+\boldsymbol{k},\boldsymbol{\beta}+\boldsymbol{k}},I^d} \Big)^{1/2}. \tag{8.191}$$

Note the difference of the above definitions with those in (8.126). It is clear that $\mathbb{K}^0_{\boldsymbol{\alpha},\boldsymbol{\beta}}(I^d) = L^2_{\boldsymbol{\omega}^{\boldsymbol{\alpha},\boldsymbol{\beta}}}(I^d)$, and

$$B^{dm}_{\boldsymbol{\alpha},\boldsymbol{\beta}}(I^d) \subset \mathbb{K}^m_{\boldsymbol{\alpha},\boldsymbol{\beta}}(I^d) \subset B^m_{\boldsymbol{\alpha},\boldsymbol{\beta}}(I^d). \tag{8.192}$$

By (8.122), the norm and semi-norm of $\mathbb{K}^m_{\boldsymbol{\alpha},\boldsymbol{\beta}}(I^d)$ can be characterized in terms of the Jacobi expansion coefficients in (8.121):

$$\|u\|_{\mathbb{K}^m_{\boldsymbol{\alpha},\boldsymbol{\beta}}(I^d)} = \Big\{ \sum_{\boldsymbol{n}\in\mathbb{N}_0^d} \Big(\sum_{0\le|\boldsymbol{k}|_\infty\le m} \chi^{\boldsymbol{\alpha},\boldsymbol{\beta}}_{\boldsymbol{n},\boldsymbol{k}} \Big) |\hat{u}^{\boldsymbol{\alpha},\boldsymbol{\beta}}_{\boldsymbol{n}}|^2 \Big\}^{1/2},$$
$$|u|_{\mathbb{K}^m_{\boldsymbol{\alpha},\boldsymbol{\beta}}(I^d)} = \Big\{ \sum_{\boldsymbol{n}\in\mathbb{N}_0^d} \Big(\sum_{|\boldsymbol{k}|_\infty=m} \chi^{\boldsymbol{\alpha},\boldsymbol{\beta}}_{\boldsymbol{n},\boldsymbol{k}} \Big) |\hat{u}^{\boldsymbol{\alpha},\boldsymbol{\beta}}_{\boldsymbol{n}}|^2 \Big\}^{1/2}. \tag{8.193}$$

Hereafter, we assume that the regularity index m is a fixed integer.

The main result on the Jacobi hyperbolic cross approximation is stated below.

Theorem 8.10. *For any $u \in \mathbb{K}^m_{\boldsymbol{\alpha},\boldsymbol{\beta}}(I^d)$,*

$$\|\partial_{\boldsymbol{x}}^{\boldsymbol{l}}(\pi_N^{\boldsymbol{\alpha},\boldsymbol{\beta}} u - u)\|_{\boldsymbol{\omega}^{\boldsymbol{l}+\boldsymbol{\alpha},\boldsymbol{l}+\boldsymbol{\beta}},I^d} \le D_1 N^{|\boldsymbol{l}|_\infty - m} |u|_{\mathbb{K}^m_{\boldsymbol{\alpha},\boldsymbol{\beta}}(I^d)}, \quad 0 \le \boldsymbol{l} \le m, \tag{8.194}$$

where $D_1 = 1$ for $m = 0$, and for $m \ge 1$,

$$D_1 := D_1(\boldsymbol{l}, m, d, \boldsymbol{\alpha}, \boldsymbol{\beta}) = m^{(d-1)(m-|\boldsymbol{l}|_\infty)} \prod_{j=1}^d \Big(\max\Big\{1, \frac{m^2}{2m+\alpha_j+\beta_j}\Big\}\Big)^{m-l_j}. \tag{8.195}$$

Proof. Since the result is trivial for $m = 0$, we assume $m \ge 1$.

By (8.120)–(8.122) and (8.124),

$$\|\partial_{\boldsymbol{x}}^{\boldsymbol{l}}(\pi_N^{\boldsymbol{\alpha},\boldsymbol{\beta}} u - u)\|^2_{\boldsymbol{\omega}^{\boldsymbol{\alpha}+\boldsymbol{l},\boldsymbol{\beta}+\boldsymbol{l}},I^d} = \sum_{\boldsymbol{n}\in\Upsilon^c_{N,\boldsymbol{l}}} \chi^{\boldsymbol{\alpha},\boldsymbol{\beta}}_{\boldsymbol{n},\boldsymbol{l}} |\hat{u}^{\boldsymbol{\alpha},\boldsymbol{\beta}}_{\boldsymbol{n}}|^2$$
$$= \sum_{\boldsymbol{n}\in\Upsilon^c_{N,m}} \chi^{\boldsymbol{\alpha},\boldsymbol{\beta}}_{\boldsymbol{n},\boldsymbol{l}} |\hat{u}^{\boldsymbol{\alpha},\boldsymbol{\beta}}_{\boldsymbol{n}}|^2 + \sum_{\boldsymbol{n}\in\Upsilon^c_{N,\boldsymbol{l}}\setminus\Upsilon^c_{N,m}} \chi^{\boldsymbol{\alpha},\boldsymbol{\beta}}_{\boldsymbol{n},\boldsymbol{l}} |\hat{u}^{\boldsymbol{\alpha},\boldsymbol{\beta}}_{\boldsymbol{n}}|^2. \tag{8.196}$$

(i) $\boldsymbol{n} \in \Upsilon^c_{N,m}$. In this case, $\boldsymbol{n} \geq m$, so we have

$$\begin{aligned}\sum_{\boldsymbol{n}\in\Upsilon^c_{N,m}} \chi^{\boldsymbol{\alpha},\boldsymbol{\beta}}_{\boldsymbol{n},\boldsymbol{l}} |\hat{u}^{\boldsymbol{\alpha},\boldsymbol{\beta}}_{\boldsymbol{n}}|^2 &\leq \max_{\boldsymbol{n}\in\Upsilon^c_{N,m}} \left\{ \frac{\chi^{\boldsymbol{\alpha},\boldsymbol{\beta}}_{\boldsymbol{n},\boldsymbol{l}}}{\chi^{\boldsymbol{\alpha},\boldsymbol{\beta}}_{\boldsymbol{n},\boldsymbol{m}}} \right\} \sum_{\boldsymbol{n}\in\Upsilon^c_{N,m}} \chi^{\boldsymbol{\alpha},\boldsymbol{\beta}}_{\boldsymbol{n},\boldsymbol{m}} |\hat{u}^{\boldsymbol{\alpha},\boldsymbol{\beta}}_{\boldsymbol{n}}|^2 \\ &\overset{(8.122)}{\leq} \max_{\boldsymbol{n}\in\Upsilon^c_{N,m}} \left\{ \frac{\chi^{\boldsymbol{\alpha},\boldsymbol{\beta}}_{\boldsymbol{n},\boldsymbol{l}}}{\chi^{\boldsymbol{\alpha},\boldsymbol{\beta}}_{\boldsymbol{n},\boldsymbol{m}}} \right\} \left\|\boldsymbol{\partial}^{\boldsymbol{m}}_{\boldsymbol{x}} u\right\|^2_{\boldsymbol{\omega}^{\boldsymbol{\alpha}+m,\boldsymbol{\beta}+m},I^d},\end{aligned} \tag{8.197}$$

where we have set $\boldsymbol{m} = (m, m, \ldots, m)$ and $\boldsymbol{\partial}^{\boldsymbol{m}}_{\boldsymbol{x}} u = \partial^{(m,m,\ldots,m)}_{\boldsymbol{x}} u$. Thus, we only need to estimate the maximum value in (8.197).

A direct calculation by using (8.113) and (8.119) yields

$$\begin{aligned}\frac{\chi^{\boldsymbol{\alpha},\boldsymbol{\beta}}_{\boldsymbol{n},\boldsymbol{l}}}{\chi^{\boldsymbol{\alpha},\boldsymbol{\beta}}_{\boldsymbol{n},\boldsymbol{m}}} &= \prod_{j=1}^{d}\prod_{i=l_j}^{m-1} n_j^{-2}\Big(1+\frac{\alpha_j+\beta_j+1}{n_j}-\frac{i(i+\alpha_j+\beta_j+1)}{n_j^2}\Big)^{-1} \\ &= \Big(\prod_{j=1}^{d} n_j^{2(l_j-m)}\Big) \prod_{j=1}^{d}\prod_{i=l_j}^{m-1} \underbrace{\Big(1+\frac{\alpha_j+\beta_j+1}{n_j}-\frac{i(i+\alpha_j+\beta_j+1)}{n_j^2}\Big)^{-1}}_{:=g(i,j)}.\end{aligned} \tag{8.198}$$

Notice that for any $\boldsymbol{n} \in \Upsilon^c_{N,m}$ and $0 \leq |\boldsymbol{l}| \leq m$,

$$\prod_{j=1}^{d} n_j^{2(l_j-m)} \leq \prod_{j=1}^{d} n_j^{2(|\boldsymbol{l}|_\infty-m)} \leq N^{2(|\boldsymbol{l}|_\infty-m)}. \tag{8.199}$$

Next, we estimate the upper bound of the second product in (8.198). Note that $g(i,j)$ is decreasing with respect to i, i.e., $g(i,j) \leq g(m-1,j)$. Hence,

$$\begin{aligned}\prod_{j=1}^{d}\prod_{i=l_j}^{m-1} g(i,j) &\leq \prod_{j=1}^{d} \big[g(m-1,j)\big]^{m-l_j} \\ &= \prod_{j=1}^{d}\Big(1+\frac{\alpha_j+\beta_j+1}{n_j}-\frac{(m-1)(m+\alpha_j+\beta_j+1)}{n_j^2}\Big)^{l_j-m}.\end{aligned} \tag{8.200}$$

To obtain an upper bound independent of N, we define

$$f_j(t) := -(m-1)(m+\alpha_j+\beta_j+1)t^2 + (\alpha_j+\beta_j+1)1 + t \quad \text{with} \quad t = \frac{1}{n_j}.$$

Assuming that $n_1 n_2 \ldots n_d = \widetilde{N} > N \gg 1$, one verifies that

$$m \leq n_j \leq \frac{\widetilde{N}}{m^{d-1}} \quad \Rightarrow \quad \frac{m^{d-1}}{\widetilde{N}} \leq t \leq \frac{1}{m}.$$

Obviously, for $m=1$, we have

$$f_j(t)=1+\frac{\alpha_j+\beta_j+1}{n_j}\geq\begin{cases}1, & \text{if} \quad \alpha_j+\beta_j+1\geq 0,\\ \alpha_j+\beta_j+2, & \text{if} \quad -2<\alpha_j+\beta_j+1<0.\end{cases} \tag{8.201}$$

For $m\geq 2$, using the properties of quadratic functions, we find that

$$f_j(t)\geq\min\left\{f_j\left(\frac{1}{m}\right),\, f_j\left(\frac{m^{d-1}}{\widetilde{N}}\right)\right\}. \tag{8.202}$$

A direct calculation leads to

$$f_j\left(\frac{1}{m}\right)=\frac{2m+\alpha_j+\beta_j}{m^2}; \quad f_j\left(\frac{m^{d-1}}{\widetilde{N}}\right)\simeq 1 \quad \text{for} \ \widetilde{N}>N\gg 1. \tag{8.203}$$

A combination of the above facts gives

$$\prod_{j=1}^{d}\prod_{i=l_j}^{m-1}g(i,j)\leq\prod_{j=1}^{d}\left(\max\left\{1,\frac{m^2}{2m+\alpha_j+\beta_j}\right\}\right)^{m-l_j}:=\tilde{c}^2, \tag{8.204}$$

which is valid for all $\boldsymbol{n}\geq m\geq 1$ and $N\gg 1$. Consequently, we derive from (8.198), (8.199) and (8.204) that

$$\max_{\boldsymbol{n}\in\Upsilon^c_{N,m}}\left\{\frac{\boldsymbol{\chi}^{\boldsymbol{\alpha},\boldsymbol{\beta}}_{\boldsymbol{n},\boldsymbol{l}}}{\boldsymbol{\chi}^{\boldsymbol{\alpha},\boldsymbol{\beta}}_{\boldsymbol{n},m}}\right\}\leq\tilde{c}^2N^{2(|\boldsymbol{l}|_\infty-m)}. \tag{8.205}$$

Now, we deal with the second summation in (8.196).

(ii) $\boldsymbol{n}\in\Upsilon^c_{N,\boldsymbol{l}}\setminus\Upsilon^c_{N,m}$. In this case, a little care has to be taken for the modes with $n_j<m$. Notice that

$$\Upsilon^c_{N,\boldsymbol{l}}\setminus\Upsilon^c_{N,m}=\left\{\boldsymbol{n}\in\mathbb{N}_0^d\ :\ |\boldsymbol{n}|_{\text{mix}}>N,\ \boldsymbol{n}\geq\boldsymbol{l},\ \exists j,\ \text{s.t.}\ l_j\leq n_j<m\right\}.$$

For clarity, we split the index set $\{1\leq j\leq d\}=\aleph\cup\aleph^c$ with

$$\aleph=\left\{j\ :\ l_j\leq n_j<m,\ 1\leq j\leq d\right\},\quad \aleph^c=\left\{j\ :\ n_j\geq m,\ 1\leq j\leq d\right\}. \tag{8.206}$$

Clearly, $\aleph\cap\aleph^c=\varnothing$ and neither of these two index sets is empty. Define

$$\tilde{\chi}^{\alpha_j,\beta_j}_{n_j,l_j,m}:=\max\left\{\chi^{\alpha_j,\beta_j}_{n_j,l_j},\chi^{\alpha_j,\beta_j}_{n_j,m}\right\}=\begin{cases}0, & \text{if}\ n_j<l_j,\\ \chi^{\alpha_j,\beta_j}_{n_j,l_j}, & \text{if}\ l_j\leq n_j<m,\\ \chi^{\alpha_j,\beta_j}_{n_j,m}, & \text{if}\ m\leq n_j.\end{cases} \tag{8.207}$$

Hence, for any $j \in \aleph$, $\tilde{\chi}_{n_j,l_j,m}^{\alpha_j,\beta_j} = \chi_{n_j,l_j}^{\alpha_j,\beta_j}$, while for any $j \in \aleph^c$, $\tilde{\chi}_{n_j,l_j,m}^{\alpha_j,\beta_j} = \chi_{n_j,m}^{\alpha_j,\beta_j}$. Moreover,

$$\tilde{\boldsymbol{\chi}}_{\boldsymbol{n},\boldsymbol{l},m}^{\boldsymbol{\alpha},\boldsymbol{\beta}} = \Big(\prod_{j\in\aleph} \chi_{n_j,l_j}^{\alpha_j,\beta_j}\Big)\Big(\prod_{k\in\aleph^c} \chi_{n_k,m}^{\alpha_k,\beta_k}\Big) = \boldsymbol{\chi}_{\boldsymbol{n},\boldsymbol{k}}^{\boldsymbol{\alpha},\boldsymbol{\beta}}, \tag{8.208}$$

where $\boldsymbol{k}$ is a $d-$dimensional index consisting of l_j for $j \in \aleph$ and m for $j \in \aleph^c$. Since $|\boldsymbol{k}|_\infty = m$, we find from (8.122) and (8.193) that

$$\sum_{\boldsymbol{n}\in\Upsilon_{N,\boldsymbol{l}}^c\setminus\Upsilon_{N,m}^c} \tilde{\boldsymbol{\chi}}_{\boldsymbol{n},\boldsymbol{l},m}^{\boldsymbol{\alpha},\boldsymbol{\beta}} |\hat{u}_{\boldsymbol{n}}^{\boldsymbol{\alpha},\boldsymbol{\beta}}|^2 = \sum_{\boldsymbol{n}\in\Upsilon_{N,\boldsymbol{l}}^c\setminus\Upsilon_{N,m}^c} \boldsymbol{\chi}_{\boldsymbol{n},\boldsymbol{k}}^{\boldsymbol{\alpha},\boldsymbol{\beta}} |\hat{u}_{\boldsymbol{n}}^{\boldsymbol{\alpha},\boldsymbol{\beta}}|^2 \le |u|^2_{\mathbb{K}^m_{\boldsymbol{\alpha},\boldsymbol{\beta}}(I^d)}. \tag{8.209}$$

We treat the second summation in (8.196) as

$$\begin{aligned}
\sum_{\boldsymbol{n}\in\Upsilon_{N,\boldsymbol{l}}^c\setminus\Upsilon_{N,m}^c} \boldsymbol{\chi}_{\boldsymbol{n},\boldsymbol{l}}^{\boldsymbol{\alpha},\boldsymbol{\beta}} |\hat{u}_{\boldsymbol{n}}^{\boldsymbol{\alpha},\boldsymbol{\beta}}|^2 &= \max_{\boldsymbol{n}\in\Upsilon_{N,\boldsymbol{l}}^c\setminus\Upsilon_{N,m}^c} \left\{ \frac{\boldsymbol{\chi}_{\boldsymbol{n},\boldsymbol{l}}^{\boldsymbol{\alpha},\boldsymbol{\beta}}}{\tilde{\boldsymbol{\chi}}_{\boldsymbol{n},\boldsymbol{l},m}^{\boldsymbol{\alpha},\boldsymbol{\beta}}} \right\} \sum_{\boldsymbol{n}\in\Upsilon_{N,\boldsymbol{l}}^c\setminus\Upsilon_{N,m}^c} \tilde{\boldsymbol{\chi}}_{\boldsymbol{n},\boldsymbol{l},m}^{\boldsymbol{\alpha},\boldsymbol{\beta}} |\hat{u}_{\boldsymbol{n}}^{\boldsymbol{\alpha},\boldsymbol{\beta}}|^2 \\
&\overset{(8.209)}{\le} \max_{\boldsymbol{n}\in\Upsilon_{N,\boldsymbol{l}}^c\setminus\Upsilon_{N,m}^c} \left\{ \frac{\boldsymbol{\chi}_{\boldsymbol{n},\boldsymbol{l}}^{\boldsymbol{\alpha},\boldsymbol{\beta}}}{\tilde{\boldsymbol{\chi}}_{\boldsymbol{n},\boldsymbol{l},m}^{\boldsymbol{\alpha},\boldsymbol{\beta}}} \right\} |u|^2_{\mathbb{K}^m_{\boldsymbol{\alpha},\boldsymbol{\beta}}(I^d)}.
\end{aligned} \tag{8.210}$$

Thus, it remains to estimate the maximum. By a direct calculation,

$$\begin{aligned}
\frac{\boldsymbol{\chi}_{\boldsymbol{n},\boldsymbol{l}}^{\boldsymbol{\alpha},\boldsymbol{\beta}}}{\tilde{\boldsymbol{\chi}}_{\boldsymbol{n},\boldsymbol{l},m}^{\boldsymbol{\alpha},\boldsymbol{\beta}}} &= \prod_{j\in\aleph^c} \frac{\chi_{n_j,l_j}^{\alpha_j,\beta_j}}{\chi_{n_j,m}^{\alpha_j,\beta_j}} \le \Big(\prod_{j\in\aleph^c} n_j^{2(l_j-m)}\Big) \prod_{j\in\aleph^c}\prod_{i=l_j}^{m-1} g(i,j) \\
&\le \Big(\prod_{j\in\aleph^c} n_j^{2(l_j-m)}\Big) \prod_{j\in\aleph^c} \left(\max\left\{1, \frac{m^2}{2m+\alpha_j+\beta_j}\right\}\right)^{m-l_j} \\
&\overset{(8.204)}{\le} \tilde{c}^2\Big(\prod_{j\in\aleph^c} n_j^{2(l_j-m)}\Big).
\end{aligned} \tag{8.211}$$

In view of $m \ge 1$ and $|\boldsymbol{n}|_{\text{mix}} = \bar{n}_1 \ldots \bar{n}_d > N$, we deduce that

$$\prod_{j\in\aleph^c} \bar{n}_j > \frac{N}{\prod_{j\in\aleph} \bar{n}_j} > \frac{N}{\prod_{j\in\aleph} m}. \tag{8.212}$$

A combination of the above estimates leads to

$$\begin{aligned}
\frac{\boldsymbol{\chi}_{\boldsymbol{n},\boldsymbol{l}}^{\boldsymbol{\alpha},\boldsymbol{\beta}}}{\tilde{\boldsymbol{\chi}}_{\boldsymbol{n},\boldsymbol{l},m}^{\boldsymbol{\alpha},\boldsymbol{\beta}}} &\le \tilde{c}^2 \prod_{j\in\aleph^c} n_j^{2(l_j-m)} \le \tilde{c}^2 \prod_{j\in\aleph^c} n_j^{2(|\boldsymbol{l}|_\infty-m)} \le \tilde{c}^2 \Big(\frac{N}{\prod_{j\in\aleph} m}\Big)^{2(|\boldsymbol{l}|_\infty-m)} \\
&\le \tilde{c}^2 m^{2(d-1)(m-|\boldsymbol{l}|_\infty)} N^{2(|\boldsymbol{l}|_\infty-m)}.
\end{aligned} \tag{8.213}$$

Finally, the estimate (8.194)–(8.195) follows from (8.196), (8.205), (8.210) and (8.213). □

By the definition of $\mathbb{K}^l_{\boldsymbol{\alpha},\boldsymbol{\beta}}(I^d)$ in (8.190)–(8.191) and Theorem 8.10, we immediately obtain the following result.

Corollary 8.1.

$$\|\boldsymbol{\pi}_N^{\boldsymbol{\alpha},\boldsymbol{\beta}} u - u\|_{\mathbb{K}^l_{\boldsymbol{\alpha},\boldsymbol{\beta}}(I^d)} \le D_2 N^{l-m} |u|_{\mathbb{K}^m_{\boldsymbol{\alpha},\boldsymbol{\beta}}(I^d)}, \quad 0 \le l \le m, \tag{8.214}$$

where the constant D_2 is given by

$$D_2 := D_2(l,m,d,\boldsymbol{\alpha},\boldsymbol{\beta}) = \Big(\sum_{0 \le |\boldsymbol{l}|_\infty \le l} D_1^2 N^{2(|\boldsymbol{l}|_\infty - l)} \Big)^{1/2}, \tag{8.215}$$

with D_1 being the same as in Theorem 8.10.

The above result clearly indicates that the Jacobi-weighted Korobov-type spaces $\mathbb{K}^l_{\boldsymbol{\alpha},\boldsymbol{\beta}}(I^d)$ are the natural functional spaces for hyperbolic cross approximations. We note that the above result is in the same form as the result in Theorem 8.1, except that Jacobi-weighted Korobov norms are used here instead of Jacobi-weighted Sobolev norms used in Theorem 8.1.

To characterize the error in terms of the dimensionality of the approximation space $X_N^{\boldsymbol{\alpha},\boldsymbol{\beta}}$, we find from Lemma 8.9 that for any $\varepsilon > 0$, and $N \gg 1$,

$$M \le C_d N^{1+\varepsilon(d-1)} \quad \Rightarrow \quad N^{-1} \le C_d^{1/(1+\varepsilon(d-1))} M^{-(1/(1+\varepsilon(d-1)))}.$$

Therefore, as a direct consequence of Corollary 8.1, we have the following estimate.

Corollary 8.2. *For any $\varepsilon > 0$ and $0 \le l \le m$,*

$$\|\boldsymbol{\pi}_N^{\boldsymbol{\alpha},\boldsymbol{\beta}} u - u\|_{\mathbb{K}^l_{\boldsymbol{\alpha},\boldsymbol{\beta}}(I^d)} \le D_2 C_d^{1/(1+\varepsilon(d-1))} M^{\frac{|\boldsymbol{l}|_\infty - m}{1+\varepsilon(d-1)}} |u|_{\mathbb{K}^m_{\boldsymbol{\alpha},\boldsymbol{\beta}}(I^d)}. \tag{8.216}$$

8.5.2 *Optimized Hyperbolic Cross Jacobi Approximations*

While the use of the regular hyperbolic cross (8.187) significantly improved the convergence rate with respect to the number of unknowns, the "curse of dimensionality" is not completely broken as the convergence rate still deteriorates, albeit very slowly, as d increases (cf. (8.216)). In order to completely break the "curse of dimensionality", we consider the following family of spaces (cf. Bungartz and Griebel (2004), Griebel and Hamaekers (2007)):

$$V_{N,\gamma}^{\boldsymbol{\alpha},\boldsymbol{\beta}} := \mathrm{span}\{ \boldsymbol{J}_{\boldsymbol{n}}^{\boldsymbol{\alpha},\boldsymbol{\beta}} \, : \, |\boldsymbol{n}|_{\mathrm{mix}} |\boldsymbol{n}|_\infty^{-\gamma} \le N^{1-\gamma} \}, \quad -\infty \le \gamma < 1. \tag{8.217}$$

In particular, we have $V_{N,0}^{\boldsymbol{\alpha},\boldsymbol{\beta}} = X_N^{\boldsymbol{\alpha},\boldsymbol{\beta}}$ in (8.187) and $V_{N,-\infty}^{\boldsymbol{\alpha},\boldsymbol{\beta}} := \mathrm{span}\{\boldsymbol{J}_{\boldsymbol{n}}^{\boldsymbol{\alpha},\boldsymbol{\beta}} : |\boldsymbol{n}|_\infty \le N\}$ (i.e., the full grid). But for $0<\gamma<1$, the trade-off between N^γ and $|\boldsymbol{n}|_\infty^\gamma$ leads to the following reduction of cardinality (see Lemma 3 in Griebel and Hamaekers (2007)):

$$\mathrm{card}\big(V_{N,\gamma}^{\boldsymbol{\alpha},\boldsymbol{\beta}}\big) = C(\gamma,d)N, \quad 0<\gamma<1. \tag{8.218}$$

Thus, the space $V_{N,\gamma}^{\boldsymbol{\alpha},\boldsymbol{\beta}}$ with $0<\gamma<1$ is referred to as the optimized hyperbolic cross space.

We plot in Fig. 8.4 the ratio $\dim\big(V_{N,\gamma}^{\boldsymbol{\alpha},\boldsymbol{\beta}}\big)/N$ for various N with $\gamma = 0.9$ and $d = 2,\ldots,10$, which indicates that the constant $C(\gamma,d)$ is independent of N, but grows as d increases.

In this case, the complement index set in (8.188) takes the form

$$\Upsilon_{N,\boldsymbol{k}}^{c} = \big\{\boldsymbol{n}\in\mathbb{N}_0^d : |\boldsymbol{n}|_{\mathrm{mix}}|\boldsymbol{n}|_\infty^{-\gamma} > N^{1-\gamma} \text{ and } \boldsymbol{n}\ge\boldsymbol{k}\big\}, \quad \forall \boldsymbol{k}\in\mathbb{N}_0^d. \tag{8.219}$$

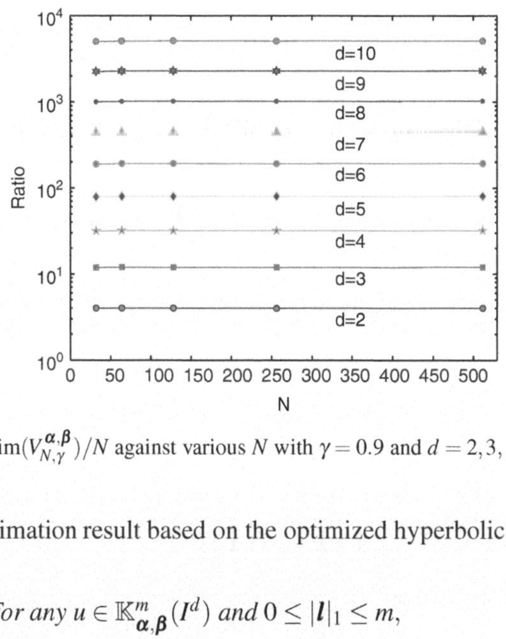

Fig. 8.4 The ratio $\dim(V_{N,\gamma}^{\boldsymbol{\alpha},\boldsymbol{\beta}})/N$ against various N with $\gamma = 0.9$ and $d = 2,3,\ldots,10$

The main approximation result based on the optimized hyperbolic cross is stated as follows.

Theorem 8.11. *For any $u\in\mathbb{K}_{\boldsymbol{\alpha},\boldsymbol{\beta}}^m(I^d)$ and $0\le|\boldsymbol{l}|_1\le m$,*

$$\big\|\partial_{\boldsymbol{x}}^{\boldsymbol{l}}(\pi_{N,\gamma}^{\boldsymbol{\alpha},\boldsymbol{\beta}}u-u)\big\|_{\boldsymbol{\omega}^{\boldsymbol{\alpha}+\boldsymbol{l},\boldsymbol{\beta}+\boldsymbol{l}},I^d} \le D_3|u|_{\mathbb{K}_{\boldsymbol{\alpha},\boldsymbol{\beta}}^m(I^d)} \times \begin{cases} N^{|\boldsymbol{l}|_1-m}, & \text{if } 0<\gamma\le\frac{|\boldsymbol{l}|_1}{m},\\ N^{|\boldsymbol{l}|_1-m+(\gamma m-|\boldsymbol{l}|_1)(1-\frac{1}{d})}, & \text{if } \frac{|\boldsymbol{l}|_1}{m}\le\gamma<1,\end{cases} \tag{8.220}$$

where

$$D_3 := \left[\prod_{j=1}^{d}\left(\max\left\{1, \frac{m^2}{2m+\alpha_j+\beta_j}\right\}\right)^{m-l_j}\right]$$
$$\times m^{(d-1)m} \times \begin{cases} m^{-\frac{(d-1)(|\boldsymbol{l}|_1-\gamma m)}{1-\gamma}}, & \text{if } 0<\gamma\le \frac{|\boldsymbol{l}|_1}{m}, \\ 1, & \text{if } \frac{|\boldsymbol{l}|_1}{m}\le\gamma<1. \end{cases}$$

Proof. The estimate is trivial for $m=0$, so we assume $m\ge 1$. Let $\tilde{\chi}^{\boldsymbol{\alpha},\boldsymbol{\beta}}_{\boldsymbol{n},\boldsymbol{l},m}$ be the same as defined in (8.208). Following the proof of Theorem 8.10, we begin by estimating (8.196) with $\Upsilon^c_{N,\boldsymbol{l}}$ defined in (8.219), and separated into two subsets: (i) $\Upsilon^c_{N,m}$ and (ii) $\Upsilon^c_{N,\boldsymbol{l}}\setminus\Upsilon^c_{N,m}$ as before, namely,

$$\begin{aligned} \left\|\partial_{\boldsymbol{x}}^{\boldsymbol{l}}\left(\boldsymbol{\pi}_N^{\boldsymbol{\alpha},\boldsymbol{\beta}}u-u\right)\right\|^2_{\boldsymbol{\omega}^{\boldsymbol{\alpha}+\boldsymbol{l},\boldsymbol{\beta}+\boldsymbol{l}},I^d} &\le \max_{\boldsymbol{n}\in\Upsilon^c_{N,m}}\left\{\frac{\chi^{\boldsymbol{\alpha},\boldsymbol{\beta}}_{\boldsymbol{n},\boldsymbol{l}}}{\tilde{\chi}^{\boldsymbol{\alpha},\boldsymbol{\beta}}_{\boldsymbol{n},\boldsymbol{l},m}}\right\}\sum_{\boldsymbol{n}\in\Upsilon^c_{N,m}}\tilde{\chi}^{\boldsymbol{\alpha},\boldsymbol{\beta}}_{\boldsymbol{n},\boldsymbol{l},m}\left|\hat{u}^{\boldsymbol{\alpha},\boldsymbol{\beta}}_{\boldsymbol{n}}\right|^2 \\ &+ \max_{\boldsymbol{n}\in\Upsilon^c_{N,\boldsymbol{l}}\setminus\Upsilon^c_{N,m}}\left\{\frac{\chi^{\boldsymbol{\alpha},\boldsymbol{\beta}}_{\boldsymbol{n},\boldsymbol{l}}}{\tilde{\chi}^{\boldsymbol{\alpha},\boldsymbol{\beta}}_{\boldsymbol{n},\boldsymbol{l},m}}\right\}\sum_{\boldsymbol{n}\in\Upsilon^c_{N,\boldsymbol{l}}\setminus\Upsilon^c_{N,m}}\tilde{\chi}^{\boldsymbol{\alpha},\boldsymbol{\beta}}_{\boldsymbol{n},\boldsymbol{l},m}\left|\hat{u}^{\boldsymbol{\alpha},\boldsymbol{\beta}}_{\boldsymbol{n}}\right|^2. \end{aligned} \tag{8.221}$$

We now estimate the first term, and consider $\boldsymbol{n}\in\Upsilon^c_{N,m}$. Like (8.198), we have

$$\frac{\chi^{\boldsymbol{\alpha},\boldsymbol{\beta}}_{\boldsymbol{n},\boldsymbol{l}}}{\chi^{\boldsymbol{\alpha},\boldsymbol{\beta}}_{\boldsymbol{n},m}} = \left(\prod_{j=1}^{d} n_j^{2(l_j-m)}\right)\left\{\prod_{j=1}^{d}\prod_{i=l_j}^{m-1} g(i,j)\right\}. \tag{8.222}$$

We first deal with the product in the parentheses. Notice that for any $\boldsymbol{n}\in\Upsilon^c_{N,m}$,

$$|\boldsymbol{n}|_{\text{mix}}|\boldsymbol{n}|_\infty^{-\gamma} > N^{1-\gamma} \implies \left(\frac{|\boldsymbol{n}|_\infty^\gamma}{|\boldsymbol{n}|_{\text{mix}}}\right)^{\frac{1}{1-\gamma}} < \frac{1}{N}. \tag{8.223}$$

Therefore,

$$\begin{aligned} \prod_{j=1}^{d} n_j^{2(l_j-m)} &= \left(\prod_{j=1}^{d} n_j^{2l_j}\right)\left(\prod_{j=1}^{d} n_j\right)^{-2m} \\ &\le \left(\prod_{j=1}^{d}|\boldsymbol{n}|_\infty^{2l_j}\right)|\boldsymbol{n}|_{\text{mix}}^{-2m} = |\boldsymbol{n}|_\infty^{2|\boldsymbol{l}|_1}|\boldsymbol{n}|_{\text{mix}}^{-2m} \\ &= \left(\frac{|\boldsymbol{n}|_\infty^\gamma}{|\boldsymbol{n}|_{\text{mix}}}\right)^{\frac{2(m-|\boldsymbol{l}|_1)}{1-\gamma}}\left(\frac{|\boldsymbol{n}|_\infty}{|\boldsymbol{n}|_{\text{mix}}}\right)^{\frac{2(|\boldsymbol{l}|_1-\gamma m)}{1-\gamma}} \\ &\overset{(8.223)}{\le} N^{2(|\boldsymbol{l}|_1-m)}\left(\frac{|\boldsymbol{n}|_\infty}{|\boldsymbol{n}|_{\text{mix}}}\right)^{\frac{2(|\boldsymbol{l}|_1-\gamma m)}{1-\gamma}}. \end{aligned} \tag{8.224}$$

One verifies readily that

$$\frac{|\boldsymbol{n}|_\infty}{|\boldsymbol{n}|_{\mathrm{mix}}} \le \frac{1}{m^{d-1}}, \quad \forall\, \boldsymbol{n} \in \Upsilon^c_{N,m}. \tag{8.225}$$

Hence, if $0 < \gamma \le \frac{|\boldsymbol{l}|_1}{m}$,

$$\begin{aligned} \prod_{j=1}^d n_j^{2(l_j-m)} &\le N^{2(|\boldsymbol{l}|_1-m)} \max_{\boldsymbol{n}\in\Upsilon^c_{N,m}} \Big(\frac{|\boldsymbol{n}|_\infty}{|\boldsymbol{n}|_{\mathrm{mix}}}\Big)^{\frac{2(|\boldsymbol{l}|_1-\gamma m)}{1-\gamma}} \\ &\le m^{-\frac{2(d-1)(|\boldsymbol{l}|_1-\gamma m)}{1-\gamma}} N^{2(|\boldsymbol{l}|_1-m)}. \end{aligned} \tag{8.226}$$

Next, for any $\boldsymbol{n} \in \Upsilon^c_{N,m}$, we have $|\boldsymbol{n}|_\infty > N^{\frac{1}{d}}$ and

$$|\boldsymbol{n}|_{\mathrm{mix}} |\boldsymbol{n}|_\infty^{-\gamma} > N^{1-\gamma} \;\Rightarrow\; \frac{|\boldsymbol{n}|_{\mathrm{mix}}}{|\boldsymbol{n}|_\infty} < N^{1-\gamma} |\boldsymbol{n}|_\infty^{\gamma-1} < N^{(1-\gamma)(1-\frac{1}{d})}, \tag{8.227}$$

which, together with (8.224), implies that if $\frac{|\boldsymbol{l}|_1}{m} \le \gamma < 1$,

$$\begin{aligned} \prod_{j=1}^d n_j^{2(l_j-m)} &\le N^{2(|\boldsymbol{l}|_1-m)} \max_{\boldsymbol{n}\in\Upsilon^c_{N,m}} \Big(\frac{|\boldsymbol{n}|_\infty}{|\boldsymbol{n}|_{\mathrm{mix}}}\Big)^{\frac{2(|\boldsymbol{l}|_1-\gamma m)}{1-\gamma}} \\ &\le N^{2(|\boldsymbol{l}|_1-m)+2(\gamma m-|\boldsymbol{l}|_1)(1-1/d)}. \end{aligned} \tag{8.228}$$

It remains to estimate the term in the braces of (8.222). Observe that for any $\boldsymbol{n} \in \Upsilon^c_{N,m}$, we have

$$|\boldsymbol{n}|_\infty \ge |\boldsymbol{n}|_{\mathrm{mix}}^{1/d} \quad\Rightarrow\quad |\boldsymbol{n}|_{\mathrm{mix}} > N^{(1-\gamma)/(1-\gamma/d)} \gg 1. \tag{8.229}$$

Hence, the product in the braces with maximum taken over $\Upsilon^c_{N,m}$ (cf. (8.219)) has the same upper bound $\tilde c^2$ as in (8.204).

Next we consider the second summation in (8.221). Defining $\aleph$ and $\aleph^c$ associated with $\Upsilon^c_{N,\boldsymbol{l}} \setminus \Upsilon^c_{N,m}$ as in (8.206), and following the derivation of the estimate (8.213), we have

$$\begin{aligned} \frac{\boldsymbol{\chi}^{\boldsymbol{\alpha},\boldsymbol{\beta}}_{\boldsymbol{n},\boldsymbol{l}}}{\tilde{\boldsymbol{\chi}}^{\boldsymbol{\alpha},\boldsymbol{\beta}}_{\boldsymbol{n},\boldsymbol{l},m}} &= \Big(\prod_{j\in\aleph} \frac{\chi^{\alpha_j,\beta_j}_{n_j,l_j}}{\tilde\chi^{\alpha_j,\beta_j}_{n_j,l_j,m}}\Big)\Big(\prod_{k\in\aleph^c} \frac{\chi^{\alpha_k,\beta_k}_{n_k,l_k}}{\tilde\chi^{\alpha_k,\beta_k}_{n_k,l_k,m}}\Big) \overset{(8.204)}{\le} \tilde c^2 \prod_{k\in\aleph^c} n_k^{2(l_k-m)} \\ &\le \tilde c^2 \Big(\prod_{k\in\aleph^c} |\boldsymbol{n}|_\infty^{2l_k}\Big)\Big(\frac{\prod_{j\in\aleph}\bar n_j}{|\boldsymbol{n}|_{\mathrm{mix}}}\Big)^{2m} \le \tilde c^2 m^{2(d-1)m} |\boldsymbol{n}|_\infty^{2|\boldsymbol{l}|_1} |\boldsymbol{n}|_{\mathrm{mix}}^{-2m} \\ &\overset{(8.224)}{\le} \tilde c^2 m^{2(d-1)m} N^{2(|\boldsymbol{l}|_1-m)} \Big(\frac{|\boldsymbol{n}|_\infty}{|\boldsymbol{n}|_{\mathrm{mix}}}\Big)^{\frac{2(|\boldsymbol{l}|_1-\gamma m)}{1-\gamma}}. \end{aligned} \tag{8.230}$$

Since the estimate (8.225) is also valid for all $\boldsymbol{n} \in \Upsilon^c_{N,\boldsymbol{l}} \setminus \Upsilon^c_{N,m}$, we can follow the derivations of (8.226)–(8.228) to obtain

$$\frac{\chi^{\boldsymbol{\alpha},\boldsymbol{\beta}}_{\boldsymbol{n},\boldsymbol{l}}}{\tilde{\chi}^{\boldsymbol{\alpha},\boldsymbol{\beta}}_{\boldsymbol{n},\boldsymbol{l},m}} \le \tilde{c}^2 m^{2(d-1)m} \times \begin{cases} m^{-\frac{2(d-1)(|\boldsymbol{l}|_1-\gamma m)}{1-\gamma}} N^{2(|\boldsymbol{l}|_1-m)}, & \text{if } 0<\gamma\le \frac{|\boldsymbol{l}|_1}{m}, \\ N^{2(|\boldsymbol{l}|_1-m)+2(\gamma m-|\boldsymbol{l}|_1)(1-1/d)}, & \text{if } \frac{|\boldsymbol{l}|_1}{m} \le \gamma < 1. \end{cases} \tag{8.231}$$

Furthermore, (8.209) holds for the optimized hyperbolic cross.

Finally, a combination of the above estimates leads to the desired result. □

Unlike the results for the regular hyperbolic approximation in Theorem 8.10, we can not replace the norm at the left-hand side of (8.220) by the norm in $\mathbb{K}^l_{\boldsymbol{\alpha},\boldsymbol{\beta}}(I^d)$ as in Corollary 8.1, due to the term "$|\boldsymbol{l}|_1$" in the power of N. Instead, we can derive immediately the following estimate in the weighted Sobolev space $B^l_{\boldsymbol{\alpha},\boldsymbol{\beta}}(I^d)$.

Corollary 8.3. *For any $u \in \mathbb{K}^m_{\boldsymbol{\alpha},\boldsymbol{\beta}}(I^d)$, $0 \le l \le m$ and $0<\gamma<1$,*

$$\|\boldsymbol{\pi}^{\boldsymbol{\alpha},\boldsymbol{\beta}}_{N,\gamma} u - u\|_{B^l_{\boldsymbol{\alpha},\boldsymbol{\beta}}(I^d)} \le D_4 N^{l-m} |u|_{\mathbb{K}^m_{\boldsymbol{\alpha},\boldsymbol{\beta}}(I^d)}, \quad 0<\gamma\le\frac{l}{m}, \tag{8.232}$$

where

$$D_4 = \Big(\sum_{0\le|\boldsymbol{l}|_1\le l} D_3^2 N^{2(|\boldsymbol{l}|_1-l)}\Big)^{1/2}, \tag{8.233}$$

and D_3 is the same as in Theorem 8.11.

The above result provides a convergence rate which is independent of dimension d for the approximation space $V^{\boldsymbol{\alpha},\boldsymbol{\beta}}_{N,\gamma}$.

8.5.3 Extensions to Generalized Jacobi Polynomials

As illustrated in Chap. 5 (also see Guo et al. (2006a, 2009)), the use of generalized Jacobi polynomials (GJPs) greatly simplifies the analysis and implementation of spectral methods. We now show that the results established in the previous section can be extended to the case of generalized Jacobi polynomials with both indexes being integers.

Let $k,l \in \mathbb{Z}$ (the set of all integers), and let $\{J^{k,l}_n : n \ge n_0\}$ be the GJPs defined in (6.1). In this context, it is more suitable to consider the normalized GJPs, denoted by $\{\tilde{J}^{k,l}_n : n \ge n_0\}$, as in (8.110). Importantly, as with the classical Jacobi polynomials (cf. (8.111)), they satisfy the derivative relation:

$$\partial_x \tilde{J}^{k,l}_n(x) = d^{k,l}_n \tilde{J}^{k+1,l+1}_{n-1}(x), \tag{8.234}$$

where the explicit expression of $d^{k,l}_n$ (behaves like $O(n)$) is given in (8.111). Hence, $\{\partial^r_x \tilde{J}^{k,l}_n\}$ are mutually orthogonal with respect to the (generalized) Jacobi weight function $\omega^{k+r,l+r}$. In view of there two important properties, we can extend the analysis and the results for the classical Jacobi polynomials to the GJPs. In particular, we can extend Theorem 8.10 to the cases with both indexes being arbitrary integers.

Theorem 8.12. *Let $\boldsymbol{\pi}_N^{\boldsymbol{k},\boldsymbol{l}}$ be the $L^2_{\boldsymbol{\omega}^{\boldsymbol{k},\boldsymbol{l}}}$-orthogonal projection upon the hyperbolic cross*

$$X_N^{\boldsymbol{k},\boldsymbol{l}} := span\{\boldsymbol{J}_{\boldsymbol{n}}^{\boldsymbol{k},\boldsymbol{l}} : |\boldsymbol{n}|_{\text{mix}} \le N; \boldsymbol{n} \ge \boldsymbol{n}_0\}, \quad \boldsymbol{k},\boldsymbol{l} \in \mathbb{Z}^d. \tag{8.235}$$

Then for any $u \in \mathbb{K}^m_{\boldsymbol{k},\boldsymbol{l}}(I^d)$,

$$\left\|\boldsymbol{\pi}_N^{\boldsymbol{k},\boldsymbol{l}} u - u\right\|_{\mathbb{K}^{\mu}_{\boldsymbol{k},\boldsymbol{l}}(I^d)} \le D_5 N^{\mu-m} |u|_{\mathbb{K}^m_{\boldsymbol{k},\boldsymbol{l}}(I^d)}, \quad 0 \le \mu \le m, \tag{8.236}$$

where D_5 is a positive constant depending on $d, \boldsymbol{k}, \boldsymbol{l}, \mu$ and m, but independent of N.

Notice that the explicit dependence of D_5 on d can be worked out as in Theorem 8.10. To avoid repetition, we leave the detail of the proof to the interested reader (see Problem 8.6 for the case: $\boldsymbol{k} = \boldsymbol{l} = -1$).

8.5.4 Sparse Spectral-Galerkin Methods

We present in this section sparse spectral-Galerkin methods for solving the model equation (8.1) in a high-dimensional cube $\Omega = (-1,1)^d$.

Given a suitable approximation space Y_N and an interpolation operator I_N from a computational grid to Y_N. A weighted spectral-Galerkin method for (8.1) with $u|_{\partial\Omega} = 0$ is

$$\begin{cases} \text{Find } u_N \in X_N = Y_N \cap H_0^1(\Omega) \text{ such that} \\ \alpha(u_N, v_N)_\omega - (\Delta u_N, v_N)_\omega = (I_N f, v_N)_\omega, \quad \forall v_N \in X_N. \end{cases} \tag{8.237}$$

Based on the previous discussions in this section, a hyperbolic cross based approximation space appears to be a good candidate for Y_N when $d \ge 4$. However, in order to construct an efficient algorithm, two main issues need to be addressed:

- The first issue is how to define Y_N and I_N such that, for any function $f \in C(\Omega)$, the transform between values of f at the computational grid and expansion coefficients of $I_N f$ in Y_N can be computed efficiently.
- The second issue is to construct a suitable basis of X_N such that the resulting linear system for (8.237) can be solved efficiently. The structure of the matrix associated to (8.237) essentially depends on the basis functions of X_N.

A particularly interesting computational grid Σ_N is the Smolyak's *sparse grid* (cf. Smolyak (1960)) constructed from the *nested* Chebyshev-Gauss-Lobatto quadrature, which enjoys two distinct properties: (a) it is *spectrally accurate* with nested points; (b) FFT can be used for the transform.

We construct below hierarchical basis functions corresponding to the Smolyak's sparse grids based on a nested one-dimensional quadrature, followed by a discussion on sparse spectral algorithms for solving (8.237).

8.5.4.1 One-Dimensional Hierarchical Basis

Let $I=(-1,1)$. Let $\mathscr{U}^i$ be a scheme ($\{\mathscr{U}^i\}$ could be a sequence of functionals for quadrature or a sequence of operators for interpolation) which uses N_i grid points $\mathscr{X}^i$ in I, namely,

$$\mathscr{X}^i=\{x_0^i,x_1^i,\ldots,x_{N_i-1}^i\},\quad i=1,2,\ldots.$$

Conventionally, we set $\mathscr{X}^0=\emptyset$ and $\mathscr{U}^0$ to be the zero functional/operator.

The grids $\{\mathscr{X}^i\}$ are called **nested grids**, if $\mathscr{X}^1\subset\mathscr{X}^2\subset\ldots$. For nested grids, we can rearrange the grid points in such a way that

$$\mathscr{X}=\mathscr{X}^1\cup(\mathscr{X}^2\backslash\mathscr{X}^1)\cup(\mathscr{X}^3\backslash\mathscr{X}^2)\cup\ldots=\{x_0,x_1,x_2,\ldots\}$$

with $\{x_j,j\in\mathscr{I}^i\}=\mathscr{X}^i$, where $\mathscr{I}^i=\{0,1,\ldots,N_i-1\}$.

Let $\omega(x)>0$ $(x\in I)$ be a weight function, $V_1\subset V_2\subset\ldots\subset V_i\ldots$ be a sequence of finite dimensional spaces in $L^2_\omega(I)$, and $\{\phi_k(x):k=0,1,\ldots\}$ be a set of basis functions of $L^2_\omega(I)$ with

$$V_i=\text{span}\{\phi_k:k\in\mathscr{I}^i\}.$$

Then, for $f\in C(\bar{I})$, we can determine a unique set of coefficients $\{b_k^i\}$ such that

$$f(x_j^i)=\sum_{k\in\mathscr{I}^i}b_k^i\phi_k(x_j^i),\quad\forall j=0,1,\ldots,N_i-1. \tag{8.238}$$

Note that fast transforms between $\{f(x_j^i),x_j^i\in\mathscr{X}^i\}$ and $\{b_k^i,k\in\mathscr{I}^i\}$ are available if the basis functions $\{\phi_k\}$ are Fourier series or Chebyshev polynomials.

For schemes with nested grids, if one can find a set of basis functions $\{\tilde{\phi}_k\}$, such that $V_i=\text{span}\{\tilde{\phi}_k:k\in\mathscr{I}^i\}$ and

$$\tilde{\phi}_k(x_j)=0,\quad\forall j\in\mathscr{I}^i,k\notin\mathscr{I}^i. \tag{8.239}$$

Then $\{\tilde{\phi}_k\}$ is called a set of **hierarchical bases**.

An important property of the hierarchical bases is that the expansion coefficients $\{b_k^i\}$ do not depend on the level index i, i.e., we can write

$$f(x_j)=\sum_{k\in\mathscr{I}^i}b_k\tilde{\phi}_k(x_j),\quad\forall j\in\mathscr{I}^i,\ i=1,2,\ldots. \tag{8.240}$$

Theorem 8.1. *Hierarchical bases always exist for nested schemes. Furthermore, a set of hierarchical bases is given by*

$$\tilde{\phi}_k(x)=\phi_k(x)+\sum_{l\in\mathscr{I}^i}c_{k,l}\phi_l(x),\quad\forall k\in\tilde{\mathscr{I}}^{i+1},\ i=0,1,2,\ldots, \tag{8.241}$$

where $\tilde{\mathscr{I}}^{i+1} = \mathscr{I}^{i+1} \backslash \mathscr{I}^i$, $\mathscr{I}^0 = \varnothing$, *and* $c_{k,l} = -\phi_k(x_j)A_{jl}$ *with* $A = (A_{jl})_{l,j \in \mathscr{I}^i}$ *being the inverse matrix of* $B = (\phi_l(x_j))_{l,j \in \mathscr{I}^i}$.

Proof. Evaluating (8.241) at $\mathscr{X}^i$, we derive by using (8.239) that

$$\sum_{l \in \mathscr{I}^i} c_{k,l}\phi_l(x_j) = -\phi_k(x_j), \quad \text{for } j \in \mathscr{I}^i. \tag{8.242}$$

Since $\{\phi_l(x), l \in \mathscr{I}^i\}$ are basis functions and $\{x_j, j \in \mathscr{I}^i\}$ are quadrature points, $(\phi_l(x_j))_{l,j \in \mathscr{I}^i}$ is a full rank matrix. We can then determine $\{c_{k,l}\}$ from (8.242), hence the hierarchical bases in (8.241). □

We now describe in some detail two practical hierarchical bases based on the Chebyshev-Gauss-Lobatto quadrature.

(CH1) Let $\mathscr{X}^i = \{x_j^i = \cos j\pi/2^i, j = 0, \ldots, 2^i\}$ for $i \geqslant 1$ be the Chebyshev-Gauss-Lobatto grid at the i-th level with the number of grid points being $N_i = 2^i + 1$ $(i \geq 1)$; for $i = 0$, we set $N_0 = 0$. We rearrange the grid points into a hierarchical order $x_{\alpha^i(j)} = x_j^i$, where $\alpha^i(j)$ is the reorder vector for grid level $i > 0$. For example, one may take

$$\alpha^i(j) - \begin{cases} 0, & j = 0, \\ \dfrac{2^i}{2^l}\left[1 + 2(2^l - j)\right], & j > 0,\ l = \min\limits_{2^s \geq j} s. \end{cases}$$

The original 1 basis functions are the Chebyshev polynomials $T_k(x) := \cos(k \arccos(x))$. The corresponding hierarchical bases are given by

$$\tilde{T}_k = T_k, \text{ for } k \in \mathscr{I}^1; \quad \tilde{T}_k = T_k - T_{2^i - k}, \text{ for } k \in \tilde{\mathscr{I}}^i,\ i > 1. \tag{8.243}$$

(CH2) Let us consider the Chebyshev scheme for functions satisfying homogeneous Dirichlet boundary conditions. Set

$$\mathscr{X}^i = \left\{x_j^i = \cos(\frac{(j+1)\pi}{2^i}), j = 0, \ldots, N_i - 1\right\}$$

with $N_i = 2^i - 1$ for $i \geqslant 1$. A set of bases with homogeneous Dirichlet boundary conditions are given by:

$$\hat{T}_{2k} = T_{2k+2} - T_0, \quad \hat{T}_{2k+1} = T_{2k+3} - T_1, \quad k = 0, 1, \ldots.$$

Then, a set of hierarchical bases is given by

$$\begin{aligned} &\tilde{T}_k = \hat{T}_k, \text{ for } k \in \mathscr{I}^1, \\ &\tilde{T}_k = \hat{T}_k - \hat{T}_{2^i - 2 - (k+2)} = T_{k+2} - T_{2^i - (k+2)}, \text{ for } k \in \tilde{\mathscr{I}}^i, i > 1. \end{aligned} \tag{8.244}$$

8.5.4.2 Multi-Dimensional Hierarchical Basis and Sparse Grid

Given a one-dimensional scheme $\mathscr{U}^i$, the d-dimensional sparse grid by Smolyak's construction is

$$\mathscr{U}_d^q = \sum_{d \leqslant |\boldsymbol{i}|_1 \leqslant q} \Delta^{i_1} \otimes \Delta^{i_2} \otimes \ldots \otimes \Delta^{i_d}, \tag{8.245}$$

where $\boldsymbol{i} = (i_1, i_2, \ldots, i_d)$ is the multiple index of grid level with all subindexes starting from 1, and $|\boldsymbol{i}|_1 = i_1 + i_2 + \ldots + i_d$, and $\Delta^i = \mathscr{U}^i - \mathscr{U}^{i-1}$ for $i = 1, 2, \ldots$.

It is clear that all the points in the sparse grid are in the set

$$\mathscr{X}_d^q = \bigcup_{d \leqslant |\boldsymbol{i}|_1 \leqslant q} \mathscr{X}^{i_1} \times \mathscr{X}^{i_2} \times \ldots \times \mathscr{X}^{i_d}.$$

It is shown (cf. Barthelmann et al. (2000)) that (8.245) is equivalent to

$$\mathscr{U}_d^q = \sum_{q-d < |\boldsymbol{i}|_1 \leqslant q} (-1)^{q-|\boldsymbol{i}|} \binom{d-1}{q-|\mathbf{i}|_1} \mathscr{U}^{i_1} \otimes \mathscr{U}^{i_2} \otimes \ldots \otimes \mathscr{U}^{i_d}, \tag{8.246}$$

where $\binom{k}{n}$ is the binomial coefficient of selecting k elements from an n-element set. In order to use (8.246) as a quadrature rule or an interpolation operator, we need to compute $\mathscr{U}^{i_1} \otimes \mathscr{U}^{i_2} \otimes \ldots \otimes \mathscr{U}^{i_d}$ for every $q-d < |\boldsymbol{i}|_1 \leqslant q$. Therefore, the computational complexity of a direct evaluation of (8.246) is about

$$C \sum_{q-d < |\boldsymbol{i}|_1 \leqslant q} N_{i_1} N_{i_2} \ldots N_{i_d},$$

regardless whether the 1-D grids are nested or not. However, for nested grids, we can use the properties of hierarchical bases to design more efficient algorithms as described below.

Given $f \in C(\bar{I}^d)$ with $d = 1$, we can write the interpolation operator $\mathscr{U}^i$ as

$$\mathscr{U}^i(f)(x) = \sum_{k \in \mathscr{I}^i} b_k^i \phi_k(x),$$

where $\{b_k^i\}$ are determined by (8.238). However, by using the hierarchical bases, we have

$$\mathscr{U}^i(f)(x) = \sum_{k \in \mathscr{I}^i} b_k \tilde{\phi}_k(x), \quad \Delta^i(f)(x) = \sum_{k \in \tilde{\mathscr{I}}^i} b_k \tilde{\phi}_k(x).$$

Therefore, (8.245) becomes

$$\begin{aligned} \mathscr{U}_d^q(f)(\boldsymbol{x}) &= \sum_{d \leqslant |i|_1 \leqslant q} \sum_{\boldsymbol{k} \in \tilde{\mathscr{I}}^{i_1} \times \ldots \times \tilde{\mathscr{I}}^{i_d}} b_{k_1,k_2,\ldots,k_d} \tilde{\phi}_{k_1}(x_1) \tilde{\phi}_{k_2}(x_2) \ldots \tilde{\phi}_{k_d}(x_d) \\ &= \sum_{\boldsymbol{k} \in \mathscr{I}_d^q} b_{\boldsymbol{k}} \tilde{\phi}_{\boldsymbol{k}}(\boldsymbol{x}), \end{aligned}$$

where $\boldsymbol{x} = (x_1, x_2, \ldots, x_d)$, $\tilde{\phi}_{\boldsymbol{k}}(\boldsymbol{x}) = \tilde{\phi}_{k_1}(x_1)\tilde{\phi}_{k_2}(x_2)\ldots\tilde{\phi}_{k_d}(x_d)$, and

$$\mathscr{I}_d^q = \bigcup_{d \le |\boldsymbol{i}|_1 \le q} \tilde{\mathscr{I}}^{i_1} \times \ldots \times \tilde{\mathscr{I}}^{i_d}. \tag{8.247}$$

The expansion coefficients $\{b_{\boldsymbol{k}}, \boldsymbol{k} \in \mathscr{I}_d^q\}$ can be determined by

$$f(\boldsymbol{x}_{\boldsymbol{j}}) = \sum_{\boldsymbol{k} \in \mathscr{I}_d^q} b_{\boldsymbol{k}} \tilde{\phi}_{\boldsymbol{k}}(\boldsymbol{x}_{\boldsymbol{j}}), \quad \forall \boldsymbol{j} \in \mathscr{I}_d^q. \tag{8.248}$$

Hence, $\mathscr{U}_d^q$ defines an interpolation operator which maps the function values on the grid $\mathscr{X}_d^q$ onto the space

$$V_d^q = \text{span}\{\tilde{\phi}_{\boldsymbol{k}}, \boldsymbol{k} \in \mathscr{I}_d^q\}. \tag{8.249}$$

Figure 8.5 shows a sparse grid $\mathscr{X}_2^5$ based on the one-dimensional Chebyshev Gauss-Lobatto quadrature and the corresponding index set $\mathscr{I}_2^5$ in the interpolation space. Note that the index set $\mathscr{I}_d^q$ is closed related to the index set of a hyperbolic cross (cf. Fig. 8.2).

By using the properties of the hierarchical basis, Shen and Yu (2010) developed a fast transform between the function values at the sparse grid based on the Chebyshev Gauss-Lobatto quadrature and the coefficients of the expansion in Chebyshev hierarchical basis.

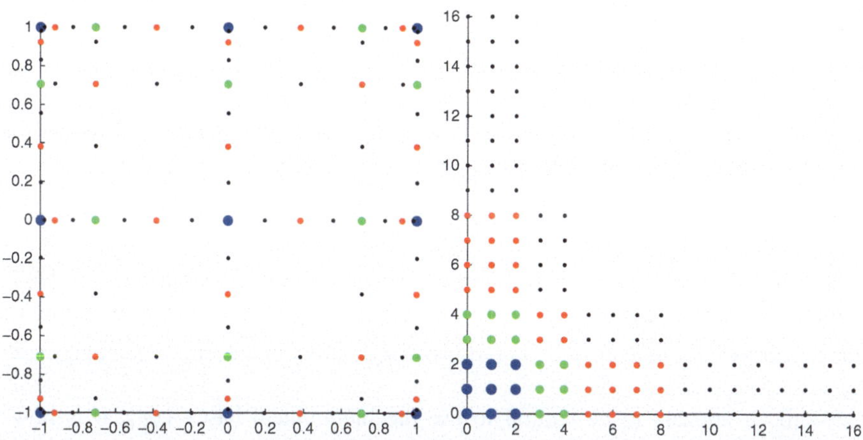

Fig. 8.5 Left: the sparse grid $\mathscr{X}_2^5$ constructed from the Chebyshev-Gauss-Lobatto quadrature; Right: the corresponding index set $\mathscr{I}_2^5$ in the frequency space. The one-to-one mapping between the points on the left and right figures is given by $\mathscr{X}_2^5 = \{(x_{j_1}, x_{j_2}), (j_1, j_2) \in \mathscr{I}_2^5\}$

8.5.4.3 Sparse Spectral-Galerkin Method

We are now in a position to address the second issue raised in the beginning of this section, namely, how to construct efficient sparse spectral algorithms for (8.237)? To this end, we first need to specify Y_N, I_N and X_N.

Given an integer $q \geq d$, let us denote by $\mathscr{I}_d^q$ and $\mathscr{J}_d^q$ the index sets associated with the schemes CH2 and CH1, respectively. More precisely,

$$\mathscr{I}_d^q = \Big\{(k_1,k_2,\ldots,k_d) : 0 \leqslant k_s < 2^{i_s} - 1, i_s \geqslant 0, \sum_{s=1}^{d} i_s = q\Big\},$$

and for CH1,

$$\mathscr{J}_d^q = \Big\{(k_1,k_2,\ldots,k_d) : 0 \leqslant k_s < 2^{i_s} + 1, i_s \geqslant 0, \sum_{s=1}^{d} i_s = q\Big\}.$$

We define the sparse approximation space

$$\begin{aligned} X_N &:= V_d^q = \mathrm{span}\{\tilde{\phi}_{\boldsymbol{k}} : \boldsymbol{k} = (k_1,\ldots,k_d) \in \mathscr{I}_d^q\}, \\ Y_N &:= W_d^q = \mathrm{span}\{\tilde{\phi}_{\boldsymbol{k}} : \boldsymbol{k} = (k_1,\ldots,k_d) \in \mathscr{J}_d^q\}, \end{aligned} \tag{8.250}$$

where $\{\tilde{\phi}_{\boldsymbol{k}}\}$ are the d-dimensional hierarchical basis functions based on (8.244) and (8.243), respectively. We also denote by $I_N := \mathscr{U}_d^q$ the interpolation operator (cf. (8.246)) associated with the scheme CH1. Then, we can rewrite (8.237) as

$$\begin{cases} \text{Find } u_d^q \in V_d^q \text{ such that} \\ \alpha(u_d^q, v)_\omega - (\Delta u_d^q, v)_\omega = (\mathscr{U}_d^q f, v)_\omega, \quad \forall v \in V_d^q, \end{cases} \tag{8.251}$$

where $(u,v)_\omega = \int_\Omega uv\omega d\boldsymbol{x}$, and $\omega(\boldsymbol{x}) = \Pi_{i=1}^d (1-x_i^2)^{-1/2}$ in the Chebyshev case and $\omega(\boldsymbol{x}) = 1$ in the Legendre case.

Given a set of basis functions $\{\phi_{\boldsymbol{k}}\}$ (not necessarily the hierarchical bases) for V_d^q, (8.251) can reduce to a linear system

$$(\alpha M + S)\bar{u} = \bar{f}, \tag{8.252}$$

where M and S are respectively the mass and stiffness matrices associated with $\{\phi_{\boldsymbol{k}}\}$, namely $M = \big((\phi_{\boldsymbol{j}}, \phi_{\boldsymbol{k}})\big)_{\boldsymbol{k},\boldsymbol{j} \in \mathscr{I}_d^q}$, $S = \big(-(\Delta\phi_{\boldsymbol{j}}, \phi_{\boldsymbol{k}})\big)_{\boldsymbol{k},\boldsymbol{j} \in \mathscr{I}_d^q}$, $\bar{u}$ is a vector consisting of the expansion coefficients of u_d^q in terms of $\{\phi_{\boldsymbol{k}}\}$, and $\bar{f}$ is a vector with component $\bar{f}_{\boldsymbol{k}} = (\mathscr{U}_d^q f, \phi_{\boldsymbol{k}})_\omega$.

While it appears to be natural to use the hierarchical bases $\{\tilde{\phi}_{\boldsymbol{k}}\}$, it is however not the most efficient choice as the number of non-zero elements in the mass and stiffness matrices increase rapidly as d increases. On the other hand, the spectral-Galerkin basis functions $\psi_k(x) := T_k(x) - T_{k+2}(x)$ in the Chebyshev case and $\varphi_k(x) := L_k(x) - L_{k+2}(x)$ in the Legendre case lead to very simple mass and stiffness matrices and enjoy many other nice properties (cf. Chap. 4 and Shen (1994, 1995)). Therefore, we shall use these as basis functions for V_d^q.

Notice that we still use the hierarchical basis (8.243) to obtain $\mathscr{U}_d^q f$, the expansion in the hierarchical basis, from the values of f at the sparse grid. Therefore, in order to compute the right-hand side vector $\boldsymbol{f}$, we first transform $\mathscr{U}_d^q f$ into an expansion based on standard Chebyshev or Legendre polynomials, then we use the orthogonality of Chebyshev or Legendre polynomials to compute $\boldsymbol{f_k} = (\mathscr{U}_d^q f, \phi_{\boldsymbol{k}})_\omega$.

We now describe briefly, for both the Chebyshev and Legendre weight functions $\boldsymbol{\omega}$, how the approximate solution u_d^q for (8.251) can be efficiently obtained.

- In the Chebyshev case, the above method leads to non-symmetric, non-sparse system which can only be solved by an iterative method. While the system matrix is not sparse, but the matrix–vector multiplication can be performed by a fast algorithm developed in Shen and Yu (2010).
- In the Legendre case, the above method leads to a symmetric, positive definite, sparse system which can be solved by an iterative method or sparse solver. However, the price we pay for the sparsity of stiffness matrix is an extra step in evaluating $\boldsymbol{f}$, namely transform the expansion of $\mathscr{U}_q^d f$ in terms of Chebyshev polynomials to the expansion in Legendre polynomials. Therefore, this method can be classified as the sparse Chebyshev-Legendre-Galerkin method.

We refer to Shen and Yu (2010) for more detail on the implementation of the sparse Chebyshev-Galerkin and Chebyshev-Legendre-Galerkin methods. In Fig. 8.6, we plot the sparse structure of the stiffness and mass matrices.

We present below two numerical examples to illustrate the convergence properties of the sparse spectral-Galerkin method and compare them with the usual spectral-Galerkin method. We consider the Poisson equation with the following two exact solutions:

$$u_1(\boldsymbol{x}) = \prod_{i=1}^{d} \sin(k\pi \frac{x_i+1}{2}),$$

$$u_2(\boldsymbol{x}) = \prod_{i=1}^{d} \left(h_k(x_i) - \frac{1+x_i}{2}\right),$$

where

$$h_k(x) = \begin{cases} 0, & x \leqslant 0, \\ x^k, & x > 0, \end{cases} \quad k = 2, 3, \ldots$$

Note that u_1 is an isotropic analytical function, while the u_2 only has a finite regularity in each direction.

We recall that the convergence rate of the sparse spectral methods depends on the regularity of solutions in the weighted Korobov spaces (cf. Theorem 8.10), while that of the usual spectral methods depends on the regularity of solutions in the Jacobi-weighted Sobolev spaces (cf. Theorem 8.1). Therefore, the sparse spectral method will be much better than the usual spectral method for functions with similar regularity index m in both weighted Sobolev and weighted Korobov spaces. However, for very smooth functions such as isotropic analytical functions, both sparse

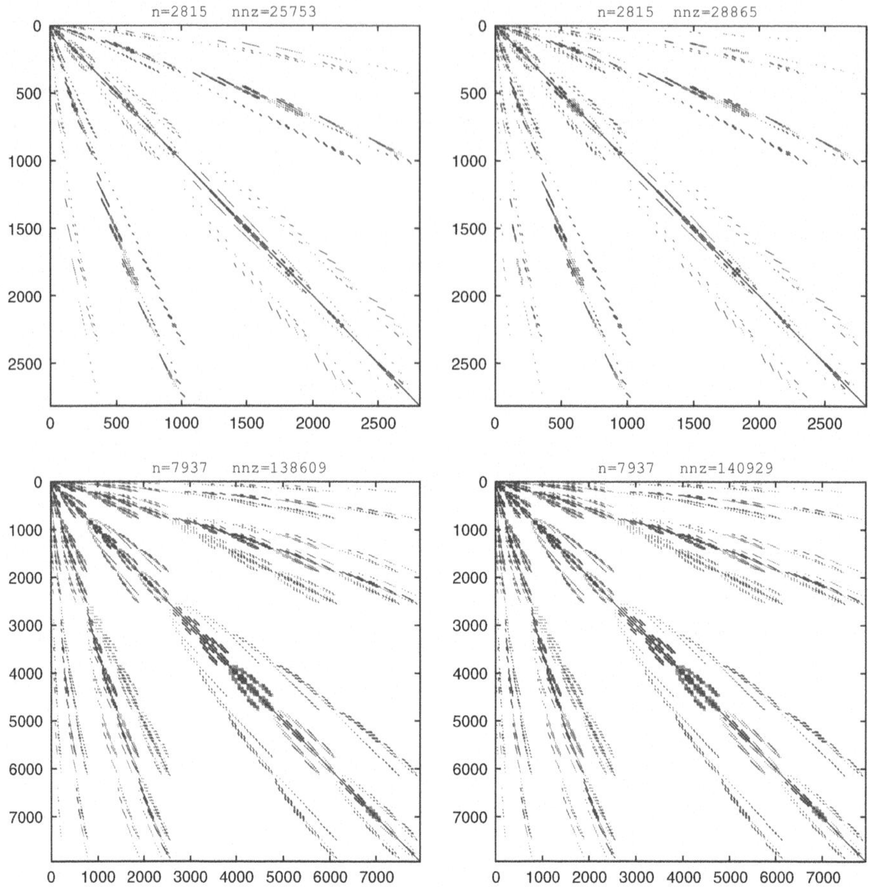

Fig. 8.6 The structure of system matrices of Chebyshev-Legendre-Galerkin sparse grid method. Left column: stiffness matrix; Right column: the sum of stiffness matrix and mass matrices. First row: $d = 3, q = 9$; second row: $d = 4, q = 10$

spectral method and full grid spectral method will converge exponentially fast (with respect to the number of unknowns) so there is not much advantage using a sparse spectral method.

Case 1: The exact solution u_1 is a tensor product of one-dimensional analytic functions. So the sparse spectral method does not have an advantage over the usual full grid spectral method. This observation is also consistent with the spectrum shown in Fig. 8.7.

In Fig. 8.8, we present the L^2_ω-error of the solutions obtained by Chebyshev-Legendre-Galerkin method on full grid and sparse grid. The results indicate that the sparse grid method has similar convergence rate as the full grid method in terms of number of unknowns, while the results with the sparse grid method are sightly better in higher dimensions.

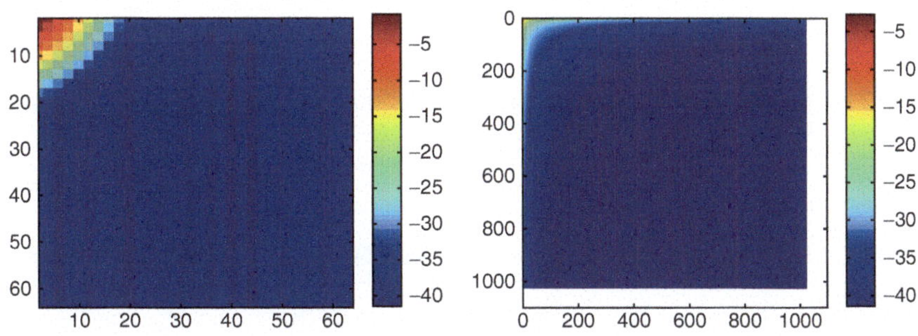

Fig. 8.7 The Chebyshev spectrum of the exact solutions u_1 and u_2 with $k = 2$ and $k = 3$, respectively

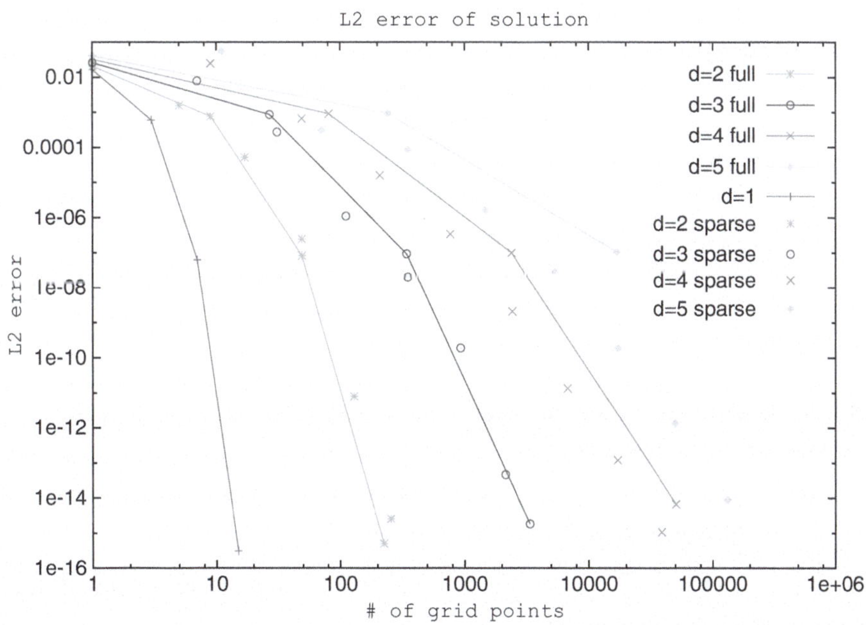

Fig. 8.8 Convergence history of the Chebyshev-Legendre-Galerkin method on full grid (*solid line*) and sparse grid (*dotted line*) for Poisson equation with the exact solution u_1 ($k = 1$)

Case 2: The exact solution u_2 is a product of one-dimensional functions in $H^{k-\frac{1}{2}+\varepsilon}(I)$ (for any $\varepsilon > 0$). It is easy to see that the solution belongs to $K^m(I^d)$ and $\tilde{H}^m_\omega(I^d)$ with the same index $m = k - \frac{1}{2} + \varepsilon$. Therefore, this is an ideal case for the sparse spectral method. Figure 8.9 shows that the sparse grid method is much more efficient than full grid method for this problem.

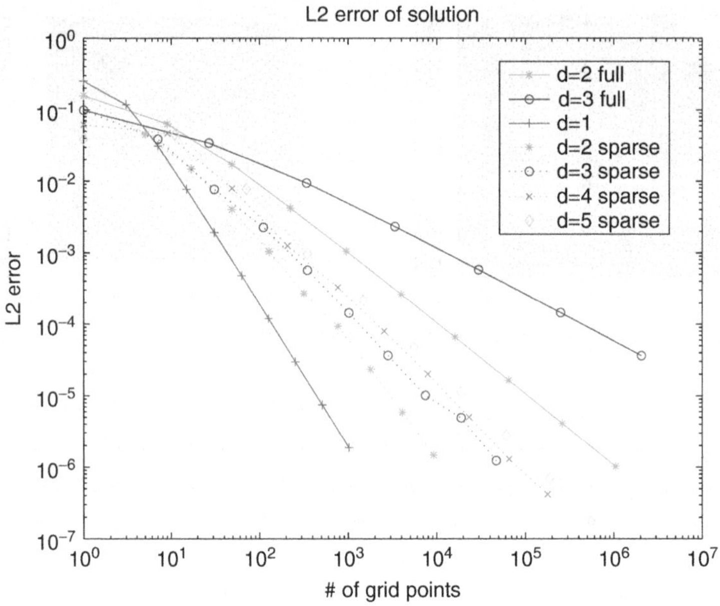

Fig. 8.9 The Chebyshev-Legendre-Galerkin method on full grid (*solid line*) and sparse grid (*dotted line*) for Poisson equation with exact solution u_2 $(k = 3)$

Problems

8.1. Implement and test the full diagonalization algorithm based on the Legendre-Galerkin method for solving (8.14) with $\Omega = (-1,1)^2$.

8.2. Write down the full diagonalization algorithm based on the Legendre-Galerkin method for (8.1) with $u|_{\partial\Omega} = 0$ and $\Omega = (-1,1)^3$.

8.3. Let $\omega = (1-t^2)^{-1/2}$ and $\phi_j(t) = (1-t^2)T_j(t)$. Determine the entries of A, B and C in (8.41) (cf. Remark 8.8).

8.4. Derive the estimate in Theorem 8.5.

8.5. Estimate the interpolation errors of the d-dimensional tensorial Legendre-Gauss-Lobatto interpolation in the weighted L^2-norm.

8.6. Prove Theorem 8.12 with $\boldsymbol{k} = \boldsymbol{l} = -1$.

Chapter 9
Applications in Multi-Dimensional Domains

We consider in this chapter several multi-dimensional problems, which (a) are of current interest; (b) are suitable for spectral approximations; and (c) can be efficiently solved by using the basic spectral algorithms developed in previous chapters. These include steady state problems: the Helmholtz equation for acoustic scattering and the Stokes equations, as well as time-dependent problems including the Allen–Cahn equation, the Cahn–Hilliard equation, the Navier–Stokes equations, and the Gross-Pitaevskii equation. For applications of spectral methods to other multi-dimensional problems in science and engineering, we refer, for instance, to Boyd (2001), Canuto et al. (2006), Hesthaven et al. (2007) and the references therein.

For time-dependent problems, we shall first present semi-discretization (in time) schemes which lead to, at each time step, elliptic type equations that can be efficiently solved by using spectral methods presented in previous chapters. Special attentions will be paid to how to design simple yet accurate and stable time discretization schemes for the Allen–Cahn equation, the Cahn–Hilliard equation, the Navier–Stokes equations, and the Gross-Pitaevskii equation.

9.1 Helmholtz Equation for Acoustic Scattering

Time harmonic wave propagations appear in many applications such as wave scattering and transmission, noise reduction, fluid-solid interaction, and sea and earthquake wave propagation. We describe in this section a spectral-Galerkin method for solving the Helmholtz equation arising from acoustic scattering problems in an exterior domain $\Omega = \mathbb{R}^d \setminus D$ $(d = 1,2,3)$ with D being a bounded obstacle.

The Helmholtz equation in exterior domains presents a great challenge to numerical analysts and computational scientists, for (a) the domain is unbounded, (b) the problem is indefinite, and (c) the solution is highly oscillatory (when the wave number is large) and decays slowly. There is an abundant literature on different numerical techniques that have been developed for this problem such as boundary element methods (cf. Ciskowski and Brebbia (1991)), infinite element methods (cf. Gerdes

J. Shen et al., *Spectral Methods: Algorithms, Analysis and Applications*,
Springer Series in Computational Mathematics 41, DOI 10.1007/978-3-540-71041-7_9,

and Demkowicz (1996)), Dirichlet-to-Neumann (DtN) methods (cf. Keller and Givoli (1989)), perfectly matched layers (PML) (cf. Berenger (1994)), among others.

In many of these approaches, an essential step is to solve the Helmholtz equation (9.8) in a finite domain with an exact or approximate non-reflecting boundary condition at the outer boundary. In particular, with a proper boundary perturbation technique (cf. Nicholls and Reitich (2003)), the Helmholtz equation in exterior domains can be reduced to a sequence of Helmholtz equations (9.8) in a separable bounded domain Ω.

In this section, we shall restrict our attention to the cases with the obstacle D being a sphere or a circle, and present in detail a very efficient spectral-Galerkin algorithm for solving the reduced problem (9.8). The algorithm presented below can be easily combined with the so-called *transformed field expansion* method (cf. Nicholls and Reitich (2003)) to treat general obstacles (cf. Nicholls and Shen (2006), Fang et al. (2007), Nicholls and Shen (2009)).

9.1.1 Time-Harmonic Wave Equations

The wave equation

$$\partial_t^2 w - c^2 \Delta w = g \tag{9.1}$$

arises from many applications, such as electromagnetic wave propagations and acoustics, where c is the speed of sound. In many situations, we may assume that the inhomogeneity g is periodic in time, i.e.,

$$g(x,t) = f(x)e^{-\mathrm{i}\omega t}. \tag{9.2}$$

In this case, the solution of (9.1) is of the form: $w(x,t) = u(x)e^{-\mathrm{i}\omega t}$, where the amplitude $u(x)$ satisfies the Helmholtz equation

$$-\Delta u - k^2 u = f. \tag{9.3}$$

The constant $k = \omega/c$ is called the wave number. To illustrate the physical meaning of k, we consider the 1-D Helmholtz equation

$$-u''(x) - k^2 u(x) = 0, \quad \text{in } (-\infty, +\infty), \tag{9.4}$$

whose general solution takes the form

$$u(x) = Ae^{\mathrm{i}kx} + Be^{-\mathrm{i}kx} \tag{9.5}$$

with arbitrary constants A and B. Hence, the time-harmonic solution of (9.1) is

$$w(x,t) = Ae^{\mathrm{i}(kx-\omega t)} + Be^{-\mathrm{i}(kx+\omega t)}. \tag{9.6}$$

We see that the first term on the right-hand side represents an *outgoing* wave (traveling from left to right) with the phase speed $c = \omega/k$, while the second indicates an *incoming* wave (traveling from right to left) with the phase speed $-c$. In practice, some conditions are imposed to eliminate the incoming wave, which will be discussed shortly.

In the case of acoustic scattering from an obstacle D, the Helmholtz equation (9.3) is set in the exterior domain $\Omega = \mathbb{R}^d \backslash D$. In order for the problem to be well-posed, we need to impose the so-called Sommerfeld radiation condition at infinity

$$\frac{\partial u}{\partial r} - iku = o\big(r^{\frac{1-d}{2}}\big) \quad \text{as } r \to \infty \quad \text{for } d = 1,2,3, \tag{9.7}$$

which ensures that waves do not reflect from far field. On the surface of the obstacle D, we may impose a Dirichlet, Neumann or Robin boundary condition, which corresponds to sound soft, sound hard or impedance surface of the obstacle, respectively.

9.1.2 Dirichlet-to-Neumann (DtN) Map

A classical approach to reduce a problem in an unbounded domain to a bounded domain is to use the so-called Dirichlet-to-Neumann map. The basic idea is to introduce a sufficiently large ball B (resp. a disk for 2-D) so that $D \subseteq B$ and $\text{supp}(f) \subseteq B$, and reduce the original problem to

$$\begin{cases} -\Delta u - k^2 u = f, & \text{in } \Omega := B \cap (\mathbb{R}^d \setminus \bar{D}), \\ u = g, & \text{on } \partial D, \\ \dfrac{\partial u}{\partial r} + T(u) = 0, & \text{on } \partial B, \end{cases} \tag{9.8}$$

where T is the DtN map. To derive the formulation of T in 3-D case, we consider the "auxiliary" problem exterior to the artificial ball B :

$$\begin{cases} -\Delta u - k^2 u = 0, & \text{in } \Omega_{\text{ext}} := \mathbb{R}^3 \setminus \bar{B}, \\ u = \Psi, & \text{on } \partial B. \end{cases} \tag{9.9}$$

This problem can be solved analytically via separation of variables in spherical coordinate (r, θ, ϕ):

$$u(r,\theta,\phi) = \sum_{l=0}^{\infty} h_l^{(1)}(kr) \sum_{m=-l}^{l} \hat{u}_{lm} Y_l^m(\theta,\phi), \tag{9.10}$$

where $r > 0$, $\theta \in [0,\pi]$, $\phi \in [0,2\pi]$, $h_l^{(1)}(z)$ is the spherical Hankel function of the first kind of order l (cf. Morse and Feshback (1953)), and Y_l^m is the spherical harmonic function defined in (8.91).

To determine the coefficients $\{\hat{u}_{lm}\}$ in (9.10), we expand the Dirichlet data in (9.9) as

$$\Psi(\theta,\phi) = \sum_{l=0}^{\infty}\sum_{m=-l}^{l} \hat{\psi}_{lm} Y_l^m(\theta,\phi). \tag{9.11}$$

Hence, letting $r = b$ in (9.10) and comparing the coefficients of two expansions, one verifies that

$$\hat{u}_{lm} = \frac{\hat{\psi}_{lm}}{h_l^{(1)}(kb)}. \tag{9.12}$$

Plugging it into (9.10) leads to the exact solution of (9.9):

$$u(r,\theta,\phi) = \sum_{l=0}^{\infty} \frac{h_l^{(1)}(kr)}{h_l^{(1)}(kb)} \sum_{m=-l}^{l} \hat{\psi}_{lm} Y_l^m(\theta,\phi). \tag{9.13}$$

Differentiating (9.13) with respect to r and setting $r = b$, we derive

$$\frac{\partial u}{\partial r}(b,\theta,\phi) = \sum_{l=0}^{\infty} k \frac{\partial_z h_l^{(1)}(kb)}{h_l^{(1)}(kb)} \sum_{m=-l}^{l} \hat{\psi}_{lm} Y_l^m(\theta,\phi). \tag{9.14}$$

The DtN map is now simply obtained by setting

$$\begin{aligned} T(u) = \frac{\partial u}{\partial n}\Big|_{\partial B} &= -\frac{\partial u}{\partial r}\Big|_{r=b} \\ &= -\sum_{l=0}^{\infty} k \frac{\partial_z h_l^{(1)}(kb)}{h_l^{(1)}(kb)} \sum_{m=-l}^{l} \hat{\psi}_{lm} Y_l^m(\theta,\phi), \end{aligned} \tag{9.15}$$

where the normal vector n of Ω_{ext} is pointing to the negative radial direction.

Similarly, in the 2-D setting, let B be a sufficiently large disk containing D and the support of f. Consider the "auxiliary" problem

$$\begin{cases} -\Delta u - k^2 u = 0, & \text{in } \Omega_{\text{ext}} := \mathbb{R}^2 \setminus \bar{B}, \\ u = \Psi, & \text{on } \partial B, \end{cases} \tag{9.16}$$

which admits the analytical solution in polar coordinate (r,ϕ) :

$$u(r,\phi) = \sum_{l=-\infty}^{\infty} \hat{u}_l H_l^{(1)}(kr) e^{\mathrm{i}l\phi}, \quad r > b,\ \phi \in [0,2\pi). \tag{9.17}$$

Here, $H_l^{(1)}$ is the Hankel function of the first kind of order l (cf. Morse and Feshback (1953)). The coefficients $\{\hat{u}_l\}$ can be determined by the boundary value $\Psi(\phi)$ with the expansion

$$\Psi(\phi) = \sum_{l=-\infty}^{\infty} \hat{\psi}_l e^{\mathrm{i}l\phi}. \tag{9.18}$$

Setting $r = b$ in (9.17), and comparing the coefficients of the above two expansions lead to $\hat{u}_l = \hat{\psi}_l / H_l^{(1)}(kb)$. Hence, the exact solution of (9.16) is

$$u(r,\phi) = \sum_{l=-\infty}^{\infty} \frac{\hat{\psi}_l}{H_l^{(1)}(kb)} H_l^{(1)}(kr) e^{il\phi}. \tag{9.19}$$

The 2-D DtN map is given by

$$T(u) = \frac{\partial u}{\partial n}\Big|_{\partial B} = -\frac{\partial u}{\partial r}\Big|_{r=b} = -\sum_{l=-\infty}^{\infty} k \frac{\partial_z H_l^{(1)}(kb)}{H_l^{(1)}(kb)} \hat{\psi}_l e^{il\phi}. \tag{9.20}$$

Finally, in the 1-D case, it is easy to show that

$$T(u) = -\mathrm{i}u. \tag{9.21}$$

9.1.3 Spectral-Galerkin Method

We now present a spectral-Galerkin method for the truncated problem (9.8) with the DtN map given by (9.15), (9.20) or (9.21). We shall only consider the obstacle D being a disk or a ball of radius a. Combining the spectral-Galerkin method for these simple separable domains with a *transformed field expansion* (cf. Nicholls and Reitich (2003)), we can then deal with general obstacles (cf. Nicholls and Shen (2006), Fang et al. (2007), Nicholls and Shen (2009)).

In the 3-D case, we expand

$$\{u,f,g\} = \sum_{l=0}^{\infty} \sum_{m=-l}^{l} \{\hat{u}_{lm}(r), \hat{f}_{lm}(r), \hat{g}_{lm}(r)\} Y_l^m(\theta,\phi), \tag{9.22}$$

and likewise for the 2-D case,

$$(u,f,g) = \sum_{l=-\infty}^{\infty} (\hat{u}_l(r), \hat{f}_l(r), \hat{g}_l(r)) e^{il\phi}. \tag{9.23}$$

For brevity, we use u to denote $\hat{u}_{lm}$ or $\hat{u}_l$, and likewise for f and g below. With the above setup, the problem of interest is reduced to the following sequence of 1-D problems

$$\begin{aligned} &-\frac{1}{r^{d-1}} \frac{d}{dr}\left(r^{d-1} \frac{du}{dr}\right) + C_l \frac{u}{r^2} - k^2 u = f, \quad r \in (a,b), \\ &u(a) = g, \quad u'(b) - kD_{l,k} u(b) = 0, \end{aligned} \tag{9.24}$$

where $C_l = 0,\ l^2,\ l(l+1)$ for $d = 1,2,3$, respectively, and the DtN kernel is defined by

$$D_{l,k} := D_{l,k}(b) = \begin{cases} \mathrm{i}, & \text{if } d=1, \\ \dfrac{\partial_z H_l^{(1)}(kb)}{H_l^{(1)}(kb)}, & \text{if } d=2, \\ \dfrac{\partial_z h_l^{(1)}(kb)}{h_l^{(1)}(kb)}, & \text{if } d=3. \end{cases} \tag{9.25}$$

Note that for $d=2$, we have $D_{l,k} = D_{-l,k}$ (cf. Shen and Wang (2007a)). It is known (see, e.g., Harari and Hughes (1992), Demkowicz and Ihlenburg (2001)) that

$$\mathrm{Re}\big(D_{l,k}\big) < 0, \quad \mathrm{Im}\big(D_{l,k}\big) > 0, \quad \text{for } d = 2,3, \tag{9.26}$$

which ensure the well-posedness of the problem (9.24).

We are now in a position to describe the spectral-Galerkin method for (9.24)-(9.25). For convenience, we make a change of variable

$$x = 2\frac{r-a}{b-a} - 1, \quad x \in (-1,1), \ r \in (a,b), \tag{9.27}$$

and denote

$$\tilde{u}(x) = u(r), \ \tilde{f}(x) = \frac{(b-a)^2}{4} f(r), \ \tilde{g} = g,$$
$$c = \frac{b+a}{b-a}, \ \tilde{k} = \frac{k(b-a)}{2}.$$

Then the problem (9.24)–(9.25) is set in $x \in (-1,1)$ and of the form

$$\begin{aligned} &-\frac{1}{(x+c)^{d-1}}\frac{d}{dx}\Big((x+c)^{d-1}\frac{d\tilde{u}}{dx}\Big) + C_l\frac{\tilde{u}}{(x+c)^2} \\ &\qquad -\tilde{k}^2\tilde{u} = \tilde{f}, \quad \text{in } (-1,1),\ d=1,2,3, \\ &\tilde{u}(-1) = \tilde{g}, \\ &\tilde{u}_x(1) - \tilde{k}D_{l,k}\tilde{u}(1) = 0. \end{aligned} \tag{9.28}$$

One verifies readily that the function

$$s(x) = \frac{\tilde{k}D_{l,k}x + (1-\tilde{k}D_{l,k})}{1-2\tilde{k}D_{l,k}}\tilde{g}$$

satisfies the boundary conditions in (9.28). Multiplying the first equation of (9.28) by $(x+c)^{2d-2}$ and setting

$$\begin{aligned} &\tilde{u}(x) = \hat{u}(x) + s(x), \quad h = \frac{\tilde{k}D_{l,k}\tilde{g}}{1-2\tilde{k}D_{l,k}}, \\ &\hat{f}(x) = (x+c)^{2d-2}\tilde{f}(x) - \big[C_l(x+c)^{2d-4} \\ &\qquad -\tilde{k}^2(x+c)^{2d-2}\big]s(x) + (d-1)(x+c)^{2d-3}h, \end{aligned}$$

we end up with the following problem with homogeneous boundary conditions:

$$
\begin{aligned}
&-(x+c)^{d-1}\frac{d}{dx}\Big[(x+c)^{d-1}\frac{d\hat{u}}{dx}\Big]+\Big(C_l(x+c)^{2d-4}\\
&\qquad -\tilde{k}^2(x+c)^{2d-2}\Big)\hat{u}=\hat{f}, \qquad x\in(-1,1),\, d=1,2,3,\\
&\hat{u}(-1)=0, \quad \hat{u}'(1)-\tilde{k}D_{l,k}\hat{u}(1)=0.
\end{aligned}
\tag{9.29}
$$

To simplify the implementation, we make a transform to convert the Robin boundary condition (at $x=1$) to the Neumann boundary condition. For this purpose, let

$$
v(x)=\hat{u}(x)e^{-\tilde{k}D_{l,k}x}, \quad f(x)=\hat{f}(x)e^{-\tilde{k}D_{l,k}x}. \tag{9.30}
$$

The problem (9.29) is converted to

$$
\begin{aligned}
&-(x+c)^{d-1}\frac{d}{dx}\Big((x+c)^{d-1}\frac{dv}{dx}\Big)+p(x)\frac{dv}{dx}+q(x)v=f,\\
&v(-1)=v'(1)=0,
\end{aligned}
\tag{9.31}
$$

where

$$
\begin{aligned}
p(x) &:= p(x;l,k,d)=-2\tilde{k}D_{l,k}(x+c)^{2d-2},\\
q(x) &:= q(x;l,k,d)=-(x+c)^{2d-2}\Big(\frac{(d-1)\tilde{k}D_{l,k}}{x+c}+\tilde{k}^2D_{l,k}^2\Big)\\
&\quad+\Big(C_l(x+c)^{2d-4}-\tilde{k}^2(x+c)^{2d-2}\Big).
\end{aligned}
\tag{9.32}
$$

Let P_N be the space of complex polynomials of degree $\le N$. Define the approximation space

$$
X_N=\{u\in P_N : u(-1)=u'(1)=0\}. \tag{9.33}
$$

The spectral-Galerkin approximation to (9.31)-(9.32) is

$$
\begin{cases}
\text{Find } v_N\in X_N \text{ such that}\\
-\displaystyle\int_{-1}^{1}(x+c)^{d-1}((x+c)^{d-1}v_N')'\bar{w}_N dx+\int_{-1}^{1}pv_N'\bar{w}_N dx\\
\qquad +\displaystyle\int_{-1}^{1}qv_N\bar{w}_N dx=\int_{-1}^{1}f\bar{w}_N dx, \quad \forall w_N\in X_N.
\end{cases}
\tag{9.34}
$$

Let $L_n(x)$ be the Legendre polynomial of degree n, and define

$$
\phi_n=(L_n+L_{n+1})-\Big(\frac{n+1}{n+2}\Big)^2(L_{n+1}+L_{n+2}). \tag{9.35}
$$

One verifies readily that $\phi_n(-1)=\phi_n'(1)=0$. Therefore, the real part (or the imaginary part) of X_N is

$$
X_N^R := \text{span}\{\phi_n : n=0,1,\dots,N-2\}.
$$

In actual computations, we split (9.31) into real and imaginary parts by setting $v_N = v_N^R + \mathrm{i}v_N^I$, and likewise for p, q and f. Consequently, the scheme (9.34) is equivalent to

$$\begin{cases} \text{Find } v_N^R, v_N^I \in X_N^R \text{ such that} \\ -\big((x+c)^{d-1}\partial_x((x+c)^{d-1}\partial_x v_N^R), \varphi\big) + \big(p^R \partial_x v_N^R - p^I \partial_x v_N^I, \varphi\big) \\ \quad + \big(q^R v_N^R - q^I v_N^I, \varphi\big) = \big(f^R, \varphi\big), \quad \forall \varphi \in X_N^R, \\ -\big((x+c)^{d-1}\partial_x((x+c)^{d-1}\partial_x v_N^I), \psi\big) + \big(p^I \partial_x v_N^R + p^R \partial_x v_N^I, \psi\big) \\ \quad + \big(q^I v_N^R + q^R v_N^I, \psi\big) = \big(f^I, \psi\big), \quad \forall \psi \in X_N^R. \end{cases} \tag{9.36}$$

Let us denote

$$\begin{aligned} &a_{ij} = -\big((x+c)^{d-1}\partial_x((x+c)^{d-1}\partial_x \phi_j), \phi_i\big), \quad A = (a_{ij}); \\ &\hat{P}_{ij}^z = \big(p^z \partial_x \phi_j, \phi_i\big), \quad P^z = (\hat{P}_{ij}^z); \\ &\hat{Q}_{ij}^z = \big(q^z \phi_j, \phi_i\big), \quad Q^z = (\hat{q}_{ij}^z); \\ &f_i^z = \big(f^z, \phi_i\big), \quad \mathbf{f}^z = \big(f_0^z, f_1^z, \ldots, f_{N-2}^z\big)^T; \\ &v_N^z = \sum_{n=0}^{N-2} v_n^z \phi_n, \quad \mathbf{v}^z = \big(v_0^z, v_1^z, \ldots, v_{N-2}^z\big)^T, \end{aligned}$$

where $z = R$ or $z = I$. Then the matrix form of (9.36) is

$$\begin{bmatrix} A + P^R + Q^R & -(P^I + Q^I) \\ P^I + Q^I & A + P^R + Q^R \end{bmatrix} \begin{bmatrix} \mathbf{v}^R \\ \mathbf{v}^I \end{bmatrix} = \begin{bmatrix} \mathbf{f}^R \\ \mathbf{f}^I \end{bmatrix}. \tag{9.37}$$

Note that using the properties of Legendre polynomials, one verifies that the above matrices are sparse so the above system can be efficiently solved.

We now present some numerical results obtained by the above spectral-Galerkin method. We consider (9.24) with the exact solution: $u(r) = H_l^{(1)}(kr)$. We fix $l = 1$, $d = 2$, and concentrate on the approximation behavior of our scheme with respect to the frequency k and the thickness of the annulus $b - a$.

In the first set of tests, we take $a = 1$ and $b = 2$. In Fig. 9.1, we present the relative L^2-error versus the number of mode $N = N_r$ for a wide range of wave numbers. We note that as soon as $N_r > k(b-a)/2$, the errors start to decay, and for moderate to large wave numbers, the errors decay slowly until about $N_r \sim k(b-a)$, and finally for $N_r > k(b-a)$, the errors decay exponentially.

In the second set of tests, we take $a = 1$ and $b = 1.25$. The results are plotted in Fig. 9.2. We observe similar behaviors as in the first set except that now we have $b - a = \frac{1}{4}$ and only about $1/4$ of the modes are needed to achieve a similar accuracy. These behaviors are consistent with the error estimates in Shen and Wang (2007a) (cf. Remark 4.2 in Shen and Wang (2007a)).

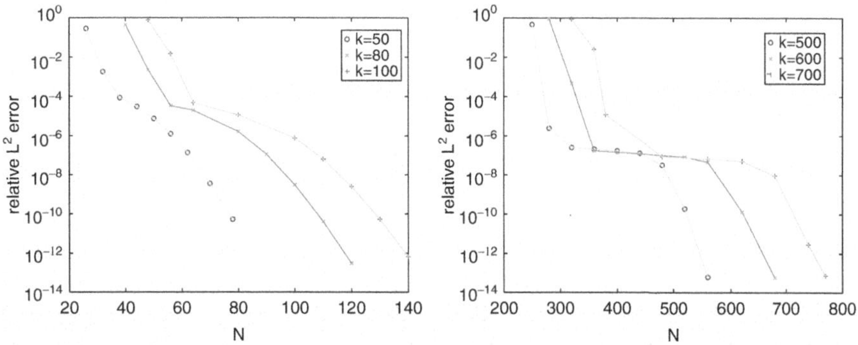

Fig. 9.1 Relative L^2-error versus N as compared to an exact solution: $a = 1, b = 2$

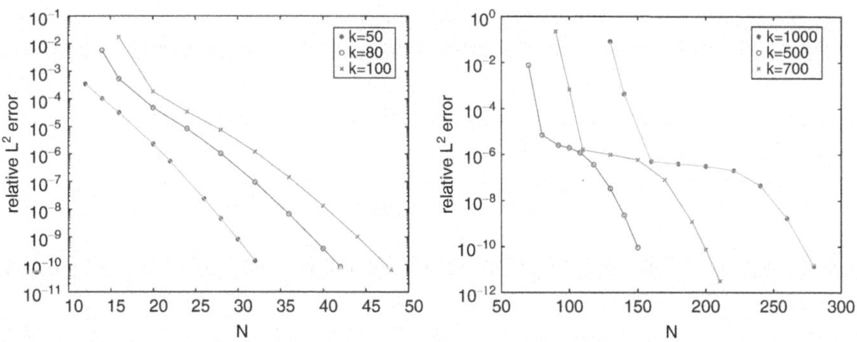

Fig. 9.2 Relative L^2-error versus N as compared to an exact solution: $a = 1, b = 1.25$

Note that error estimates with explicit dependence on k for the 1-D problem were derived in Sect. 4.5.4. Complete error analysis for the method presented above can be found in Shen and Wang (2007a).

9.2 Stokes Equations

The Stokes equations play an important role in fluid mechanics and solid mechanics. Numerical approximation of Stokes equations has attracted considerable attention in the last few decades and is still an active research direction (cf. Girault and Raviart (1986), Brezzi and Fortin (1991), Bernardi and Maday (1997), Elman et al. (2005) and the references therein).

In this section, we shall restrict our attention to the Stokes equations in primitive variables and consider iterative algorithms for which only Possion type equations have to be solved in each iteration.

9.2.1 Stokes Equations and Uzawa Operator

We consider the generalized Stokes equations in primitive variables:

$$\begin{cases} \alpha \boldsymbol{u} - \nu \Delta \boldsymbol{u} + \nabla p = \boldsymbol{f}, & \text{in } \Omega \subset R^d, \\ \nabla \cdot \boldsymbol{u} = 0, \quad \text{in } \Omega; & \boldsymbol{u}|_{\partial\Omega} = 0. \end{cases} \tag{9.38}$$

In the above, the unknowns are the velocity vector $\boldsymbol{u}$ and the pressure p, $\boldsymbol{f}$ is a given body force, ν is the viscosity coefficient, and α is a non-negative number. When $\alpha = 0$, the above reduces to the standard Stokes equations. The case $\alpha > 0$ often arises from a coupled semi-implicit time discretization for the unsteady Navier–Stokes equations (9.101). For the sake of simplicity, the homogeneous Dirichlet boundary condition is assumed, and other admissible boundary conditions can be treated similarly (cf. Girault and Raviart (1986)).

One main difficulty for solving (9.38) is that the velocity $\boldsymbol{u}$ and the pressure p are coupled by the incompressibility constraint $\nabla \cdot \boldsymbol{u} = 0$. However, one can formally decouple the pressure from the velocity as follows:

Let $\boldsymbol{X} = H_0^1(\Omega)^d$ and denote by $A : \boldsymbol{X} \to \boldsymbol{X}'$ the operator defined by

$$\langle A\boldsymbol{u}, \boldsymbol{v} \rangle_{\boldsymbol{X}',\boldsymbol{X}} = \alpha(\boldsymbol{u}, \boldsymbol{v}) + \nu(\nabla \boldsymbol{u}, \nabla \boldsymbol{v}), \quad \forall \boldsymbol{u}, \boldsymbol{v} \in \boldsymbol{X}. \tag{9.39}$$

Then, applying the operator $\nabla \cdot A^{-1}$ to (9.38), we find that the pressure can be determined by

$$Bp := -\nabla \cdot A^{-1} \nabla p = -\nabla \cdot A^{-1} \boldsymbol{f}. \tag{9.40}$$

Once p is obtained from (9.40), we can obtain $\boldsymbol{u}$ from (9.38) by inverting the operator A, namely,

$$\boldsymbol{u} = A^{-1}(\boldsymbol{f} - \nabla p). \tag{9.41}$$

Let

$$M = \Big\{ q \in L^2(\Omega) : \int_\Omega q dx = 0 \Big\}. \tag{9.42}$$

The operator $B := -\nabla \cdot A^{-1} \nabla : M \to M$ is usually referred to as the Uzawa operator or the Schur complement associated with the Stokes operator. We have

$$(Bp, q) := -(\nabla \cdot A^{-1} \nabla p, q) = (A^{-1} \nabla p, \nabla q) = (p, Bq). \tag{9.43}$$

Therefore, B is a self-adjoint positive definite operator.

9.2.2 Galerkin Method for the Stokes Problem

The weak formulation for (9.38) is

$$\begin{cases} \text{Find } \boldsymbol{u} \in \boldsymbol{X} \text{ and } p \in M \text{ such that} \\ \alpha(\boldsymbol{u},\boldsymbol{v}) + \nu(\nabla \boldsymbol{u}, \nabla \boldsymbol{v}) - (\nabla\cdot \boldsymbol{v}, p) = \langle \boldsymbol{f}, \boldsymbol{v}\rangle_{\boldsymbol{X}',\boldsymbol{X}}, \quad \forall \boldsymbol{v} \in \boldsymbol{X}, \\ (\nabla\cdot\boldsymbol{u}, q) = 0, \quad \forall q \in M. \end{cases} \tag{9.44}$$

Let $\boldsymbol{X}_N$ and M_N be a suitable pair of finite dimensional approximation spaces for $\boldsymbol{X}$ and M. The corresponding Galerkin method for (9.44) is

$$\begin{cases} \text{Find } (\boldsymbol{u}_N, p_N) \in \boldsymbol{X}_N \times M_N \text{ such that} \\ \alpha(\boldsymbol{u}_N,\boldsymbol{v}_N) + \nu(\nabla \boldsymbol{u}_N, \nabla \boldsymbol{v}_N) - (p_N, \nabla\cdot \boldsymbol{v}_N) = (\boldsymbol{f}, \boldsymbol{v}_N), \quad \forall \boldsymbol{v}_N \in \boldsymbol{X}_N, \\ (\nabla\cdot\boldsymbol{u}_N, q_N) = 0, \quad \forall q_N \in M_N. \end{cases} \tag{9.45}$$

It is well-known (see, e.g., Girault and Raviart (1986)) that the discrete problem (9.45) admits a unique solution if and only if there exists a positive constant β_N such that

$$\inf_{q_N \in M_N} \sup_{0 \neq \boldsymbol{v}_N \in \boldsymbol{X}_N} \frac{(q_N, \nabla\cdot \boldsymbol{v}_N)}{\|q_N\| |||\boldsymbol{v}_N|||} \geq \beta_N, \tag{9.46}$$

where $|||\boldsymbol{v}|||^2 = \alpha\|\boldsymbol{v}\|^2 + \nu\|\nabla\boldsymbol{v}\|^2$. The above condition is referred to as Brezzi-Babuška inf-sup condition (cf. Babuška (1973), Brezzi (1974) and Theorem 1.1) and β_N is referred to as the inf-sup constant.

Let $\{\boldsymbol{\phi}_k\}_{k=1}^{N_u}$ and $\{\psi_k\}_{k=1}^{N_p}$ be respectively the basis functions of $\boldsymbol{X}_N$ and M_N. Then we can write

$$\boldsymbol{u}_N = \sum_{k=1}^{N_u} \tilde{u}_k \boldsymbol{\phi}_k, \quad p_N = \sum_{k=1}^{N_p} \tilde{p}_k \psi_k. \tag{9.47}$$

Set

$$\begin{aligned} & a_{ij} = \alpha(\boldsymbol{\phi}_j, \boldsymbol{\phi}_i) + \nu(\nabla\boldsymbol{\phi}_j, \nabla\boldsymbol{\phi}_i), \quad A_N = (a_{ij})_{i,j=1,\dots,N_u}, \\ & b_{ij} = -(\psi_i, \nabla\cdot\boldsymbol{\phi}_j), \quad B_N = (b_{ij})_{i=1,\dots,N_p, j=1,\dots,N_u}, \\ & \mathbf{u} = (\tilde{u}_1, \dots, \tilde{u}_{N_u})^T, \quad \mathbf{p} = (\tilde{p}_1, \dots, \tilde{p}_{N_p})^T, \\ & f_i = (I_N \boldsymbol{f}, \boldsymbol{\phi}_i), \quad \mathbf{f} = (f_1, \dots, f_{N_u})^T. \end{aligned} \tag{9.48}$$

Then the problem (9.45) reduces to

$$A_N \mathbf{u} + B_N^T \mathbf{p} = \mathbf{f}; \quad B_N \mathbf{u} = 0. \tag{9.49}$$

The main difficulty in numerically solving the above linear system is that it is indefinite so standard iterative algorithms (cf. Appendix C) would not be efficient.

As in the space continuous case, $\mathbf{p}$ can be obtained by inverting the discrete Uzawa operator:

$$B_N A_N^{-1} B_N^T \mathbf{p} = B_N A_N^{-1} \mathbf{f}. \tag{9.50}$$

If there exists $\beta_N > 0$ such that the inf-sup condition is satisfied, then it is easy to show that the above problem is well-posed and the discrete Uzawa operator is symmetric positive definite. Therefore, one can apply a suitable iterative method for solving (9.50). It is essential that the iterative method can be performed without explicitly forming the full matrix $B_N A_N^{-1} B_N^T$. It is also expected the convergence rate of such iterative method will depend essentially on the condition number of the matrix $B_N A_N^{-1} B_N^T$. It is shown (see, e.g., Maday et al. (1993)) that

$$\mathrm{cond}(B_N A_N^{-1} B_N^T) = \beta_N^{-2}. \tag{9.51}$$

Therefore, the effectiveness of these iterative methods is directly related to the magnitude of β_N. It is well-known that there exist many finite element pairs of $(\boldsymbol{X}_N, M_N)$ such that $\beta_N = \beta > 0$ (cf. Girault and Raviart (1986), Brezzi and Fortin (1991)). In the literature, such pairs are referred to as *stable* Stokes pairs, as they lead to optimal error estimates for both the velocity and pressure (cf. Theorem 9.1).

We now consider how to choose $\boldsymbol{X}_N$ and M_N in a spectral method. To simplify the presentation, we shall consider only $\Omega := (-1,1)^d$ with $d = 2$ or 3. In this case, the obvious choice for $\boldsymbol{X}_N$ in a spectral method is $\boldsymbol{X}_N = (P_N \cap H_0^1(\Omega))^d$. However, how to choose M_N is not a trivial question. For any given M_N, let us define

$$Z_N = \big\{ q_N \in M_N : (q_N, \nabla \cdot \boldsymbol{v}_N) = 0, \ \forall \boldsymbol{v}_N \in \boldsymbol{X}_N \big\}. \tag{9.52}$$

Obviously if $(\boldsymbol{u}_N, p_N)$ is a solution of (9.45), then so is $(\boldsymbol{u}_N, p_N + q_N)$ for any $q_N \in Z_N$. Hence, any mode in Z_N is called a spurious mode. For the most convenient choice $M_N = \{q_N \in P_N : \int_\Omega q_N dx = 0\}$, it is shown that Z_N spans a seven-dimensional space if $d = 2$ and $12N+3$-dimensional space if $d = 3$, and that the corresponding (after filtering the spurious mode) inf-sup constant β_N behaves like $O(N^{-1})$ (cf. Bernardi and Maday (1992a, 1997)). Therefore, it is not a good choice for the pressure space. On the other hand, if we set

$$M_N = \Big\{ q_N \in P_{N-2} : \int_\Omega q_N dx = 0 \Big\}, \tag{9.53}$$

then the corresponding Z_N is empty, and this leads to a well-posed problem (9.45) with the inf-sup constant (see, e.g., Bernardi and Maday (1997))

$$\beta_N \geq C(\alpha, \nu) N^{-(d-1)/2} \qquad (d = 2 \text{ or } 3). \tag{9.54}$$

This pair of spaces is known as the $P_N \times P_{N-2}$ method in the literature, and is the most commonly used pair for spectral/spectral-element approximations of Stokes and Navier–Stokes equations.

Remark 9.1. *It is shown in* Bernardi and Maday (1999) *that for any given* $0 < \lambda < 1$,

$$M_N^{(\lambda)} = \Big\{ q \in P_{[\lambda N]} : \int_\Omega q dx = 0 \Big\}$$

leads to a well-posed problem (9.45) *with an inf-sup constant which is independent of N but is of course dependent on λ in such a way that $\beta_N \to 0$ as $\lambda \to 1^-$.*

We now present a simple iterative algorithm, known as the Uzawa algorithm (cf. Arrow et al. (1958)) for solving (9.49): Given an arbitrary $\mathbf{p}^0$, compute $(\mathbf{u}^{k+1}, \mathbf{p}^{k+1})$ recursively by

$$\begin{aligned} A_N\mathbf{u}^{k+1} + B_N^T\mathbf{p}^k &= \mathbf{f}; \\ \mathbf{p}^{k+1} &= \mathbf{p}^k - \rho_k B_N\mathbf{u}^{k+1}. \end{aligned} \tag{9.55}$$

It is clear that, for each k, the above system can be solved by the efficient Legendre-Galerkin algorithm developed in Sect. 8.1.

As for the convergence, it can be shown (cf., for instance, Marion and Temam (1998)) that, in the case of $\alpha = 0$ and $\rho_k = \nu$, we have

$$\|\boldsymbol{u}_N^k - \boldsymbol{u}_N\|_1 + \|p_N^k - p_N\| \lesssim (1-\beta_N^2)^k, \tag{9.56}$$

where $(\boldsymbol{u}_N, p_N)$ is the solution of (9.45) and β_N is the inf-sup constant defined in (9.46). Thus, for a given tolerance ε, the number of Uzawa steps needed is proportional to $\frac{\log\varepsilon}{\beta_N^2}$. On the other hand, we can also use the Conjugate Gradient (CG) iteration to solve (9.50). The cost of each CG iteration is essentially the same as one step of (9.55). However, the number of the CG steps needed for the same tolerance, thanks to Theorem C.1, is proportional to $\frac{\log\varepsilon}{\beta_N}$. Therefore, applying the CG method to (9.50) is preferred over using the iterative Uzawa algorithm (9.55) for (9.49).

9.2.3 Error Analysis

The inf-sup constant β_N not only plays an important role in the implementation of the approximation (9.45), but also it is of paramount importance in the error analysis. Let us denote

$$\boldsymbol{V}_N = \{\boldsymbol{v}_N \in \boldsymbol{X}_N : (q_N, \nabla\cdot\boldsymbol{v}_N) = 0, \quad \forall q_N \in M_N\}. \tag{9.57}$$

Then, with respect to the error analysis, we have

Theorem 9.1. *Assuming* (9.46)*, the following error estimates hold:*

$$\begin{aligned} &|||\boldsymbol{u} - \boldsymbol{u}_N||| \lesssim \inf_{\boldsymbol{v}_N\in\boldsymbol{V}_N} |||\boldsymbol{u} - \boldsymbol{v}_N|||, \\ &\beta_N\|p - p_N\| \lesssim \inf_{\boldsymbol{v}_N\in\boldsymbol{V}_N} |||\boldsymbol{u} - \boldsymbol{v}_N||| + \inf_{q_N\in M_N} \|p - q_N\|, \end{aligned} \tag{9.58}$$

where $(\boldsymbol{u}, p)$ *and* $(\boldsymbol{u}_N, p_N)$ *are respectively the solution of* (9.38) *and* (9.45).

Proof. Let us denote

$$\boldsymbol{V} = \{\boldsymbol{v} \in \boldsymbol{X} : (q, \nabla\cdot\boldsymbol{v}) = 0, \quad \forall q \in M\}. \tag{9.59}$$

Then, by the definition of $\boldsymbol{V}$ and $\boldsymbol{V}_N$,

$$\begin{aligned} &\alpha(\boldsymbol{u},\boldsymbol{v}) + \nu(\nabla\boldsymbol{u},\nabla\boldsymbol{v}) = (\boldsymbol{f},\boldsymbol{v}), \quad \forall \boldsymbol{v}\in\boldsymbol{V}, \\ &\alpha(\boldsymbol{u}_N,\boldsymbol{v}_N) + \nu(\nabla\boldsymbol{u}_N,\nabla\boldsymbol{v}_N) = (f,\boldsymbol{v}_N), \quad \forall \boldsymbol{v}_N\in\boldsymbol{V}_N. \end{aligned} \tag{9.60}$$

Since $\boldsymbol{V}_N \subset \boldsymbol{V}$, we have

$$\alpha(\boldsymbol{u}-\boldsymbol{u}_N,\boldsymbol{v}_N)+\nu(\nabla(\boldsymbol{u}-\boldsymbol{u}_N),\nabla\boldsymbol{v}_N)=0, \quad \forall \boldsymbol{v}_N \in \boldsymbol{V}_N.$$

Hence,

$$\begin{aligned}|||\boldsymbol{u}-\boldsymbol{u}_N|||^2 &= \alpha(\boldsymbol{u}-\boldsymbol{u}_N,\boldsymbol{u}-\boldsymbol{u}_N)+\nu(\nabla(\boldsymbol{u}-\boldsymbol{u}_N),\nabla(\boldsymbol{u}-\boldsymbol{u}_N))\\ &= \inf_{\boldsymbol{v}_N\in\boldsymbol{V}_N}\{\alpha(\boldsymbol{u}-\boldsymbol{u}_N,\boldsymbol{u}-\boldsymbol{v}_N)+\nu(\nabla(\boldsymbol{u}-\boldsymbol{u}_N),\nabla(\boldsymbol{u}-\boldsymbol{v}_N))\},\end{aligned}$$

which implies immediately

$$|||\boldsymbol{u}-\boldsymbol{u}_N||| \le \inf_{\boldsymbol{v}_N\in\boldsymbol{V}_N}|||\boldsymbol{u}-\boldsymbol{v}_N|||.$$

Next, we derive from (9.38)-(9.45) the identity

$$\alpha(\boldsymbol{u}-\boldsymbol{u}_N,\boldsymbol{v}_N)+\nu(\nabla(\boldsymbol{u}-\boldsymbol{u}_N),\nabla\boldsymbol{v}_N)-(p-p_N,\nabla\cdot\boldsymbol{v}_N)=0, \quad \forall \boldsymbol{v}_N\in\boldsymbol{X}_N. \tag{9.61}$$

Hence, by using (9.46) and the above identity, we find that for any $q_N \in M_N$,

$$\begin{aligned}\beta_N\|q_N-p_N\| &\le \sup_{\boldsymbol{v}_N\in\boldsymbol{X}_N}\frac{(q_N-p_N,\nabla\cdot\boldsymbol{v}_N)}{|||\boldsymbol{v}_N|||}\\ &= \sup_{\boldsymbol{v}_N\in\boldsymbol{X}_N}\frac{\alpha(\boldsymbol{u}-\boldsymbol{u}_N,\boldsymbol{v}_N)+\nu(\nabla(\boldsymbol{u}-\boldsymbol{u}_N),\nabla\boldsymbol{v}_N)-(p-q_N,\nabla\cdot\boldsymbol{v}_N)}{|||\boldsymbol{v}_N|||}.\end{aligned}$$

It follows from the identity $\|\nabla\boldsymbol{v}\|=\|\nabla\times\boldsymbol{v}\|+\|\nabla\cdot\boldsymbol{v}\|$, $\forall\boldsymbol{v}\in\boldsymbol{X}$, and the Cauchy–Schwarz inequality that

$$\beta_N\|q_N-p_N\| \le |||\boldsymbol{u}-\boldsymbol{u}_N|||+\frac{1}{\nu}\|p-q_N\|, \quad \forall q_N\in M_N.$$

Therefore,

$$\begin{aligned}\beta_N\|p-p_N\| &\le \beta_N \inf_{q_N\in M_N}\{\|p-q_N\|+\|q_N-p_N\|\}\\ &\lesssim |||\boldsymbol{u}-\boldsymbol{u}_N|||+\inf_{q_N\in M_N}\|p-q_N\|\\ &\lesssim \inf_{\boldsymbol{v}_N\in\boldsymbol{V}_N}|||\boldsymbol{u}-\boldsymbol{v}_N|||+\inf_{q_N\in M_N}\|p-q_N\|.\end{aligned}$$

This completes the proof of this theorem. □

Consider now the $P_N\times P_{N-2}$ method, it can be shown (cf. Bernardi and Maday (1997)) that $\inf_{\boldsymbol{v}_N\in\boldsymbol{V}_N}|||\boldsymbol{u}-\boldsymbol{v}_N|||\lesssim N^{1-m}\|\boldsymbol{u}\|_m$. Therefore, the error estimate (9.58) becomes:

$$|\boldsymbol{u}-\boldsymbol{u}_N|_1+N^{-\frac{d-1}{2}}\|p-p_N\|\lesssim\left(N^{1-m}\|\boldsymbol{u}\|_m+N^{-s}\|p\|_s\right). \tag{9.62}$$

We observe that the convergence rate of velocity approximation is optimal, while that of the pressure approximation is only sub-optimal. However, such a loss of accuracy for the pressure is considered acceptable in the case of spectral methods thanks to their high-order of convergence when the solution $(\boldsymbol{u}, p)$ is sufficiently smooth.

9.3 Allen–Cahn and Cahn–Hilliard Equations

We consider in this section the Allen–Cahn equation:

$$\begin{cases} u_t - \Delta u + \dfrac{1}{\varepsilon^2} f(u) = 0, \quad \text{in } \Omega \times (0,T], \\ \dfrac{\partial u}{\partial n}\Big|_{\partial\Omega} = 0, \\ u|_{t=0} = u_0; \end{cases} \tag{9.63}$$

and the Cahn–Hilliard equation:

$$\begin{cases} u_t - \Delta\Big(- \Delta u + \dfrac{1}{\varepsilon^2} f(u)\Big) = 0, \quad \text{in } \Omega \times (0,T], \\ \dfrac{\partial u}{\partial n}\Big|_{\partial\Omega} = 0, \quad \dfrac{\partial}{\partial n}\Big(\Delta u - \dfrac{1}{\varepsilon^2} f(u)\Big)\Big|_{\partial\Omega} = 0, \\ u|_{t=0} = u_0. \end{cases} \tag{9.64}$$

In the above, Ω is a bounded domain in $\mathbb{R}^d$ $(d = 2,3)$, n is its outward normal, and $f(u) = F'(u)$ with $F(u)$ being a given energy potential.

We recall that the one-dimensional Allen–Cahn and Cahn–Hilliard equations were considered in Sects. 2.3.3 and 6.6.1, respectively. The Allen–Cahn (cf. Allen and Cahn (1979)) and Cahn–Hilliard (cf. Cahn and Hilliard (1958)) were originally developed as models for some material science applications. They have also been widely used in fluid dynamics to describe moving interfaces through a phase-field approach (see, e.g., Anderson et al. (1998), Lowengrub and Truskinovsky (1998), Liu and Shen (2003) and the references therein).

The Allen–Cahn and Cahn–Hilliard equations are often associated with periodic boundary conditions in many materials science applications (cf. Chen and Shen (1998), Chen (2002)). The discussion below on designing stable time discretization schemes applies directly to the periodic case, and the resulting linear/nonlinear systems at each time step can be easily solved by using a Fourier spectral method. Therefore, we shall not treat the periodic case separately.

An important feature of the Allen–Cahn (resp. Cahn–Hilliard) equation is that it can be viewed as the gradient flow in L^2 (resp. H^{-1}) of the Lyapunov energy functional

$$E(u) := \int_\Omega \Big(\frac{1}{2}|\nabla u|^2 + \frac{1}{\varepsilon^2} F(u)\Big)\, dx. \tag{9.65}$$

More precisely, by taking the inner product of (9.63) and (9.64) with $-\Delta u+\frac{1}{\varepsilon^2}f(u)$, we immediately find the energy law for (9.63):

$$\frac{\partial}{\partial t}E(u(t)) = -\int_\Omega |-\Delta u+\frac{1}{\varepsilon^2}f(u)|^2 dx; \tag{9.66}$$

and for (9.64):

$$\frac{\partial}{\partial t}E(u(t)) = -\int_\Omega |\nabla(-\Delta u+\frac{1}{\varepsilon^2}f(u))|^2 dx. \tag{9.67}$$

It is desirable for a numerical scheme to obey a similar discrete energy law.

9.3.1 Simple Semi-Implicit Schemes

We consider a usual first-order semi-implicit method for (9.63):

$$\begin{cases} \text{Find } u^{n+1} \in H^1(\Omega) \text{ such that} \\ \frac{1}{\delta t}(u^{n+1}-u^n,\psi)+(\nabla u^{n+1},\nabla\psi)+\frac{1}{\varepsilon^2}(f(u^n),\psi)=0, \quad \forall \psi \in H^1(\Omega), \end{cases} \tag{9.68}$$

where δt is the time step size and u^n is the approximation of the solution at $n\delta t$. At each time step, the above system leads to a second-order problem with constant coefficients that can be easily solved. On the other hand, a direct extension of the above scheme for (9.64) will lead to, at each time step, a fourth-order equation that is more difficult to deal with. To avoid solving a multi-dimensional fourth-order equation at each time step, we rewrite (9.64) as a mixed formulation:

$$\begin{cases} \text{Find } u, w \in H^1(\Omega) \text{ such that} \\ (u_t,q)+(\nabla w,\nabla q)=0, \quad \forall q \in H^1(\Omega), \\ (\nabla u,\nabla\psi)+\frac{1}{\varepsilon^2}(f(u),\psi)=(w,\psi), \quad \forall \psi \in H^1(\Omega). \end{cases} \tag{9.69}$$

Then, a first-order semi-implicit method for (9.69) is:

$$\begin{cases} \text{Find } u^{n+1}, w^{n+1} \in H^1(\Omega) \text{ such that} \\ \frac{1}{\delta t}(u^{n+1}-u^n,q)+(\nabla w^{n+1},\nabla q)=0, \quad \forall q \in H^1(\Omega), \\ (\nabla u^{n+1},\nabla\psi)+\frac{1}{\varepsilon^2}(f(u^n),\psi)=(w^{n+1},\psi), \quad \forall \psi \in H^1(\Omega). \end{cases} \tag{9.70}$$

We shall assume that the potential function $F(u)$ satisfies the following condition: there exists a constant L such that

$$\max_{u\in\mathbb{R}} |f'(u)| \le L, \tag{9.71}$$

where $f(u)=F'(u)$.

Note that the condition (9.71) is satisfied by many physically relevant potentials by restricting the growth of $F(u)$ to be quadratic for $|u| \geq M$. Consider, for example, the Ginzburg-Landau double-well potential $F(u) = \frac{1}{4}(u^2 - 1)^2$. Its quartic growth at infinity introduces various technical difficulties in the analysis and approximation of Allen–Cahn and Cahn–Hilliard equations. Since the Allen–Cahn equation satisfies the maximum principle, and it has been shown in Caffarelli and Muler (1995) that for a truncated potential $F(u)$ with quadratic growth at infinity, the maximum norm of the solution for the Cahn–Hilliard equation is bounded, it has been a common practice (cf. Kessler et al. (2004), Condette et al. (2011)) to consider the Allen–Cahn and Cahn–Hilliard equations with a truncated double-well potential $F(u)$ satisfying (9.71).

Theorem 9.2. *Assume that the condition* (9.71)*, and*

$$\delta t \leq \frac{2\varepsilon^2}{L} \quad \textit{for the scheme } (9.68), \tag{9.72}$$

$$\delta t \leq \frac{4\varepsilon^4}{L^2} \quad \textit{for the scheme } (9.70), \tag{9.73}$$

hold. Then, the solutions of (9.68) *and* (9.70) *satisfy*

$$E(u^{n+1}) \leq E(u^n), \quad \forall n \geq 0.$$

Proof. We shall only prove the result for (9.70). The proof for (9.68) is similar and simpler.

Taking $q = \delta t w^{n+1}$ and $\psi = u^{n+1} - u^n$ in (9.70), and using the Taylor expansion

$$F(u^{n+1}) - F(u^n) = f(u^n)(u^{n+1} - u^n) + \frac{f'(\xi^n)}{2}(u^{n+1} - u^n)^2, \tag{9.74}$$

we find

$$(u^{n+1} - u^n, w^{n+1}) + \delta t \|\nabla w^{n+1}\|^2 = 0, \tag{9.75}$$

and

$$\begin{aligned}\frac{1}{2}\big(\|\nabla u^{n+1}\|^2 - \|\nabla u^n\|^2 + \|\nabla(u^{n+1} - u^n)\|^2\big) + \frac{1}{\varepsilon^2}(F(u^{n+1}) - F(u^n), 1)\\ + \frac{1}{2\varepsilon^2}(f'(\xi^n)(u^{n+1} - u^n), u^{n+1} - u^n) = (w^{n+1}, u^{n+1} - u^n).\end{aligned} \tag{9.76}$$

On the other hand, taking $q = \sqrt{\delta t}(u^{n+1} - u^n)$ in (9.70), we obtain

$$\begin{aligned}\frac{1}{\sqrt{\delta t}}\|u^{n+1} - u^n\|^2 &= -\sqrt{\delta t}(\nabla w^{n+1}, \nabla(u^{n+1} - u^n))\\ &\leq \frac{\delta t}{2}\|\nabla w^{n+1}\|^2 + \frac{1}{2}\|\nabla(u^{n+1} - u^n)\|^2.\end{aligned} \tag{9.77}$$

Summing up the above three relations and using (9.71), we arrive at

$$\begin{aligned}\frac{1}{\sqrt{\delta t}}\|u^{n+1}-u^n\|^2+\frac{\delta t}{2}\|\nabla w^{n+1}\|^2+\frac{1}{2}(\|\nabla u^{n+1}\|^2-\|\nabla u^n\|^2)\\+\frac{1}{\varepsilon^2}(F(u^{n+1})-F(u^n),1)\\\leq-\frac{1}{2\varepsilon^2}(f'(\xi^n)(u^{n+1}-u^n),u^{n+1}-u^n)\\\leq\frac{L}{2\varepsilon^2}\|u^{n+1}-u^n\|^2.\end{aligned}$$

We then conclude that the desired result holds under the condition (9.73). □

Notice that, due to the explicit treatment for the nonlinear term, the stability conditions (9.72) and (9.73) are very severe when $\varepsilon \ll 1$. On the other hand, a fully implicit treatment, namely replacing $f(u^n)$ by $f(u^{n+1})$ in (9.68) and (9.70), will not be very helpful as conditions similar to (9.72) and (9.73) are needed for the nonlinear systems at each time step to have a unique solution. Therefore, it is highly desirable to construct simple schemes which admit a unique solution with a much relaxed stability condition, or ideally, unconditionally stable. We shall present two different approaches below.

9.3.2 Convex Splitting Schemes

The first approach is the so-called convex splitting originally proposed by Eyre (1998). Recently, the idea has been applied to various gradient flows (cf. Hu et al. (2009), Wang et al. (2010)). Assume that we can split the potential function $F(u)$ as the difference of two convex functions, i.e.,

$$F(u)=F_c(u)-F_e(u) \quad \text{with} \quad F_c''(u),\, F_e''(u)\geq 0. \tag{9.78}$$

For example, we can split the usual Ginzburg-Landau potential as $F(u)=\frac{1}{4}(u^4+1)-\frac{1}{2}u^2$. Then, a first-order convex splitting scheme for (9.63) reads

$$\begin{cases}\text{Find } u^{n+1}\in H^1(\Omega) \text{ such that}\\ \dfrac{1}{\delta t}(u^{n+1}-u^n,\psi)+(\nabla u^{n+1},\nabla\psi)\\ \qquad+\dfrac{1}{\varepsilon^2}(f_c(u^{n+1})-f_e(u^n),\psi)=0, \quad \forall\psi\in H^1(\Omega),\end{cases} \tag{9.79}$$

where $f_c(u)=F_c'(u)$ and $f_e(u)=F_e'(u)$.

Theorem 9.3. *The scheme* (9.79) *is unconditionally stable. More precisely, we have*

$$E(u^{n+1})\leq E(u^n)-\frac{1}{\delta t}\|u^{n+1}-u^n\|^2-\frac{1}{2}\|\nabla(u^{n+1}-u^n)\|^2. \tag{9.80}$$

Furthermore, the solution u^{n+1} of the nonlinear equation (9.79) *is the unique minimizer of the convex functional*

$$Q(u)=\int_\Omega\Big(\frac{1}{\delta t}|u|^2+\frac{1}{2}|\nabla u|^2+\frac{1}{\varepsilon^2}F_c(u)+g(u^n)\Big)dx, \tag{9.81}$$

where $g(u^n)=-\frac{1}{\delta t}u^n-\frac{1}{\varepsilon^2}f_e(u^n)$.

Proof. Taking $\psi=u^{n+1}-u^n$ in (9.79) and using the Taylor expansions

$$F_c(u^n)-F_c(u^{n+1})=f_c(u^{n+1})(u^n-u^{n+1})+\frac{F_c''(\xi_n)}{2}(u^n-u^{n+1})^2,$$

$$F_e(u^{n+1})-F_e(u^n)=f_e(u^n)(u^{n+1}-u^n)+\frac{F_e''(\eta_n)}{2}(u^{n+1}-u^n)^2,$$

thanks to (9.78), we find that

$$\begin{aligned}\frac{1}{\delta t}\|u^{n+1}-u^n\|^2&+\frac{1}{2}(\|\nabla u^{n+1}\|^2-\|\nabla u^n\|^2+\|\nabla(u^{n+1}-u^n)\|^2)\\&+\frac{1}{\varepsilon^2}\{(F(u^{n+1}),1)-(F(u^n),1)\}\\&=-\frac{1}{2\varepsilon^2}(F_c''(\xi_n)+F_e''(\eta_n),(u^n-u^{n+1})^2)\le 0,\end{aligned}$$

which implies (9.80).

On the other hand, it is clear that (9.79) is the Euler-Lagrange equation of $\min_{u\in H^1(\Omega)}Q(u)$. Since $Q(u)$ is convex, u^{n+1} is its unique minimizer. □

It is also possible to construct second-order convex splitting schemes. For the Ginzburg-Landau potential $F(u)=\frac{1}{4}(u^2-1)^2$ and $f(u)=F'(u)=u^3-u$, a second-order convex splitting scheme is as follows (cf. Hu et al. (2009)):

$$\begin{cases}\text{Find } u^{n+1}\in H^1(\Omega) \text{ such that}\\ \dfrac{1}{\delta t}(u^{n+1}-u^n,\psi)+\Big(\nabla\dfrac{u^{n+1}+u^n}{2},\nabla\psi\Big)\\ \qquad+\dfrac{1}{4\varepsilon^2}(((u^{n+1})^2+(u^n)^2)(u^{n+1}+u^n),\psi)\\ \qquad-\dfrac{1}{2\varepsilon^2}(3u^n-u^{n-1},\psi)=0,\quad\forall\psi\in H^1(\Omega).\end{cases} \tag{9.82}$$

It is not hard to show that the scheme is unconditionally stable (cf. Hu et al. (2009)).

One can easily extend the above convex splitting schemes for the Cahn–Hilliard equation. For example, a first-order convex splitting scheme for (9.69) is

$$\begin{cases} \text{Find } u^{n+1}, w^{n+1} \in H^1(\Omega) \text{ such that} \\ \frac{1}{\delta t}(u^{n+1} - u^n, q) + (\nabla w^{n+1}, \nabla q) = 0, \quad \forall q \in H^1(\Omega), \\ (\nabla u^{n+1}, \nabla \psi) + \frac{1}{\varepsilon^2}(f_c(u^{n+1}) - f_e(u^n), \psi) = (w^{n+1}, \psi), \quad \forall \psi \in H^1(\Omega). \end{cases} \tag{9.83}$$

One can show that the results in Theorem 9.3 can be extended to the above scheme (see Problem 9.1).

The convex splitting schemes have very attractive properties. However, they require solving a nonlinear system at each time step.

9.3.3 Stabilized Semi-Implicit Schemes

In order to avoid solving a nonlinear system at each time step, we consider the following first-order stabilized semi-implicit method for (9.63):

$$\begin{cases} \text{Find } u^{n+1} \in H^1(\Omega) \text{ such that} \\ \left(\frac{1}{\delta t} + \frac{S}{\varepsilon^2}\right)(u^{n+1} - u^n, \psi) + (\nabla u^{n+1}, \nabla \psi) \\ \qquad + \frac{1}{\varepsilon^2}(f(u^n), \psi) = 0, \quad \forall \psi \in H^1(\Omega), \end{cases} \tag{9.84}$$

where S is a stabilizing parameter to be specified.

Similarly, a first-order stabilized semi-implicit method for (9.69) is:

$$\begin{cases} \text{Find } u^{n+1}, w^{n+1} \in H^1(\Omega) \text{ such that} \\ \frac{1}{\delta t}(u^{n+1} - u^n, q) + (\nabla w^{n+1}, \nabla q) = 0, \quad \forall q \in H^1(\Omega), \\ (\nabla u^{n+1}, \nabla \psi) + \frac{S}{\varepsilon^2}(u^{n+1} - u^n, \psi) + \frac{1}{\varepsilon^2}(f(u^n), \psi) = (w^{n+1}, \psi), \ \forall \psi \in H^1(\Omega). \end{cases} \tag{9.85}$$

The stabilizing term $\frac{S}{\varepsilon^2}(\phi^{n+1} - \phi^n)$ in the above schemes introduces an extra consistency error of order $\frac{S\delta t}{\varepsilon^2} u_t(\xi_n)$. We note however that this error is of the same order as the error introduced by the explicit treatment for the term f_e in the convex splitting schemes (9.79) and (9.83), which is,

$$\frac{1}{\varepsilon^2}\left(f_e(u(t^{n+1})) - f_e(u(t^n))\right) = \frac{\delta t}{\varepsilon^2} f_e'(\eta_n) u_t(\gamma_n).$$

The stability and error analysis of the above stabilized schemes were studied in Shen and Yang (2010). In particular, we have

Theorem 9.4. *Under the condition* (9.71)*, the stabilized schemes* (9.84) *and* (9.85) *with* $S \geq \frac{L}{2}$ *are unconditionally table, and the following energy law holds for any* δt*:*

$$E(u^{n+1}) \leq E(u^n), \quad \forall n \geq 0. \tag{9.86}$$

Proof. Once again, we shall only provide the proof for (9.85), and leave the proof for (9.84) to the interested readers.

As in the proof of Theorem 9.2, taking $q = \delta t w^{n+1}$ and $\psi = u^{n+1} - u^n$ in (9.85), we obtain (9.75) and (9.76) with an extra term $\frac{S}{\varepsilon^2}\|u^{n+1} - u^n\|^2$ in the left hand side of (9.76). Therefore, summing up (9.75) and (9.76) with this extra term, we immediately derive the desired result. □

One can also easily construct second-order stabilized schemes for (9.63) and (9.69). However, it appears not possible for such a scheme to be unconditionally stable. But ample numerical experiments indicate that the maximum allowable time step of such stabilized schemes can be orders of magnitude larger than that of standard semi-implicit schemes.

9.3.4 Spectral-Galerkin Discretizations in Space

To fix the idea, we set $\Omega = (-1,1)^d$, and

$$X_N = \{u \in P_N : u'(\pm 1) = 0\}, \quad Y_N = X_N^d.$$

The Legendre-Galerkin method for the first-order stabilized scheme (9.84) reads

$$\begin{cases} \text{Given } u_N^0 = I_N u_0 \text{ for } n \ge 0, \text{ find } u_N^{n+1} \in Y_N \text{ such that} \\ \left(\frac{1}{\delta t} + \frac{S}{\varepsilon^2}\right)(u_N^{n+1} - u_N^n, \psi_N) + (\nabla u_N^{n+1}, \nabla \psi_N) \\ \qquad + \frac{1}{\varepsilon^2}\langle f(u_N^n), \psi_N\rangle_N = 0, \quad \forall \psi_N \in Y_N, \end{cases} \tag{9.87}$$

where $\langle \cdot, \cdot \rangle_N$ is the d-dimensional discrete inner product based on the Legendre-Gauss-Lobatto points, and I_N is the corresponding interpolation operator.

Similarly, the Legendre-Galerkin method for the first-order stabilized scheme (9.85) reads

$$\begin{cases} \text{Given } u_N^0 = I_N u_0 \text{ for } n \ge 0, \text{ find } u_N^{n+1}, w_N^{n+1} \in Y_N \text{ such that} \\ \frac{1}{\delta t}(u_N^{n+1} - u_N^n, q_N) + (\nabla w_N^{n+1}, \nabla q_N) = 0, \quad \forall q_N \in Y_N, \\ (\nabla u_N^{n+1}, \nabla \psi_N) + \frac{S}{\varepsilon^2}(u_N^{n+1} - u_N^n, \psi_N) \\ \qquad + \frac{1}{\varepsilon^2}\langle f(u_N^n), \psi_N\rangle_N = (w_N^{n+1}, \psi_N), \quad \forall \psi_N \in Y_N. \end{cases} \tag{9.88}$$

The above scheme is a coupled linear system with constant coefficients for u_N^{n+1}, w_N^{n+1}. While the method of matrix decomposition presented in Sect. 8.1 does not directly apply to this system, Chen and Shen (2011) recently developed an algorithm, based on the idea of matrix decomposition, which can solve this coupled system at twice the cost of solving a second-order equation.

Let us define the discrete energy functional

$$E_N(u)=\frac{1}{2}\|\nabla u\|^2+\frac{1}{\varepsilon^2}\langle F(u),1\rangle_N. \tag{9.89}$$

Then, by proceeding as in the proof of Theorem 9.4, and noticing that the Taylor expansion (9.74) holds for each collocation point, we can prove the following:

Theorem 9.5. *Under the condition* (9.71), *the fully discrete stabilized schemes* (9.87) *and* (9.88) *with* $S\geq\frac{L}{2}$ *are unconditionally stable, and the following energy law holds for any* δt*:*

$$E_N(u^{n+1})\leq E_N(u^n),\quad \forall n\geq 0. \tag{9.90}$$

9.3.5 Error Analysis

To simplify the presentation, we shall carry out an error analysis for the Galerkin version of (9.88). Let $\bar{\Pi}_N^1$ be the projection operator defined in (8.155), and $Y_N=X_N^d$.

$$\begin{cases}\text{Given } u_N^0=\bar{\Pi}_N^1 u_0 \text{ for } n\geq 0, \text{ find } (u_N^{n+1},w_N^{n+1})\in Y_N\times Y_N \text{ such that}\\ \dfrac{1}{\delta t}(u_N^{n+1}-u_N^n,q_N)+(\nabla w_N^{n+1},\nabla q_N)=0,\quad \forall q_N\in Y_N,\\ (\nabla u_N^{n+1},\nabla\psi_N)+\dfrac{S}{\varepsilon^2}(u_N^{n+1}-u_N^n,\psi_N)\\ \qquad\qquad +\dfrac{1}{\varepsilon^2}(f(u_N^n),\psi_N)=(w_N^{n+1},\psi_N),\quad \forall\psi_N\in Y_N.\end{cases} \tag{9.91}$$

Denote

$$\begin{aligned}&\tilde{e}_N^{n+1}=\bar{\Pi}_N^1 u(t^{n+1})-u_N^{n+1},\quad \hat{e}_N^{n+1}=u(t^{n+1})-\bar{\Pi}_N^1 u(t^{n+1}),\\ &\bar{e}_N^{n+1}=\bar{\Pi}_N^1 w(t^{n+1})-w_N^{n+1},\quad \check{e}_N^{n+1}=w(t^{n+1})-\bar{\Pi}_N^1 w(t^{n+1}).\end{aligned} \tag{9.92}$$

The following results were established in Shen and Yang (2010):

Theorem 9.6. *Given* $T>0$, *we assume that for some* $m\geq 1$, $u,w\in C(0,T;H^m(\Omega))$, $u_t\in L^2(0,T;H^m(\Omega))$ *and* $u_{tt}\in L^2(0,T;L^2(\Omega))$. *Then for* $S>\frac{L}{2}$, *the solution of* (9.91) *satisfies*

$$E(u_N^{n+1})\leq E(u_N^n),$$

and the following error estimate holds

$$\begin{aligned}&\|u(t^{k+1})-u_N^{k+1}\|+\Big(\delta t\sum_{n=0}^{k}\|w(t_{n+1})-w_N^{n+1}\|^2\Big)^{/2}\\ &\qquad\leq C(\varepsilon,T)(K_1(u,\varepsilon)\delta t+K_2(u,\varepsilon)N^{-m}),\quad \forall 0\leq k\leq\frac{T}{\delta t}-1,\end{aligned} \tag{9.93}$$

where

$$
\begin{aligned}
&C(\varepsilon,T)\sim \exp(T/\varepsilon^4);\\
&K_1(u,\varepsilon)=\varepsilon^2\|u_{tt}\|_{L^2(0,T;L^2)}+\frac{1}{\varepsilon^2}\|u_t\|_{L^2(0,T;L^2)};\\
&K_2(u,\varepsilon)=\|u_0\|_m+(\varepsilon^2+\frac{\delta t}{\varepsilon^2})\|u_t\|_{L^2(0,T;H^m)}+\frac{1}{\varepsilon^2}\|u\|_{C(0,T;H^m)}+\|w\|_{C(0,T;H^m)}.
\end{aligned}
$$

Proof. Obviously, the proof of Theorem 9.4 is also valid for the fully discrete scheme (9.91). We now turn to the error estimates.

Let us define

$$
R^{n+1}:=\frac{u(t^{n+1})-u(t^n)}{\delta t}-u_t(t^{n+1}). \tag{9.94}
$$

By using the Taylor expansion with integral residuals and the Cauchy–Schwarz inequality, we derive easily

$$
\|R^{n+1}\|_s^2\le\frac{1}{\delta t^2}\left\|\int_{t^n}^{t^{n+1}}(t-t^n)u_{tt}(t)dt\right\|_s^2\le\frac{\delta t}{3}\int_{t^n}^{t^{n+1}}\|u_{tt}(t)\|_s^2dt, \tag{9.95}
$$

for $s=-1,0$.

Subtracting (9.91) from (9.69), we obtain

$$
\begin{aligned}
&\frac{1}{\delta t}(\tilde e_N^{n+1}-\tilde e_N^n,q_N)+(\nabla\tilde e_N^{n+1},\nabla q_N)\\
&\qquad=(R^{n+1},q_N)-\frac{1}{\delta t}((I-\bar\Pi_N^1)(u(t^{n+1})-u(t^n)),q_N),\\
&(\nabla\tilde e_N^{n+1},\nabla\psi_N)+\frac{S}{\varepsilon^2}(\tilde e_N^{n+1}-\tilde e_N^n,\psi_N)+\frac{1}{\varepsilon^2}(f(u(t^{n+1}))-f(u_N^n),\psi_N)\\
&\qquad=(\tilde e_N^{n+1}+\breve e_N^{n+1},\psi_N)+\frac{S}{\varepsilon^2}(\bar\Pi_N^1u(t^{n+1})-\bar\Pi_N^1u(t^n),\psi_N).
\end{aligned} \tag{9.96}
$$

Taking $q_N=2\delta t\breve e_N^{n+1}$ and $\psi_N=-2\delta t\tilde e_N^{n+1}$ and summing up the two identities, we derive

$$
\begin{aligned}
&\|\tilde e_N^{n+1}\|^2-\|\tilde e_N^n\|^2+\|\tilde e_N^{n+1}-\tilde e_N^n\|^2+2\delta t\|\breve e_N^{n+1}\|^2=2\delta t(R^{n+1},\breve e_N^{n+1})\\
&\quad-2((I-\bar\Pi_N^1)(u(t^{n+1})-u(t^n)),\breve e_N^{n+1})+\frac{2\delta tS}{\varepsilon^2}(\tilde e_N^{n+1}-\tilde e_N^n,\tilde e_N^{n+1})\\
&\quad+\frac{2\delta t}{\varepsilon^2}(f(u(t^{n+1}))-f(u_N^n),\tilde e_N^{n+1})-2\delta t(\breve e_N^{n+1},\tilde e_N^{n+1})\\
&\quad-\frac{2S\delta t}{\varepsilon^2}(\bar\Pi_N^1(u(t^{n+1})-u(t^n)),\tilde e_N^{n+1}):=\mathbf{I}+\mathbf{II}+\mathbf{III}+\mathbf{IV}+\mathbf{V}+\mathbf{VI}.
\end{aligned} \tag{9.97}
$$

Using the Cauchy–Schwarz inequality and (9.98), we derive

$$\mathbf{I} \le \varepsilon^4 \delta t \|R^{n+1}\|^2 + \frac{\delta t}{\varepsilon^4}\|\tilde{e}_N^{n+1}\|^2;$$

$$\mathbf{II} \le 2\varepsilon^4 \int_{t^n}^{t^{n+1}} \|(I-\bar{\Pi}_N^1)u_t(t)\|^2 dt + \frac{\delta t}{2\varepsilon^4}\|\tilde{e}_N^{n+1}\|^2;$$

$$\mathbf{III} \le \frac{\delta t}{4}\|\bar{e}_N^{n+1}\|^2 + \frac{4\delta t S^2}{\varepsilon^4}\|\tilde{e}_N^{n+1} - \tilde{e}_N^n\|^2;$$

$$\mathbf{V} \le \frac{\delta t}{4}\|\bar{e}_N^{n+1}\|^2 + 4\delta t\|\breve{e}_N^{n+1}\|^2;$$

$$\mathbf{VI} \le \frac{\delta t}{4}\|\bar{e}_N^{n+1}\|^2 + \frac{4S^2\delta t^2}{\varepsilon^4}\Big(\int_{t^n}^{t^{n+1}} \|(I-\bar{\Pi}_N^1)u_t(t)\|^2 dt + \int_{t^n}^{t^{n+1}} \|u_t(t)\|^2 dt\Big).$$

For the fourth term **IV**, we use (9.71) to derive

$$\begin{aligned}\|f(u_N^n) - f(u(t^{n+1}))\| &\le \|f(u_N^n) - f(u_N^{n+1})\| + \|f(u_N^{n+1}) - f(u(t^{n+1}))\| \\ &\le L\|u_N^n - u_N^{n+1}\| + L(\|\tilde{e}_N^{n+1}\| + \|\hat{e}_N^{n+1}\|) \\ &\le L(\|\tilde{e}_N^{n+1} - \tilde{e}_N^n\| + \|(I-\bar{\Pi}_N^1)(u(t^{n+1}) - u(t^n))\| \\ &+ \|u(t^{n+1}) - u(t^n)\|) + L(\|\tilde{e}_N^{n+1}\| + \|\hat{e}_N^{n+1}\|).\end{aligned} \tag{9.98}$$

We can derive from the above that

$$\begin{aligned}\mathbf{IV} \le \frac{\delta t}{4}\|\bar{e}_N^{n+1}\|^2 + C_7\Big(&\frac{\delta t L^2}{\varepsilon^4}\|\tilde{e}_N^{n+1} - \tilde{e}_N^n\|^2 + \frac{\delta t^2 L^2}{\varepsilon^4}\int_{t^n}^{t^{n+1}} \|(I-\bar{\Pi}_N^1)u_t(t)\|^2 dt \\ &+ \frac{\delta t^2 L^2}{\varepsilon^4}\int_{t^n}^{t^{n+1}} \|u_t(\cdot,t)\|^2 dt + \frac{\delta t L^2}{\varepsilon^4}(\|\tilde{e}_N^{n+1}\|^2 + \|\hat{e}_N^{n+1}\|^2)\Big).\end{aligned} \tag{9.99}$$

Combining the above inequalities into (9.97), we arrive at

$$\begin{aligned}&\|\tilde{e}_N^{n+1}\|^2 - \|\tilde{e}_N^n\|^2 + \|\tilde{e}_N^{n+1} - \tilde{e}_N^n\|^2 + \delta t\|\bar{e}_N^{n+1}\|^2 \\ &\quad\le \delta t\varepsilon^4\|R^{n+1}\|^2 + 4\delta t\|\breve{e}_N^{n+1}\|^2 + \frac{C_8\delta t}{\varepsilon^4}(\|\tilde{e}_N^{n+1}\|^2 + \|\tilde{e}_N^n\|^2 + \|\hat{e}_N^{n+1}\|^2) \\ &\quad+ 2\varepsilon^4\int_{t^n}^{t^{n+1}} \|(I-\bar{\Pi}_N^1)u_t(\cdot,t)\|^2 dt \\ &\quad+ \frac{C_9\delta t^2}{\varepsilon^4}\Big(\int_{t^n}^{t^{n+1}} \|(I-\bar{\Pi}_N^1)u_t(\cdot,t)\|^2 dt + \int_{t^n}^{t^{n+1}} \|u_t(\cdot,t)\|^2 dt\Big).\end{aligned} \tag{9.100}$$

Summing up the above inequality for $n = 0, 1, \ldots, k$ ($k \le \frac{T}{\delta t} - 1$) and using (8.158) and (9.95), we obtain

$$
\begin{aligned}
&\|\tilde{e}_N^{k+1}\|^2 + \delta t \sum_{n=0}^{k} \|\bar{e}_N^{n+1}\|^2 \\
&\quad \le 4\delta t \sum_{n=0}^{k} \|\breve{e}_N^{n+1}\|^2 + \delta t^2 \Big(\varepsilon^4 \|u_{tt}\|^2_{L^2(0,T;L^2)} + \frac{C_9}{\varepsilon^4}\|u_t\|^2_{L^2(0,T;L^2)}\Big) \\
&\quad + \frac{C_8 \delta t}{\varepsilon^4} \sum_{n=0}^{k} \big(\|\tilde{e}_N^{n+1}\|^2 + \|\tilde{e}_N^{n}\|^2 + \|\hat{e}_N^{n+1}\|^2\big) \\
&\quad + \varepsilon^4 N^{-2m} \|u_t\|^2_{L^2(0,T;H^m)} + \frac{C_9 \delta t^2}{\varepsilon^4} N^{-2m} \|u_t\|_{L^2(0,T;H^m)}.
\end{aligned}
$$

Applying the discrete Gronwall Lemma B.10 to the above inequality, we can then conclude by using the triangle inequality and (8.158). □

9.3.6 *Effect of Spatial Accuracy*

It has been observed that for interface problems governed by the Allen–Cahn or the Cahn–Hilliard type equation, spectral methods usually provide much more accurate results using fewer points than lower order methods like finite elements or finite differences. We now give a heuristic argument based on our error estimates.

To fix the idea, let us consider the Cahn–Hilliard equation (9.63) and its error estimate in (9.93). It is well-known that the solution of the Cahn–Hilliard equation will develop an interface with thickness of order ε. Therefore, it is reasonable to assume that $\partial_x^m u \sim \varepsilon^{-m}, \forall m \ge 0$. Hence, the error estimate (9.93) indicates that

$$\|u(t^n) - u_N^n\| \lesssim C(\varepsilon, T)(K_1(u,\varepsilon)\delta t + N^{-m}\varepsilon^{-1-m}).$$

Since the solution is usually smooth around the interfacial area, it can be expected that the above estimate is valid for all m. Let us ignore for the moment $C(\varepsilon, T)$. Then, as soon as $N > O(\varepsilon^{-1})$, it can be expected that the error due to the spatial discretization will decay very fast, in fact faster than any algebraic order, as N increases. In practice, it has been found that having 5-8 points inside the transitional region is sufficient to represent the interface accurately. On the other hand, for a lower order method, the corresponding error estimate is similar to (9.93) with N replaced by h^{-1}, but only with a fixed m, e.g., $m = 2$ for piece-wise linear finite elements or second-order finite differences. Hence, for $m = 2$, one needs to have $h \ll \varepsilon^{3/2}$ for the scheme to be convergent, and $h \sim \varepsilon^3$ for the spatial error to be of order $O(h)$. Therefore, an adaptive procedure is almost necessary for low-order methods to have a desirable accuracy with reasonable cost.

It is clear that similar arguments can be applied to other schemes for the Allen–Cahn and Cahn–Hilliard equations.

9.4 Unsteady Navier–Stokes Equations

The Navier–Stokes equations describe the motion of an incompressible flow. Numerical approximations of Navier–Stokes equations play an important role in many applications. There has been an enormous amount of research work, and still growing, on mathematical and numerical analysis of the Navier–Stokes equations. We refer to the books by Temam (1984), Karniadakis and Sherwin (1999), Deville et al. (2002), Glowinski (2003) for more details on the approximation of Navier–Stokes equations by the finite element, spectral and spectral element methods. In this section, we briefly describe two robust and accurate projection type schemes and the related full discretization schemes with a spectral-Galerkin discretization in space.

The unsteady Navier–Stokes equations are as follows:

$$\begin{cases} \boldsymbol{u}_t - \nu\Delta\boldsymbol{u} + \boldsymbol{u}\cdot\nabla\boldsymbol{u} + \nabla p = \boldsymbol{f}, & \text{in } \Omega\times(0,T], \\ \nabla\cdot\boldsymbol{u} = 0, & \text{in } \Omega\times[0,T], \end{cases} \tag{9.101}$$

subject to appropriate initial and boundary conditions for $\boldsymbol{u}$. In the above, the unknowns are the velocity vector $\boldsymbol{u}$ and the pressure p; $\boldsymbol{f}$ is a given body force, ν is the kinematic viscosity, Ω is an open and bounded domain in $\mathbb{R}^d$ ($d=2$ or 3 in practical situations), and $[0,T]$ is the time interval.

As for the Stokes equations, a main difficulty in approximating (9.101) is that the velocity and the pressure are coupled by the incompressibility constraint $\nabla\cdot\boldsymbol{u}=0$. A straightforward linearly-implicit time discretization of (9.101) would lead to a generalized Stokes problem (9.38) at each time step. Although the iterative algorithms presented in the previous section are acceptable for the steady Stokes problem, it is in general very costly to apply an iterative algorithm to solve a Stokes problem at each time step, particularly for a usual spectral discretization of the Stokes problem due to the non-optimality of its inf-sup constant.

A popular and effective strategy is to use a fractional step scheme to decouple the computation of the pressure from that of the velocity. This approach was first introduced by Chorin (1968) and Temam (1969) in the late 60's, and its countless variants have played and are still playing a major role in computational fluid dynamics, especially for large three-dimensional numerical simulations. We refer to Guermond et al. (2006) for an up-to-date review on this subject.

9.4.1 Second-Order Rotational Pressure-Correction Scheme

We first present a rotational pressure-correction scheme (see, for instance, Timmermans et al. (1996), Guermond and Shen (2004)) which has been widely used in practice.

Assuming $(\boldsymbol{u}^k, \boldsymbol{u}^{k-1}, p^k)$ are known, in the first substep, we look for $\tilde{\boldsymbol{u}}^{k+1}$ such that

$$\begin{cases} \frac{1}{2\delta t}(3\tilde{\boldsymbol{u}}^{k+1} - 4\boldsymbol{u}^k + \boldsymbol{u}^{k-1}) - \nu\Delta\tilde{\boldsymbol{u}}^{k+1} + \nabla p^k = \boldsymbol{g}(t_{k+1}), \\ \tilde{\boldsymbol{u}}^{k+1}|_{\partial\Omega} = 0, \end{cases} \tag{9.102}$$

where

$$\boldsymbol{g}(t_{k+1}) = \boldsymbol{f}(t_{k+1}) - (2(\boldsymbol{u}^k \cdot \nabla)\boldsymbol{u}^k - (\boldsymbol{u}^{k-1} \cdot \nabla)\boldsymbol{u}^{k-1}).$$

Then, in the second substep, we determine $(\boldsymbol{u}^{k+1}, \phi^{k+1})$ such that

$$\begin{cases} \frac{1}{2\delta t}(3\boldsymbol{u}^{k+1} - 3\tilde{\boldsymbol{u}}^{k+1}) + \nabla\phi^{k+1} = 0, \\ \nabla\cdot\boldsymbol{u}^{k+1} = 0, \\ \boldsymbol{u}^{k+1}\cdot\boldsymbol{n}|_{\partial\Omega} = 0. \end{cases} \tag{9.103}$$

The remaining task is to define a suitable p^{k+1} so that we can advance to the next time step. To this end, we first notice from (9.103) that

$$\Delta\tilde{\boldsymbol{u}}^{k+1} = \Delta\boldsymbol{u}^{k+1} + \frac{2\delta t}{3}\nabla\Delta\phi^{k+1} = \Delta\boldsymbol{u}^{k+1} + \nabla\nabla\cdot\tilde{\boldsymbol{u}}^{k+1}.$$

We then sum up the two substeps and use the above identity to obtain:

$$\begin{cases} \frac{1}{2\delta t}(3\boldsymbol{u}^{k+1} - 4\boldsymbol{u}^k + \boldsymbol{u}^{k-1}) - \nu\Delta\boldsymbol{u}^{k+1} + \nabla(\phi^{k+1} + p^k - \nu\nabla\cdot\tilde{\boldsymbol{u}}^{k+1}) = \boldsymbol{g}(t_{k+1}), \\ \nabla\cdot\boldsymbol{u}^{k+1} = 0, \\ \boldsymbol{u}^{k+1}\cdot\boldsymbol{n}|_{\partial\Omega} = 0. \end{cases} \tag{9.104}$$

Therefore, it is clear that we should set

$$p^{k+1} = \phi^{k+1} + p^k - \nu\nabla\cdot\tilde{\boldsymbol{u}}^{k+1}. \tag{9.105}$$

We note that the only difference between (9.104)-(9.105) and a *coupled* second-order scheme is that

$$\boldsymbol{u}^{k+1}\cdot\boldsymbol{\tau}|_{\partial\Omega} = -\frac{2\delta t}{3}\nabla\phi^{k+1}\cdot\boldsymbol{\tau}|_{\partial\Omega} \neq 0$$

(where $\boldsymbol{\tau}$ is the unit tangential vector) but "small". Hence, it is expected that the scheme (9.102), (9.103) and (9.105) provides a good approximation to the Navier–Stokes equations. Indeed, it is shown in Guermond and Shen (2004) (see also E and Liu (1996)) that

$$\|\boldsymbol{u}(t_k) - \boldsymbol{u}^k\| + \sqrt{\delta t}(\|\boldsymbol{u}(t_k) - \boldsymbol{u}^k\|_1 + \|p(t_k) - p^k\|) \lesssim \delta t^2. \tag{9.106}$$

We note that in the special case where only one direction is non-periodic, it is shown in Brown et al. (2001) that the factor of $\sqrt{\delta t}$ in the above estimate can be removed.

In practice, the coupled system (9.103) is decoupled by taking the divergence of the first equation in (9.103), leading to

$$\begin{aligned} &\Delta\phi^{k+1} = \frac{3}{2\delta t}\nabla\cdot\tilde{\boldsymbol{u}}^{k+1} \quad \text{in } \Omega, \quad \left.\frac{\partial\phi^{k+1}}{\partial\boldsymbol{n}}\right|_{\partial\Omega} = 0; \\ &\boldsymbol{u}^{k+1} = \tilde{\boldsymbol{u}}^{k+1} - \frac{2\delta t}{3}\nabla\phi^{k+1}. \end{aligned} \tag{9.107}$$

Hence, at each time step, the scheme (9.102)-(9.103)-(9.105) only involves inverting a Poisson-type equation for each of the velocity component $\tilde{\boldsymbol{u}}^{k+1}$ in (9.102) and a Poisson equation for ϕ^{k+1} in (9.107).

Remark 9.2. *If part of the boundary is open, i.e., the problem is prescribed with the following boundary conditions:*

$$\boldsymbol{u}|_{\Gamma_1} = \boldsymbol{h}_1, \ \boldsymbol{n}^T(\nu\nabla\boldsymbol{u} - pI)|_{\Gamma_2} = \boldsymbol{h}_2, \ \partial\Omega = \Gamma_1 \cup \Gamma_2, \tag{9.108}$$

the above scheme should be modified as follows Guermond et al. (2006):

$$\begin{cases} \frac{1}{2\delta t}(3\tilde{\boldsymbol{u}}^{k+1} - 4\boldsymbol{u}^k + \boldsymbol{u}^{k-1}) - \nu\Delta\tilde{\boldsymbol{u}}^{k+1} + \nabla p^k = \boldsymbol{g}(t_{k+1}), \\ \tilde{\boldsymbol{u}}^{k+1}|_{\Gamma_1} = \boldsymbol{h}_1^{k+1}, \ \boldsymbol{n}^T(\nu\nabla\tilde{\boldsymbol{u}}^{k+1} - p^k I)|_{\Gamma_2} = \boldsymbol{h}_2^{k+1}, \end{cases} \tag{9.109}$$

$$\begin{cases} \frac{1}{2\delta t}(3\boldsymbol{u}^{k+1} - 3\tilde{\boldsymbol{u}}^{k+1}) + \nabla\phi^{k+1} = 0; \quad \nabla\cdot\boldsymbol{u}^{k+1} = 0, \\ \boldsymbol{u}^{k+1}\cdot\boldsymbol{n}|_{\Gamma_1} = \boldsymbol{h}_1^{k+1}\cdot\boldsymbol{n}, \ \phi^{k+1}|_{\Gamma_2} = 0; \end{cases} \tag{9.110}$$

and

$$p^{k+1} = \phi^{k+1} + p^k - \nu\nabla\cdot\tilde{\boldsymbol{u}}^{k+1}. \tag{9.111}$$

9.4.2 Second-Order Consistent Splitting Scheme

Although the rotational pressure-correction scheme is quite accurate, it still suffers from a splitting error of order $\delta t^{\frac{3}{2}}$ for the H^1-norm of the velocity and L^2-norm of the pressure. We present below a consistent splitting scheme (cf. Guermond and Shen (2003)) which removes this splitting error. The key idea behind the consistent splitting schemes is to evaluate the pressure by testing the momentum equation against gradients. By taking the L^2-inner product of the momentum equation in (9.101) with ∇q and noticing that $(\boldsymbol{u}_t, \nabla q) = -(\nabla\cdot\boldsymbol{u}_t, q)$, we obtain

$$\int_\Omega \nabla p\cdot\nabla q = \int_\Omega (\boldsymbol{f} + \nu\Delta\boldsymbol{u} - \boldsymbol{u}\cdot\nabla\boldsymbol{u})\cdot\nabla q, \quad \forall q \in H^1(\Omega). \tag{9.112}$$

Note that if $\boldsymbol{u}$ is known, (9.112) is simply the weak form of a Poisson equation for the pressure. So the principle we shall follow is to compute the velocity and the pressure in two consecutive steps: First, we evaluate the velocity by making explicit the pressure, then we evaluate the pressure by making use of (9.112).

Denoting

$$\boldsymbol{g}^{k+1} = \boldsymbol{f}^{k+1} - \left(2\boldsymbol{u}^n\cdot\nabla\boldsymbol{u}^n - \boldsymbol{u}^{n-1}\cdot\nabla\boldsymbol{u}^{n-1}\right),$$

a second-order semi-implicit splitting scheme can be constructed as follows: find $\boldsymbol{u}^{k+1}$ and p^{k+1} such that

$$\frac{3\boldsymbol{u}^{k+1}-4\boldsymbol{u}^{k}+\boldsymbol{u}^{k-1}}{2\delta t}-\nu\Delta\boldsymbol{u}^{k+1}+\nabla(2p^{k}-p^{k-1})=\boldsymbol{g}^{k+1},\quad \boldsymbol{u}^{k+1}|_{\partial\Omega}=0, \tag{9.113}$$

$$(\nabla p^{k+1},\nabla q)=(\boldsymbol{g}^{k+1}+\nu\Delta\boldsymbol{u}^{k+1},\nabla q),\quad \forall q\in H^{1}(\Omega). \tag{9.114}$$

Notice that we can use (9.113) to replace $\boldsymbol{g}^{k+1}+\nu\Delta\boldsymbol{u}^{k+1}$ in (9.114) by $(3\boldsymbol{u}^{k+1}-4\boldsymbol{u}^{k}+\boldsymbol{u}^{k-1})/(2\delta t)+\nabla(2p^{k}-p^{k-1})$, leading to an equivalent formulation of (9.114):

$$(\nabla(p^{k+1}-2p^{k}+p^{k-1}),\nabla q)=\Big(\frac{3\boldsymbol{u}^{k+1}-4\boldsymbol{u}^{k}+\boldsymbol{u}^{k-1}}{2\delta t},\nabla q\Big),\quad \forall q\in H^{1}(\Omega). \tag{9.115}$$

We observe that if the domain Ω is sufficiently smooth, the solution of the above problem satisfies the following Poisson equation:

$$\begin{aligned}&-\Delta(p^{k+1}-2p^{k}+p^{k-1})=-\nabla\cdot\Big(\frac{3\boldsymbol{u}^{k+1}-4\boldsymbol{u}^{k}+\boldsymbol{u}^{k-1}}{2\delta t}\Big);\\&\frac{\partial}{\partial\boldsymbol{n}}(p^{k+1}-2p^{k}+p^{k-1})|_{\partial\Omega}=0.\end{aligned} \tag{9.116}$$

Since the exact pressure does not satisfy any prescribed boundary condition, it is clear that the pressure approximation from (9.116) is plagued by the artificial Neumann boundary condition which limits its accuracy. However, this defect can be easily overcome by using the identity $\Delta\boldsymbol{u}^{k+1}=\nabla\nabla\cdot\boldsymbol{u}^{k+1}-\nabla\times\nabla\times\boldsymbol{u}^{k+1}$, and replacing $\Delta\boldsymbol{u}^{k+1}$ in (9.114) by $-\nabla\times\nabla\times\boldsymbol{u}^{k+1}$. This procedure amounts to removing in (9.114) the term $\nabla\nabla\cdot\boldsymbol{u}^{k+1}$. It is clear that this is a consistent procedure since the exact velocity is divergence-free. Thus, (9.114) should be replaced by

$$(\nabla p^{k+1},\nabla q)=(\boldsymbol{g}^{k+1}-\nu\nabla\times\nabla\times\boldsymbol{u}^{k+1},\nabla q),\quad \forall q\in H^{1}(\Omega). \tag{9.117}$$

Once again, we can use (9.113) to reformulate (9.117) by replacing $\boldsymbol{g}^{k+1}-\nu\nabla\times\nabla\times\boldsymbol{u}^{k+1}$ by $(3\boldsymbol{u}^{k+1}-4\boldsymbol{u}^{k}+\boldsymbol{u}^{k-1})/2\delta t+\nabla(2p^{k}-p^{k-1})-\nu\nabla\nabla\cdot\boldsymbol{u}^{k+1}$. Thus, the second-order consistent splitting scheme takes the form

$$\begin{aligned}&\frac{3\boldsymbol{u}^{k+1}-4\boldsymbol{u}^{k}+\boldsymbol{u}^{k-1}}{2\delta t}-\nu\Delta\boldsymbol{u}^{k+1}+\nabla(2p^{k}-p^{k-1})=\boldsymbol{g}^{k+1},\ \boldsymbol{u}^{k+1}|_{\partial\Omega}=0,\\&(\nabla\psi^{k+1},\nabla q)=\Big(\frac{3\boldsymbol{u}^{k+1}-4\boldsymbol{u}^{k}+\boldsymbol{u}^{k-1}}{2\delta t},\nabla q\Big),\quad \forall q\in H^{1}(\Omega),\end{aligned} \tag{9.118}$$

with

$$p^{k+1}=\psi^{k+1}+(2p^{k}-p^{k-1})-\nu\nabla\cdot\boldsymbol{u}^{k+1}. \tag{9.119}$$

Ample numerical results presented in Guermond and Shen (2003) (see also Johnston and Liu (2004)) indicate that this scheme provides truly second-order accurate approximation for both the velocity and the pressure. However, a rigorous proof of this statement is still not available (cf. Guermond et al. (2006) and Liu et al. (2007)).

9.4.3 Full Discretization

It is straightforward to discretize in space the two schemes presented above. For a rectangular domain, we can use, for instance, the spectral-Galerkin method described in Sect. 8.1. To fix the idea, let $\Omega=(-1,1)^d$ and set

$$X_N=P_N^d\cap H_0^1(\Omega)^d,\quad M_N=\Big\{q\in P_{N-2}:\int_\Omega q=0\Big\}. \tag{9.120}$$

Then, the scheme (9.102)-(9.103)-(9.105) can be implemented as follows:

- **Step 1:** Find $\tilde{\boldsymbol{u}}_N^{k+1}\in X_N$ such that

$$\begin{aligned}\frac{3}{2\delta t}(\tilde{\boldsymbol{u}}_N^{k+1},\boldsymbol{v}_N)+\nu(\nabla\tilde{\boldsymbol{u}}_N^{k+1},\nabla\boldsymbol{v}_N)&=\frac{1}{2\delta t}(4\boldsymbol{u}_N^k-\boldsymbol{u}_N^{k-1}-\nabla(2p_N^k-p_N^{k-1}),\boldsymbol{v}_N)\\&+(I_N(\boldsymbol{f}^{k+1}-2\boldsymbol{u}_N^k\cdot\nabla\boldsymbol{u}_N^k+\boldsymbol{u}_N^{k-1}\cdot\nabla\boldsymbol{u}_N^{k-1}),\boldsymbol{v}_N),\quad\forall\boldsymbol{v}_N\in X_N;\end{aligned} \tag{9.121}$$

- **Step 2:** Find $\phi_N^{k+1}\in M_N$ such that

$$(\nabla\phi_N^{k+1},\nabla q_N)=\frac{3}{2\delta t}(\tilde{\boldsymbol{u}}_N^{k+1},\nabla q_N),\quad\forall q_N\in M_N; \tag{9.122}$$

- **Step 3:** Set

$$\begin{aligned}\boldsymbol{u}_N^{k+1}&=\tilde{\boldsymbol{u}}_N^{k+1}-\frac{2\delta t}{3}\nabla\phi_N^{k+1},\\p_N^{k+1}&=\phi_N^{k+1}+p_N^k-\nu\nabla\cdot\tilde{\boldsymbol{u}}_N^{k+1}.\end{aligned} \tag{9.123}$$

The scheme (9.118)-(9.119) can be implemented in a similar way:

- **Step 1:** Find $\boldsymbol{u}_N^{k+1}\in X_N$ such that

$$\begin{aligned}\frac{3}{2\delta t}(\boldsymbol{u}_N^{k+1},\boldsymbol{v}_N)+\nu(\nabla\boldsymbol{u}_N^{k+1},\nabla\boldsymbol{v}_N)&=\frac{1}{2\delta t}(4\boldsymbol{u}_N^k-\boldsymbol{u}_N^{k-1}-\nabla(2p_N^k-p_N^{k-1}),\boldsymbol{v}_N)\\&+(I_N(\boldsymbol{f}^{k+1}-2\boldsymbol{u}_N^k\cdot\nabla\boldsymbol{u}_N^k+\boldsymbol{u}_N^{k-1}\cdot\nabla\boldsymbol{u}_N^{k-1}),\boldsymbol{v}_N),\quad\forall\boldsymbol{v}_N\in X_N;\end{aligned} \tag{9.124}$$

- **Step 2:** Find $\phi_N^{k+1}\in M_N$ such that

$$(\nabla\phi_N^{k+1},\nabla q_N)=\frac{1}{2\delta t}(3\boldsymbol{u}_N^{k+1}-4\boldsymbol{u}_N^k+\boldsymbol{u}_N^{k-1},\nabla q_N),\ \forall q_N\in M_N; \tag{9.125}$$

- **Step 3:** Set

$$p_N^{k+1}=\phi_N^{k+1}+2p_N^k-p_N^{k-1}-\nu\Pi_{N-2}\nabla\cdot\boldsymbol{u}_N^{k+1}, \tag{9.126}$$

  where Π_{N-2} is the L^2-projection operator onto M_{N-2}.

Hence, at each time step, the two spectral-projection schemes presented above only involve a Poisson equation for the velocity and a Poisson equation for the pressure.

9.5 Axisymmetric Flows in a Cylinder

In this section, we apply the spectral-projection method presented above to simulate an incompressible flow inside a cylinder. To simplify the presentation, we assume that the flow is axisymmetric, so we are effectively dealing with a two-dimensional problem. We refer to Lopez et al. (2002) for the extension to full three-dimensional incompressible flows in a cylinder. For more details on the physical background of this problem and its numerical simulations, we refer to Lopez and Perry (1992), Lopez and Shen (1998) and the references therein.

9.5.1 Governing Equations and the Time Discretization

Consider a flow in an enclosed cylinder with the height H and radius R. The flow is driven by a bottom rotation rate of Ω rad s^{-1}. We shall non-dimensionalize the governing equations with the radius of the cylinder R as the length scale and $1/\Omega$ as the time scale. The Reynolds number is then $Re = \Omega R^2/\nu$, where ν is the kinematic viscosity. The flow is governed by another non-dimensional parameter, the aspect ratio of the cylinder $\Lambda = H/R$. Therefore, the domain for the space variables (r,z) is the rectangle

$$\mathscr{D} = \{(r,z) : r \in (0,1) \text{ and } z \in (0,\Lambda)\}.$$

Let (u,v,w) be the velocity field in the cylindrical polar coordinates (r,θ,z) and assume the flow is axisymmetric, i.e., independent of the azimuthal θ direction. The Navier–Stokes equation (9.101) governing this axisymmetric flow in the cylindrical polar coordinates read (cf. Lopez and Shen (1998))

$$u_t + uu_r + wu_z - \frac{1}{r}v^2 = -p_r + \frac{1}{Re}\left(\tilde{\nabla}^2 u - \frac{1}{r^2}u\right), \tag{9.127}$$

$$v_t + uv_r + wv_z + \frac{1}{r}uv = \frac{1}{Re}\left(\tilde{\nabla}^2 v - \frac{1}{r^2}v\right), \tag{9.128}$$

$$w_t + uw_r + ww_z = -p_z + \frac{1}{Re}\tilde{\nabla}^2 w, \tag{9.129}$$

$$\frac{1}{r}(ru)_r + w_z = 0, \tag{9.130}$$

where

$$\tilde{\nabla}^2 = \partial_r^2 + \frac{1}{r}\partial_r + \partial_z^2 \tag{9.131}$$

is the Laplace operator in axisymmetric cylindrical coordinates. The boundary conditions for the velocity components are zero everywhere except that (a) $v = r$ at $\{z = 0\}$ which is the bottom of the cylinder, and (b) $w_r = 0$ at $\partial\mathscr{D}\backslash\{z = 0\}$.

To simplify the presentation, we introduce the following notation:

$$\tilde{\Delta} = \begin{pmatrix} \tilde{\nabla}^2 - 1/r^2, & 0, & 0 \\ 0, & \tilde{\nabla}^2 - 1/r^2, & 0 \\ 0, & 0, & \tilde{\nabla}^2 \end{pmatrix}, \quad \tilde{\nabla} = \begin{pmatrix} \partial_r \\ 0 \\ \partial_z \end{pmatrix},$$

$$\Gamma_1 = \{(r,z) : r \in (0,1) \text{ and } z = 0\}, \quad \Gamma_2 = \{(r,z) : r = 0 \text{ and } z \in (0,\Lambda)\},$$

and rewrite (9.127)–(9.130) in vector form,

$$\begin{aligned} &\boldsymbol{u}_t + \boldsymbol{N}(\boldsymbol{u}) = -\tilde{\nabla} p + \frac{1}{Re}\tilde{\Delta}\boldsymbol{u}, \\ &\tilde{\nabla}\cdot\boldsymbol{u} := \frac{1}{r}(ru)_r + w_z = 0, \\ &\boldsymbol{u}|_{\partial\mathscr{D}\setminus(\Gamma_1\cup\Gamma_2)} = \boldsymbol{0}, \quad \boldsymbol{u}|_{\Gamma_1} = (0,r,0)^T, \quad (u,v,w_r)^T|_{\Gamma_2} = \boldsymbol{0}, \end{aligned} \tag{9.132}$$

where $\boldsymbol{u} = (u,v,w)^T$ and $\boldsymbol{N}(\boldsymbol{u})$ is the vector containing the nonlinear terms in (9.127)-(9.129).

To overcome the difficulties associated with the nonlinearity and the coupling of velocity components and the pressure, we adapt the following semi-implicit second-order rotational pressure-correction scheme (cf. Sect. 9.4) for the system of equations (9.132):

$$\begin{cases} \frac{1}{2\delta t}(3\tilde{\boldsymbol{u}}^{K+1}\mathbf{u} - 4\boldsymbol{u}^k + \boldsymbol{u}^{k-1}) - \frac{1}{Re}\tilde{\Delta}\tilde{\boldsymbol{u}} \\ \qquad = -\tilde{\nabla}p^k - (2\boldsymbol{N}(\boldsymbol{u}^k) - \boldsymbol{N}(\boldsymbol{u}^{k-1})), \\ \tilde{\boldsymbol{u}}^{k+1}|_{\partial\mathscr{D}\setminus(\Gamma_1\cup\Gamma_2)} = \boldsymbol{0}, \quad \tilde{\boldsymbol{u}}^{k+1}|_{\Gamma_1} = (0,r,0)^T, \\ (\tilde{u}^{k+1},\tilde{v}^{k+1},\tilde{w}_r^{k+1})^T|_{\Gamma_2} = \boldsymbol{0}, \end{cases} \tag{9.133}$$

$$\begin{cases} \frac{3}{2\delta t}(\boldsymbol{u}^{k+1} - \tilde{\boldsymbol{u}}^{k+1}) + \tilde{\nabla}\phi^{k+1} = \boldsymbol{0}, \\ \tilde{\nabla}\cdot\boldsymbol{u}^{k+1} = 0, \\ (\boldsymbol{u}^{k+1} - \tilde{\boldsymbol{u}}^{k+1})\cdot\boldsymbol{n}|_{\partial\mathscr{D}} = 0, \end{cases} \tag{9.134}$$

and

$$p^{k+1} = p^k + \phi^{k+1} - \frac{1}{Re}\tilde{\nabla}\cdot\boldsymbol{u}^{k+1}, \tag{9.135}$$

where δt is the time step, $\boldsymbol{n}$ is the outward normal at the boundary, and $\tilde{\boldsymbol{u}}^{k+1} = (\tilde{u}^{k+1},\tilde{v}^{k+1},\tilde{w}^{k+1})^T$ and $\boldsymbol{u}^{k+1} = (u^{k+1},v^{k+1},w^{k+1})^T$ are respectively the intermediate and final approximations of $\mathbf{u}$ at time $t = (k+1)\delta t$.

It is easy to see that $\tilde{\boldsymbol{u}}^{k+1}$ can be determined from (9.133) by solving three Helmholtz-type equations. Instead of solving for $(\boldsymbol{u}^{k+1},\phi^{k+1})$ from the coupled first-order differential equation (9.134), we apply the operator "$\tilde{\nabla}\cdot$" (see the definition in (9.132)) to the first equation in (9.134) to obtain an equivalent system

$$\begin{aligned} &\tilde{\nabla}^2\phi^{k+1} = \frac{3}{2\delta t}\tilde{\nabla}\cdot\tilde{\boldsymbol{u}}^{k+1}, \\ &\partial_{\boldsymbol{n}}\phi^{k+1}|_{\partial\mathscr{D}} = 0, \end{aligned} \tag{9.136}$$

and

$$\boldsymbol{u}^{k+1} = \tilde{\boldsymbol{u}}^{k+1} - \frac{2\delta t}{3}\tilde{\nabla}\phi^{k+1}. \tag{9.137}$$

Thus, $(\boldsymbol{u}^{k+1}, \phi^{k+1})$ can be obtained by solving an additional Poisson equation (9.136).

Next, we apply the spectral-Galerkin method for solving these equations.

9.5.1.1 Spatial Discretization

We first transform the domain $\mathscr{D}$ to the unit square $\mathscr{D}^* = (-1,1)\times(-1,1)$ by using the transformations $r = (y+1)/2$ and $z = \Lambda(x+1)/2$. Then, at each time step, the systems (9.133) and (9.136) lead to the following four Helmholtz-type equations:

$$\begin{aligned} &\alpha u - \beta u_{xx} - \frac{1}{y+1}((y+1)u_y)_y + \frac{\gamma}{(y+1)^2}u = f \quad \text{in } \mathscr{D}^*,\\ &u|_{\partial\mathscr{D}^*} = 0; \end{aligned} \tag{9.138}$$

$$\begin{aligned} &\alpha v - \beta v_{xx} - \frac{1}{y+1}((y+1)v_y)_y + \frac{\gamma}{(y+1)^2}v = g \quad \text{in } \mathscr{D}^*,\\ &v|_{\partial\mathscr{D}^*\setminus\Gamma_1^*} = 0, \quad v|_{\Gamma_1^*} = \frac{1}{2}(y+1); \end{aligned} \tag{9.139}$$

$$\begin{aligned} &\alpha w - \beta w_{xx} - \frac{1}{y+1}((y+1)w_y)_y = h, \quad \text{in } \mathscr{D}^*,\\ &w|_{\partial\mathscr{D}^*\setminus\Gamma_2^*} = 0, \quad w_r|_{\Gamma_2^*} = 0; \end{aligned} \tag{9.140}$$

and

$$\begin{aligned} &-\beta p_{xx} - \frac{1}{y+1}((y+1)p_y)_y = q \quad \text{in } \mathscr{D}^*,\\ &\partial_{\boldsymbol{n}} p|_{\partial\mathscr{D}^*} = 0. \end{aligned} \tag{9.141}$$

In the above,

$$\begin{aligned} &\Gamma_1^* = \{(x,y) : x = -1,\ y\in(-1,1)\}, \quad \Gamma_2^* = \{(x,y) : x\in(-1,1), y=-1\},\\ &\alpha = \frac{3}{8}Re/\delta t, \quad \beta = \Lambda^{-2}, \quad \gamma = 1, \end{aligned}$$

and f, g, h, q are known functions depending on the solutions at the two previous time steps.

The spectral-Galerkin method presented in Sect. 8.2 (cf. Shen (1997)) can be directly applied to (9.138)-(9.141). We next discuss the method for solving (9.138) in some detail. The other three equations can be treated similarly.

Let P_K be the space of all polynomials of degree $\le K$, and set $P_{NM} = P_N \times P_M$. Define

$$X_{NM} = \{w \in P_{NM} : w|_{\partial \mathscr{D}^*} = 0\}.$$

Then the spectral-Galerkin method for (9.138) is

$$\begin{cases} \text{Find } u_{NM} \in X_{NM} \text{ such that} \\ \alpha((y+1)u_{NM}, v)_{\tilde{\omega}} - \beta((y+1)\partial_x^2 u_{NM}, v)_{\tilde{\omega}} - (((y+1)\partial_y u_{NM})_y, v)_{\tilde{\omega}} \\ +\gamma(\frac{1}{y+1} u_{NM}, v)_{\tilde{\omega}} = ((y+1)f, v)_{\tilde{\omega}}, \quad \forall v \in X_{NM}, \end{cases} \tag{9.142}$$

where $\tilde{\omega} = \omega(x)\omega(y)$ with $\omega(s)$ being 1 or $(1-s^2)^{-1/2}$, depending on whether the Legendre or Chebyshev polynomials are used. Equation (9.142) is derived by first multiplying (9.138) by $(y+1)\omega(x)\omega(y)$ and then integrating over $\mathscr{D}^*$. The multiplication by $(y+1)$ is natural since the Jacobian of the transformation from the Cartesian coordinates to cylindrical coordinates is $r = (y+1)/2$ in the axisymmetric case. Since $u_{NM} = 0$ at $y = -1$, we see that all terms in (9.142) are well defined and that no singularity is present.

For this problem, it is easy to verify that

$$X_{NM} = \text{span}\{\phi_i(x)\rho_j(y) : i = 0,1,\ldots,N-2;\ j = 0,1,\ldots,M-2\},$$

with $\phi_l(s) = \rho_l(s) = p_l(s) - p_{l+2}(s)$ where $p_l(s)$ is either the l-th degree Legendre or Chebyshev polynomial. Set

$$u_{NM} = \sum_{i=0}^{N-2}\sum_{j=0}^{M-2} u_{ij}\phi_i(x)\rho_j(y),$$

and

$$\begin{aligned} a_{ij} &= \int_{-1}^{1} \phi_j(x)\,\phi_i(x)\,\omega(x)\,dx, \quad b_{ij} = -\int_{-1}^{1} \phi_j''(x)\,\phi_i(x)\,\omega(x)\,dx, \\ c_{ij} &= \int_{-1}^{1} (y+1)\,\rho_j(y)\,\rho_i(y)\,\omega(y)\,dy, \\ d_{ij} &= -\int_{-1}^{1} ((y+1)\,\rho_j'(y))'\,\rho_i(y)\,\omega(y)\,dy, \\ e_{ij} &= \int_{-1}^{1} \frac{1}{y+1}\rho_j(y)\,\rho_i(y)\,\omega(y)\,dy, \\ f_{ij} &= \int_{\mathscr{D}^*} (y+1) f \rho_j(y)\phi_i(x)\,\omega(x)\,\omega(y)\,dxdy, \end{aligned} \tag{9.143}$$

and let A, B, C, D, E, F and U be the corresponding matrices with entries given above. Then (9.142) is equivalent to the matrix system

$$\alpha AUC + \beta BUC + AUD + \gamma AUE = F. \tag{9.144}$$

Note that e_{ij} is well defined in spite of the term $\frac{1}{y+1}$, since $\rho_i(-1) = 0$. In the Legendre case, the matrices A, B, C, D, and E are all symmetric and sparsely banded. The entries of these matrices are either given in Sect. 8.2 and/or can be easily computed by using the properties of Legendre and Chebyshev polynomials.

9.5.2 Treatment for the Singular Boundary Condition

The boundary condition for v is discontinuous at the lower right corner $(r = 1, z = 0)$. This singular boundary condition is a mathematical idealization of the physical situation, where there is a thin gap over which v adjusts from 1 on the edge of the rotating endwall to 0 on the sidewall. Therefore, it is appropriate to use a regularized boundary condition (so that v is continuous) which is representative of the actual gap between the rotating endwall and the stationary sidewall in experiments.

In finite difference or finite element schemes, the singularity is usually regularized over a few grid spacings in the neighborhood of the corner in an *ad hoc* manner. However, this simple treatment leads to a mesh-dependent boundary condition which in turn results in mesh-dependent solutions which prevents a sensible comparison between solutions with different meshes. Essentially, the grid spacing represents the physical gap size.

The singular boundary condition at $r = 1$ is

$$v(z) = 1 \text{ at } z = 0, \quad v(z) = 0 \quad \text{for } 0 < z \leq \Lambda,$$

Table 9.1 Largest negative values of the angular momentum $\Gamma = rv$ on the grid points of a 201×501 uniform mesh, corresponding to the solutions for Stokes flow shown in Fig. 9.4

N, M	$\min(\Gamma)$ with $\varepsilon = 0.006$	$\min(\Gamma)$ with *ad hoc* B.C.
56, 80	-2.472×10^{-6}	-4.786×10^{-3}
48, 64	-9.002×10^{-6}	-6.510×10^{-3}
40, 48	-1.633×10^{-4}	-6.444×10^{-3}

which is similar to that of the driven cavity problem. Unless this singularity is treated appropriately, spectral methods may have severe difficulty dealing with it. In the past, most computations with spectral methods avoided this difficulty by using regularized boundary conditions which, unfortunately, do not approximate the physical boundary condition (see, e.g., Shen (1991), Demaret and Deville (1991)). A sensible approach is to use the boundary layer function

$$v_\varepsilon(z) = \exp\left(-\frac{2z}{\Lambda\varepsilon}\right),$$

which has the ability to approximate the singular boundary condition to within any prescribed accuracy. Outside a boundary layer of width $O(\varepsilon)$, $v_\varepsilon(z)$ converges to $v(z)$ exponentially as $\varepsilon \to 0$. However, for a given ε, approximately $\varepsilon^{-1/2}$ colloca-

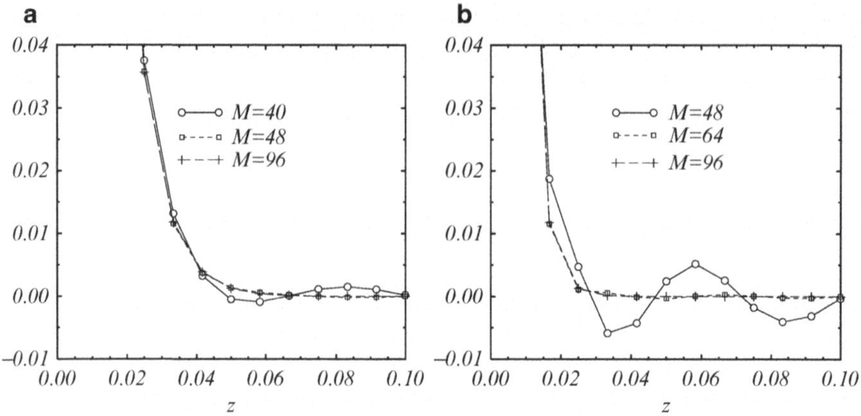

Fig. 9.3 Variation of $I_M v_\varepsilon$ (with $\Lambda = 2.5$) in the vicinity of the singularity at $z = 0$ for (**a**) $\varepsilon = 0.006$ and (**b**) $\varepsilon = 0.003$, and various M as indicated

tion points are needed to represent the boundary layer function v_ε. In other words, for a fixed number of modes M, we can only use $\varepsilon \geq \varepsilon(M)$ where $\varepsilon(M)$ can be approximately determined by comparing $I_M v_\varepsilon$ and v_ε, where $I_M v_\varepsilon$ is the polynomial interpolant of v_ε at the Gauss-Lobatto points.

Although it is virtually impossible to match the exact physical condition in the experimental gap region, the function v_ε with $\varepsilon = 0.006$ does provide a reasonable representation of the experimental gap. The function v_ε can be resolved spectrally with $M \geq M_\varepsilon$ modes, where M_ε is such that $I_M v_\varepsilon$ for a given ε is non-oscillatory. Due to the nonlinear term v^2/r in (9.127), we also require $I_M v_{\varepsilon/2}$ to be non-oscillatory (since $(v_\varepsilon)^2 = v_{\varepsilon/2}$). Figure 9.3a shows $I_M v_{0.006}$ for various M. It is clear that $I_{48} v_{0.006}$ is non-oscillatory. However, from Fig. 9.3b we see that $I_{48} v_{0.003}$ is oscillatory near $z = 0$, while $I_{64} v_{0.003}$ is not. Thus, $M \approx 64$ is required for $\varepsilon = 0.006$.

Figure 9.4 shows plots of the solution for Stokes flow ($Re = 0$) for this problem. The governing equations (9.127)-(9.130) in the case $Re = 0$ reduce to

$$\tilde{\nabla}^2 v - \frac{1}{r^2} v = \tilde{\nabla}^2 \Gamma = 0,$$

with $\Gamma = 0$ on the axis, top endwall and sidewall, and $\Gamma = r^2$ on the rotating bottom endwall. The singular boundary condition on the sidewall has been regularized in Fig. 9.4a with $v_{0.006}$ and in Fig. 9.4b with the *ad hoc* method. For the solution of the Stokes problem with $\varepsilon = 0.006$, we judge that the error is acceptably small at $M = 64$ and is very small at $M = 80$. The measure of error used here is the largest value of negative Γ of the computed solution at the grid points of a uniform 201×501 mesh; the true solution has $\Gamma \geq 0$. These values are listed in Table 9.1. In contrast, with the *ad hoc* method the error does not decrease as M increases and the computed solutions exhibit large errors for all values of M considered. We refer to Lopez and Shen (1998) for more detail on this problem.

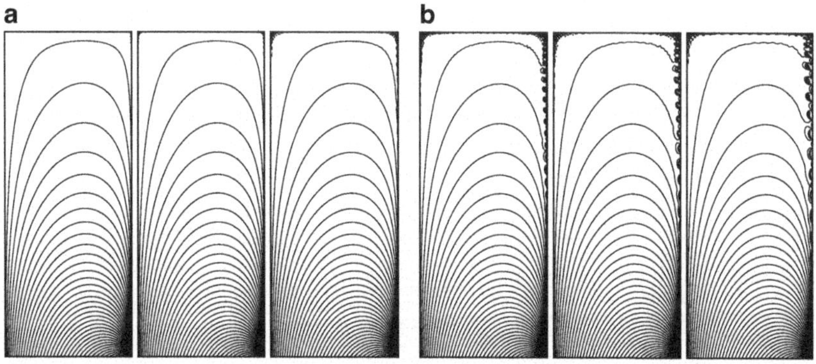

Fig. 9.4 Contours of the angular momentum $\Gamma = rv$ for Stokes flow ($Re = 0$), using $v_{0.006}$ (**a**) and the *ad hoc* (**b**) regularization of the corner singularity. The leftmost plot in each set has $N = 56$, $M = 80$, the middle plots have $N = 48$, $M = 64$, and the right plots have $N = 40$, $M = 48$. All have been projected on to 201 uniform radial locations and 501 uniform axial locations

9.6 Gross-Pitaevskii Equation

The nonlinear Schrödinger equation plays an important role in many fields of mathematical physics. In particular, when the temperature T is much smaller than the critical temperature T_c, a Bose–Einstein condensate (BEC) is well described by the macroscopic wave function $\psi = \psi(\mathbf{x},t)$ whose evolution is governed by a self-consistent, mean field nonlinear Schrödinger equation (NLSE) known as the Gross-Pitaevskii equation (GPE) (cf. Gross (1961), Pitaevskii (1961)). We present in this section a fourth-order time-splitting spectral method, developed in Bao and Shen (2005), for the numerical simulation of BEC. The scheme preserves all essential features of the GPE, such as conservative, time reversible and time transverse invariants, and it is explicit, unconditionally stable, and spectrally accurate in space and fourth-order accurate in time.

9.6.1 GPE and Its Time Discretization

We consider the non-dimensional Gross-Pitaevskii equation of the form

$$\mathrm{i}\,\frac{\partial \psi(\mathbf{x},t)}{\partial t} = -\frac{1}{2}\nabla^2\psi(\mathbf{x},t) + V(\mathbf{x})\psi(\mathbf{x},t) + \beta\,|\psi(\mathbf{x},t)|^2\psi(\mathbf{x},t), \tag{9.145}$$

where the unknown is the complex wave function ψ, $\mathrm{i} = \sqrt{-1}$, β is a positive constant and

$$V(\mathbf{x}) = \left(\gamma_x^2 x^2 + \gamma_y^2 y^2 + \gamma_z^2 z^2\right)/2 \tag{9.146}$$

is the trapping potential. There are two typical extreme regimes between the trap frequencies: (a) $\gamma_x = 1$, $\gamma_y \approx 1$ and $\gamma_z \gg 1$, it is a disk-shaped condensation; (b) $\gamma_x \gg 1$, $\gamma_y \gg 1$ and $\gamma_z = 1$, it is a cigar-shaped condensation. Following the procedure used in Bao et al. (2003b) and Leboeuf and Pavloff (2001), the disk-shaped condensation can be effectively modeled by a 2-D GPE. Similarly, a cigar-shaped condensation can be reduced to a 1-D GPE.

In general, we consider the GPE in d-dimension ($d = 1,2,3$):

$$\begin{aligned} &\mathrm{i}\,\frac{\partial \psi(\mathbf{x},t)}{\partial t} = -\frac{1}{2}\nabla^2\psi + V_d(\mathbf{x})\psi + \beta_d\,|\psi|^2\psi, \quad \mathbf{x}\in\mathbb{R}^d, \\ &\psi(\mathbf{x},0) = \psi_0(\mathbf{x}), \quad \mathbf{x}\in\mathbb{R}^d, \end{aligned} \tag{9.147}$$

with

$$\beta_d = \begin{cases} \sqrt{\gamma_x\gamma_y/2\pi}, \\ \sqrt{\gamma_z/2\pi}, \\ 1, \end{cases} \quad V_d(\mathbf{x}) = \begin{cases} \gamma_z^2 z^2/2, & d=1, \\ \left(\gamma_x^2x^2+\gamma_y^2y^2\right)/2, & d=2, \\ \left(\gamma_x^2x^2+\gamma_y^2y^2+\gamma_z^2z^2\right)/2, & d=3, \end{cases} \tag{9.148}$$

where γ_x, γ_y and γ_z are positive constants. It is easy to check the conservation:

$$\|\psi(\cdot,t)\|^2 := \int_{\mathbb{R}^d} |\psi(\mathbf{x},t)|^2\,d\mathbf{x} \equiv \int_{\mathbb{R}^d} |\psi_0(\mathbf{x})|^2\,d\mathbf{x}. \tag{9.149}$$

For convenience, we normalize the initial condition as

$$\int_{\mathbb{R}^d} |\psi_0(\mathbf{x})|^2\,d\mathbf{x} = 1. \tag{9.150}$$

Since the GPE is time reversible and time transverse invariant (cf. Bao et al. (2003b)), it is desirable to design the numerical scheme that preserves these properties as well.

For the time discretization, we shall use the fourth-order splitting scheme (D.31). For this purpose, we rewrite the GPE (9.147) in the form

$$\psi_t = f(\psi) := -\mathrm{i}A\psi - \mathrm{i}B\psi \quad \text{with} \quad \psi(\mathbf{x},0) = \psi_0(\mathbf{x}), \tag{9.151}$$

where

$$A\psi = \beta_d\,|\psi(\mathbf{x},t)|^2\psi(\mathbf{x},t); \quad B\psi = -\frac{1}{2}\nabla^2\psi(\mathbf{x},t) + V_d(\mathbf{x})\psi(\mathbf{x},t). \tag{9.152}$$

The key idea is to efficiently solve the following two sub-problems:

$$\mathrm{i}\frac{\partial \psi(\mathbf{x},t)}{\partial t} = A\psi(\mathbf{x},t), \quad \mathbf{x}\in\mathbb{R}^d, \tag{9.153}$$

and

$$\mathrm{i}\frac{\partial \psi(\mathbf{x},t)}{\partial t} = B\psi(\mathbf{x},t), \quad \mathbf{x}\in\mathbb{R}^d; \qquad \lim_{|\mathbf{x}|\to+\infty} \psi(\mathbf{x},t) = 0, \tag{9.154}$$

where the operators A and B are defined by (9.152).

Multiplying (9.153) by $\overline{\psi}$, and taking the imaginary part from both sides of the resulting equation, we find that $\partial_t|\psi|^2 = 0$, which implies that $|\psi(\mathbf{x},t)|$ is invariant in t. Hence, for $t \geq t_s$ (for any given t_s), (9.153) becomes

$$\mathrm{i}\frac{\partial \psi(\mathbf{x},t)}{\partial t} = \beta_d\, |\psi(\mathbf{x},t_s)|^2 \psi(\mathbf{x},t), \quad \mathbf{x} \in \mathbb{R}^d, \tag{9.155}$$

which can be integrated **exactly**. More precisely, we have

$$\psi(\mathbf{x},t) = e^{-\mathrm{i}\beta_d |\psi(\mathbf{x},t_s)|^2 (t-t_s)} \psi(\mathbf{x},t_s), \quad t \geq t_s, \ \mathbf{x} \in \mathbb{R}^d. \tag{9.156}$$

Now, let $= \{\mathbf{x}_k : k \in \Sigma_K\}$ be a set of collocation points, τ be a time step size. Let $\psi(\mathbf{x},t)$ be the exact solution of (9.154) with $\psi(\mathbf{x},0) = \psi_0(\mathbf{x})$. Then, the fourth-order time-splitting spectral-collocation method for the GPE (9.151) is as follows:
Let ψ_k^n be the approximation of $\psi(\mathbf{x}_k,t_n)$, and $\mathscr{F}_\tau(w,\psi_0)(\mathbf{x})$ be a spectral approximation (to be specified below) of $\psi(\mathbf{x}, w\tau)$. We compute ψ_k^{n+1} by

$$\begin{aligned}
&\psi_k^{(1)} = e^{-2\mathrm{i}w_1\tau\beta_d|\psi_k^n|^2}\psi_k^n, && \psi_k^{(2)} = \mathscr{F}_\tau(w_2,\psi^{(1)})(\mathbf{x}_k),\\
&\psi_k^{(3)} = e^{-2\mathrm{i}w_3\tau\beta_d|\psi_k^{(2)}|^2}\psi_k^{(2)}, && \psi_k^{(4)} = \mathscr{F}_\tau(w_4,\psi^{(3)})(\mathbf{x}_k),\\
&\psi_k^{(5)} = e^{-2\mathrm{i}w_3\tau\beta_d|\psi_k^{(4)}|^2}\psi_k^{(4)}, && \psi_k^{(6)} = \mathscr{F}_\tau(w_2,\psi^{(5)})(\mathbf{x}_k),\\
&\psi_k^{n+1} = e^{-2\mathrm{i}w_1\tau\beta_d|\psi_k^{(6)}|^2}\psi_k^{(6)}, && k \in \Sigma_K,
\end{aligned} \tag{9.157}$$

where w_i, $i = 1,2,3,4$ are given in (D.33).

It remains to construct an efficient and accurate scheme to obtain $\mathscr{F}_\tau(w,\psi_0)$ for (9.154). We shall construct below suitable spectral basis functions which are eigenfunctions of B so that $e^{-\mathrm{i}B\Delta t}\psi$ can be evaluated exactly (which is necessary for the full scheme to be time reversible and time transverse invariant).

9.6.2 Hermite-Collocation Method for the 1-D GPE

In one-dimensional case, (9.154) is reduced to

$$\begin{aligned}
&\mathrm{i}\,\frac{\partial \psi}{\partial t} = B\psi = -\frac{1}{2}\frac{\partial^2 \psi}{\partial z^2} + \frac{\gamma_z^2 z^2}{2}\psi, \quad z \in \mathbb{R}, \ t > 0;\\
&\lim_{|z|\to+\infty} \psi(z,t) = 0, \ t \geq 0,
\end{aligned} \tag{9.158}$$

with the normalization (9.150):

$$\|\psi(\cdot,t)\|^2 = \int_{-\infty}^{\infty} |\psi(z,t)|^2 dz \equiv \int_{-\infty}^{\infty} |\psi_0(z)|^2 dz = 1. \tag{9.159}$$

Since the problem (9.158) is posed on the whole line, it is natural to use a spectral method based on Hermite functions. Although the standard Hermite functions could be used as basis functions here, they are not the most appropriate ones. Below, we construct properly scaled Hermite functions which are eigenfunctions of B. Let $\{H_l(z)\}$ be the standard Hermite polynomials (cf. (7.58)). Define the scaled Hermite function

$$h_l(z) = \frac{\gamma_z^{1/4}}{\pi^{1/4}\sqrt{2^l\, l!}} H_l\big(\sqrt{\gamma_z} z\big) e^{-\gamma_z z^2/2}, \quad z \in \mathbb{R}. \tag{9.160}$$

It follows from (7.57) and (7.58) that

$$\int_{-\infty}^{\infty} h_l(z) h_n(z)\, dz = \frac{1}{\sqrt{\pi 2^l l! 2^n n!}} \int_{-\infty}^{\infty} H_l(z) H_n(z) e^{-z^2}\, dz = \delta_{ln}, \tag{9.161}$$

and

$$B h_l(z) = -\frac{1}{2} h_l''(z) + \frac{\gamma_z^2 z^2}{2} h_l(z) = \mu_l^z h_l(z), \quad \mu_l^z = \frac{2l+1}{2}\gamma_z. \tag{9.162}$$

Hence, $\{h_l\}$ are the eigenfunctions of B defined in (9.158).

We now describe the Gauss quadrature associated with the scaled Hermite functions. Let $\{x_k, \omega_k\}_{k=0}^N$ be the Hermite-Gauss points and weights given in Theorem 7.3. We define the scaled Hermite-Gauss points and weights by

$$z_k = x_k/\sqrt{\gamma_z}, \quad \omega_k^z = \omega_k\, e^{x_k^2}/\sqrt{\gamma_z}, \quad 0 \le k \le N. \tag{9.163}$$

We then derive from (7.80) and (9.160) the (discrete) orthogonality:

$$\sum_{k=0}^{N} h_l(z_k)\, h_n(z_k) \omega_k^z = \sum_{k=0}^{N} \frac{H_l(x_k)}{\pi^{1/4}\sqrt{2^l\, l!}} \frac{H_n(x_k)}{\pi^{1/4}\sqrt{2^n n!}} \omega_k = \delta_{ln}, \quad 0 \le l, n \le N. \tag{9.164}$$

Define $X_N = \mathrm{span}\{h_l : l = 0, 1, \ldots, N\}$. The Hermite-collocation method for (9.158) is:
Find $\psi_N(z,t) \in X_N$, i.e.,

$$\psi_N(z,t) = \sum_{l=0}^{N} \hat{\psi}_l(t)\, h_l(z), \tag{9.165}$$

such that

$$\mathrm{i}\frac{\partial \psi_N}{\partial t}(z_k,t) = B\psi(z_k,t) = -\frac{1}{2}\frac{\partial^2 \psi_N}{\partial z^2}(z_k,t) + \frac{\gamma_z^2 z_k^2}{2}\psi_N(z_k,t), \quad 0 \le k \le N. \tag{9.166}$$

Note that $\lim_{|z|\to+\infty} h_l(z) = 0$ (cf. Sect. 7.2), so the decay condition $\lim_{|z|\to+\infty} \psi_N(z,t) = 0$ is automatically satisfied.

Plugging (9.165) into (9.166), thanks to (9.164) and (9.162), we find

$$\mathrm{i}\frac{d\hat{\psi}_l(t)}{dt} = \mu_l^z\, \hat{\psi}_l(t) = \frac{2l+1}{2}\gamma_z\, \hat{\psi}_l(t), \quad l = 0, 1, \ldots, N. \tag{9.167}$$

Hence,

$$\hat{\psi}_l(t) = e^{-\mathrm{i}\mu_l^z(t-t_s)}\hat{\psi}_l(t_s), \quad t \geq t_s. \tag{9.168}$$

In other words, let $\phi(z) = \sum_{l=0}^{N} \hat{\phi}_l \, h_l(z)$, we have

$$\mathscr{F}_\tau(w,\phi)(z) = \sum_{l=0}^{N} e^{-\mathrm{i}\mu_l^z w\tau}\hat{\phi}_l \, h_l(z). \tag{9.169}$$

Since each of the sub-problems (9.153) and (9.154) is conservative and our numerical scheme (9.157) with (9.169) solves the two sub-problems exactly in the discrete space, one can easily establish the following result (cf. Bao and Shen (2005)).

Lemma 9.1. *The time-splitting Hermite-collocation method* (9.157) *with* (9.169) *preserves the conservation* (9.149), *i.e.,*

$$\|\psi^n\|_{l^2}^2 = \sum_{k=0}^{N} \omega_k^z |\psi_k^n|^2 = \sum_{k=0}^{N} \omega_k^z |\psi_0(z_k)|^2 = \|\psi_0\|_{l^2}^2, \quad n = 0, 1, \ldots. \tag{9.170}$$

We leave the proof of this lemma as an exercise (see Problem 9.6).

9.6.3 Laguerre Method for the 2-D GPE with Radial Symmetry

In the 2-D case with radial symmetry, i.e., $d = 2$, $\gamma_x = \gamma_y$ and $\psi_0(x,y) = \psi_0(r)$ (with $r = \sqrt{x^2+y^2}$) in (9.147)–(9.148), we can write the solution of (9.147)–(9.148) as $\psi(x,y,t) = \psi(r,t)$. Therefore, (9.154) becomes

$$\begin{aligned}
&\mathrm{i}\frac{\partial \psi(r,t)}{\partial t} = B\psi(r,t) = -\frac{1}{2r}\frac{\partial}{\partial r}\Big(r\frac{\partial \psi(r,t)}{\partial r}\Big) + \frac{\gamma_r^2 r^2}{2}\psi(r,t), \\
&\lim_{r\to\infty} \psi(r,t) = 0,
\end{aligned} \tag{9.171}$$

where $\gamma_r = \gamma_x = \gamma_y$. The normalization (9.150) reduces to

$$\|\psi(\cdot,t)\|^2 = 2\pi\int_0^\infty |\psi(r,t)|^2 r\,dr \equiv 2\pi\int_0^\infty |\psi_0(r)|^2 r\,dr = 1. \tag{9.172}$$

Note that it can be shown, similarly as for the Poisson equation in a 2-D disk (cf. Shen (1997)), that the problem (9.171) admits a unique solution without any condition at the pole $r = 0$.

Since (9.171) is posed on a semi-infinite interval, it is natural to consider Laguerre functions. Again the standard Laguerre functions need to be properly scaled so that they are the eigenfunctions of B.

Let $\{\mathscr{L}_m\}$ be the usual Laguerre polynomials as defined in (7.3). Recall the properties:

$$\begin{aligned} &\int_0^\infty \mathscr{L}_m(r)\mathscr{L}_n(r)e^{-r}dr = \delta_{mn}, \\ &r\mathscr{L}''_m(r) + (1-r)\mathscr{L}'_m(r) + m\mathscr{L}_m(r) = 0. \end{aligned} \tag{9.173}$$

We define the scaled Laguerre functions by

$$\hat{L}_m(r) = \sqrt{\frac{\gamma_r}{\pi}}\, e^{-\gamma_r r^2/2}\, \mathscr{L}_m(\gamma_r r^2), \quad 0 \le r < \infty. \tag{9.174}$$

It follows from (9.173) and (9.174) that

$$2\pi \int_0^\infty \hat{L}_m(r)\hat{L}_n(r) r\, dr = \int_0^\infty \mathscr{L}_m(r)\mathscr{L}_n(r)e^{-r}\, dr = \delta_{mn}, \tag{9.175}$$

and

$$-\frac{1}{2r}\frac{\partial}{\partial r}\Big(r\frac{\partial \hat{L}_m(r)}{\partial r}\Big) + \frac{1}{2}\gamma_r^2 r^2 \hat{L}_m(r) = \mu_m^r \hat{L}_m(r), \;\; \mu_m^r = \gamma_r(2m+1). \tag{9.176}$$

Hence, $\{\hat{L}_m\}$ are the eigenfunctions of B defined in (9.171).

We now introduce the Gauss-Radau quadrature associated with the scaled Laguerre functions. Let $\{x_j^{(0)}, \omega_j^{(0)}\}_{j=0}^M$ be the Laguerre-Gauss-Radau points and weights given in Theorem 7.1. We have from (7.1) and (7.29) that

$$\sum_{j=0}^M \mathscr{L}_m(x_j^{(0)})\mathscr{L}_n(x_j^{(0)})\,\omega_j^{(0)} = \delta_{mn}, \quad 0 \le n, m \le M. \tag{9.177}$$

Define the corresponding scaled Laguerre-Gauss-Radau points and weights by

$$r_j = \sqrt{x_j^{(0)}/\gamma_r}, \quad \omega_j^r = \pi\omega_j^{(0)} e^{x_j^{(0)}}/\gamma_r, \quad 0 \le j \le M. \tag{9.178}$$

Hence, we have from (9.174) and (9.177) that

$$\sum_{j=0}^M \hat{L}_m(r_j)\hat{L}_n(r_j)\omega_j^r = \sum_{j=0}^M \mathscr{L}_m(x_j^{(0)})\mathscr{L}_n(x_j^{(0)})\,\omega_j^{(0)} = \delta_{mn}, \tag{9.179}$$

for all $0 \le m, n \le M$.

Let $Y_M = \text{span}\{\hat{L}_m : m = 0, 1, \ldots, M\}$. The Laguerre-collocation method for (9.171) is:
Find $\psi_M(r,t) \in Y_M$, i.e.,

$$\psi_M(r,t) = \sum_{m=0}^M \hat{\psi}_m(t)\hat{L}_m(r), \quad 0 \le r < \infty, \tag{9.180}$$

such that

$$\mathrm{i}\,\frac{\partial \psi_M}{\partial t}(r_j,t) = B\psi_M(r_j,t) = -\frac{1}{2r}\frac{\partial}{\partial r}\Big(r\frac{\partial \psi_M}{\partial r}\Big)(r_j,t) + \frac{\gamma_r^2 r_j^2}{2}\psi_M(r_j,t), \tag{9.181}$$

for $0 \le j \le M$. Note that $\lim_{|r|\to\infty} \hat{L}_m(r) = 0$ (cf. (7.22)), so ψ_M automatically meets $\psi_M \to 0$ as $|r| \to \infty$. Plugging (9.180) into (9.181), we find from (9.179) and (9.176) that

$$\mathrm{i}\frac{d\hat{\psi}_m(t)}{dt} = \mu_m^r \hat{\psi}_m(t) = \gamma_r(2m+1)\hat{\psi}_m(t), \quad 0 \le m \le M. \tag{9.182}$$

Hence,

$$\hat{\psi}_m(t) = e^{-\mathrm{i}\mu_m^r(t-t_s)}\hat{\psi}_m(t_s), \quad t \ge t_s. \tag{9.183}$$

In other words, let $\phi(r) = \sum_{m=0}^{M} \hat{\phi}_m \hat{L}_m(r)$, we have

$$\mathscr{F}_\tau(w,\phi)(r) = \sum_{m=0}^{M} e^{-\mathrm{i}\mu_m^r w\tau}\hat{\phi}_m \hat{L}_m(r). \tag{9.184}$$

Similar to Lemma 9.1, we have the following stability result.

Lemma 9.2. *The time-splitting Laguerre-collocation method* (9.157) *with* (9.184) *preserves the conservation* (9.149)*, i.e.,*

$$\|\psi^n\|_{l^2}^2 = \sum_{j=0}^{M} \omega_j^r |\psi_j^n|^2 = \sum_{j=0}^{M} \omega_j^r |\psi_0(r_j)|^2 = \|\psi_0\|_{l^2}^2, \quad n \ge 0.$$

9.6.4 Laguerre-Hermite Method for the 3-D GPE with Cylindrical Symmetry

Consider now the 3-D case with cylindrical symmetry, i.e., $d = 3, \gamma_x = \gamma_y$ and $\psi_0(x,y,z) = \psi_0(r,z)$ in (9.147)-(9.148), its solution with $d = 3$ is of the form $\psi(x,y,z,t) = \psi(r,z,t)$. Therefore, (9.154) becomes

$$\begin{aligned} \mathrm{i}\frac{\partial\psi(r,z,t)}{\partial t} &= B\psi(r,z,t) = -\frac{1}{2}\left[\frac{1}{r}\frac{\partial}{\partial r}\left(r\frac{\partial\psi}{\partial r}\right) + \frac{\partial^2\psi}{\partial z^2}\right] \\ &+ \frac{1}{2}\left(\gamma_r^2 r^2 + \gamma_z^2 z^2\right)\psi, \quad 0 < r < \infty, \ -\infty < z < \infty, \ t > 0, \\ &\lim_{r\to\infty}\psi(r,z,t) = 0, \quad \lim_{|z|\to\infty}\psi(r,z,t) = 0, \qquad t \ge 0, \end{aligned} \tag{9.185}$$

where $\gamma_r = \gamma_x = \gamma_y$. The normalization (9.150) now is

$$\|\psi(\cdot,t)\|^2 = 2\pi\int_0^\infty\int_{-\infty}^\infty |\psi(r,z,t)|^2 r\,dzdr \equiv \|\psi_0\|^2 = 1. \tag{9.186}$$

Since the two-dimensional computational domain here is a tensor product of a semi-infinite interval and the whole line, it is natural to combine the Hermite-collocation and Laguerre-collocation methods. In particular, the product of scaled Hermite and

Laguerre functions $\{\hat{L}_m(r)h_l(z)\}$ are eigenfunctions of B defined in (9.185), since we derive from (9.162) and (9.176) that

$$\begin{aligned}
-\frac{1}{2}\left[\frac{1}{r}\frac{\partial}{\partial r}\left(r\frac{\partial}{\partial r}\right)+\frac{\partial^2}{\partial z^2}\right](\hat{L}_m(r)\,h_l(z))+\frac{1}{2}\left(\gamma_r^2r^2+\gamma_z^2z^2\right)(\hat{L}_m(r)\,h_l(z))\\
=\left[-\frac{1}{2r}\frac{d}{dr}\left(r\frac{d\hat{L}_m(r)}{dr}\right)+\frac{1}{2}\gamma_r^2r^2\hat{L}_m(r)\right]h_l(z)\\
+\left[-\frac{1}{2}\frac{d^2h_l(z)}{dz^2}+\frac{1}{2}\gamma_z^2z^2h_l(z)\right]\hat{L}_m(r)\\
=\mu_m^r\hat{L}_m(r)h_l(z)+\mu_l^z h_l(z)\hat{L}_m(r)=(\mu_m^r+\mu_l^z)\hat{L}_m(r)h_l(z).
\end{aligned}\tag{9.187}$$

Now, let $X_{MN}=\text{span}\{\hat{L}_m(r)h_l(z):\,m=0,1,\ldots,M,\ l=0,1,\ldots,N\}$. The Laguerre-Hermite collocation method for (9.185) is:
Find $\psi_{MN}(r,z,t)\in X_{MN}$, i.e.,

$$\psi_{MN}(r,z,t)=\sum_{m=0}^{M}\sum_{l=0}^{N}\tilde{\psi}_{ml}(t)\,\hat{L}_m(r)\,h_l(z),\tag{9.188}$$

such that, for all $0\le j\le M,\ 0\le k\le N$,

$$\begin{aligned}
\mathrm{i}\frac{\partial\psi_{MN}}{\partial t}(r_j,z_k,t)=B\psi_{MN}(r_j,z_k,t)\\
=-\frac{1}{2}\left[\frac{1}{r}\frac{\partial}{\partial r}\left(r\frac{\partial\psi_{MN}}{\partial r}\right)+\frac{\partial^2\psi_{MN}}{\partial z^2}\right](r_j,z_k,t)\\
+\frac{1}{2}\left(\gamma_r^2r_j^2+\gamma_z^2z_k^2\right)\psi_{MN}(r_j,z_k,t).
\end{aligned}\tag{9.189}$$

Inserting (9.188) into (9.189), we find from (9.162), (9.176) and (9.187) that

$$\mathrm{i}\frac{d\tilde{\psi}_{ml}(t)}{dt}=\left(\mu_m^r+\mu_l^z\right)\tilde{\psi}_{ml}(t),\quad 0\le m\le M,\ 0\le l\le N.\tag{9.190}$$

Hence,

$$\tilde{\psi}_{ml}(t)=e^{-\mathrm{i}(\mu_m^r+\mu_l^z)(t-t_s)}\,\tilde{\psi}_{ml}(t_s),\quad t\ge t_s.\tag{9.191}$$

In other words, let $\phi(r,z)=\sum_{m=0}^{M}\sum_{l=0}^{N}\hat{\phi}_{ml}\,\hat{L}_m(r)h_l(z)$, we have

$$\mathscr{F}_\tau(w,\phi)(r,z)=\sum_{m=0}^{M}\sum_{l=0}^{N}e^{-\mathrm{i}(\mu_m^r+\mu_l^z)w\tau}\hat{\phi}_{ml}\,\hat{L}_m(r)h_l(z).\tag{9.192}$$

We have the following stability result.

Lemma 9.3. *The time-splitting Laguerre-Hermite collocation method* (9.157) *with* (9.192) *preserves the conservation* (9.149), *i.e.,*

$$\begin{aligned}
\|\psi^n\|_{l^2}^2=\sum_{j=0}^{M}\sum_{k=0}^{N}|\psi_{jk}^n|^2\omega_j^r\omega_k^z=\sum_{j=0}^{M}\sum_{k=0}^{N}|\psi_0(r_j,z_k)|^2\omega_j^r\omega_k^z\\
=\|\psi_0\|_{l^2}^2,\quad n\ge 0.
\end{aligned}\tag{9.193}$$

9.6.5 Numerical Results

We now present some numerical results. We define the condensate width along the r- and z-axis as

$$\sigma_\alpha^2 = \int_{\mathbb{R}^d} \alpha^2 |\psi(\mathbf{x},t)|\, d\mathbf{x}, \quad \alpha = x,y,z, \quad \sigma_r^2 = \sigma_x^2 + \sigma_y^2.$$

Example 9.1. **The 1-D Gross-Pitaevskii equation**. We choose $d = 1$, $\gamma_z = 2$, and $\beta_1 = 50$ in (9.147). The initial data $\psi_0(z)$ is chosen as the ground state of the 1-D GPE (9.147) with $d = 1$, $\gamma_z = 1$ and $\beta_1 = 50$. This corresponds to an experimental setup where initially the condensate is assumed to be in its ground state, and the trap frequency is doubled at $t = 0$.

We solve this problem by using (9.157) with $N = 31$ and time step $k = 0.001$. Figure 9.5 plots the condensate width and central density $|\psi(0,t)|^2$ as functions of time. Our numerical experiments also show that the scheme (9.157) with $N = 31$ gives similar numerical results as the TSSP method (cf. Bao et al. (2003a)) for this example, with 513 grid points over the interval $[-12,12]$ and time step $\tau = 0.001$.

In order to test the 4th-order accuracy in time of (9.157), we compute a numerical solution with a very fine mesh, e.g., $N = 81$, and a very small time step, e.g., $\tau = 0.0001$, as the 'exact' solution ψ. Let ψ^τ denote the numerical solution under $N = 81$ and time step τ. Since N is large enough, the truncation error from space discretization is negligible compared to that from time discretization. Table 9.2 shows the errors $\max|\psi(t) - \psi^\tau(t)|$ and $\|\psi(t) - \psi^\tau(t)\|_{l^2}$ at $t = 2.0$ for different time steps τ. The results in Table 9.2 demonstrate the 4th-order accuracy in time of (9.157).

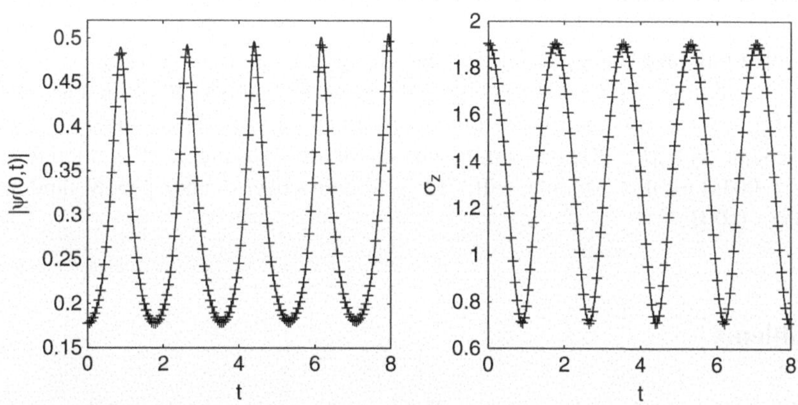

Fig. 9.5 Evolution of central density and condensate width in Example 9.1. '–': 'exact solutions' obtained by the TSSP in Bao et al. (2003a) with 513 grid points over an interval $[-12,12]$; '+++': Numerical results by (9.157) with 31 grid points on the whole z-axis. (**a**). Central density $|\psi(0,t)|^2$; (**b**). Condensate width σ_z

Table 9.2 Time discretization errors of (9.157) at $t = 2$ with $N = 81$

τ	1/40	1/80	1/160	1/320
$\max\lvert\psi(t)-\psi^{\tau}(t)\rvert$	0.1619	4.715E-6	3.180E-7	2.036E-8
$\lVert\psi(t)-\psi^{\tau}(t)\rVert_{l^2}$	0.2289	7.379E-6	4.925E-7	3.215E-8

Example 9.2. **The 2-D Gross-Pitaevskii equation with radial symmetry**. We choose $d = 2$, $\gamma_r = \gamma_x = \gamma_y = 2$, $\beta_2 = 50$ in (9.147). The initial data $\psi_0(r)$ is chosen as the ground state of the 2-D GPE (9.147) with $d = 2$, $\gamma_r = \gamma_x = \gamma_y = 1$ and $\beta_2 = 50$. Again this corresponds to an experimental setup where initially the condensate is assumed to be in its ground state, and the trap frequency is doubled at $t = 0$.

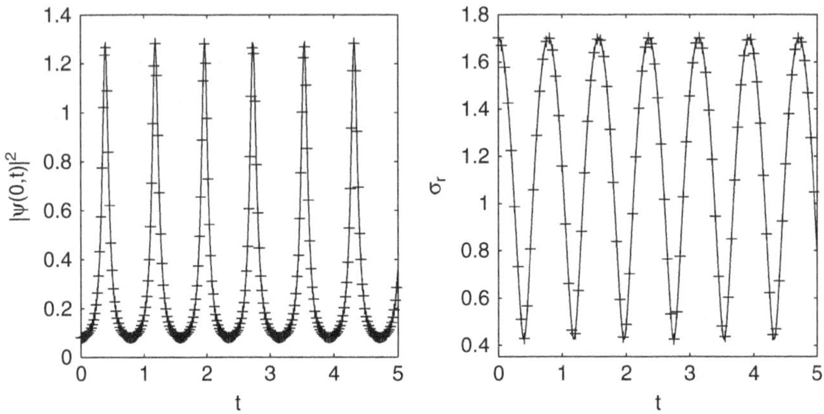

Fig. 9.6 Evolution of central density and condensate width '–': 'exact solutions' obtained by TSSP in Bao et al. (2003a) with 513^2 grid points over a box $[-8,8]^2$; '+ + + ': Numerical results by our scheme with 30 grid points on the semi-infinite interval $[0,\infty)$. (**a**). Central density $|\psi(0,t)|^2$; (**b**). Condensate width σ_r

We solve this problem by using the time splitting Laguerre-spectral method with $M = 30$ and time step $k = 0.001$. Figure 9.6 plots the condensate width and central density $|\psi(0,t)|^2$ as functions of time. Our numerical experiments also show that our scheme with $M = 30$ gives similar numerical results as the TSSP method in Bao et al. (2003a) for this example, with 513^2 grid points over the box $[-8,8]^2$ and time step $k = 0.001$.

Problems

9.1. State and prove results similar to Theorem 9.3 for the scheme (9.83).

9.2. Implement the Uzawa algorithm for solving the Stokes problem using $M_N = \{q \in P_{N-2} : (q,1) = 0\}$ and $M_N = \{q \in P_{[\lambda N]} : (q,1) = 0\}$ for $\lambda = 0.7, 0.8, 0.9$ with $N = 16, 32, 64, 128$. Explain your results.

9.3. Prove the statement (9.56).

9.4. Write a program implementing the rotational pressure-correction scheme and consistent splitting scheme using P_N for the velocity and P_{N-2} for the pressure. Let the exact solution $(\boldsymbol{u}, p)$ of (9.101) be

$$\boldsymbol{u}(x,y,t) = \pi \sin t(\sin 2\pi y \sin^2 \pi x, -\sin 2\pi x \sin^2 \pi y), \quad p(x,y,t) = \sin t \cos \pi x \sin \pi y.$$

Compare the errors of the velocity and pressure at time $t = 1$ in both the L^2-norm and H^1-norm using the two schemes with $N = 32$ for $\delta t = 0.1, 0.05, 0.025, 0.0125$. Explain your results.

9.5. Use the rotational pressure correction scheme to compute the steady state solution of the regularized driven cavity problem, i.e., $\Omega = (0,1)^2$ with the boundary condition

$$\boldsymbol{u}|_{y=1} = (16x^2(1-x^2), 0), \quad \boldsymbol{u}|_{\partial\Omega \setminus \{y=1\}} = 0.$$

Take $N = 32$ and $Re = 1/\nu = 400$. Compare your results with the benchmark results in Shen (1991).

9.6. Show the stability result in Lemma 9.1.

Appendix A
Properties of the Gamma Functions

We list here some basic properties of the Gamma function (see, e.g., Abramowitz and Stegun (1964)), defined by

$$\Gamma(z) = \int_0^{\infty} t^{z-1} e^{-t} dt, \quad \forall z \in \mathbb{C} \text{ with } \mathrm{Re}(z) > 0. \tag{A.1}$$

In particular, we have $\Gamma(1) = 1$ and $\Gamma(1/2) = \sqrt{\pi}$.

- Recursion formula:

$$\Gamma(z+1) = z\Gamma(z), \quad \Gamma(n+1) = n!, \tag{A.2}$$

  and

$$\Gamma(2z) = (2\pi)^{-1/2} 2^{2z-1/2} \Gamma(z)\Gamma(z+1/2). \tag{A.3}$$

- Connection with binomial coefficient:

$$\binom{z}{w} = \frac{\Gamma(z+1)}{\Gamma(w+1)\Gamma(z-w+1)}. \tag{A.4}$$

- Relation with Beta function:

$$B(x,y) = \int_0^1 t^{x-1}(1-t)^{y-1} dt = \frac{\Gamma(x)\Gamma(y)}{\Gamma(x+y)}, \quad x,y > 0. \tag{A.5}$$

  In particular, for $\alpha, \beta > -1$,

$$\begin{aligned} \int_{-1}^1 (1-x)^{\alpha}(1+x)^{\beta} dx &= 2^{\alpha+\beta+1} \int_0^1 t^{\beta}(1-t)^{\alpha} dt \\ &= 2^{\alpha+\beta+1} \frac{\Gamma(\alpha+1)\Gamma(\beta+1)}{\Gamma(\alpha+\beta+2)}. \end{aligned} \tag{A.6}$$

J. Shen et al., *Spectral Methods: Algorithms, Analysis and Applications*, Springer Series in Computational Mathematics 41, DOI 10.1007/978-3-540-71041-7,

- Stirling's formula:

$$\Gamma(x) = \sqrt{2\pi} x^{x-1/2} e^{-x} \left\{ 1 + \frac{1}{12x} + \frac{1}{288x^2} + O(x^{-3}) \right\}, \quad x \gg 1. \tag{A.7}$$

Moreover, we have

$$\sqrt{2\pi} n^{n+1/2} < n! e^n < \sqrt{2\pi} n^{n+1/2} \left(1 + \frac{1}{4n} \right), \quad n \geq 1. \tag{A.8}$$

Appendix B
Essential Mathematical Concepts

We provide here some essential mathematical concepts which have been used in the mathematical analysis throughout the book. For a more comprehensive presentation, we refer to Yosida (1980) and Adams (1975).

Let $(X;d)$ be a metric space. A sequence $\{x_k\}$ in X is called a *Cauchy sequence*, if

$$d(x_k, x_l) \to 0 \quad \text{as} \quad k, l \to \infty.$$

The space $(X;d)$ is said to be a *complete space* if every Cauchy sequence in X converges to an element in X.

B.1 Banach Space

Definition B.1. *Given a (real) vector space X, a* norm *on X is a function $\|\cdot\| : X \to \mathbb{R}$ such that*

- $\|u+v\| \le \|u\| + \|v\|, \quad \forall u, v \in X;$
- $\|\alpha u\| = |\alpha|\|u\|, \quad \forall u \in X$ *and* $\forall \alpha \in \mathbb{R};$
- $\|u\| \ge 0, \quad \forall u \in X;$
- $\|u\| = 0$ *if and only if* $u = 0.$

In particular, a semi-norm *on X is a function $|\cdot| : X \to \mathbb{R}$ satisfying the first three conditions.*

The space $(X, \|\cdot\|)$ is called a normed vector space. A Banach space is a normed vector space which is complete with respect to the metric:

$$d(u, v) = \|u - v\|, \quad \forall u, v \in X.$$

J. Shen et al., *Spectral Methods: Algorithms, Analysis and Applications*,
Springer Series in Computational Mathematics 41, DOI 10.1007/978-3-540-71041-7,

B.2 Hilbert Space

Definition B.2. *Let X be a real vector space. An inner product on X is a function $(u,v): X \times X \to \mathbb{R}$ such that*

- $(u,v) = (v,u), \quad \forall u,v \in X;$
- $(\alpha u + \beta v, w) = \alpha(u,w) + \beta(v,w), \quad \forall u,v,w \in X$ *and* $\forall \alpha, \beta \in \mathbb{R};$
- $(u,u) \geq 0, \quad \forall u \in X;$
- $(u,u) = 0$ *if and only if* $u = 0$.

Two elements $u,v \in X$ are said to be *orthogonal* in X, if $(u,v) = 0$. The inner product $(\cdot,\cdot)$ induces a *norm* on X, given by

$$\|u\| = \sqrt{(u,u)}, \quad \forall u \in X.$$

Correspondingly, the metric on X can be defined by $d(u,v) = \|u - v\|$.

A *Hilbert space* is a Banach space endowed with an inner product (i.e., every Cauchy sequence in X is convergent with respect to the induced norm).

In a Hilbert space, the Cauchy–Schwarz inequality holds:

$$|(u,v)| \leq \|u\| \|v\|, \quad \forall u,v \in X. \tag{B.1}$$

Remark B.1. *If X is a complex vector space, the inner product (u,v) is a complex valued function. In the Definition B.2, the first condition should be replaced by*

$$(u,v) = \overline{(v,u)}, \quad \forall u,v \in X.$$

Next, we introduce the dual space of a Banach/Hilbert space X.

Definition B.3. *A functional $F: X \to \mathbb{R}$ is said to be* linear *or* continuous, *if there exists a constant $c > 0$ such that*

$$|F(u)| \leq c\|u\|, \quad \forall u \in X. \tag{B.2}$$

Let X' be the set of all linear functionals on X, and define the norm

$$\|F\|_{X'} = \sup_{u \in X; u \neq 0} \frac{|F(u)|}{\|u\|}.$$

Then the space X' is a Banach space, which is called the *dual space* of X.

The bilinear form $F(u) = \langle F, u \rangle : X' \times X \to \mathbb{R}$, is called the *duality pairing* on $X' \times X$. If X is a Hilbert space, then its dual space X' is a Hilbert space as well. Moreover, according to the *Riesz Representation Theorem*, X and X' are isometric, and X' can be canonically identified to X. More precisely, for any linear functional $F \in X'$, there exists a unique $u \in X$ such that

$$F(v) = \langle F, v \rangle = (u,v), \quad \forall v \in X \quad \text{and} \quad \|F\|_{X'} = \|u\|.$$

In a normed space X, a sequence $\{v_n\} \subset X$ is (strongly) convergent to $v \in X$, if $\|v_n - v\| \to 0$ as $n \to \infty$. It is possible to introduce another type of convergence in a weaker sense.

Definition B.4. *A sequence $\{v_n\}$ in X is called weakly convergent to $v \in X$, if $F(v_n) \to F(v)$ in $\mathbb{R}$ for all $F \in X'$.*

If a sequence $\{v_n\}$ converges to v in X, it is also weakly convergent. The converse is not true unless X is a finite dimensional space.

In a dual space X', a sequence of functional $\{F_n\}$ in X' is called *weakly* convergent* to $F \in X'$, if $\{F_n(v)\}$ converges to $F(v)$ for all $v \in X$. The weak convergence implies the weak* convergence.

B.3 Lax-Milgram Lemma

Definition B.5. *Let X be a Hilbert space with norm $\|\cdot\|$. A functional $a(\cdot,\cdot) : X \times X \to \mathbb{R}$ defines a* bilinear form, *if for any $u, v, w \in X$ and $\alpha, \beta \in \mathbb{R}$,*

$$a(\alpha u + \beta v, w) = \alpha a(u,w) + \beta a(v,w),$$
$$a(u, \alpha v + \beta w) = \alpha a(u,w) + \beta a(v,w).$$

That is, for any fixed u, both the functionals $a(u,\cdot) : X \to \mathbb{R}$ and $a(\cdot,u) : X \to \mathbb{R}$ are linear. The bilinear form is symmetric, *if $a(u,v) = a(v,u)$ for any $u, v \in X$.*

Definition B.6. *A bilinear form $a(\cdot,\cdot)$ on a Hilbert space X is said to be* continuous, *if there exists a constant $C > 0$ such that*

$$|a(u,v)| \le C\|u\|\,\|v\|, \quad \forall u, v \in X, \tag{B.3}$$

and coercive *on X, if there exists a constant $\alpha > 0$ such that*

$$a(u,u) \ge \alpha \|u\|^2, \quad \forall u \in X. \tag{B.4}$$

It is clear that if $a(\cdot,\cdot)$ is symmetric, continuous and coercive on the Hilbert space X, then $a(\cdot,\cdot)$ defines an inner product on X.

Theorem B.1. (Lax-Milgram lemma). *Let X be a Hilbert space, let $a(\cdot,\cdot) : X \times X \to \mathbb{R}$ be a continuous and coercive bilinear form, and let $F : X \to \mathbb{R}$ be a linear functional in X'. Then the variational problem:*

$$\begin{cases} \text{Find } u \in X \text{ such that} \\ a(u,v) = F(v), \quad \forall v \in X, \end{cases} \tag{B.5}$$

has a unique solution. Moreover, we have

$$\|u\| \le \frac{1}{\alpha}\|F\|_{X'}. \tag{B.6}$$

B.4 L^p-Space

Let Ω be a Lebesgue-measurable subset of $\mathbb{R}^d$ ($d = 1,2,3$) with non-empty interior, and let u be a Lebesgue measurable function on Ω. In what follows, the integration is in the Lebesgue sense.

Definition B.7. *For $1 \le p \le \infty$, let*

$$L^p(\Omega) := \left\{ u \;:\; u \text{ is measurable on } \Omega \text{ and } \|u\|_{L^p(\Omega)} < \infty \right\}, \tag{B.7}$$

where for $1 \le p < \infty$,

$$\|u\|_{L^p(\Omega)} := \left(\int_\Omega |u(x)|^p dx \right)^{1/p}, \tag{B.8}$$

and

$$\|u\|_{L^\infty(\Omega)} := \operatorname{ess}\sup_{x\in\Omega} |u(x)|. \tag{B.9}$$

Remark B.2. *Some remarks are in order.*

(i) *The space $L^\infty(\Omega)$ consists all functions that are essentially bounded on Ω. A function u is said to be* essentially bounded *on Ω, if there exists a constant K such that $|u(x)| \le K$ a.e. on Ω. The greatest lower bound of such constants K is called the* essential supremum *of $|u(x)|$ on Ω, denoted by $\|u\|_{L^\infty(\Omega)}$.*
(ii) *We identify functions in $L^p(\Omega)$ that are equal almost everywhere on Ω. The elements of $L^p(\Omega)$ are equivalence classes of measurable functions that satisfy (B.7) with the equivalence relation: $u \equiv v$, if they only differ on a measurable subset of measure zero.*

Equipped with the norm $\|\cdot\|_{L^p(\Omega)}$, the space $L^p(\Omega)$ with $1 \le p \le \infty$ is a Banach space. In particular, the space $L^2(\Omega)$ is a Hilbert space equipped with the inner product

$$(u,v)_{L^2(\Omega)} = \int_\Omega u(x)v(x)dx, \quad \forall u,v \in L^2(\Omega). \tag{B.10}$$

Definition B.8. *If p and q are positive real numbers such that*

$$p + q = pq \quad \text{or} \quad \frac{1}{p} + \frac{1}{q} = 1, \tag{B.11}$$

then we call (p,q) a pair of conjugate exponents. *As $p \to 1$, (B.11) forces $q \to \infty$. Consequently, $(1,\infty)$ is also regarded as a pair of conjugate exponents.*

Theorem B.2.

- **Minkowski's inequality**. *If $u,v \in L^p(\Omega)$ with $1 \le p \le \infty$, then $u+v \in L^p(\Omega)$, and*

$$\|u+v\|_{L^p(\Omega)} \le \|u\|_{L^p(\Omega)} + \|v\|_{L^p(\Omega)}. \tag{B.12}$$

- **Hölder's inequality**. *Let p and q be conjugate exponents with $1 \le p \le \infty$. If $u \in L^p(\Omega)$ and $v \in L^q(\Omega)$, then $uv \in L^1(\Omega)$, and*

$$\int_\Omega |u(x)v(x)|dx \le \|u\|_{L^p(\Omega)}\|v\|_{L^q(\Omega)}. \tag{B.13}$$

In particular, if $p = 2$, the Hölder's inequality reduces to the Cauchy–Schwarz inequality (B.1).

It follows from (B.13) that $L^q(\Omega) \subset L^p(\Omega)$, if $p \le q$ and Ω has a finite measure.

As a final remark, given a weight function $\omega(x)$, which is almost everywhere positive and Lebesgue integrable on Ω, $\omega(x)dx$ also defines a Lebesgue measure on Ω. Replacing dx in (B.8) by $\omega(x)dx$, we define the norm $\|\cdot\|_{L^p_\omega(\Omega)}$ and the space $L^p_\omega(\Omega)$ with $1 \le p < \infty$, which is a Banach space. In particular, the space $L^2_\omega(\Omega)$ is a Hilbert space with the inner product and norm given by

$$(u,v)_\omega = \int_\Omega u(x)v(x)\omega(x)dx, \quad \|u\|_\omega = \sqrt{(u,u)_\omega}.$$

One verifies that the inequalities (B.12) and (B.13) hold in the weighted norms.

B.5 Distributions and Weak Derivatives

A multi-index $\boldsymbol{\alpha} = (\alpha_1, \ldots, \alpha_d)$ is a d-tuple of non-negative integers $\{\alpha_i\}$. Denote $|\boldsymbol{\alpha}| = \sum_{i=1}^d \alpha_i$, and define the partial derivative operator

$$D^{\boldsymbol{\alpha}} = \frac{\partial^{|\boldsymbol{\alpha}|}}{\partial x_1^{\alpha_1} \ldots \partial x_d^{\alpha_d}}.$$

For any $x = (x_1, \ldots, x_d) \in \mathbb{R}^d$, we define $x^{\boldsymbol{\alpha}} = x_1^{\alpha_1} \ldots x_d^{\alpha_d}$.

For any $\Omega \subset \mathbb{R}^d$, let $\mathscr{D}(\Omega)$ (or $C_0^\infty(\Omega)$) be the set of all infinitely differentiable functions with compact support in Ω. With the aid of $\mathscr{D}(\Omega)$, we can extend the conventional derivatives to the notion of generalized (weak) derivatives.

We first recall the topology on the vector space $C_0^\infty(\Omega)$.

Definition B.9. *A sequence of functions $\{v_n\}$ in $C_0^\infty(\Omega)$ is said to be* convergent in the sense of $\mathscr{D}(\Omega)$ *to the function $v \in C_0^\infty(\Omega)$ provided that*

(i) *there exists $K \subset\subset \Omega$, which means the closure $\bar{K} \subset \Omega$ and $\bar{K}$ is compact (i.e., closed and bounded), such that the support of $v_n - v \subset K$ for every n;*
(ii) $\lim_{n\to\infty} D^{\boldsymbol{\alpha}} v_n(x) = D^{\boldsymbol{\alpha}} v(x)$ *uniformly on K for each multi-index $\boldsymbol{\alpha}$.*

Definition B.10. *The dual space $\mathscr{D}'(\Omega)$ of $\mathscr{D}(\Omega)$ is called the space of (Schwarz)* distributions. *A sequence of distributions $\{T_n\}$ in $\mathscr{D}'(\Omega)$ is called weakly* convergent to a distribution $T \in \mathscr{D}'(\Omega)$, if $T_n(v) \to T(v)$ in $\mathbb{R}$ for every $v \in \mathscr{D}(\Omega)$.*

For example, we consider the distributions induced by *locally integrable* functions.

Definition B.11. *Given a domain $\Omega \subset \mathbb{R}^d$, the set of all* locally integrable *functions is denoted by*

$$L^1_{\text{loc}}(\Omega) = \{u : u \in L^1(K), \forall \text{ compact } K \subset \text{interior } \Omega\}. \qquad \text{(B.14)}$$

Corresponding to every $u \in L^1_{\text{loc}}(\Omega)$, there is a distribution $T_u \in \mathscr{D}'(\Omega)$ defined by

$$T_u(v) = \int_\Omega u(x)v(x)dx, \quad \forall v \in \mathscr{D}(\Omega). \qquad \text{(B.15)}$$

However, not every distribution $T \in \mathscr{D}'(\Omega)$ is of the form (B.15), and for instance, the Delta function δ is such a distribution.

Definition B.12. *Let T be a distribution in $\mathscr{D}'(\Omega)$, and let $\boldsymbol{\alpha}$ be a multi-index. Then $D^{\boldsymbol{\alpha}}T$ is also a distribution in $\mathscr{D}'(\Omega)$, defined as follows*

$$\langle D^{\boldsymbol{\alpha}}T, v\rangle = (-1)^{|\boldsymbol{\alpha}|}\langle T, D^{\boldsymbol{\alpha}}v\rangle, \quad \forall v \in \mathscr{D}(\Omega), \qquad \text{(B.16)}$$

where $\langle\cdot,\cdot\rangle$ is the duality paring of $\mathscr{D}'(\Omega)$ and $\mathscr{D}(\Omega)$.

Notice that by definition, a distribution is infinitely differentiable. Moreover, if T is a smooth function, its generalized derivative coincides with the usual derivative.

Definition B.13. *A given function $u \in L^1_{\text{loc}}(\Omega)$ has a weak derivative $D^{\boldsymbol{\alpha}}u$, if there exists a function $w \in L^1_{\text{loc}}(\Omega)$ such that*

$$\int_\Omega w(x)v(x)dx = (-1)^{|\boldsymbol{\alpha}|}\int_\Omega u(x)D^{\boldsymbol{\alpha}}v(x)dx, \quad \forall v \in \mathscr{D}(\Omega). \qquad \text{(B.17)}$$

If such a w exists, we define $D^{\boldsymbol{\alpha}}u = w$.

We can extend the above discussion to periodic distributions. Let $\Omega = (0,2\pi)^d$, and define the space $C_p^\infty(\bar{\Omega})$ as the vector space of functions that are infinitely differentiable with all derivatives being 2π-periodic in each space direction. A sequence $\{\phi_n\}$ in $C_p^\infty(\bar{\Omega})$ converges to a function ϕ in $C_p^\infty(\bar{\Omega})$, if $D^{\boldsymbol{\alpha}}\phi_n \to D^{\boldsymbol{\alpha}}\phi$ uniformly on $\bar{\Omega}$ for every multi-index $\boldsymbol{\alpha}$. Similarly, a *periodic distribution* is a continuous linear form $T : C_p^\infty(\bar{\Omega}) \to \mathbb{C}$, that is, $\langle T, \phi_n\rangle \to \langle T, \phi\rangle$ in $\mathbb{C}$ whenever $\phi_n \to \phi$ in $C_p^\infty(\bar{\Omega})$. The derivative of a periodic distribution T can be defined by (B.16) with $C_p^\infty(\bar{\Omega})$ in place of $\mathscr{D}(\Omega)$.

B.6 Sobolev Spaces

Using the notion of weak derivatives, we define the Sobolev spaces on the L^p-spaces. Such spaces are most often used for the variational theory of partial differential equations. In what follows, we restrict the discussions to the Hilbert spaces (i.e., with $p = 2$), and refer to Adams (1975) for a comprehensive presentation of general Sobolev spaces.

Definition B.14. *The Sobolev space $H^m(\Omega)$ with $m \in \mathbb{N}$ is the space of functions $u \in L^2(\Omega)$ such that all the distributional derivatives of order up to m can be represented by functions in $L^2(\Omega)$. That is,*

$$H^m(\Omega) = \{u \in L^2(\Omega) \,:\, D^{\boldsymbol{\alpha}} u \in L^2(\Omega) \text{ for } 0 \le |\boldsymbol{\alpha}| \le m\}, \tag{B.18}$$

equipped with the norm and semi-norm

$$\|u\|_{m,\Omega} = \Big(\sum_{|\boldsymbol{\alpha}|=0}^{m} \|D^{\boldsymbol{\alpha}} u\|^2_{L^2(\Omega)} \Big)^{1/2}, \quad |u|_{m,\Omega} = \Big(\sum_{|\boldsymbol{\alpha}|=m} \|D^{\boldsymbol{\alpha}} u\|^2_{L^2(\Omega)} \Big)^{1/2}. \tag{B.19}$$

The space $H^m(\Omega)$ is a Hilbert space endowed with the inner product

$$(u,v)_{m,\Omega} = \sum_{|\boldsymbol{\alpha}|=0}^{m} \int_\Omega D^{\boldsymbol{\alpha}} u(x) D^{\boldsymbol{\alpha}} v(x) dx.$$

The following density property holds (see, e.g., Brenner and Scott (2008)).

Theorem B.3. *For any $\Omega \in \mathbb{R}^d$, $C^\infty(\bar{\Omega})$ is dense in $H^m(\Omega)$ for any integer $m \ge 0$.*

Definition B.15. *For any positive integer m, the space $H_0^m(\Omega)$ is the closure of $C_0^\infty(\Omega)$ with respect to the norm $\|\cdot\|_{m,\Omega}$. The dual space of $H_0^m(\Omega)$ is denoted by $H^{-m}(\Omega)$ with the norm*

$$\|u\|_{-m,\Omega} = \sup_{0 \ne v \in H_0^m(\Omega)} \frac{\langle u, v\rangle}{\|v\|_{m,\Omega}}. \tag{B.20}$$

We have the following *Poincaré-Friedrichs inequality* (see, e.g., Ciarlet (1978)).

Theorem B.4. *Let Ω be a bounded open subset of $\mathbb{R}^d$. Then there exists a positive constant $c(\Omega)$ such that*

$$\|u\|_{0,\Omega} \le c(\Omega)|u|_{1,\Omega}, \quad \forall u \in H_0^1(\Omega), \tag{B.21}$$

which implies that the semi-norm $|\cdot|_{m,\Omega}$ is a norm of $H_0^m(\Omega)$, equivalent to the norm $\|\cdot\|_{m,\Omega}$.

For any real $r > 0$, the Sobolev space $H^r(\Omega)$ can be defined by space interpolation (see, e.g., Bergh and Löfström (1976), Adams (1975) and Lions and Magenes (1968)).

An important property of the Sobolev spaces is the *embedding result*, which indicates the close connection with continuous functions.

Theorem B.5. *Let Ω be a domain in $\mathbb{R}^d$ with Lipschitz boundary $\partial\Omega$, and let $r > n/2$. Then there exists a positive constant C such that*

$$\|u\|_{L^\infty(\Omega)} \leq C\|u\|_{r,\Omega}. \tag{B.22}$$

Moreover, there is a continuous function in the $L^\infty(\Omega)$ equivalence class of u.

Another important result of Sobolev spaces is the so-called *trace theorem*. The trace of a function $u \in H^r(\Omega)$ on the boundary $\partial\Omega$ is meaningful by defining it as the restriction of $\tilde{u}$ on the boundary, where $\tilde{u} \in C^0(\bar{\Omega})$ is among the equivalence class of u.

Theorem B.6. *Let Ω be a domain in $\mathbb{R}^d$ with Lipschitz boundary $\partial\Omega$, and $r > 1/2$.*

(i) *There exits a unique linear continuous map $\gamma_0 : H^r(\Omega) \to H^{r-1/2}(\partial\Omega)$ such that $\gamma_0 v = v|_{\partial\Omega}$ for each $v \in H^r(\Omega) \cap C^0(\bar{\Omega})$.*
(ii) *There exists a linear continuous map $\tau_0 : H^{r-1/2}(\partial\Omega) \to H^r(\Omega)$ such that $\gamma_0\tau_0\phi = \phi$ for each $\phi \in H^{r-1/2}(\partial\Omega)$.*

Analogous results also hold if we consider the trace γ_Γ over a Lipschitz continuous subset Γ of the boundary $\partial\Omega$.

We see that any function in $H^{r-1/2}(\partial\Omega), r > 1/2$, is the trace on $\partial\Omega$ of a function in $H^r(\Omega)$. This provides a characterization of the space $H^{r-1/2}(\partial\Omega)$. In particular, the proceeding theorem indicates that there exists a positive constant c such that

$$\|v\|_{L^2(\partial\Omega)} \leq c\|v\|_{1,\Omega}, \quad \forall v \in H^1(\Omega). \tag{B.23}$$

In addition, if the boundary $\partial\Omega$ is Lipschitz continuous, we can characterize the space $H_0^1(\Omega)$ by

$$H_0^1(\Omega) = \{v \in H^1(\Omega) \ : \ \gamma_0 v = 0\}. \tag{B.24}$$

If Γ is part of $\partial\Omega$, we define

$$H_\Gamma^1(\Omega) = \{v \in H^1(\Omega) \ : \ \gamma_\Gamma v = 0\}. \tag{B.25}$$

It is worthwhile to point out that the Poincaré-Friedrichs inequality (B.21) is also valid for functions in $H_\Gamma^1(\Omega)$, provided that Γ is non-empty.

The *interpolation theorem* is also found useful in the analysis.

Theorem B.7. *Assume that Ω is an open subset of $\mathbb{R}^d$ with a Lipschitz boundary $\partial\Omega$. Let $r_1 < r_2$ be two real numbers and set $r = (1-\theta)r_1 + \theta r_2$ with $0 \leq \theta \leq 1$. Then there exists a constant $c > 0$ such that*

$$\|u\|_{r,\Omega} \leq c\|u\|_{r_1,\Omega}^{1-\theta}\|u\|_{r_2,\Omega}^{\theta}, \quad \forall u \in H^{r_2}(\Omega). \tag{B.26}$$

In the Definition B.14 of the Sobolev space, one can require the functions as well as its distributional derivatives to be square integrable with respect to the measure $\omega(x)dx$ on Ω. This provides a natural framework to deal with the Chebyshev and

Jacobi spectral methods. In a very similar fashion, we define the space $H^m_\omega(\Omega)$, the norm $\|\cdot\|_{m,\omega,\Omega}$ and the semi-norm $|\cdot|_{m,\omega,\Omega}$ by replacing the L^2-space and the norm $\|\cdot\|_{L^2(\Omega)}$ in (B.18) and (B.19) by the weighted L^2-space and the weighted norm $\|\cdot\|_{L^2_\omega(\Omega)}$, respectively. For real $r>0$, the space $H^r_\omega(\Omega)$ is defined by space interpolation as usual. Moreover, its subspace $H^r_{0,\omega}(\Omega)$ can be defined as the closure of $C_0^\infty(\Omega)$ in $H^r_\omega(\Omega)$ as before.

In the analysis of Fourier methods, it is necessary to define Sobolev space of periodic functions. In this framework, the functions are complex-valued and their weak derivatives are in the sense of periodic distribution. In particular, for $\Omega=(0,2\pi)$ and for any integer $m>0$,

$$H^m_p(0,2\pi)=\left\{u\in H^m(0,2\pi)\;:\;u^{(k)}(0)=u^{(k)}(2\pi),\,0\le k\le m-1\right\}. \tag{B.27}$$

In this context, the norm can be characterized by the Fourier coefficients of the underlying function (cf. (2.46) in Chap. 2).

B.7 Integral Identities: Divergence Theorem and Green's Formula

We collect some integral identities of advanced calculus in the setting of Sobolev spaces, which are useful in the formulation of multi-dimensional variational problems.

In what follows, let $\boldsymbol{u}=(u_1,\ldots,u_d)$ be a vector, and let $\nabla=(\partial_{x_1},\ldots,\partial_{x_d})$ be the d-dimensional gradient operator. Assume that Ω is a domain with Lipschitz boundary, and $\boldsymbol{\nu}$ denotes its unit outward normal to $\partial\Omega$.

Lemma B.1. (Divergence Theorem). *Let $\boldsymbol{u}$ be a Lebesgue integrable function on Ω. Then*

$$\int_\Omega \nabla\cdot\boldsymbol{u}\,dx=\int_{\partial\Omega}\boldsymbol{u}\cdot\boldsymbol{\nu}\,d\gamma. \tag{B.28}$$

Applying (B.28) with $\boldsymbol{u}=uw\boldsymbol{e}_i$ with $\boldsymbol{e}_i$ is the ith unit coordinate vector, leads to the following identity.

Lemma B.2. (Green's formula). *Let $u,w\in H^1(\Omega)$. Then for $i=1,\ldots,n$,*

$$\int_\Omega \partial_{x_i}uw\,dx=-\int_\Omega u\partial_{x_i}w\,dx+\int_{\partial\Omega}uw\nu_i\,d\gamma. \tag{B.29}$$

In a vector form, if $\boldsymbol{u}\in(H^1(\Omega))^d$ and $w\in H^1(\Omega)$, we have

$$\int_\Omega \nabla\cdot\boldsymbol{u}w\,dx=-\int_\Omega\boldsymbol{u}\cdot\nabla w\,dx+\int_{\partial\Omega}\boldsymbol{u}\cdot\boldsymbol{\nu}\,w\,d\gamma. \tag{B.30}$$

Applying (B.30) with $\boldsymbol{u}=\nabla\phi$ yields the following formulas.

Lemma B.3. *Let $\phi \in H^2(\Omega)$ and $w \in H^1(\Omega)$. Then*

$$\int_\Omega (-\Delta\phi) w \, dx = \int_\Omega \nabla\phi \cdot \nabla w \, dx - \int_{\partial\Omega} \partial_\nu \phi \, w \, d\gamma. \tag{B.31}$$

If, in addition, $w \in H^2(\Omega)$, we have

$$\int_\Omega \big(w\Delta\phi - \phi\Delta w\big) dx = \int_{\partial\Omega} \big(w\partial_\nu\phi - \phi\partial_\nu w\big) d\gamma. \tag{B.32}$$

B.8 Some Useful Inequalities

We present below some useful embedding inequalities on finite/infinite intervals.

B.8.1 Sobolev-Type Inequalities

Let (a,b) be a finite interval. There holds the *Sobolev inequality*:

$$\max_{x\in[a,b]} |u(x)| \le \Big(\frac{1}{b-a} + 2\Big)^{1/2} \|u\|^{1/2}_{L^2(a,b)} \|u\|^{1/2}_{H^1(a,b)}, \quad \forall u \in H^1(a,b), \tag{B.33}$$

which is also known as the *Gagliardo-Nirenberg interpolation inequality*.

This inequality may take the form.

Lemma B.4. *For any $u \in H^1(a,b)$,*

$$\max_{x\in[a,b]} |u(x)| \le \frac{1}{\sqrt{b-a}} \|u\|_{L^2(a,b)} + \sqrt{b-a}\|u'\|_{L^2(a,b)}. \tag{B.34}$$

Proof. For any $x_1, x_2 \in [a,b]$,

$$|u(x_1) - u(x_2)| \le \int_{x_1}^{x_2} |u'(x)| dx \le \sqrt{b-a}\|u'\|_{L^2(a,b)},$$

which implies $u \in C[a,b]$. Denote $|u(x_*)| = \min_{x\in[a,b]} |u(x)|$, and we have

$$|u(x)| - |u(x_*)| \le \sqrt{b-a}\|u'\|_{L^2(a,b)}.$$

Moreover,

$$|u(x_*)| \le \frac{1}{b-a}\int_a^b |u(x)| dx \le \frac{1}{\sqrt{b-a}} \|u\|_{L^2(a,b)}.$$

A combination of the above two inequalities leads to (B.34). □

Lemma B.5. *Let $\omega = e^{-x}$. For any $u \in H^1_\omega(0,\infty)$ with $u(0)=0$, we have*

$$\|e^{-x/2}u\|_{L^\infty(0,\infty)} \le \sqrt{2}\|u\|_\omega^{1/2}|u|_{1,\omega}^{1/2}, \tag{B.35a}$$

$$\|u\|_\omega \le 2|u|_{1,\omega}. \tag{B.35b}$$

Proof. Since $u(0)=0$, we have

$$\begin{aligned} e^{-x}u^2(x) &= \int_0^x \partial_y(e^{-y}u^2(y))dy \\ &= 2\int_0^x e^{-y}u(y)u'(y)dy - \int_0^x e^{-y}u^2(y)dy, \quad \forall x \in (0,\infty), \end{aligned}$$

from which we derive

$$\begin{aligned} e^{-x}u^2(x) + \int_0^x e^{-y}u^2(y)dy &\le 2\int_0^\infty e^{-y}|u(y)u'(y)|dy \\ &\le 2\|u\|_\omega |u|_{1,\omega}. \end{aligned}$$

This implies the first inequality, and letting $x \to \infty$ leads to the second one. □

Lemma B.6. *Let $\omega(x) = e^{-x^2}$. Then for any $u \in H^1_\omega(-\infty,\infty)$, we have*

$$\|e^{-x^2/2}u\|_{L^\infty(-\infty,\infty)} \le 2\|u\|_\omega^{1/2}|u|_{1,\omega}^{1/2}, \tag{B.36a}$$

$$\|xu\|_\omega \le \|u\|_{1,\omega}. \tag{B.36b}$$

Proof. Applying integration by parts and the Schwarz inequality yields

$$\int_{-\infty}^\infty xu^2(x)\omega(x)dx = \int_{-\infty}^\infty u(x)u'(x)\omega(x)dx \le \|u\|_\omega |u|_{1,\omega}, \tag{B.37}$$

which implies $xu^2(x)\omega(x) \to 0$ as $|x| \to \infty$. Therefore, we have

$$\begin{aligned} \int_{-\infty}^\infty (xu(x))^2\omega(x)dx &= -\frac{1}{2}\int_{-\infty}^\infty xu^2(x)d\omega(x) \\ &= \frac{1}{2}\int_{-\infty}^\infty u^2(x)\omega(x)dx + \int_{-\infty}^\infty xu(x)u'(x)\omega(x)dx \\ &\le \frac{1}{2}\|u\|_\omega^2 + \frac{1}{2}\|xu\|_\omega^2 + \frac{1}{2}|u|_{1,\omega}^2 \\ &= \frac{1}{2}\|u\|_{1,\omega}^2 + \frac{1}{2}\|xu\|_\omega^2, \end{aligned}$$

which gives (B.36b).

Next, since

$$\begin{aligned} e^{-x^2}u^2(x) &= \int_{-\infty}^x \partial_y(e^{-y^2}u^2(y))dy \\ &= 2\int_{-\infty}^x u(y)u'(y)\omega(y)dy - 2\int_{-\infty}^x yu^2(y)\omega(y)dy, \end{aligned}$$

we deduce from the Schwarz inequality that

$$e^{-x^2}u^2(x)+2\int_{-\infty}^{x} yu^2(y)\omega(y)dy \le 2\|u\|_\omega |u|_{1,\omega}.$$

Thus, by (B.37),

$$e^{-x^2}u^2(x) \le 4\|u\|_\omega |u|_{1,\omega},$$

which yields (B.36a). □

B.8.2 Hardy-Type Inequalities

Let $a<b$ be two real numbers, and let $\alpha<1$. Then for any $\phi \in L^2_\omega(a,b)$ with $\omega=(x-a)^\alpha$, we have the following Hardy inequality (see Hardy et al. (1952)):

$$\int_a^b \Big(\frac{1}{x-a}\int_a^x \phi(y)dy\Big)^2 (x-a)^\alpha dt \le \frac{4}{1-\alpha}\int_a^b \phi^2(x)(x-a)^\alpha dx. \tag{B.38}$$

Similarly, for any $\phi \in L^2_\omega(a,b)$ with $\omega=(b-x)^\alpha$, we have

$$\int_a^b \Big(\frac{1}{b-x}\int_x^b \phi(y)dy\Big)^2 (b-x)^\alpha dt \le \frac{4}{1-\alpha}\int_a^b \phi^2(x)(b-x)^\alpha dx. \tag{B.39}$$

Next, we apply the above Hardy inequality to derive some useful inequalities associated with the Jacobi weight function $\omega^{\alpha,\beta}(x)=(1-x)^\alpha(1+x)^\beta$ with $x\in I:=(-1,1)$.

Lemma B.7. *If* $-1<\alpha,\beta<1$, *then*

$$\|u\|_{\omega^{\alpha-2,\beta-2}} \le c\|u'\|_{\omega^{\alpha,\beta}}, \quad \forall u \in H^1_{0,\omega^{\alpha,\beta}}(I), \tag{B.40}$$

which implies the Poincaré-type inequality:

$$\|u\|_{\omega^{\alpha,\beta}} \le c\|u'\|_{\omega^{\alpha,\beta}}, \quad \forall u \in H^1_{0,\omega^{\alpha,\beta}}(I). \tag{B.41}$$

Proof. Taking $a=-1, b=1$ and $\phi=\partial_x u$ in (B.39) yields that for $\alpha<1$,

$$\int_0^1 u^2(x)(1-x)^{\alpha-2}dx \le c\int_0^1 (u'(x))^2(1-x)^\alpha dx.$$

Hence,

$$\begin{aligned}\int_0^1 u^2(x)(1-x)^{\alpha-2}(1+x)^{\beta-2}dx &\le c\int_0^1 u^2(x)(1-x)^{\alpha-2}dx\\ &\le c\int_0^1 (u'(x))^2(1-x)^\alpha dx \le c\int_0^1 (u'(x))^2(1-x)^\alpha(1+x)^\beta dx.\end{aligned}$$

Similarly, for $\beta < 1$, we use (B.39) to derive

$$\int_{-1}^{0} u^2(x)(1-x)^{\alpha-2}(1+x)^{\beta-2}dx \le c\int_{-1}^{0}(u'(x))^2(1-x)^{\alpha}(1+x)^{\beta}dx.$$

A combination of the above two inequalities leads to (B.40).

In view of $\omega^{\alpha,\beta}(x) < \omega^{\alpha-2,\beta-2}(x)$, (B.41) follows from (B.40). □

A consequence of Lemma B.7 is the following result.

Corollary B.1. *If $-1<\alpha,\beta<1$, then for any $u\in H^1_{0,\omega^{\alpha,\beta}}(I)$, we have $u\omega^{\alpha,\beta}\in H^1_{0,\omega^{-\alpha,-\beta}}(I)$.*

Proof. A direct calculation shows that

$$|u\omega^{\alpha,\beta}|^2_{1,\omega^{-\alpha,-\beta}} \le c\big(|u|^2_{1,\omega^{\alpha,\beta}} + \|u\|^2_{\omega^{\alpha-2,\beta-2}}\big) \overset{\text{(B.40)}}{\le} c|u|^2_{1,\omega^{\alpha,\beta}}.$$

On the other hand, since $\|u\omega^{\alpha,\beta}\|_{\omega^{-\alpha,-\beta}} = \|u\|_{\omega^{\alpha,\beta}}$, we have

$$\|u\omega^{\alpha,\beta}\|_{1,\omega^{-\alpha,-\beta}} \le c\|u\|_{1,\omega^{\alpha,\beta}}.$$

This ends the proof. □

The following inequalities can be found in Guo (2000).

Lemma B.8. *Let $\alpha,\beta>-1$. Then for any function $u\in H^1_{\omega^{\alpha+2,\beta+2}}(I)$ with $u(x_0)=0$ for some $x_0\in(-1,1)$, we have*

$$\|u\|_{\omega^{\alpha,\beta}} \le c\|u'\|_{\omega^{\alpha+2,\beta+2}}, \tag{B.42}$$

which implies

$$\|u\|_{\omega^{\alpha,\beta}} \le c\|u'\|_{\omega^{\alpha,\beta}}. \tag{B.43}$$

Proof. The inequality (B.43) follows directly from (B.42), so it suffices to prove the first one. For any $x\in[x_0,1]$,

$$u^2(x)(1-x)^{\alpha+1} = \int_{x_0}^{x} \partial_y\big(u^2(y)(1-y)^{\alpha+1}\big)dy.$$

Hence, by the Schwarz inequality,

$$\begin{aligned}
&u^2(x)(1-x)^{\alpha+1} + (\alpha+1)\int_{x_0}^{x} u^2(y)(1-y)^{\alpha}dy\\
&= 2\int_{x_0}^{x} u(y)u'(y)(1-y)^{\alpha+1}dy\\
&\le 2\Big(\int_{x_0}^{x} u^2(y)(1-y)^{\alpha}dy\Big)^{1/2}\Big(\int_{x_0}^{x}(u'(y))^2(1-y)^{\alpha+2}dy\Big)^{1/2},
\end{aligned}$$

which implies

$$\int_{x_0}^{x} u^2(y)(1-y)^{\alpha}dy \le \frac{4}{(\alpha+1)^2}\int_{x_0}^{x}\big(u'(y)\big)^2(1-y)^{\alpha+2}dy.$$

Letting $x \to 1$ in the above inequality leads to

$$\begin{aligned}\int_{x_0}^{1} u^2(x)(1-x)^{\alpha}dx &\le \frac{4}{(\alpha+1)^2}\int_{x_0}^{1}\big(u'(x)\big)^2(1-x)^{\alpha+2}dx\\ &\le c\int_{x_0}^{1}\big(u'(x)\big)^2(1-x)^{\alpha+2}(1+x)^{\beta+2}dx.\end{aligned}$$

Similarly, we can derive

$$\int_{-1}^{x_0} u^2(x)(1+x)^{\beta}dx \le c\int_{-1}^{x_0}\big(u'(x)\big)^2(1-x)^{\alpha+2}(1+x)^{\beta+2}dx.$$

Finally, (B.42) follows from the above two inequalities. □

A direct consequence of (B.43) is as follows.

Corollary B.2. *For any $u \in H^1(a,b)$ with $u(x_0) = 0$ for some $x_0 \in (a,b)$, the following Poincaré inequality holds:*

$$\|u\|_{L^2(a,b)} \le c\|u'\|_{L^2(a,b)}, \quad \forall u \in H^1(a,b). \tag{B.44}$$

Remark B.3. *In fact, the Poincaré inequality* (B.44) *holds, when the condition in Corollary B.2 is replaced by*

$$\int_a^b u(x)dx = 0. \tag{B.45}$$

B.8.3 Gronwall Inequalities

The Gronwall type inequalities are very useful in the stability and convergence analysis of initial-boundary value problems. The following is a typical Gronwall inequality:

Lemma B.9. *Let $f(t)$ be a non-negative integrable function over $(t_0,T]$, and let $g(t)$ and $E(t)$ be continuous functions on $[t_0,T]$. If $E(t)$ satisfies*

$$E(t) \le g(t) + \int_{t_0}^{t} f(\tau)E(\tau)d\tau, \quad \forall t \in [t_0,T], \tag{B.46}$$

then we have

$$E(t) \le g(t) + \int_{t_0}^{t} f(s)g(s)\exp\Big(\int_s^t f(\tau)d\tau\Big)ds, \quad \forall t \in [t_0,T]. \tag{B.47}$$

If, in addition, g is non-decreasing, then

$$E(t) \leq g(t)\exp\Big(\int_{t_0}^{t} f(\tau)d\tau\Big), \quad \forall t \in [t_0, T]. \tag{B.48}$$

On the other hand, discrete Gronwall inequalities are often used in the stability and convergence analysis of time discretization schemes. In particular, a useful discrete analogue of Lemma B.9 is:

Lemma B.10. *Let y^n, h^n, g^n, f^n be four nonnegative sequences satisfying*

$$y^m + k\sum_{n=0}^{m} h^n \leq B + k\sum_{n=0}^{m}(g^n y^n + f^n), \text{ with } k\sum_{n=0}^{T/k} g^n \leq M, \ \forall\, 0 \leq m \leq T/k.$$

We assume $kg^n < 1$ and let $\sigma = max_{0\leq n\leq T/k}(1-kg^n)^{-1}$. Then

$$y^m + k\sum_{n=1}^{m} h^n \leq \exp(\sigma M)(B + k\sum_{n=0}^{m} f^n), \ \forall m \leq T/k.$$

We refer to, for instance, Quarteroni and Valli (2008) for a proof of the above two lemmas.

Appendix C
Basic Iterative Methods and Preconditioning

We review below some basic iterative methods for solving the linear system

$$Ax = b, \tag{C.1}$$

where $A \in \mathbb{R}^{n\times n}$ is an invertible matrix and $b \in \mathbb{R}^n$ is a given vector. We refer to Barrett et al. (1994), Golub and Van Loan (1996) and Saad (2003) for more detailed presentation in this matter.

C.1 Krylov Subspace Methods

The basic idea of Krylov subspace methods is to form an orthogonal basis of the sequence of successive matrix powers times the initial residual (the Krylov sequence), and then look for the approximation to the solution by minimizing the residual over the subspace formed by these orthogonal basis.

In what follows, we shall mainly discuss two proto-type of Krylov subspace methods: the *Conjugate Gradient (CG)* method and the *Generalized Minimal Residual (GMRES)* method. The CG method of Hestenes and Stiefel (1952) is the *method of choice* for solving large *symmetric positive definite* linear systems, while the GMRES method proposed by Saad and Schultz (1986) is popular for solving nonsymmetric linear systems.

C.1.1 Conjugate Gradient (CG) Method

Throughout this section, let $A \in \mathbb{R}^{n\times n}$ be a symmetric positive definite matrix. It can be verified that $\hat{x}$ is the solution of $Ax = b$ if and only if $\hat{x}$ minimizes the quadratic functional

$$J(x) = \frac{1}{2}x^T Ax - x^T b. \tag{C.2}$$

J. Shen et al., *Spectral Methods: Algorithms, Analysis and Applications*,
Springer Series in Computational Mathematics 41, DOI 10.1007/978-3-540-71041-7,

Suppose that $x^{(k)}$ has been obtained. Then $x^{(k+1)}$ can be found by

$$x^{(k+1)} = x^{(k)} + \alpha_k p^{(k)}, \tag{C.3}$$

where the scalar α_k is called the step size factor and the vector $p^{(k)}$ is called the search direction. The coefficient α_k in (C.3) is selected such that

$$\alpha_k = \arg\min_{\alpha} J\big(x^{(k)} + \alpha p^{(k)}\big). \tag{C.4}$$

A simple calculation shows that

$$\alpha_k = \frac{(r^{(k)}, p^{(k)})}{(Ap^{(k)}, p^{(k)})}, \tag{C.5}$$

where the inner product of two column vectors is defined by $(u,v) = u^T v$, and the residual is given by

$$r^{(k)} := b - Ax^{(k)}. \tag{C.6}$$

Notice that the residual at the $(k+1)$th step is updated by

$$\begin{aligned} r^{(k+1)} &:= b - Ax^{(k+1)} = b - A\big(x^{(k)} + \alpha_k p^{(k)}\big) \\ &= b - Ax^{(k)} - \alpha_k Ap^{(k)} = r^{(k)} - \alpha_k Ap^{(k)}. \end{aligned} \tag{C.7}$$

In the conjugate gradient method, we select the *next search direction* $p^{(k+1)}$ satisfying the orthogonality

$$(p^{(k+1)}, Ap^{(k)}) = 0, \tag{C.8}$$

i.e.,

$$p^{(k+1)} = r^{(k+1)} + \beta_k p^{(k)}, \tag{C.9}$$

One verifies the orthogonality

$$\beta_k = -\frac{(Ap^{(k)}, r^{(k+1)})}{(Ap^{(k)}, p^{(k)})}. \tag{C.10}$$

It is important to notice the orthogonality

$$\big(p^{(i)}, Ap^{(j)}\big) = 0, \quad \big(r^{(i)}, r^{(j)}\big) = 0, \quad i \neq j. \tag{C.11}$$

Moreover, it can be shown that if A is a real $n \times n$ positive definite matrix, then, assuming exact arithmetic, the iteration converges in at most n steps, i.e., $x^{(m)} = \hat{x}$ for some $m \leq n$. Furthermore, the residual vectors satisfy

$$\big(r^{(k)}, p^{(j)}\big) = 0 \quad \text{for each} \quad j = 1, \ldots, k-1. \tag{C.12}$$

Therefore, we can reformulate the scalers α_k and β_k as

$$\alpha_k = \frac{(r^{(k)}, r^{(k)})}{(Ap^{(k)}, p^{(k)})}, \quad \beta_k = \frac{(r^{(k+1)}, r^{(k+1)})}{(r^{(k)}, r^{(k)})}. \tag{C.13}$$

We summarize the **CG Algorithm** below.

CG Algorithm

1. Initialization: choose $x^{(0)}$, compute $r^{(0)} = b - Ax^{(0)}$ and set $p^{(0)} = r^{(0)}$.
2. For $k = 0, 1, \ldots,$

   (i) Compute
   $$\alpha_k = (r^{(k)}, r^{(k)})/(Ap^{(k)}, p^{(k)}).$$
   (ii) Set
   $$x^{(k+1)} = x^{(k)} + \alpha_k p^{(k)}.$$
   (iii) Compute
   $$r^{(k+1)} = r^{(k)} - \alpha_k A p^{(k)}.$$
   (iv) If the stopping rule does not apply, continue.
   (v) Compute
   $$\beta_k = (r^{(k+1)}, r^{(k+1)})/(r^{(k)}, r^{(k)}).$$
   (vi) Set
   $$p^{(k+1)} = r^{(k+1)} + \beta_k p^{(k)}.$$
3. endFor

The following theorem on the rate of convergence of the CG method can be found in e.g., Golub and Van Loan (1996).

Theorem C.1. *Suppose that $A \in \mathbb{R}^{n \times n}$ is symmetric positive definite matrix and $b \in \mathbb{R}^n$. Then the CG Algorithm produces iterates $\{x^{(k)}\}$ satisfying*

$$\|\hat{x} - x^{(k)}\|_A \leq 2\left(\frac{\sqrt{\kappa}-1}{\sqrt{\kappa}+1}\right)^k \|\hat{x} - x^{(0)}\|_A, \tag{C.14}$$

where $\hat{x}$ is the exact solution of $Ax = b$, $\|x\|_A = \sqrt{x^T A x}$, and $\kappa = \|A\|_2 \|A^{-1}\|_2$ (the condition number of A).

Some remarks are in order.

(i) For a symmetric positive definite matrix, we have $\|A\|_2 = \lambda_n$ and $\|A^{-1}\|_2 = \lambda_1^{-1}$, where λ_n and λ_1 are the largest and smallest eigenvalues of A, respectively. One derives from Theorem C.1 the estimate in the 2-norm:

$$\|\hat{x} - x^{(k)}\|_2 \leq 2\sqrt{\kappa}\left(\frac{\sqrt{\kappa}-1}{\sqrt{\kappa}+1}\right)^k \|\hat{x} - x^{(0)}\|_2. \tag{C.15}$$

(ii) The CG method involves one matrix–vector multiplication, three vector updates, and two inner products per iteration. If the matrix is sparse or has a special structure, these operators can be performed efficiently.
(iii) Unlike the traditional SOR type method, there is no free parameter to choose in the CG algorithm.

C.1.2 BiConjugate Gradient (BiCG) Method

The Conjugate Gradient method is not suitable for non-symmetric systems because the residual vectors can not be made orthogonal with short recurrences (see Faber and Manteuffel (1984) for the proof). The BiConjugate Gradient (BiCG) method takes another approach, replacing the orthogonal sequence of residuals by two mutually orthogonal sequences, at the price of no longer providing a minimization.

The BiCG method augments the update relations for residuals in the CG method by relations based on both A and A^T. More precisely, given two pairs: $(p^{(j)}, \tilde{p}^{(j)})$ and $(r^{(j)}, \tilde{r}^{(j)})$, we update

$$x^{(j+1)} = x^{(j)} + \alpha_j p^{(j)}, \tag{C.16}$$

and the two sequences of residuals

$$r^{(j+1)} = r^{(j)} - \alpha_j A p^{(j)}, \quad \tilde{r}^{(j+1)} = \tilde{r}^{(j)} - \alpha_j A^T \tilde{p}^{(j)}. \tag{C.17}$$

Require that $(r^{(j+1)}, \tilde{r}^{(j)}) = 0$ and $(r^{(j)}, \tilde{r}^{(j+1)}) = 0$ for all j. This leads to

$$\alpha_j = (r^{(j)}, \tilde{r}^{(j)}) / (Ap^{(j)}, \tilde{p}^{(j)}). \tag{C.18}$$

The two sequences of search directions are updated by

$$p^{(j+1)} = r^{(j+1)} + \beta_j p^{(j)}, \quad \tilde{p}^{(j+1)} = \tilde{r}^{(j+1)} + \beta_j \tilde{p}^{(j)}. \tag{C.19}$$

By requiring that $(Ap^{(j+1)}, \tilde{p}^{(j)}) = 0$ and $(Ap^{(j)}, \tilde{p}^{(j+1)}) = 0$, we obtain

$$\beta_j = (r^{(j+1)}, \tilde{r}^{(j+1)}) / (r^{(j)}, \tilde{r}^{(j)}). \tag{C.20}$$

The above derivations lead to the **BiCG Algorithm** outlined below.
Some remarks are in order.

(i) The BiCG algorithm is particularly suitable for matrices which are positive definite, i.e., $(Ax,x) > 0$ for all $x \neq 0$, but not necessary to be symmetric.
(ii) If A is symmetric positive definite and $\tilde{r}^{(0)} = r^{(0)}$, then the BiCG algorithm delivers the same results as the CG method, but at twice of the cost per iteration.
(iii) The algorithm breaks down if $(Ap_j, \tilde{p}_j) = 0$. Otherwise, the amount of work and storage is of the same order as the CG algorithm.

BiCG Algorithm

1. Initialization: choose $x^{(0)}$, compute $r^{(0)} = b - Ax^{(0)}$ and set $p^{(0)} = r^{(0)}$; choose $\tilde{r}^{(0)}$ (such that $(r^{(0)}, \tilde{r}^{(0)}) \neq 0$, e.g., $\tilde{r}^{(0)} = r^{(0)}$).
2. For $j = 0, 1, \ldots,$

   (i) Compute
   $$\alpha_j = (r^{(j)}, \tilde{r}^{(j)})/(Ap^{(j)}, \tilde{p}^{(j)}).$$
   (ii) Set
   $$x^{(j+1)} = x^{(j)} + \alpha_j p^{(j)}.$$
   (iii) Compute
   $$r^{(j+1)} = r^{(j)} - \alpha_j A p^{(j)}, \quad \tilde{r}^{(j+1)} = \tilde{r}^{(j)} - \alpha_j A^T \tilde{p}^{(j)}.$$
   (iv) If the stopping rule does not apply, continue.
   (v) Compute
   $$\beta_j = (r^{(j+1)}, \tilde{r}^{(j+1)})/(r^{(j)}, \tilde{r}^{(j)}).$$
   (vi) Set
   $$p^{(j+1)} = r^{(j+1)} + \beta_j p^{(j)}, \quad \tilde{p}^{(j+1)} = \tilde{r}^{(j+1)} + \beta_j \tilde{p}^{(j)}.$$
3. endFor

C.1.3 Conjugate Gradient Squared (CGS) Method

The BiCG algorithm requires multiplication by both A and A^T at each iteration. Obviously, this demands extra work, and in addition, it is sometimes cumbersome to multiply by A^T than by A. For example, there may be a special formula for the product of A with a given vector when A represents, say, a Jacobian, but a corresponding formula for the product of A^T with a given vector may not be available. In other cases, data may be stored on a parallel machine in such a way that multiplication by A is efficient but multiplication by A^T involves extra communication between processors. For these reasons it is desirable to have an iterative method that requires multiplication only by A and that generates good approximate solutions. A method for such purposes is the Conjugate Gradient Squared (CGS) method.

From the recurrence relations of BiCG algorithms, we see that

$$r^{(j)} = \Phi_j^a(A) r^{(0)} + \Phi_j^b(A) p^{(0)},$$

where $\Phi_j^a(A)$ and $\Phi_j^b(A)$ are polynomials of degree j of A. Choosing $p^{(0)} = r^{(0)}$ gives

$$r^{(j)} = \Phi_j(A) r^{(0)} \quad \text{where} \quad \Phi_j = \Phi_j^a + \Phi_j^b,$$

with $\Phi_0 \equiv 1$. Similarly,

$$p^{(j)} = \pi_j(A) r^{(0)},$$

where $\pi_j(A)$ is a polynomial of degree j of A. As $\tilde{r}^{(j)}$ and $\tilde{p}^{(j)}$ are updated, using the same recurrence relation as for $r^{(j)}$ and $p^{(j)}$, we have

$$\tilde{r}^{(j)} = \Phi_j(A^T) \tilde{r}^{(0)}, \quad \tilde{p}^{(j)} = \pi_j(A^T) \tilde{r}^{(0)}. \tag{C.21}$$

Hence,

$$\alpha_j = \frac{(\Phi_j(A)r^{(0)}, \Phi_j(A^T)\tilde{r}^{(0)})}{(A\pi_j(A)r^{(0)}, \pi_j(A^T)\tilde{r}^{(0)})} = \frac{(\Phi_j^2(A)r^{(0)}, \tilde{r}^{(0)})}{(A\pi_j^2(A)r^{(0)}, \tilde{r}^{(0)})}. \tag{C.22}$$

From the BiCG algorithm,

$$\Phi_{j+1}(t) = \Phi_j(t) - \alpha_j t\pi_j(t), \quad \pi_{j+1}(t) = \Phi_{j+1}(t) + \beta_j\pi_j(t). \tag{C.23}$$

Observe that

$$\Phi_j\pi_j = \Phi_j(\Phi_j + \beta_{j-1}\pi_{j-1}) = \Phi_j^2 + \beta_{j-1}\Phi_j\pi_{j-1}. \tag{C.24}$$

It follows from the above results that

$$\begin{aligned}
&\Phi_{j+1}^2 = \Phi_j^2 - 2\alpha_j t(\Phi_j^2 + \beta_{j-1}\Phi_j\pi_{j-1}) + \alpha_j^2 t^2\pi_j^2,\\
&\Phi_{j+1}\pi_j = \Phi_j\pi_j - \alpha_j t\pi_j^2 = \Phi_j^2 + \beta_{j-1}\Phi_j\pi_{j-1} - \alpha_j t\pi_j^2,\\
&\pi_{j+1}^2 = \Phi_{j+1}^2 + 2\beta_j\Phi_{j+1}\pi_j + \beta_j^2\pi_j^2.
\end{aligned} \tag{C.25}$$

Define

$$\begin{aligned}
&r^{(j)} = \Phi_j^2(A)r^{(0)}, \quad p^{(j)} = \pi_j^2(A)r^{(0)},\\
&q^{(j)} = \Phi_{j+1}(A)\pi_j(A)r^{(0)}, \quad d^{(j)} = 2r^{(j)} + 2\beta_{j-1}q^{(j-1)} - \alpha_j Ap^{(j)}.
\end{aligned}$$

It can be verified that

$$\begin{aligned}
&r^{(j+1)} = r^{(j)} - \alpha_j Ad^{(j)},\\
&q^{(j)} = r^{(j)} + \beta_{j-1}q^{(j-1)} - \alpha_j Ap^{(j)},\\
&p^{(j+1)} = r^{(j+1)} + 2\beta_j q^{(j)} + \beta_j^2 p^{(j)},\\
&d^{(j)} = 2r^{(j)} + 2\beta_{j-1}q^{(j-1)} - \alpha_j Ap^{(j)}.
\end{aligned}$$

Correspondingly,

$$x^{(j+1)} = x^{(j)} + \alpha_j d^{(j)}. \tag{C.26}$$

This gives the **CGS Algorithm** as summarized below.

The CGS method requires two matrix–vector multiplications at each step but no multiplications by the transpose. For problems where the BiCG method converges well, the CGS method typically requires only about half as many steps and, therefore, half the work of the BiCG method (assuming that multiplication by A or A^T requires the same amount of work). When the norm of the BiCG residual increases at a step, however, that of the CGS residual usually increases by approximately the square of the increase of the BiCG residual norm. The CGS algorithm convergence curve may therefore show wild oscillations that can sometimes lead to numerical instability and break down.

CGS Algorithm

1. Initialization: choose $x^{(0)}$, compute $r^{(0)} = b - Ax^{(0)}$ and set $p^{(0)} = r^{(0)} = u^{(0)}, q^{(0)} = 0$; choose $\tilde{r}^{(0)}$ such that $(r^{(0)}, \tilde{r}^{(0)}) \neq 0$.
2. For $j = 0, 1, \ldots,$

   (i) Compute
   $$\alpha_j = (r^{(j)}, \tilde{r}^{(0)})/(Ap^{(j)}, \tilde{r}^{(0)}),$$
   and
   $$q^{(j+1)} = u^{(j)} - \alpha_j Ap^{(j)}.$$
   (ii) Set
   $$x^{(j+1)} = x^{(j)} + \alpha_j (u^{(j)} + q^{(j+1)}).$$
   (iii) Compute
   $$r^{(j+1)} = r^{(j)} - \alpha_j A(u^{(j)} + q^{(j+1)}).$$
   (iv) If the stopping rule does not apply, continue.
   (v) Compute
   $$\beta_j = (r^{(j+1)}, \tilde{r}^{(0)})/(r^{(j)}, \tilde{r}^{(0)}),$$
   and
   $$u^{(j+1)} = r^{(j+1)} + \beta_j q^{(j+1)}.$$
   (vi) Set
   $$p^{(j+1)} = u^{(j+1)} + \beta_j (q^{(j+1)} + \beta_j p^{(j)}).$$
3. endFor

C.1.4 BiConjugate Gradient Stabilized (BiCGStab) Method

The BiConjugate Gradient Stabilized (BiCGStab) method was developed by Van der Vorst (1992) to solve non-symmetric linear systems while avoiding the irregular convergence patterns of the CGS method. The main idea is to produce a residual of the form

$$r^{(j)} = \Psi_j(A)\Phi_j(A)r^{(0)}, \tag{C.27}$$

where Φ_j is again the BiCG polynomial but Ψ_j is chosen to keep the residual norm small at each step while retaining the rapid overall convergence of the CGS method. For example, $\Psi_j(t)$ could be of the form

$$\Psi_{j+1}(t) = (1 - w_j t)\Psi_j(t). \tag{C.28}$$

In the BiCGStab algorithm, the solution is updated in such a way that $r^{(j)}$ is of the form (C.27), where $\Psi_j(A)$ is a polynomial of degree j satisfying (C.28). Then

$$\begin{aligned} \Psi_{j+1}\Phi_{j+1} &= (1 - w_j t)\Psi_j(\Phi_j - \alpha_j t \pi_j) \\ &= (1 - w_j t)(\Psi_j \Phi_j - \alpha_j t \Psi_j \pi_j), \end{aligned} \tag{C.29}$$

and

$$\begin{aligned} \Psi_j \pi_j &= \Psi_j(\Phi_j + \beta_{j-1}\pi_{j-1}) \\ &= \Psi_j \Phi_j + \beta_{j-1}(1 - w_{j-1}t)\Psi_{j-1}\pi_{j-1}. \end{aligned} \tag{C.30}$$

Let $r^{(j)} = \Phi_j(A)\Psi_j(A)r^{(0)}$ and $p^{(j)} = \Psi_j(A)\pi_j(A)r^{(0)}$. It can be verified that

$$\begin{aligned} r^{(j+1)} &= (I - w_j A)(r^{(j)} - \alpha_j A p^{(j)}), \\ p^{(j+1)} &= r^{(j+1)} + \beta_j (I - w_j A) p^{(j)}. \end{aligned} \tag{C.31}$$

Letting $s^{(j)} = r^{(j)} - \alpha_j A p^{(j)}$, we obtain

$$r^{(j+1)} = (I - w_j A) s^{(j)}. \tag{C.32}$$

The parameter w_j is chosen to minimize the 2-norm of $r^{(j+1)}$, i.e.,

$$w_j = \frac{(As^{(j)}, s^{(j)})}{(As^{(j)}, As^{(j)})}. \tag{C.33}$$

We also need to find an updating formula for α_j and β_j, which ideally only involves $r^{(k)}, p^{(k)}$ and $s^{(k)}$. This seems to be rather complicated, so we omit the derivation here.

The **BiCGStab Algorithm** is summarized below.

BiCGStab Algorithm

1. Initialization: choose $x^{(0)}$, compute $r^{(0)} = b - Ax^{(0)}$ and set $p^{(0)} = r^{(0)}$; choose $\tilde{r}^{(0)}$ such that $(r^{(0)}, \tilde{r}^{(0)}) \neq 0$.
2. For $j = 0, 1, \ldots,$

   (i) Compute
   $$\alpha_j = \frac{(r^{(j)}, \tilde{r}^{(0)})}{(Ap^{(j)}, \tilde{r}^{(0)})}.$$

   (ii) Set
   $$s^{(j)} = r^{(j)} - \alpha_j A p^{(j)},$$
   and compute
   $$w_j = \frac{(As^{(j)}, s^{(j)})}{(As^{(j)}, As^{(j)})}.$$

   (iii) Set
   $$x^{(j+1)} = x^{(j)} + \alpha_j p^{(j)} + w_j s^{(j)}; \quad r^{(j+1)} = s^{(j)} - w_j A s^{(j)}.$$

   (iv) If the stopping rule does not apply, continue.

   (v) Compute
   $$\beta_j = \frac{\alpha_j}{w_j} \frac{(r^{(j+1)}, \tilde{r}^{(0)})}{(r^{(j)}, \tilde{r}^{(0)})}.$$

   (vi) Set
   $$p^{(j+1)} = r^{(j+1)} + \beta_j (p^{(j)} - w_j A p^{(j)}).$$

3. endFor

In general, the BiCGStab method often converges about as fast as the CGS algorithm. We also notice that the BiCGStab method requires two matrix–vector products and four inner products, i.e., two inner products more than the BiCG and CGS methods.

C.1.5 Generalized Minimal Residual (GMRES) Method

The Generalized Minimal Residual method proposed by Saad and Schultz (1986) is one of the most important tools for solving general *non-symmetric* system: $Ax = b$. In the k-th iteration of the GMRES method, we need to find $x^{(k)}$ that minimizes $\|b - Ax\|_2$ over the set

$$S_k := x^{(0)} + \text{span}\{r^{(0)}, Ar^{(0)}, \ldots, A^{k-1}r^{(0)}\}, \tag{C.34}$$

where $r^{(0)} = b - Ax^{(0)}$. In other words, for any $x \in S_k$, we have

$$x = x^{(0)} + \sum_{j=0}^{k-1} \gamma_j A^j r^{(0)}. \tag{C.35}$$

Moreover, it can be shown that

$$r = b - Ax = r^{(0)} - \sum_{j=1}^{k} \gamma_{j-1} A^j r^{(0)}. \tag{C.36}$$

Like the CG algorithm, the GMRES method will obtain the *exact* solution of $Ax = b$ within n iterations. Moreover, if b is a linear combination of k eigenvectors of A, say $b = \sum_{p=1}^{k} \gamma_p u_{i_p}$, then the GMRES method will terminate within at most k iterations.

The first important issue is to find a basis for S_k. Suppose that we have a matrix $V_k = [v_1^k, v_2^k, \ldots, v_k^k]$, whose columns form an orthogonal basis of S_k. Then any $z \in S_k$ can be expressed as

$$z = \sum_{p=1}^{k} u_p v_p^k = V_k u, \tag{C.37}$$

where $u = (u_1, \ldots, u_k) \in \mathbb{R}^k$. Once we have found V_k, we can convert the original least-squares problem: $\min_{x \in S_k} \|b - Ax\|_2$ into a least-squares problem in $\mathbb{R}^k$. More precisely, let $x^{(k)}$ be the solution after the k-th iteration. Then we have $x^{(k)} = x^{(0)} + V_k y^{(k)}$, where the vector $y^{(k)}$ minimizes

$$\min_{y \in \mathbb{R}^k} \|b - A(x^{(0)} + V_k y)\|_2 = \min_{y \in \mathbb{R}^k} \|r^{(0)} - AV_k y\|_2. \tag{C.38}$$

This is a standard linear least-squares problem that can be solved by a QR decomposition.

To find an orthonormal basis of S_k, one can use the **modified Gram-Schmidt orthogonalization** as highlighted below.
This algorithm produces the columns of the matrix V_k, which also form an orthonormal basis for S_k. Note that the algorithm breaks down when a division by zero occurs.

If the modified Gram-Schmidt process does not break down, we can use it to carry out the GMRES method (i.e., to solve the minimization problem (C.38)) in an efficient way. More precisely, define

Modified Gram-Schmidt Orthogonalization

1. Initialization: choose $x^{(0)}$, and set $r^{(0)} = b - Ax^{(0)}$ and $v^{(1)} = r^{(0)}/\|r^{(0)}\|_2$.
2. For $j = 1, 2, \ldots, k$,

   Compute

$$v^{(j+1)} = \frac{Av^{(j)} - \sum_{l=1}^{j} \left(Av^{(j)}, v^{(l)}\right) v^{(l)}}{\|Av^{(j)} - \sum_{l=1}^{j} \left(Av^{(j)}, v^{(l)}\right) v^{(l)}\|_2}.$$

3. endFor

$$h_{ij} = \left(Av^{(j)}, v^{(i)}\right), \quad 1 \le i \le j \le k.$$

From the modified Gram-Schmidt algorithm, we obtain a $k \times k$ matrix $H_k = (h_{ij})$, which is upper Hessenberg, i.e., its entries satisfy $h_{ij} = 0$ if $i > j + 1$. Moreover, this process produces a matrix $V_k = [v_1^k, v_2^k, \ldots, v_k^k]$, whose columns form an orthonormal basis for S_k, and we have

$$AV_k = V_{k+1} \tilde{H}_k, \tag{C.39}$$

where $\tilde{H}_k$ is generated by H_k (see P. 548 of Golub and Van Loan (1996)). This allows us to convert the problem: updating $x^{(k)} = x^{(0)} + V_k y^{(k)}$ by solving (C.38), into an alternative formulation. Using the fact $r^{(0)} = b - Ax^{(0)}$, and (C.39), one verifies that

$$\begin{aligned} r^{(k)} &= b - Ax^{(k)} = r^{(0)} - A\left(x^{(k)} - x^{(0)}\right) \\ &= \beta V_{k+1} e_1 - AV_k y^{(k)} = V_{k+1}\left(\beta e_1 - \tilde{H}_k y^{(k)}\right), \end{aligned} \tag{C.40}$$

where e_1 is the first unit k-vector $(1, 0, \ldots, 0)^T$, and $\beta = \|r^{(0)}\|_2$. Therefore, the problem (C.38) becomes

$$\min_{y \in \mathbb{R}^k} \left\| \beta e_1 - \tilde{H}_k y \right\|_2. \tag{C.41}$$

To find the minimizer $y^{(k)}$ of (C.41), it is necessary to we look at the linear system $\tilde{H}_k y = \beta e_1$, which can be solved by using rotation matrices to perform Gauss-elimination for $\tilde{H}_k$ (see, e.g., Saad (2003)). Here, we skip the details.

The pseudocode of the **GMRES Algorithm** for solving $Ax = b$ with A being a non-symmetric matrix is given below.

GMRES Algorithm

1. Initialization: choose $x^{(0)}$, and set $r^{(0)} = b - Ax^{(0)}, \beta = \|r^{(0)}\|_2$ and $v^{(1)} = r^{(0)}/\|r^{(0)}\|_2$.
2. For a given k, find the basis for S_k by e.g., the modified Gram-Schmidt orthogonalization process.
3. Form $\tilde{H}_k$, and solve (C.41) to find $y^{(k)}$.
4. Set $x^{(k)} = x^{(0)} + V_k y^{(k)}$.
5. Check convergence; if necessary, set $x^{(0)} = x^{(k)}, v^{(1)} = r^{(k)}/\|r^{(k)}\|_2$, and go to Step 2.

C.2 Preconditioning

The convergence rate of iterative methods depends on spectral properties of the coefficient matrix. Hence, one may attempt to transform the linear system into an equivalent system that has more favorable spectral properties. A *preconditioner* is a matrix for such a transformation. A good preconditioner is a matrix M that is easy to invert and the condition number of $M^{-1}A$ is small. In other words, the preconditioned system $M^{-1}Ax = M^{-1}b$ can be solved efficiently by an appropriate iterative method.

C.2.1 Preconditioned Conjugate Gradient (PCG) Method

Based on this idea, the Preconditioned Conjugate Gradient (PCG) method can be derived with a slight modification of the CG method as described below.

In this algorithm, we need to solve the system $M\bar{r} = r$, which might be as complicated as the original system. The idea for reducing the condition number of $M^{-1}A$ is to choose M such that M^{-1} is close to A^{-1}, while the system $M\bar{r} = r$ is easy to solve. The following theorem provides a choice of M.

Theorem C.2. *Let A be an $n \times n$ nonsingular matrix, and let $A = P - Q$ be a splitting of A such that P is nonsingular. If $H = P^{-1}Q$ and $\rho(H) < 1$, then*

$$A^{-1} = \left(\sum_{k=0}^{\infty} H^k\right)P^{-1}. \tag{C.42}$$

Based on this theorem, we can regard the matrices

$$\begin{aligned} M &= P(I + H + \ldots + H^{m-1})^{-1}, \\ M^{-1} &= (I + H + \ldots + H^{m-1})P^{-1}, \end{aligned} \tag{C.43}$$

PCG Algorithm

1. Initialization: choose $x^{(0)}$, compute $r^{(0)} = b - Ax^{(0)}$ and solve $M\bar{r}^{(0)} = r^{(0)}$. Set $p^{(0)} = \bar{r}^{(0)}$.
2. For $k = 0, 1, \ldots,$

   (i) Compute
   $$\alpha_k = (\bar{r}^{(k)}, r^{(k)})/(Ap^{(k)}, p^{(k)}).$$
   (ii) Set
   $$x^{(k+1)} = x^{(k)} + \alpha_k p^{(k)}.$$
   (iii) Compute
   $$r^{(k+1)} = r^{(k)} - \alpha_k A p^{(k)}.$$
   (iv) If the stopping rule does not apply, continue.
   (v) Solve
   $$M\bar{r}^{(k+1)} = r^{(k+1)}.$$
   (vi) Compute
   $$\beta_k = (\bar{r}^{(k+1)}, r^{(k+1)})/(\bar{r}^{(k)}, r^{(k)}).$$
   (vii) Set
   $$p^{(k+1)} = \bar{r}^{(k+1)} + \beta_k p^{(k)}.$$
3. endFor

as the approximations of A and A^{-1}, respectively. Thus the solution of the system $M\bar{r} = r$ becomes
$$\bar{r} = M^{-1}r = \big(I + H + \ldots + H^{m-1}\big)P^{-1}r.$$
Equivalently, the solution $\bar{r} = r_m$ is the result of applying m steps of the iterative scheme
$$Pr^{(i+1)} = Qr^{(i)} + r, \quad i = 0, 1, \ldots, m-1; \quad r_0 = 0.$$
If $P = D$ and $Q = L + U$, the above iteration is the standard Jacobi method. Then in the PCG method, we replace the system $M\bar{r}^{(k+1)} = r^{(k+1)}$ with *do m Jacobi iterations on* $Ar = r^{(k+1)}$ *to obtain* $\bar{r}^{(k+1)}$. The resulting method is called the **m-step Jacobi PCG** method.

In practice, we may just use the one-step Jacobi PCG method, i.e., $M = D$. Similarly, the symmetric Gauss-Seidel and symmetric Successive Over-Relaxation (SSOR) methods can also be used as preconditioners:

- Symmetric Gauss-Seidel preconditioner:
$$M = (D-L)D^{-1}(D-U), \quad M^{-1} = (D-U)^{-1}D(D-L)^{-1}.$$
- SSOR preconditioner:
$$M = \frac{\omega}{2-\omega}(\omega^{-1}D - L)D^{-1}(\omega^{-1}D - U),$$
$$M^{-1} = \omega(2-\omega)(D-\omega U)^{-1}D(D-\omega L)^{-1}.$$

C.2.2 Preconditioned GMRES Method

If we use M as a left preconditioner for the GMRES method, then we are trying to minimize the residual in the space:

$$S_m(A, r^{(0)}) = \text{span}\{r^{(0)}, M^{-1}Ar^{(0)}, \ldots, (M^{-1}A)^{m-1}r^{(0)}\}. \tag{C.44}$$

The resulting algorithm is the very same as the original GMRES method.

If M is used as a right preconditioner, we just need to replace A in the original GMRES algorithm by AM^{-1}. Also we need to update $x^{(k)}$ by

$$x^{(k)} = x^{(0)} + M^{-1}V_k y^{(k)}. \tag{C.45}$$

In practice, for the GMRES method, the Gauss-Seidel and SOR methods can also be used as preconditioners:

- Gauss-Seidel preconditioner:

$$M = D - L, \quad M^{-1} = (D - L)^{-1}.$$

- SOR preconditioner:

$$M = \omega^{-1}D - L, \, M^{-1} = \omega(D - \omega L)^{-1}.$$

The preconditioned CGS or BiCGStab algorithms can be constructed similarly. In general, to use preconditioners for the CGS method or the BiCGStab method, it is only necessary to replace the matrix A in the original algorithms by $M^{-1}A$ or AM^{-1}.

Appendix D
Basic Time Discretization Schemes

We describe below several standard methods for ordinary differential equations (ODEs), and present some popular time discretization schemes which are widely used in conjunction with spectral methods for partial differential equations (PDEs). We refer to by Gear (1971), Lambert (1991), Hairer et al. (1993), Hairer and Wanner (1996), LeVeque (2007), and Butcher (2008) for thorough discussions on numerical ODEs.

D.1 Standard Methods for Initial-Valued ODEs

Consider the initial value problem (IVP):

$$\frac{dU}{dt} = F(U,t), \quad t>0; \quad U(0)=U_0. \tag{D.1}$$

In general, it may represent a system of ODEs, i.e., $U,F,U_0 \in \mathbb{R}^d$. To this end, let τ be the time step size, and let U^n be the approximation of U at $t_n = n\tau, n \geq 1$.

The simplest method is to approximate dU/dt by the finite difference quotient $U'(t) \approx (U(t+\tau)-U(t))/\tau$. This gives the *forward Euler's method*:

$$U^{n+1} = U^n + \tau F(U^n,t_n), \quad n \geq 0; \quad U^0 = U_0. \tag{D.2}$$

From the initial data U^0, we can compute U^1, then U^2, and so on. Accordingly, it is called a *time marching scheme.*

The *backward Euler's method* is similar, but it is based on approximating $U'(t_{n+1})$ by the backward difference:

$$U^{n+1} = U^n + \tau F(U^{n+1},t_{n+1}), \quad n \geq 0; \quad U^0 = U_0. \tag{D.3}$$

J. Shen et al., *Spectral Methods: Algorithms, Analysis and Applications*,
Springer Series in Computational Mathematics 41, DOI 10.1007/978-3-540-71041-7,

In contrast with (D.2), to march from U^n to U^{n+1}, (D.3) requires to solve for U^{n+1}. It can be viewed as looking for a zero of the function:

$$g(u) = u - \tau F(u, t_{n+1}) - U^n,$$

which can be located by using an iterative method such as the *Newton's method.*

In view of this, the backward scheme (D.3) is an *implicit* method as U^{n+1} must be solved at each iteration, whereas the forward Euler method (D.2) is an *explicit method*.

Another implicit method is the *trapezoidal method*, obtained by averaging two Euler's methods as follows:

$$\frac{U^{n+1} - U^n}{\tau} = \frac{1}{2}\{F(U^n, t_n) + F(U^{n+1}, t_{n+1})\}, \quad n \geq 0; \quad U^0 = U_0. \tag{D.4}$$

As one might expect, this symmetric approximation is second-order accurate, whereas the Euler's methods are only first-order accurate.

The conceptually simplest approach to construct higher-order methods is to use more terms in the Taylor expansion. For example, we consider

$$U(t_{n+1}) \approx U(t_n) + \tau U'(t_n) + \frac{\tau^2}{2} U''(t_n), \tag{D.5}$$

where the remainder of $O(\tau^3)$ has been dropped. In view of (D.1), $U'(t_n)$ can be replaced by $F(U^n, t_n)$, and notice that

$$U''(t) = \frac{d}{dt} F(U(t), t) = F_U(U, t) U'(t) + F_t(U, t),$$

which motivates the approximation:

$$U''(t_n) \approx F_U(U^n, t_n) F(U^n, t_n) + F_t(U^n, t_n).$$

Consequently, we obtain the scheme:

$$U^{n+1} = U^n + \tau F(U^n, t_n) + \frac{\tau^2}{2}\{F_t(U^n, t_n) + F_U(U^n, t_n) F(U^n, t_n)\}. \tag{D.6}$$

It can be shown the above scheme has a second-order accuracy provided that F and the underlying solution U are smooth. However, this can result in very messy and problematic expressions that must be worked out for each equation.

D.1.1 Runge–Kutta Methods

The Taylor method outlined in the previous part has the desirable property of high-order accuracy, but the disadvantage of requiring the computation and evaluation of the derivatives of F, makes it less attractive in practice. One important class of higher-order methods without derivative computations is known as the Runge–Kutta methods.

The second-order Runge–Kutta method is of the form:

$$\begin{aligned} &K_1 = F(U^n, t_n), \quad K_2 = F(U^n + a\tau K_1, t_n + b\tau), \\ &U^{n+1} = U^n + \tau(\alpha K_1 + \beta K_2), \end{aligned} \tag{D.7}$$

where the parameters satisfy

$$\alpha + \beta = 1, \quad a\beta = b\beta = \frac{1}{2}. \tag{D.8}$$

Notice that there are three equations for four unknowns, so we are free to choose one parameter. This results in different schemes.

- If $\alpha = 0, \beta = 1$ and $a = b = 1/2$, we have the *midpoint method*:

$$U^{n+1} = U^n + \tau F\Big(U^n + \frac{\tau}{2} F(U^n, t_n), t_n + \frac{\tau}{2}\Big). \tag{D.9}$$

- If $\alpha = \beta = 1/2$ and $a = b = 1$, the scheme (D.7) is known as the *modified Euler's method*:

$$U^{n+1} = U^n + \frac{\tau}{2}\Big\{F(U^n, t_n) + F\big(U^n + \tau F(U^n, t_n), t_{n+1}\big)\Big\}. \tag{D.10}$$

- If $\alpha = 1/4, \beta = 3/4$ and $a = b = 2/3$, the scheme (D.7) is known as the *Heun method*:

$$U^{n+1} = U^n + \frac{\tau}{4}\Big\{F(U^n, t_n) + 3F\Big(U^n + \frac{2}{3}\tau F(U^n, t_n), t_n + \frac{2}{3}\tau\Big)\Big\}. \tag{D.11}$$

The third-order Runge–Kutta method is given by:

$$\begin{cases} K_1 = F(U^n, t_n), \\ K_2 = F\Big(U^n + \frac{\tau}{2} K_1, t_n + \frac{\tau}{2}\Big), \\ K_3 = F\big(U^n - \tau K_1 + 2\tau K_2, t_n + \tau\big), \\ U^{n+1} = U^n + \frac{\tau}{6}\big(K_1 + 4K_2 + K_3\big). \end{cases} \tag{D.12}$$

The classical fourth-order Runge–Kutta (RK4) method is

$$\begin{cases} K_1 = F(U^n, t_n), \\ K_2 = F\Big(U^n + \frac{\tau}{2} K_1, t_n + \frac{\tau}{2}\Big), \\ K_3 = F\Big(U^n + \frac{\tau}{2} K_2, t_n + \frac{\tau}{2}\Big), \\ K_4 = F(U^n + \tau K_3, t_{n+1}), \\ U^{n+1} = U^n + \frac{\tau}{6}(K_1 + 2K_2 + 2K_3 + K_4). \end{cases} \tag{D.13}$$

The above formula requires four levels of storage, i.e., K_1, K_2, K_3 and K_4. An equivalent formulation is:

$$\begin{cases} U = U^n, \quad G = U, \quad P = F(U, t_n), \\ U = U + \tau P/2, \quad G = P, \quad P = F\big(U, t_n + \tau/2\big), \\ U = U + \tau(P - G)/2, \quad G = G/6, \quad P = F\big(U, t_n + \tau/2\big) - P/2, \\ U = U + \tau P, \quad G = G - P, \quad P = F(U, t_{n+1}) + 2P, \\ U^{n+1} = U + \tau(G + P/6). \end{cases} \tag{D.14}$$

This version of the RK4 method requires only three levels (U, G and P) of storage.

As we saw in the derivation of the Runge–Kutta method of order 2, a number of parameters must be selected. A similar situation occurs in finding higher-order Runge–Kutta methods. Consequently, there is not just one Runge–Kutta method for each order, but a family of methods. It is worthwhile to point out that the number of required *function evaluations* increases more rapidly than the order of the Runge–Kutta methods. This makes the higher-order Runge–Kutta methods less attractive than some other classical fourth-order methods.

D.1.2 Multi-Step Methods

The methods discussed to this point are called *one-step methods* because the approximation at time t_{n+1} involves the information from only one of the previous time t_n. Although these methods might use functional evaluation information at points between t_n and t_{n+1}, they do not retain that information for direct use in future approximations.

We next review some methods using approximations at more than one previous approximations, which are called *multi-step methods.* In general, the r-step linear multi-step methods (LLMs) takes the form

$$\sum_{j=0}^{r} \alpha_j U^{n+j} = \tau \sum_{j=0}^{r} \beta_j F(U^{n+j}, t_{n+j}), \tag{D.15}$$

where U^{n+r} is computed from the equation in terms of the previous approximations $U^{n+r-1}, U^{n+r-2}, \ldots, U^n$ and F at these points (which can be stored and reused if F is expensive to evaluate).

If $\beta_r = 0$, the method (D.15) is explicit; otherwise it is implicit. Note that we can multiply both sides by any non-zero constant and have essentially the same method, so the normalization $\alpha_r = 1$ is often assumed.

The *leap frog method* is a second-order, two-step scheme given by

$$U^{n+1} = U^{n-1} + 2\tau F(U^n, t_n). \tag{D.16}$$

Some special classes of methods are particularly useful with distinctive names. We list a few of them.

The *Adams methods* take the form

$$U^{n+r} = U^{n+r-1} + \tau \sum_{j=0}^{r} \beta_j F(U^{n+j}, t_{n+j}). \tag{D.17}$$

These methods all have

$$\alpha_r = 1, \quad \alpha_{r-1} = -1, \quad \alpha_j = 0 \text{ for all } j < r-1. \tag{D.18}$$

The coefficients β_j are chosen to maximize the order of accuracy. If we require $\beta_r = 0$, the method is explicit and the r coefficients $\{\beta_j\}_{j=0}^{r-1}$ can be chosen so that the method has order r. This can be done by using Taylor expansion of the local truncation error and then choosing the coefficients to eliminate as many terms as possible. This process leads to explicit *Adams-Bashforth methods*.

Another way to derive the Adams-Bashforth methods is by writing

$$U(t_{n+1}) - U(t_n) = \int_{t_n}^{t_{n+1}} U'(t)dt = \int_{t_n}^{t_{n+1}} F(U(t),t)dt, \tag{D.19}$$

and by applying a quadrature formula to approximate

$$\int_{t_n}^{t_{n+1}} F(U(t),t)dt \approx \int_{t_n}^{t_{n+1}} L_{n,r-1}(t)dt, \tag{D.20}$$

where $L_{n,r-1}(t)$ is the Lagrange interpolating polynomial of degree $r-1$ at the points $t_n, t_{n-1}, \ldots, t_{n+r-1}$. The first few schemes obtained from this procedure are listed below, where we denote $F_k := F(U^k, t_k)$.

Explicit Adams-Bashforth methods

1-step:

$$U^{n+1} = U^n + \tau F_n \quad \text{(forward Euler)}$$

2-step:

$$U^{n+2} = U^{n+1} + \frac{\tau}{2}\{3F_{n+1} - F_n\} \quad \text{(AB2)}$$

3-step:

$$U^{n+3} = U^{n+2} + \frac{\tau}{12}\{23F_{n+2} - 16F_{n+1} + 5F_n\} \quad \text{(AB3)}$$

4-step:

$$U^{n+4} = U^{n+3} + \frac{\tau}{24}\{55F_{n+3} - 59F_{n+2} + 37F_{n+1} - 9F_n\} \quad \text{(AB4)}$$

If $\beta_r \neq 0$, then we have one more free parameter. This allows us to derive the r-step *Adams-Moulton methods* of order $r+1$. They are implicit, and we list several such schemes in the box.

One difficulty with LLMs if $r > 1$ is that we have to provide the r initial values $U^0, U^1, \ldots, U^{r-1}$ before we apply the multi-step methods. The initial value U^0 is given, but the other values are not and typically must be generated by some other methods, such as the Runge–Kutta methods.

Implicit Adams-Moulton methods

1-step:

$$U^{n+1} = U^n + \frac{\tau}{2}\{F_{n+1} + F_n\} \quad \text{(Crank-Nicolson method)}$$

2-step:

$$U^{n+2} = U^{n+1} + \frac{\tau}{12}\{5F_{n+2} + 8F_{n+1} - F_n\} \quad \text{(AM3)}$$

3-step:

$$U^{n+3} = U^{n+2} + \frac{\tau}{24}\{9F_{n+3} + 19F_{n+2} - 5F_{n+1} + F_n\} \quad \text{(AM4)}$$

4-step:

$$U^{n+4} = U^{n+3} + \frac{\tau}{720}\{251F_{n+4} + 646F_{n+3} - 264F_{n+2} + 106F_{n+1} - 19F_n\}$$

We have not touched on the theoretical issues of convergence and stability of these methods, and refer the readers to the books of Gear (1971), Lambert (1991), LeVeque (2007), and Butcher (2008) for more detail.

D.1.3 Backward Difference Methods (BDF)

The Adams-Bashforth methods might be unstable due to the fact they are obtained by integrating the interpolating polynomial outside the interval of the data that define the polynomial. This can be remedied by using multilevel implicit methods:

- Second-order backward difference method (BDF2):

$$\frac{1}{2\tau}\left(3U^{n+1} - 4U^n + U^{n-1}\right) = F(U^{n+1}, t_{n+1}). \tag{D.21}$$

- Third-order backward difference method (BDF3):

$$\frac{1}{6\tau}\left(11U^{n+1} - 18U^n + 9U^{n-1} - 2U^{n-2}\right) = F(U^{n+1}, t_{n+1}). \tag{D.22}$$

- Fourth-order backward difference method (BDF4):

$$\frac{1}{12\tau}\left(25U^{n+1} - 48U^n + 36U^{n-1} - 16U^{n-2} + 3U^{n-3}\right) = F(U^{n+1}, t_{n+1}). \tag{D.23}$$

In some applications, $F(u,t)$ is often the sum of linear and nonlinear terms. In this case, some combination of the backward difference method and extrapolation method can be used. To fix the idea, let us consider

$$U_t = L(U) + N(U), \tag{D.24}$$

where L is a linear operator and N is a nonlinear operator. By combining a second-order backward differentiation (BDF2) for the time derivative term and a second-order extrapolation (EP2) for the explicit treatment of the nonlinear term, we obtain a second-order scheme (BDF2/EP2) for (D.24):

$$\frac{1}{2\tau}\left(3U^{n+1} - 4U^n + U^{n-1}\right) = L(U^{n+1}) + N(2U^n - U^{n-1}). \tag{D.25}$$

A third-order scheme for solving (D.24) can be constructed in a similar manner, which leads to the so-called BDF3/EP3 scheme:

$$\begin{aligned} &\frac{1}{6\tau}\left(11U^{n+1} - 18U^n + 9U^{n-1} - 2U^{n-2}\right) \\ &\qquad = L(U^{n+1}) + N(3U^n - 3U^{n-1} + U^{n-2}). \end{aligned} \tag{D.26}$$

D.2 Operator Splitting Methods

In some situations, $F(u,t)$ is the sum of several terms with different nature. It is oftentimes advisable to use an operator splitting method (also called fractional step method) (cf. Yanenko (1971), Marchuk (1974), Godunov (1959), Strang (1968), Goldman and Kaper (1996)). To demonstrate the main idea, we consider

$$\frac{\partial u}{\partial t} = F(u) = Au + Bu; \quad u(t_0) = u_0, \tag{D.27}$$

where $F(u) = Au + Bu$ is frequently split according to physical components, such as density, velocity, energy or dimension.

We first consider the Strang's operator splitting method. For a given time step size $\tau > 0$, let $t_n = n\tau$, and let u^n be the approximation of $u(t_n)$. Let us formally write the solution $u(x,t)$ of (D.27) as

$$u(t) = e^{t(A+B)}u_0 =: S(t)u_0. \tag{D.28}$$

Similarly, denote by $S_1(t) := e^{tA}$ the solution operator for $u_t = Au$, and by $S_2(t) := e^{tB}$ the solution operator for $u_t = Bu$. Then the first-order operator splitting is based on the approximation

$$u^{n+1} \approx S_2(\tau)S_1(\tau)u^n, \tag{D.29}$$

or on the one with the roles of S_2 and S_1 reversed. To construct a second-order scheme, the Strang splitting (cf. Strang (1968)) can be used, in which the solution $S(t_n)u_0$ is approximated by

$$u^{n+1} \approx S_2(\tau/2)S_1(\tau)S_2(\tau/2)u^n, \tag{D.30}$$

or by the one with the roles of S_2 and S_1 reversed.

A fourth-order symplectic time integrator (cf. Yoshida (1990), Lee and Fornberg (2003)) for (D.27) is as follows:

$$\begin{aligned} &u^{(1)} = e^{2w_1A\tau}\,u^n, \quad u^{(2)} = e^{2w_2B\tau}\,u^{(1)}, \quad u^{(3)} = e^{2w_3A\tau}\,u^{(2)},\\ &u^{(4)} = e^{2w_4B\tau}\,u^{(3)}, \quad u^{(5)} = e^{2w_3A\tau}\,u^{(4)}, \quad u^{(6)} = e^{2w_2B\tau}\,u^{(5)},\\ &u^{n+1} = e^{2w_1A\tau}\,u^{(6)}; \end{aligned} \tag{D.31}$$

or equivalently,

$$\begin{aligned} u^{n+1} \approx\, &S_1(2w_1\tau)S_2(2w_2\tau)S_1(2w_3\tau)S_2(2w_4\tau)\\ &S_1(2w_3\tau)S_2(2w_2\tau)S_1(2w_1\tau)u^n, \end{aligned} \tag{D.32}$$

where

$$\begin{aligned} w_1 &= 0.33780\,17979\,89914\,40851,\\ w_2 &= 0.67560\,35959\,79828\,81702,\\ w_3 &= -0.08780\,17979\,89914\,40851,\\ w_4 &= -0.85120\,71979\,59657\,63405. \end{aligned} \tag{D.33}$$

References

Abramowitz M, Stegun I (1964) Handbook of Mathematical Functions with Formulas, Graphs, and Mathematical Tables. Dover publications

Adams R (1975) Sobolov Spaces. Acadmic Press, New York

Aguirre J, Rivas J (2005) Hermite pseudospectral approximations. An error estimate. Journal of Mathematical Analysis and Applications 304(1):189–197

Ali I, Brunner H, Tang T (2009) Spectral methods for pantograph-type differential and integral equations with multiple delays. Front Math China 4:49–61

Allen S, Cahn J (1979) A microscopic theory for antiphase boundary motion and its application to antiphase domain coarsening. Acta Metall Mater 27:1085–1095

Alpert B, Rokhlin V (1991) A fast algorithm for the evaluation of Legendre expansions. SIAM J Sci Stat Comput 12:158–179

Anderson D, McFadden G, Wheeler A (1998) Diffuse-interface methods in fluid mechanics. Annual review of fluid mechanics, Vol 30 30:139–165

Arrow K, Hurwicz L, Uzawa H (1958) Studies in Nonlinear Programming. Stanford University Press

Ascher U, Mattheij R, Russell R (1995) Numerical solution of boundary value problems for ordinary differential equations, Classics in Applied Mathematics, vol 13. Society for Industrial and Applied Mathematics (SIAM), Philadelphia, PA, corrected reprint of the 1988 original

Auteri F, Quartapelle L (2000) Galerkin-Legendre spectral method for the 3D Helmholtz equation. J Comput Phys 161(2):454–483

Azaiez M, Shen J, Xu C, Zhuang Q (2008) A Laguerre-Legendre spectral method for the Stokes problem in a semi-infinite channel. SIAM J Numer Anal 47:271–292

Babuška I (1973) The finite element method with Lagrangian multipliers. Numer Math 20:179–192

Babuška I, Aziz A (1972) Survey lectures on the mathematical foundations of the finite element method. In: The mathematical foundations of the finite element method with applications to partial differential equations (Proc. Sympos., Univ. Maryland, Baltimore, Md., 1972), Academic Press, New York, pp 1–359, with the collaboration of G. Fix and R. B. Kellogg

Bao W, Shen J (2005) A Fourth-order time-splitting Laguerre-Hermite pseudo-spectral method for Bose-Einstein condensates. SIAM J Sci Comput 26:2110–2028

Bao W, Jaksch D, Markowich P (2003a) Numerical solution of the Gross-Pitaevskii equation for Bose-Einstein condensation. J Comput Phys 187(1):318–342

Bao W, Jin S, Markowich P (2003b) Numerical study of time-splitting spectral discretizations of nonlinear Schrödinger equations in the semiclassical regimes. SIAM J Sci Comput 25(1):27–64 (electronic)

Barrett R, Berry M, Chan T, et al (1994) Templates for the Solution of Linear Systems: Building Blocks for Iterative Methods. Society for Industrial and Applied Mathematics (SIAM), Philadelphia, PA

J. Shen et al., *Spectral Methods: Algorithms, Analysis and Applications*, Springer Series in Computational Mathematics 41, DOI 10.1007/978-3-540-71041-7,

Barthelmann V, Novak E, Ritter K (2000) High dimensional polynomial interpolation on sparse grids. Adv Comput Math 12:273–288

Bellen A, Zennaro M (2003) Numerical Methods for Delay Differential Equations. Oxford University Press, Oxford

Bellman R (1961) Adaptive Control Processes: A Guided Tour. Princeton University Press, Princeton, N.J.

Berenger J (1994) A perfectly matched layer for the absorption of electromagnetic waves. J Comput Phys 114(2):185–200

Bergh J, Löfström J (1976) Interpolation Spaces. An Introduction. Springer-Verlag, Berlin, grundlehren der Mathematischen Wissenschaften, No. 223

Bernardi C, Maday Y (1992a) Approximations Spectrales de Problèmes aux Limites Elliptiques. Springer-Verlag, Paris

Bernardi C, Maday Y (1992b) Polynomial interpolation results in Sobolev spaces. Journal of Computational and Applied Mathematics 43(1-2):53–80

Bernardi C, Maday Y (1997) Spectral Method. In: Ciarlet P, Lions L (eds) Handbook of Numerical Analysis, V. 5 (Part 2), North-Holland

Bernardi C, Maday Y (1999) Uniform inf-sup conditions for the spectral discretization of the Stokes problem. Math Models Methods Appl Sci 9(3):395–414

Bernardi C, Coppoletta G, Maday Y (1992) Some spectral approximations of two-dimensional fourth-order problems. Math Comp 59:63–76

Bernardi C, Dauge M, Maday Y (1999) Spectral Methods for Axisymmetric Domains. Gauthier-Villars, Éditions Scientifiques et Médicales Elsevier, Paris

Bjørstad P, Tjostheim B (1997) Efficient algorithms for solving a fourth order equation with the spectral-galerkin method. SIAM J Sci Comput 18:621–632

Boyd J (1978) The choice of spectral functions on a sphere for boundary and eigenvalue problems: a comparison of Chebyshev, Fourier and associated Legendre expansions. Monthly Weather Rev 106:1184–1191

Boyd J (1980) The rate of convergence of Hermite function series. Math Comp 35(152):1309–1316

Boyd J (1982) The optimization of convergence for Chebyshev polynomial methods in an unbounded domain. Journal of Computational Physics 45(1):43–79

Boyd J (1987a) Orthogonal rational functions on a semi-infinite interval. J Comput Phys 70:63–88

Boyd J (1987b) Spectral methods using rational basis functions on an infinite interval. J Comput Phys 69:112–142

Boyd J (2001) Chebyshev and Fourier Spectral Methods, 2nd edn. Dover Publications Inc., Mineola, NY

Boyd J (2004) Prolate spheroidal wavefunctions as an alternative to Chebyshev and Legendre polynomials for spectral element and pseudospectral algorithms. J Comput Phys 199(2):688–716

Brenner S, Scott L (2008) The Mathematical Theory of Finite Element Methods, 3rd edn. Springer Verlag

Brezzi F (1974) On the existence, uniqueness and approximation of saddle-point problems arising from Lagrangian multipliers. Rev Française Automat Informat Recherche Opérationnelle Sér Rouge 8(R-2):129–151

Brezzi F, Fortin M (1991) Mixed and Hybrid Finite Element Methods. Springer-Verlag, New York

Brown DL, Cortez R, Minion ML (2001) Accurate projection methods for the incompressible Navier-Stokes equations. J Comput Phys 168(2):464–499

Brunner H (2004) Collocation Methods for Volterra Integral and Related Functional Differential Equations, Cambridge Monographs on Applied and Computational Mathematics, vol 15. Cambridge University Press, Cambridge

Brunner H, Hu Q (2007) Optimal superconvergence results for delay integro-differential equations of pantograph type. SIAM J Numer Anal 45:986–1004

Brunner H, Tang T (1989) Polynomial spline collocation methods for the nonlinear Basset equation. Comput Math Appl 18(5):449–457

Bungartz H, Griebel M (2004) Sparse grids. Acta Numer 13:147–269

Butcher J (2008) Numerical Methods for Ordinary Differential Equations, 2nd edn. John Wiley & Sons Ltd., Chichester

Butzer P, Nessel R (1971) Fourier Analysis and Approximation. Academic Press, New York, volume 1: One-dimensional theory, Pure and Applied Mathematics, Vol. 40

Caffarelli L, Muler N (1995) An L^∞ bound for solutions of the Cahn-Hilliard equation. Arch Rational Mech Anal 133(2):129–144

Cahn J, Hilliard J (1958) Free energy of a nonuniform system, I: Interfacial free energy. J Chem Phys 28:258

Canuto C, Quarteroni A (1982) Approximation results for orthogonal polynomials in Sobolev spaces. Math Comp 38:67–86

Canuto C, Quarteroni A (1985) Preconditioner minimal residual methods for Chebyshev spectral calculations. J Comput Phys 60:315–337

Canuto C, Quarteroni A (1994) Variational methods in the theoretical analysis of spectral approximations. In: Voigt R, Gottlieb D, Hussaini M (eds) Spectral Methods for Partial Differential Equations, SIAM, pp 55–78

Canuto C, Hussaini M, Quarteroni A, Zang T (1987) Spectral Methods in Fluid Dynamics. Springer-Verlag

Canuto C, Hussaini M, Quarteroni A, Zang T (2006) Spectral Methods. Scientific Computation, Springer-Verlag, Berlin, fundamentals in single domains

Canuto C, Hussaini M, Quarteroni A, Zang T (2007) Spectral Methods. Scientific Computation, Springer, Berlin, evolution to complex geometries and applications to fluid dynamics

Céa J (1964) Approximation variationnelle des problèmes aux limites. Ann Inst Fourier (Grenoble) 14(fasc. 2):345–444

Chen F, Shen J (2011) Efficient spectral Galerkin methods for coupled systems of elliptic equations with separable boundary conditions. Preprint

Chen L (2002) Phase-field models for microstructure evolution. Annual Review of Material Research 32:113

Chen L, Shen J (1998) Applications of semi-implicit Fourier-spectral method to phase-field equations. Comput Phys Comm 108:147–158

Chen Q, Gottlieb D, Hesthaven J (2005) Spectral methods based on prolate spheroidal wave functions for hyperbolic PDEs. SIAM J Numer Anal 43(5):1912–1933 (electronic)

Chen Y, Tang T (2010) Convergence analysis of the Jacobi spectral-collocation methods for Volterra integral equations with a weakly singular kernel. Math Comp 79:147–167

Cheney E (1998) Introduction to Approximation Theory. AMS Chelsea Publishing, Providence, RI, reprint of the second (1982) edition

Chorin A (1968) Numerical solution of the Navier-Stokes equations. Math Comp 22:745–762

Christov C (1982) A complete orthogonal system of functions in $l^2(-\infty,\infty)$ space. SIAM J Appl Math 42:1337–1344

Ciarlet P (1978) The Finite Element Method for Elliptic Problems. North-Holland

Ciskowski R, Brebbia C (1991) Boundary Element Methods in Acoutics. Kluwer Academic Publishers

Cloot A, Weideman J (1992) An adaptive algorithm for spectral computations on unbounded domains. J Comput Phys 102(2):398–406

Colin T, Ghidaglia J (2001) An initial-boundary value problem for the Korteweg-de Vries equation posed on a finite interval. Adv Differential Equations 6(12):1463–1492

Condette N, Melcher C, Süli E (2011) Spectral approximation of pattern-forming nonlinear evolution equations with double-well potentials of quadratic growth. Math Comp 80(273):205–223

Cooley J, Tukey J (1965) An algorithm for the machine calculation of complex Fourier series. Math Comp 19:297–301

Cummings P, Feng X (2006) Sharp regularity coefficient estimates for complex-valued acoustic and elastic Helmholtz equations. Math Models Methods Appl Sci 16(1):139–160

Davis P (1975) Interpolation and Approximation. Dover Publications, Inc, New Year

Davis P, Rabinowitz P (1984) Methods of Numerical Integration, 2nd edn. Computer Science and Applied Mathematics, Academic Press Inc., Orlando, FL

Delves L, Mohanmed J (1985) Computational Methods for Integral Equations. Cambridge University Press

Demaret P, Deville M (1991) Chebyshev collocation solution of the Navier-Stokes equations using multi-domain decomposition and finite element preconditioning. J Comput Phys 95:359–386

Demkowicz L, Ihlenburg F (2001) Analysis of a coupled finite-infinite element method for exterior Helmholtz problems. Numer Math 88(1):43–73

Deville M, Mund E (1985) Chebyshev pseudospectral solution of second-order elliptic equations with finite element preconditioning. J Comput Phys 60:517–533

Deville M, Fischer P, Mund E (2002) High-Order Methods for Incompressible Fluid Flow, Cambridge Monographs on Applied and Computational Mathematics, vol 9. Cambridge University Press, Cambridge

Djidjeli K, Price W, Twizell E, Wang Y (1995) Numerical methods for the solution of the third- and fifth-order dispersive Korteweg-de-Vries equations. J Comput Appl Math 58:307–336

Dobrovol'skiĭ N, Roshchenya A (1998) On the number of points in a lattice in a hyperbolic cross. Mat Zametki 63(3):363–369

Don W, Gottlieb D (1994) The Chebyshev-Legendre method: implementing Legendre methods on Chebyshev points. SIAM J Numer Anal 31:1519–1534

Douglas J, Santos J, Sheen D, Bennethum L (1993) Frequency domain treatment of one-dimensional scalar waves. Math Models Methods Appl Sci 3(2):171–194

E W, Liu JG (1996) Projection method. II. Godunov-Ryabenki analysis. SIAM J Numer Anal 33(4):1597–1621

Eisen H, Heinrichs W, Witsch K (1991) Spectral collocation methods and polar coordinate singularities. J Comput Phys 96:241–257

El-Daou M, Ortiz E (1998) The tau method as an analytic tool in the discussion of equivalence results across numerical methods. Computing 60(4):365–376

Elman H, Silvester D, Wathen A (2005) Finite elements and fast iterative solvers: with applications in incompressible fluid dynamics. Numerical Mathematics and Scientific Computation, Oxford University Press, New York

Elnagar G, Kazemi M (1996) Chebyshev spectral solution of nonlinear Volterra-Hammerstein integral equations. J Comput Appl Math 76(1-2):147–158

Engquist B, Majda A (1977) Absorbing boundary conditions for the numerical simulation of waves. Math Comp 31(139):629–651

Eyre D (1998) Unconditionally gradient stable time marching the Cahn-Hilliard equation. In: Computational and mathematical models of microstructural evolution (San Francisco, CA, 1998), Mater. Res. Soc. Sympos. Proc., vol 529, MRS, Warrendale, PA, pp 39–46

Faber V, Manteuffel T (1984) Necessary and sufficient conditions for the existence of a conjugate gradient method. SIAM J Numer Anal 21(2):352–362

Fang Q, Nicholls D, Shen J (2007) A stable, high-order method for three-dimensional, bounded-obstacle, acoustic scattering. J Comput Phys 224(2):1145–1169

Fang Q, Shen J, Wang L (2009) An efficient and accurate spectral method for acoustic scattering in elliptic domains. Numer Math: Theory, Methods Appl 2:258–274

Finlayson B (1972) The Method of Weighted Residuals and Variational Principles. Academic Press, New York

Fok J, Guo B, Tang T (2002) Combined Hermite spectral-finite difference method for the Fokker-Planck equation. Math Comp 71(240):1497–1528 (electronic)

Folland G (1992) Fourier Analysis and Its Applications. The Wadsworth & Brooks/Cole Mathematics Series, Wadsworth & Brooks/Cole Advanced Books & Software, Pacific Grove, CA

Fornberg B (1995) A pseudospectral approach for polar and spherical geometries. SIAM J Sci Comput 16:1071–1081

Fox L, Mayers D, Ockendon J, Taylor A (1971) On a functional differential equation. J Inst Math Appl 8:271–307

Franken P, Deville M, Mund E (1990) On the spectrum of the iteration operator associated to the finite element preconditioning of chebyshev collocation calculations. Comput Methods in Appl Mech Eng 80:295–304

Fujiwara H (2006) High-accurate numerical method for integral equations of the first kind under multiple-precision arithmetic. Preprint, RIMS, Kyoto University

Funaro D (1992) Polynomial Approxiamtions of Differential Equations. Springer-Verlag

Funaro D, Kavian O (1990) Approximations of some diffusion evolution equations in unbounded domains by Hermite functions. Math Comp 57:597–619

Gautschi W (2004) Orthogonal polynomials: computation and approximation. Numerical Mathematics and Scientific Computation, Oxford University Press, New York, oxford Science Publications

Gear C (1971) Numerical Initial Value Problems in Ordinary Differential Equations. Prentice Hall, NJ, USA

Gerdes K, Demkowicz L (1996) Solution of 3D-Laplace and Helmholtz equations in exterior domains using $hp-$infinite elements. Comput Methods Appl Mech Engrg 137:239–273

Girault V, Raviart P (1986) Finite Element Methods for Navier-Stokes Equations. Springer-Verlag

Glowinski R (2003) Finite Element Methods for Incompressible Viscous Flow. In: Handbook of Numerical Analysis, Vol. IX, Handb. Numer. Anal., IX, North-Holland, Amsterdam, pp 3–1176

Godunov S (1959) Finite difference methods for numerical computations of discontinuous solutions of the equations of fluid dynamics. Mat Sb 47(3):271–295, in Russian

Goldman D, Kaper T (1996) Nth-order operator splitting schemes and nonreversible systems. SIAM J Numer Anal 33(1):349–367

Golub G, Van Loan C (1996) Matrix Computations, 3rd edn. Johns Hopkins Studies in the Mathematical Sciences, Johns Hopkins University Press, Baltimore, MD

Gottlieb D, Lustman L (1983) The spectrum of the Chebyshev collocation operator for the heat equation. SIAM J Numer Anal 20:909–921

Gottlieb D, Orszag S (1977) Numerical Analysis of Spectral Methods: Theory and Applications. SIAM-CBMS, Philadelphia

Gottlieb D, Shu C (1997) On the Gibbs phenomenon and its resolution. SIAM Rev 39(4):644–668

Gottlieb D, Hussaini MY, Orszag SA (1984) Theory and applications of spectral methods. In: Spectral methods for partial differential equations (Hampton, Va., 1982), SIAM, Philadelphia, PA, pp 1–54

Gottlieb D, Shu C, Solomonoff A, Vandeven H (1992) On the Gibbs phenomenon I: recovering exponential accuracy from the Fourier partial sum of a nonperiodic analytic function. Journal of Computational and Applied Mathematics 43(1-2):81–98

Goubet O, Shen J (2007) On the dual Petrov-Galerkin formulation of the KdV equation on a finite interval. Adv Differential Equations 12(2):221–239

Greengard L, Rokhlin V (1987) A fast algorithm for particle simulations. J Comput Phys 73:325–348

Griebel M, Hamaekers J (2007) Sparse grids for the Schrödinger equation. M2AN Math Model Numer Anal 41(2):215–247

Grosch C, Orszag S (1977) Numerical solution of problems in unbounded regions: coordinates transforms. J Comput Phys 25:273–296

Gross E (1961) Structure of a quantized vortex in boson systems. Nuovo Cimento (10) 20:454–477

Grote M, Keller J (1995) On nonreflecting boundary conditions. J Comput Phys 122(2):231–243

Guermond J, Shen J (2003) A new class of truly consistent splitting schemes for incompressible flows. J Comput Phys 192(1):262–276

Guermond J, Shen J (2004) On the error estimates of rotational pressure-correction projection methods. Math Comp 73:1719–1737

Guermond J, Minev P, Shen J (2006) An overview of projection methods for incompressible flows. Comput Methods Appl Mech Engrg 195:6011–6045

Guo B (1998a) Gegenbauer approximation and its applications to differential equations on the whole line. J Math Anal Appl 226:180–206

Guo B (1998b) Spectral Methods and Their Applications. World Scientific Publishing Co. Inc., River Edge, NJ

Guo B (1999) Error estimation of Hermite spectral method for nonlinear partial differential equations. Math Comp 68(227):1067–1078

Guo B (2000) Jacobi approximations in certain Hilbert spaces and their applications to singular differential equations. J Math Anal Appl 243:373–408

Guo B, Shen J (2001) On spectral approximations using modified Legendre rational functions: application to the Korteweg-de Vries equation on the half line. Indiana Univ Math J 50:181–204

Guo B, Shen J (2008) Irrational approximations and their applications to partial differential equations in exterior domains. Adv Comput Math 28(3):237–267

Guo B, Wang L (2001) Jacobi interpolation approximations and their applications to singular differential equations. Advances in Computational Mathematics 14:227–276

Guo B, Wang L (2004) Jacobi approximations in non-uniformly Jacobi-weighted Sobolev spaces. J Approx Theory 128(1):1–41

Guo B, Xu C (2000) On two-dimensional unsteady incompressible fluid flow in an infinite strip. Math Methods Appl Sci 23(18):1617–1636

Guo B, Shen J, Wang Z (2000) A rational approximation and its applications to differential equations on the half line. J Sci Comp 15:117–147

Guo B, Shen J, Wang Z (2002) Chebyshev rational spectral and pseudospectral methods on a semi-infinite interval. Internat J Numer Methods Engrg 53(1):65–84, *p* and *hp* finite element methods: mathematics and engineering practice (St. Louis, MO, 2000)

Guo B, Shen J, Xu C (2003) Spectral and pseudospectral approximations using Hermite functions: Application to the Dirac equation. Adv Comput Math 19:35–55

Guo B, Shen J, Xu C (2005) Generalized Laguerre approximation and its applications to exterior problems. J Comput Math 23(2):113–130

Guo B, Shen J, Wang L (2006a) Optimal spectral-Galerkin methods using generalized Jacobi polynomials. J Sci Comput 27(1-3):305–322

Guo B, Wang L, Wang Z (2006b) Generalized Laguerre interpolation and pseudospectral method for unbounded domains. SIAM J Numer Anal 43(6):2567–2589 (electronic)

Guo B, Shen J, Wang L (2009) Generalized Jacobi polynomials/functions and their applications. Appl Numer Math 59(5):1011–1028

Haidvogel D, Zang T (1979) The accurate solution of Poisson's equation by expansion in Chebyshev polynomials. J Comput Phys 30:167–180

Hairer E, Wanner G (1996) Solving Ordinary Differential Equations. II, Springer Series in Computational Mathematics, vol 14, 2nd edn. Springer-Verlag, Berlin, stiff and differential-algebraic problems

Hairer E, Nørsett S, Wanner G (1993) Solving Ordinary Differential Equations. I. Nonstiff Problems, 2nd edition, Springer Series in Computational Mathematics, vol 8. Springer-Verlag, Berlin

Haldenwang P, Labrosse G, Abboudi S, Deville M (1984) Chebyshev 3-D spectral and 2-D pseudospectral solvers for the Helmholtz equation. J Comput Phys 55:115–128

Harari I, Hughes T (1992) Analysis of continuous formulations underlying the computation of time-harmonic acoustics in exterior domains. Comput Methods Appl Mech Engrg 97(1): 103–124

Hardy G, Littlewood J, Pólya G (1952) Inequalities. Cambridge University Press, UK

Heinrichs W (1989) Spectral methods with sparse matrices. Numer Math 56:25–41

Hestenes M, Stiefel E (1952) Methods of conjugate gradients for solving linear systems. J Res Nat Bur Stand 49(6):409–436

Hesthaven J, Warburton T (2008) Nodal Discontinuous Galerkin Methods, Texts in Applied Mathematics, vol 54. Springer, New York, algorithms, analysis, and applications

Hesthaven J, Gottlieb S, Gottlieb D (2007) Spectral Methods for Time-Dependent Problems. Cambridge Monographs on Applied and Computational Mathematics, Cambridge

Hu Z, Wise S, Wang C, Lowengrub J (2009) Stable and efficient finite-difference nonlinear-multigrid schemes for the phase field crystal equation. J Comput Phys 228(15):5323–5339

Huang W, Sloan D (1992) The pseudospectral method for third-order differential equations. SIAM J Numer Anal 29(6):1626–1647

Huang W, Sloan D (1993) Pole condition for singular problems: the pseudospectral approximation. J Comput Phys 107:254–261

Hyman J, Nicolaenko B (1986) The Kuramoto-Sivashinsky equation: a bridge between PDEs and dynamical systems. Phys D 18(1-3):113–126

Ihlenburg F, Babuška I (1995) Finite element solution of the Helmholtz equation with high wave number, part I: the h-version of FEM. Computers Math Applic 30:9–37

Ihlenburg F, Babuška I (1997) Finite element solution of the Helmholtz equation with high wave number. II. The *h-p* version of the FEM. SIAM J Numer Anal 34(1):315–358

Iserles A (1993) On the generalized pantograph functional differential equation. Europ J Appl Math 4:1–38

Johnston H, Liu JG (2004) Accurate, stable and efficient Navier-Stokes solvers based on explicit treatment of the pressure term. J Comput Phys 199(1):221–259

Karniadakis G, Sherwin S (1999) Spectral/hp Element Methods for CFD. Oxford UniversityPress

Karniadakis G, Sherwin S (2005) Spectral/*hp* Element Methods for Computational Fluid Dynamics, 2nd edn. Numerical Mathematics and Scientific Computation, Oxford University Press, New York

Kassam A, Trefethen L (2005) Fourth-order time stepping for stiff PDEs. SIAM J Sci Comput 26(4):1214–1233

Kawahara R (1972) Oscillatory solitary waves in dispersive media. J Phys Soc Japan 33:260–264

Keller J, Givoli D (1989) Exact non-reflecting boundary conditions. J Comput Phys 82:172–192

Kessler D, Nochetto R, Schmidt A (2004) A posteriori error control for the Allen-Cahn problem: circumventing Gronwall's inequality. M2AN Math Model Numer Anal 38(1):129–142

Kichenassamy S, Olver P (1992) Existence and nonexistence of solitary wave solutions to higher-order model evolution equations. SIAM J Math Anal 23(5):1141–1166

Kim S, Parter S (1997) Preconditioning Chebyshev spectral collocation by finite difference operators. SIAM J Numer Anal 34, No. 3:939–958

Körner T (1988) Fourier Analysis. Cambridge University Press, Cambridge

Korobov N (1992) Exponential Sums and Their Applications, Mathematics and its Applications (Soviet Series), vol 80. Kluwer Academic Publishers Group, Dordrecht, translated from the 1989 Russian original by Y.N. Shakhov

Korteweg D, de Vries G (1895) On the change of form of long waves advancing in a rectangular canal, and on a new type of long stationary waves. Philosophical Magazine 39:422–443

Kreiss HO, Oliger J (1979) Stability of the Fourier method. SIAM J Numer Anal 16(3):421–433

Krylov V (1962) Approximate Calculation of Integrals. Macmillan New York

Kuramoto Y, Tsuzuki T (1976) Persistent propagation of concentration waves in dissipative media far from thermal equilibrium. Prog Theoret Phys 55(2):356–369

Lambert J (1991) Numerical Methods for Ordinary Differential Systems: The initial value problem. John Wiley & Sons Ltd., Chichester

Le Maître O, Knio O (2010) Spectral Methods for Uncertainty Quantification. Scientific Computation, Springer, New York, with applications to computational fluid dynamics

Leboeuf P, Pavloff N (2001) Bose-Einstein beams: coherent propagation through a guide. Phys Rev A 64:aritcle 033,602

Lee J, Fornberg B (2003) A split step approach for the 3-d Maxwell's equations. J Comput and Appl Math 158(2):485–505

Lether F (1978) On the construction of Gauss-Legndre quadrature rules. J Comp Appl Math 4(1):47–52

LeVeque R (2007) Finite Difference Methods for Ordinary and Partial Differential Equations: Steady-State and Time-Dependent Problems. Society for Industrial and Applied Mathematics (SIAM), Philadelphia, PA

Levin A, Lubinsky D (1992) Christoffel functions, orthogonal polynomials, and Nevai's conjecture for Freud weights. Constr Approx 8(4):461–533

Lions J, Magenes E (1968) Problemes aux limites non homogenes et applications. Vol. 1-2. Dunod, Paris

Liu C, Shen J (2003) A phase field model for the mixture of two incompressible fluids and its approximation by a Fourier-spectral method. Physica D 179(3-4):211–228

Liu JG, Liu J, Pego RL (2007) Stability and convergence of efficient Navier-Stokes solvers via a commutator estimate. Comm Pure Appl Math 60(10):1443–1487

Liu Y, Liu L, Tang T (1994) The numerical computation of connecting orbits in dynamical systems: a rational spectral approach. J Comput Phys 111(2):373–380

Lopez J, Perry A (1992) Axisymmetric vortex breakdown. III. Onset of periodic flow and chaotic advection. J Fluid Mech 234:449–471

Lopez J, Shen J (1998) An efficient spectral-projection method for the Navier-Stokes equations in cylindrical geometries I. axisymmetric cases. J Comput Phys 139:308–326

Lopez J, Marques F, Shen J (2002) An efficient spectral-projection method for the Navier-Stokes equations in cylindrical geometries. II. Three-dimensional cases. J Comput Phys 176(2):384–401

Lowengrub J, Truskinovsky L (1998) Quasi-incompressible Cahn-Hilliard fluids and topological transitions. R Soc Lond Proc Ser A Math Phys Eng Sci 454(1978):2617–2654

Lynch R, Rice J, Thomas D (1964) Direct solution of partial differential equations by tensor product methods. Numer Math 6:185–199

Ma H, Sun W (2000) A Legendre-Petrov-Galerkin method and Chebyshev-collocation method for the third-order differential equations. SIAM J Numer Anal 38(5):1425–1438

Ma H, Sun W (2001) Optimal error estimates of the Legendre-Petrov-Galerkin method for the Korteweg-de Vries equation. SIAM J Numer Anal 39(4):1380–1394 (electronic)

Ma H, Sun W, Tang T (2005) Hermite spectral methods with a time-dependent scaling for parabolic equations in unbounded domains. SIAM J Numer Anal 43(1):58–75

Maday Y, Pernaud-Thomas B, Vandeven H (1985) Reappraisal of Laguerre type spectral methods. La Recherche Aerospatiale 6:13–35

Maday Y, Meiron D, Patera A, Rønquist E (1993) Analysis of iterative methods for the steady and unsteady Stokes problem: application to spectral element discretizations. SIAM J Sci Comput 14(2):310–337

Marchuk G (1974) Numerical Methods in Weather Prediction. Academic Press, New York

Marion M, Temam R (1998) Navier-Stokes Equations: Theory and Approximation. In: Handbook of Numerical Analysis, Vol. VI, Handb. Numer. Anal., VI, North-Holland, Amsterdam, pp 503–688

Mastroianni G, Occorsio D (2001a) Lagrange interpolation at Laguerre zeros in some weighted uniform spaces. Acta Math Hungar 91(1-2):27–52

Mastroianni G, Occorsio D (2001b) Optimal systems of nodes for lagrange interpolation on bounded intervals. a survey. Journal of Computational and Applied Mathematics 134:325–341

McLachlan N (1951) Theory and Applications of Mathieu Functions. Oxford Press, London

Melenk J (1995) On generalized finite element methods. PhD thesis, University of Maryland, College Park

Merryfield W, Shizgal B (1993) Properties of collocation third-derivative operators. J Comput Phys 105(1):182–185

Morse P, Feshback H (1953) Methods of Theoretical Physics. McGraw-Hill, New York

Nagashima H, Kawahara M (1981) Computer simulation of solitary waves of the nonlinear wave equations. J Phys Soc Japan 50:3792–3800

Nicholls D, Reitich F (2003) Analytic continuation of Dirichlet-Neumann operators. Numer Math 94(1):107–146

Nicholls D, Shen J (2006) A stable, high–order method for two–dimensional bounded–obstacle scattering. SIAM J Sci Comput 28:1398–1419

Nicholls D, Shen J (2009) A rigorous numerical analysis of the transformed field expansion method. SIAM J Numer Anal 47(4):2708–2734

Nicolaenko B, Scheurer B, Temam R (1985) Some global dynamical properties of the Kuramoto-Sivashinsky equations: nonlinear stability and attractors. Phys D 16(2):155–183

Orszag S (1971) Accurate solution of the Orr-Sommerfeld equation. Journal of Fluid Mechanics 50:689–703, [Combines Chebyshev method with QR or QZ matrix eigensolver to solve a linear stability problem.]

Orszag S (1974) Fourier series on spheres. Monthly Weather Rev 102:56–75

Orszag S (1980) Spectral methods for complex geometries. J Comput Phys 37:70–92

Orszag S, Patera A (1983) Secondary instability of wall-bounded shear flows. J Fluid Mech 128:347–385

Pan V (1997) Solving a polynomial equation: some history and recent progress. SIAM Rev 39(2):187–220

Parkes E, Zhu Z, Duffy B, Huang H (1998) Sech-polynomial travelling solitary-wave solutions of odd-order generalized KdV equations. Physics Letters A 248(2-4):219–224

Parter S, Rothman E (1995) Preconditioning lengendre spectral collocation approximations to elliptic problems. SIAM J Numer Anal 32, No. 2:333–385

Pitaevskii L (1961) Vortex lines in an imperfect bose gase. Sov Phys JETP 13:451–454

Potts D, Steidl G, Tasche M (1998) Fast algorithms for discrete polynomial transforms. Math Comp 67(224):1577–1590

Qu C, Wong R (1988) Szego's conjecture on Lebesgue constants for Legendre series. Pacific J Math 135(1):157–188

Quarteroni A, Valli A (2008) Numerical Approximation of Partial Differential Equations. Springer Verlag

Ragozin D (1970) Polynomial approximation on compact manifolds and homogeneous spaces. Trans Amer Math Soc 150:41–53

Ragozin D (1971) Constructive polynomial approximation on spheres and projective spaces. Trans Amer Math Soc 162:157–170

Reddy S, Weideman J (2005) The accuracy of the Chebyshev differencing method for analytic functions. SIAM J Numer Anal 42(5):2176–2187

Rivlin T (1974) The Chebyshev Polynomials. Jong Wiley and Sons

Rokhlin V, Tygert M (2006) Fast algorithms for spherical harmonic expansions. SIAM J Sci Comput 27(6):1903–1928 (electronic)

Saad Y (2003) Iterative Methods for Sparse Linear Systems, 2nd edn. Society for Industrial and Applied Mathematics, Philadelphia, PA

Saad Y, Schultz M (1986) GMRES: a generalized minimal residual algorithm for solving nonsymmetric linear systems. SIAM J Sci Statist Comput 7(3):856–869

Shen J (1991) Hopf bifurcation of the unsteady regularized driven cavity flows. J Comput Phys 95:228–245

Shen J (1994) Efficient spectral-Galerkin method I. direct solvers for second- and fourth-order equations by using Legendre polynomials. SIAM J Sci Comput 15:1489–1505

Shen J (1995) Efficient spectral-Galerkin method II. direct solvers for second- and fourth-order equations by using Chebyshev polynomials. SIAM J Sci Comput 16:74–87

Shen J (1996) Efficient Chebyshev-Legendre Galerkin methods for elliptic problems. In: Ilin AV, Scott R (eds) Proceedings of ICOSAHOM'95, Houston J. Math., pp 233–240

Shen J (1997) Efficient spectral-Galerkin methods III. polar and cylindrical geometries. SIAM J Sci Comput 18:1583–1604

Shen J (1999) Efficient spectral-Galerkin methods IV. spherical geometries. SIAM J Sci Comput 20:1438–1455

Shen J (2000) A new fast Chebyshev-Fourier algorithm for the Poisson-type equations in polar geometries. Appl Numer Math 33:183–190

Shen J (2003) A new dual-Petrov-Galerkin method for third and higher odd-order differential equations: application to the KDV equation. SIAM J Numer Anal 41:1595–1619

Shen J, Tang T (2006) Spectral and High-Order Methods with Applications. Science Press, Beijing

Shen J, Wang L (2004) Error analysis for mapped Legendre spectral and pseudospectral methods. SIAM J Numer Anal 42:326–349

Shen J, Wang L (2005) Spectral approximation of the Helmholtz equation with high wave numbers. SIAM J Numer Anal 43(2):623–644

Shen J, Wang L (2006) Laguerre and composite Legendre-Laguerre dual-Petrov-Galerkin methods for third-order equations. Discrete Contin Dyn Syst Ser B 6(6):1381–1402 (electronic)

Shen J, Wang L (2007a) Analysis of a spectral-Galerkin approximation to the Helmholtz equation in exterior domains. SIAM J Numer Anal 45:1954–1978

Shen J, Wang L (2007b) Fourierization of the Legendre-Galerkin method and a new space-time spectral method. Appl Numer Math 57(5-7):710–720

Shen J, Wang L (2007c) Legendre and Chebyshev dual-Petrov-Galerkin methods for hyperbolic equations. Computer Methods in Applied Mechanics and Engineering 196(37-40):3785–3797

Shen J, Wang L (2008) On spectral approximations in elliptical geometries using Mathieu functions. Math Comp 78(266):815–844

Shen J, Wang L (2009) Some recent advances on spectral methods for unbounded domains. Communcations in Computational Physics 5(2-4):195–241

Shen J, Wang L (2010) Sparse spectral approximations of high-dimensional problems based on hyperbolic cross. SIAM J Numer Anal 48:1087–1109

Shen J, Yang X (2010) Numerical Approximations of Allen-Cahn and Cahn-Hilliard Equations. Discrete and Continuous Dynamical Systems Series A 28:1669–1691

Shen J, Yu H (2010) Efficient spectral sparse grid methods and applications to high dimensional elliptic problems. SIAM J Sci Comput 32:3228–3250

Slepian D, Pollak H (1961) Prolate spheroidal wave functions, Fourier analysis and uncertainty. I. Bell System Tech J 40:43–63

Smolyak S (1960) Quadrature and interpolation formulas the classes W_s^a and E_s^a. Dokl Akad Nauk SSSR 131:1028–1031, russian, Engl. Transl.: Soviet Math. Dokl. 1:384–387, 1963

Stein E, Shakarchi R (2003) Fourier Analysis: An Introduction, Princeton Lectures in Analysis, vol 1. Princeton University Press, Princeton, NJ

Strang G (1968) On the construction and comparison of difference schemes. SIAM J Numer Anal 5(3):506–517

Strang G, Fix G (1973) An analysis of the finite element method. Prentice-Hall Inc., Englewood Cliffs, N. J., prentice-Hall Series in Automatic Computation

Swarztrauber P, Spotz W (2000) Generalized discrete spherical harmonic transforms. J Comput Phys 159(2):213–230

Szegö G (1975) Orthogonal Polynomials (fourth edition), vol 23. AMS Coll. Publ.

Tadmor E (1986) The exponential accuracy of Fourier and Chebyshev differencing methods. SIAM J Numer Anal 23(1):1–10

Tang T (1993) The Hermite spectral method for Gaussian-type functions. SIAM J Sci Comput 14:594–606

Tang T, Xu X (2009) Accuracy enhancement using spectral postprocessing for differential equations and integral equations. Commun Comput Phys 5(2-4):779–792

Temam R (1969) Sur l'approximation de la solution des équations de Navier-Stokes par la méthode des pas fractionnaires ii. Arch Rat Mech Anal 33:377–385

Temam R (1984) Navier-Stokes Equations: Theory and Numerical Analysis. North-Holland, Amsterdam

Timan A (1994) Theory of Approximation of Functions of a Real Variable. Dover Publications Inc., New York, translated from the Russian by J. Berry, Translation edited and with a preface by J. Cossar, Reprint of the 1963 English translation

Timmermans L, Minev P, Van De Vosse F (1996) An approximate projection scheme for incompressible flow using spectral elements. Int J Numer Methods Fluids 22:673–688

Trefethen L (2000) Spectral Methods in MATLAB, Software, Environments, and Tools, vol 10. Society for Industrial and Applied Mathematics (SIAM), Philadelphia, PA

Trott M (1999) Graphica I: The World of Mathematica Graphics. The Imaginary Made Real: The Images of Michael Trott. Wolfram Media, Champaign, IL

Tygert M (2010) Recurrence relations and fast algorithms. Appl Comput Harmon Anal 28(1): 121–128

Van der Vorst H (1992) Bi-CGSTAB: a fast and smoothly converging variant of Bi-CG for the solution of nonsymmetric linear systems. SIAM J Sci Statist Comput 13(2):631–644

Wang C, Wang X, Wise S (2010) Unconditionally stable schemes for equations of thin film epitaxy. Discrete Contin Dyn Syst 28(1):405–423

Wang L (2010) Analysis of spectral approximations using prolate spheroidal wave functions. Math Comp 79(270):807–827

Wang L, Guo B (2009) Interpolation approximations based on Gauss-Lobatto-Legendre-Birkhoff quadrature. J Approx Theory 161(1):142–173

Wang L, Shen J (2005) Error analysis for mapped Jacobi spectral methods. J Sci Comput 24(2):183–218

Wang L, Guo B, Wang Z (2002) A generalized Gauss-type quadrature formula and its applications to pseudospectral method. Numer Math J Chinese Univ (English Ser) 11(2):179–196

Wang Z, Guo B (2002) A rational approximation and its applications to nonlinear partial differential equations on the whole line. J Math Anal Appl 274(1):374–403

Wang Z, Guo B (2004) Modified Legendre rational spectral method for the whole line. J Comput Math 22(3):457–474

Watson G (1966) A Treatise on the Theory of Bessel Functions. Cambridge University Press

Wu H, Ma H, Li H (2003) Optimal error estimates of the Chebyshev-Legendre spectral method for solving the generalized Burgers equation. SIAM J Numer Anal 41(2):659–672 (electronic)

Xiu D (2010) Numerical Methods for Stochastic Computations. Princeton University Press, Princeton, NJ, a spectral method approach

Xu C, Guo B (2002) Mixed Laguerre-Legendre spectral method for incompressible flow in an infinite strip. Adv Comput Math 16(1):77–96

Yanenko N (1971) The Method of Fractional Steps: The Solution of Problems of Mathematical Physics In Several Variables. Springer-Verlag, New York

Yoshida H (1990) Construction of higher order symplectic integrators. Phys Lett A 150(5-7):262–268

Yosida K (1980) Functional Analysis. Springer-Verlag, Berlin

Yuan J, Shen J, Wu J (2008) A dual-Petrov-Galerkin method for the Kawahara-type equations. J Sci Comput 34(1):48–63

Zabusky N, Galvin C (1971) Shallow water waves, the Korteveg-de Vries equation and solitons. J Fluid Mech 47:811–824

Zhang X, Guo B (2006) Spherical harmonic-generalized Laguerre spectral method for exterior problems. J Sci Comput 27(1-3):523–537

Zygmund A (2002) Trigonometric Series, 3rd edn. Cambridge Mathematical Library, Cambridge University Press

Index

J. Shen et al., *Spectral Methods: Algorithms, Analysis and Applications*,
Springer Series in Computational Mathematics 41, DOI 10.1007/978-3-540-71041-7,